The ecology and physiology of the fungal mycelium

'Largely invisible, little studied the vegetative mycelium of fungi still provides an almost endless series of problems whose investigation is long overdue'.
J. H. Burnett *Fundamentals of Mycology*, 1976

The ecology and physiology of the fungal mycelium

SYMPOSIUM OF
THE BRITISH MYCOLOGICAL SOCIETY
HELD AT BATH UNIVERSITY
11–15 APRIL 1983

EDITED BY
D.H.JENNINGS & A.D.M.RAYNER

CAMBRIDGE UNIVERSITY PRESS
CAMBRIDGE
LONDON NEW YORK NEW ROCHELLE
MELBOURNE SYDNEY

CAMBRIDGE UNIVERSITY PRESS
Cambridge, New York, Melbourne, Madrid, Cape Town,
Singapore, São Paulo, Delhi, Tokyo, Mexico City

Cambridge University Press
The Edinburgh Building, Cambridge CB2 8RU, UK

Published in the United States of America by Cambridge University Press, New York

www.cambridge.org
Information on this title: www.cambridge.org/9780521106269

First published 1984
Reprinted 1986
First paperback edition 2011

A catalogue record for this publication is available from the British Library

Library of Congress Catalogue Card Number: 83-24011

ISBN 978-0-521-25413-7 Hardback
ISBN 978-0-521-10626-9 Paperback

Contents

Contributors

T. J. H. Adams, *Department of Biological Sciences, University of Exeter, Washington Singer Laboratories, Perry Road, Exeter, EX4 4PS, UK*

A. M. Ainsworth, *School of Biological Sciences, University of Bath, Claverton Down, Bath, BA2 7AY, UK*

S. L. Anagnostakis, *Department of Plant Pathology and Botany, The Connecticut Agricultural Experiment Station, P.O. Box 1106, New Haven, CT 06504, USA*

N. A. Anderson, *Department of Plant Pathology, University of Minnesota, St Paul, MN 55108, USA*

R. C. Aylmore, *Department of Biological Sciences, University of Exeter, Washington Singer Laboratories, Perry Road, Exeter, EX4 4PS, UK*

L. Boddy, *Department of Microbiology, University College, Newport Road, Cardiff, CF2 1TA, UK*

C. M. Brasier, *Forest Research Station, Alice Holt Lodge, Wrecclesham, Farnham, Surrey, GU10 4LH, UK*

G. M. Butler, *Department of Plant Biology, University of Birmingham, Birmingham, B15 2TT, UK*

D. Coates, *School of Biological Sciences, University of Bath, Claverton Down, Bath, BA2 7AY, UK*

J. H. Croft, *Department of Genetics, University of Birmingham, P.O. Box 363, Birmingham, B15 2TT, UK*

K. Esser, *Lehrstuhl für Allgemeine Botanik, Ruhr-Universität, Postfach 102148, D-4630 Bochum 1, Germany*

J. C. Frankland, *Institute of Terrestrial Ecology, Merlewood Research Station, Grange-over-Sands, Cumbria, LA11 6JU, UK*

P. H. Gregory, *Rothamsted Experimental Station, Harpenden, AL5 2QJ, UK*

D. H. Jennings, *Botany Department, The University, P.O. Box 147, Liverpool, L69 3BX, UK*

U. Kück, *Lehrstuhl für Allgemeine Botanik, Ruhr-Universität, Postfach 102148, D-4630 Bochum 1, Germany*

G. Lysek, *Institute for Systematic Botany and Plant Geography FU, Altensteinstrasse 6, D-1000 Berlin, F. R. Germany*

M. O. Moss, *Department of Microbiology, University of Surrey, Guildford, Surrey, GU2 5XH, UK*

T. T. Nguyen, *Department of Microbiology and Immunology, Indiana University School of Medicine, 635 Barnhill Drive, Indianapolis, Indiana 46223, USA*

D. J. Niederpruem, *Department of Microbiology and Immunology, Indiana University School of Medicine, 635 Barnhill Drive, Indianapolis, Indiana 46223, USA*

B. Nordbring-Hertz, *Department of Microbial Ecology, University of Lund, Ecology Building, Helgonavägen 5, S-223 62 Lund, Sweden*

M. Raudaskoski, *Department of Botany, University of Helsinki, Unioninkatu 44, SF-00170 Helsinki 17, Finland*

A. D. M. Rayner, *School of Biological Sciences, University of Bath, Claverton Down, Bath, BA2 7AY, UK*

D. J. Read, *Department of Botany, The University of Sheffield, S10 2TN, UK*

M. Salonen, *Department of Botany, University of Helsinki, Unioninkatu 44, SF-00170 Helsinki 17, Finland*

U. Stahl, *Lehrstuhl für Allgemeine Botanik, Ruhr-Universität, Postfach 102148, D-4630 Bochum 1, Germany*

W. Thompson, *Botany Department, The University, Liverpool, L69 3BX, UK*

N. K. Todd, *Department of Biological Sciences, University of Exeter, Washington Singer Laboratories, Perry Road, Exeter, EX4 4PS, UK*

A. P. J. Trinci, *Cryptogamic Botany Laboratories, Botany Department, University of Manchester, Manchester, M13 9PL, UK*

P. Tudzynski, *Lehrstuhl für Allgemeine Botanik, Ruhr-Universität, Postfach 102148, D-4630 Bochum 1, Germany*

S. C. Watkinson, *Botany Department, University of Oxford, South Parks Road, Oxford, OX1 3RA, UK*

J. F. Webber, *Forest Research Station, Alice Holt Lodge, Wrecclesham, Farnham, Surrey, GU10 4LH, UK*

E. N. D. Williams, *Department of Biological Sciences, University of Exeter, Washington Singer Laboratories, Perry Road, Exeter, EX4 4PS, UK*

Preface

The present volume comprises the invited papers given at the first in a new series of 'General Meetings' of the British Mycological Society held at the University of Bath from the 12th to the 15th April 1983. Behind the programme which we planned for the meeting lay a deep concern that, if mycologists are to achieve a workable conceptual basis for an understanding of the ecology and physiology of true fungi which can be placed within the context of what is known or believed to apply to other groups of organisms, there must be a proper appreciation of the peculiarities of the mycelial habit. Nowhere was the need for this appreciation, nor the slowness of its coming, made clearer than in the first Benefactor's Lecture given to the Society by Philip Gregory which opened the Symposium and is the first chapter in this volume. It is here that the full diversity of operations for which the mycelium is responsible during the life of a fungus can be most vividly appreciated within Dr Gregory's concept of 'modes'. The occurrence, often in overlapping sequence, of such operational phases and the transitions between them during the life of a mycelium was a consistent theme of the Symposium and underlies the sequence of chapters in the present volume.

The primary function of the mycelium is that of establishment and spread in or on a suitable medium with concomitant absorption of nutrients and water. This process of gaining access to and influence over those materials required for maintenance of metabolism is encapsulated by the phrase 'primary resource capture' which is used by several authors. Rayner and Webber emphasise that the process represents an important phase in competitive interactions between mycelia. The particular feature of mycelia is that primary resource capture is effected by production of a spreading thallus, potentially indefinite in extent, and especially suitable for invasion of solid or semi-solid substrata possessing a defined outer surface from which colonisation is effected. Production of such a thallus is governed principally by the growth properties and branching characteristics of component hyphae within the colony, and these features are considered in chapters 2 and 3 by Trinci and Butler respectively. One feature which became apparent at the meeting during both these papers and also in a thought-provoking paper offered by Dr J. I. Prosser, was the distinction between a phase of growth immediately following spore germination, in which branches of similar diameter are often produced at right angles, and subsequent expansion of the colony margin by a zone of hyphae, often of wide diameter bearing branches at acute angles. The transition between these phases appears to correspond to a change from establishment to exploration, but how it occurs is an intriguing question. The presence of such a transition, and whether it occurs uniformly in the colony, could be related to point-growth, sectoring phenomena, rhythmicity, and the occurrence of constitutively dense, slow-growing colonies, reported by several of the authors as well as members of the audience. Dr Butler's paper raised the old question of the nature and function of the clamp-connection in Basidiomycotina, at last putting to rest the simplistic view, familiar in elementary texts, that it is merely a device to ensure maintenance of the dikaryotic state. The situation is clearly much more complex than that, seemingly having much to do with the particular mode in which the mycelium is growing.

The clamp-connection is, of course, a specialised type of hyphal anastomosis. Such anastomoses are a particularly remarkable feature of mycelia of higher fungi where they have the effect of converting a primary radiate system into a network, and in Basidiomycotina are an essential feature in sexual outcrossing. Until recently, the cytological processes involved have been much neglected, but this is now being rectified as is evident in chapters 4 and 5 by Nguygen & Niederpruem and Aylmore & Todd. The former consider anastomoses associated with sexual interchange in particular, while the latter concentrate on vegetative anastomoses and describe, for the first time, the results of a remarkable combined light and electron microscope study of the fusion process.

Following early phases of establishment and expansion of the colony margin (trophophase), further differentiation and biochemical changes occur in the established mycelium. These are associated with such functions as migration and interchange between food bases, autolysis and recycling of nutrients, interactions with abiotic environmental variables and with other organisms, and reproductive differentiation. These functions are the concern of chapters 6 to 14 which can be regarded as coming under the general heading of differentiation, distribution and maintenance of the fungal colony. During these chapters the emphasis is not upon that image of mycelia as collections of hyphal tips growing under at least initially homogeneous conditions, but upon mycelia as complex, intercommunicating systems. This reflects the fact that, whilst many laboratory studies of mycelial growth and development are made with homogeneous media, in nature the environment to which many mycelia are exposed is highly heterogeneous, perhaps particularly in relation to nutrient and water supplies. Within these chapters there is considerable emphasis placed on the mycelia of Basidiomycotina. This is more than mere coincidence but it is not a consequence of the personal interests of those responsible for organising the programme! Rather we point to the fact that the Basidiomycotina as a group probably produce amongst the most extensive and highly differentiated mycelia under natural conditions, and hence have attracted some of the most detailed studies of mycelial biology. We hope that the principles emerging from such studies are of general value to mycologists, and that they may encourage further study of other fungi.

In chapter 6 Moss draws attention to the range of secondary metabolites produced by fungi and raises intriguing possibilities as to how this may be related to the mycelial habit. In chapter 7, Jennings reviews the mechanisms and significance of water flow through mycelia, hopefully in an effort to throw light on a number of, perhaps sometimes unexpected, aspects of mycelial biology. This discussion is complemented by that of Watkinson on differentiation in *Serpula lacrimans*, which focuses in particular on nitrogen economy. Thompson's account of saprotrophic mycelial-cord-forming fungi in deciduous woodlands provides much new information about a hitherto neglected group and some exciting insights into the behaviour of mycelia in which such functions as exploration, branching, and interactions with other mycelia are mediated by aggregated rather than individual hyphae. Read then considers in chapter 10, how the development of vegetative mycelia of mycorrhizal fungi is related to soil type and in so doing produces a rational framework which may prove very helpful in understanding the distributional ecology of these symbionts. The importance of understanding mycelial distribution in autecology and the wide range of approaches which may be brought to bear on the problem are emphasised in Frankland's case study of *Mycena galopus* in chapter 11. Then Boddy provides a valuable attempt to consider, in general terms, the relationship between mycelial distribution and micro-environmental factors, using basidiomycetes in temperate woodlands as the basis for discussion. The importance of such factors in differentiation or 'mode transitions' is made very apparent in chapters 13 and 14 by Raudaskoski & Salonen and Lysek, as is the complementarity between vegetative and reproductive development. The former chapter provides a valuable account of morphogenetic events in the mycelium associated with initiation of basidiocarps, and the latter a wide-ranging review of rhythmic growth and its physiological and ecological significance.

No student of the mycelium can afford to ignore the role of cytoplasmic and nuclear genetic factors in producing variation in mycelial characteristics. Esser and his colleagues describe the discovery of how a mitochondrial plasmid is responsible for the senescence phenomenon in *Podospora anserina*, and, after noting how similar plasmids have been detected in the fungi, consider how this is of potential significance for genetic engineering. The theme of practical implications is continued in the succeeding chapters by Anagnostakis & Anderson who consider the important plant pathogens *Endothia parasitica*, and *Rhizoctonia solani* respectively. The former emphasises, amongst other things, the role which transmissible cytoplasmic factors may have in reducing pathogenicity whilst Anderson sets the scene for following chapters by considering how heterokaryosis is responsible for the generation of variation and how it is influenced by homogenic and heterogenic incompatibility systems.

The next six chapters consider how various types of recognition systems control the interactions both between mycelia (of the same and differing species) and with other organisms. In chapter 18, Rayner & Webber provide an overview of interspecific mycelial interactions, their types, mechanisms and implications. These authors emphasise that it is important to consider the peculiarities of the mycelial habit and the various life strategies which can be exhibited by fungi if such interactions are to be placed within a coherent conceptual framework. Accordingly they introduce a system of terminology which they hope will aid production of such a framework.

The possible significance of lectin–carbohydrate associations in mediating mycelial interactions with other organisms is brought out in chapter 19 by Nordbring-Hertz's detailed case study of nematophagous fungi. Continuing the theme of interspecific interactions, but in this case considering closely related species, Croft & Dales discuss mycelial interactions and mitochondrial inheritance in *Aspergillus* spp. and provide some intriguing ideas of how variation and speciation may occur in this group based on studies using the relatively recently available techniques of protoplast fusion. The next three chapters consider intra-specific interactions, and practical implications are brought out by studies of *Ceratocystis ulmi* and *Endothia parastica* by Brasier & Anagnostakis respectively. The chapter by Brasier in particular contains a wealth of previously unpublished information about sexual and vegetative incompatibility and their implication for understanding the present and future population structure of *C. ulmi* and the devastating world-wide epidemics it has caused. In chapter 23 Rayner *et al.* bring together existing information, including much that is alluded to in previous chapters, about somatic incompatibility in higher fungi in a general discussion, based on their studies of wood-decay species, of the implications of fungal individualism in understanding several aspects of the ecology, physiology and evolution of true fungi.

Alan Rayner, David Jennings

Note on nomenclature and use of terms

In any dynamic subject, there are always problems with usage of terms based on ever-continuing changes in knowledge and philosophical outlook, and the inevitably different preferences of those involved. We have therefore not been too strict with authors over whether they should use such terms as 'strands' or 'cords', 'saprophytes' or 'saprotrophs', or 'vegetative', 'somatic', sexual', 'homogenic' or 'heterogenic' incompatibility. Where possible, however, we have tried to indicate where synonyms exist. However in a few cases we felt consistency was necessary to prevent confusion, particularly regarding the terms habitat, substratum, substrate, resource, outcrossing, outbreeding, and hyphal compartments. Habitat is used to describe the place where a fungus lives, substratum the medium within the habitat which physically supports the fungus during development, and substrate a specific biochemical constituent of the substratum. Resource is any material which sustains fungal growth. Outcrossing involves sexual conjugation between genetically different lines from the same or different parents (as opposed to self-fertilisation), whilst outbreeding is defined as outcrossing between progeny from different parental lines, i.e. non-sibs. Finally we have preferred to use the word compartments to describe those segments of a hypha which are delimited by septa and not the word cells.

With respect to taxonomic nomenclature, we have tried to use the most up-to-date name for species, regardless of that used by the original author. In some cases, where the original author may have misidentified the organism in question, this has been indicated as for example with *Coprinus lagopus*. For clarity, and except where including it in some way adds to the argument, we have not included the authorities with species names.

1

The fungal mycelium – an historical perspective

P. H. GREGORY
Rothamsted Experimental Station, Harpenden, England AL5 2JQ, UK

I will start by expressing my grateful thanks to the Council of The British Mycological Society for the invitation to give the first Benefactors Lecture. I feel this both an honour and a responsibility.

The beginner in oil painting is advised to avoid getting involved with detail, to use a large brush and hold it by the end of the handle at arm's length from the canvas. I intend to use this technique for sketching the history of ideas about the fungal mycelium. Frankly, I have left out almost everybody and everything!

Biologically the vegetative mycelium is the most important part of the fungus: it does all the hard work of exploring and exploiting the substratum and feeding off it. Yet most textbooks dismiss the mycelium in two or three pages. We write about fungi like those old-fashioned historians who treat history in terms of the Napoleons and Cleopatras, but ignore Tom, Dick and Sally. We seem blind to the vegetative mycelium.

This blindness is illustrated by the otherwise excellent and stimulating report of the Second Kananaskis Conference. In spite of its title: '*The Whole Fungus*', it ignores the vegetative mycelium (which it takes for granted) except that around page 463 it gives 60 lines to the taxonomic relationships of the Mycelia Sterilia. Our neglect of the mycelium is curious because sale of mushroom spawn has been a thriving industry for generations, and today much capital is invested in processes for growing mycelial pellets in submerged culture to produce antibiotics such as penicillin.

Discovery of the main features of the mycelium

There is no need to define the term 'mycelium': without a shrewd idea already you would not be at this Symposium. Basic facts about the mycelium were summarized by De Bary (1887) in his *Comparative Morphology*. But the mycelium is a dynamic entity, and to understand its activity we must take account of three important phenomena: (1) hyphal fusions or anastomoses; (2) perforate septa; and (3) cytoplasmic and nuclear migration.

Credit for discovering hyphal fusions probably goes to the brothers Tulasne, who depict hyphal fusions in five species in their great '*Carpologia*' (1861–5): in Grove's translation of their Latin the phenomenon is twice referred to as 'wonderful'.

Marshall Ward (1888) in his paper on the lily disease, may have been the first actually to watch vegetative fusion in progress under the microscope. In his three- or four-day-old cultures of *Botrytis elliptica* fusions were numerous and of constant occurrence.

The fusion process started with action at a distance between two neighbouring hyphae which evidently signalled across a gap equivalent to several hyphal diameters (Fig. 1). Forty minutes from the start, a lateral hypha was emerging from one of the two hyphae; in another 50 min this had elicited growth of a branch from the neighbouring hypha and had changed course to meet it; 25 min later they had almost met; in another 15 min they had met and fused so that protoplasmic continuity (the final test of true fusion) had been established between them. Like the Tulasnes, Marshall Ward was impressed and called this attraction phenomenon 'remarkable'. He also recognised that the mycelium must be in some special but undefined state for fusions to occur, because young hyphae could grow across one another without fusing.

We now know that vegetative hyphal fusions are very common in the higher fungi (Meyer, 1902), but in contrast mycelium of the zygomycetes generally lacks *vegetative* hyphal fusions (*Syncephalis* seems to be exceptional in this respect). However, in zygomycetes obvious *sexual* fusions occur between gametangia: the glamour of this sexual process in the lower fungi has eclipsed the work-a-day vegetative fusions of the higher fungi.

Septation

The living zygomycete mycelium is generally non-septate and can be described as *coenocytic-filamentous* (although some Trichomycetes can be strongly septate).

In contrast the outstanding characteristic of the higher fungi (i.e. the ascomycetes, basidiomycetes and their imperfect states) is that their

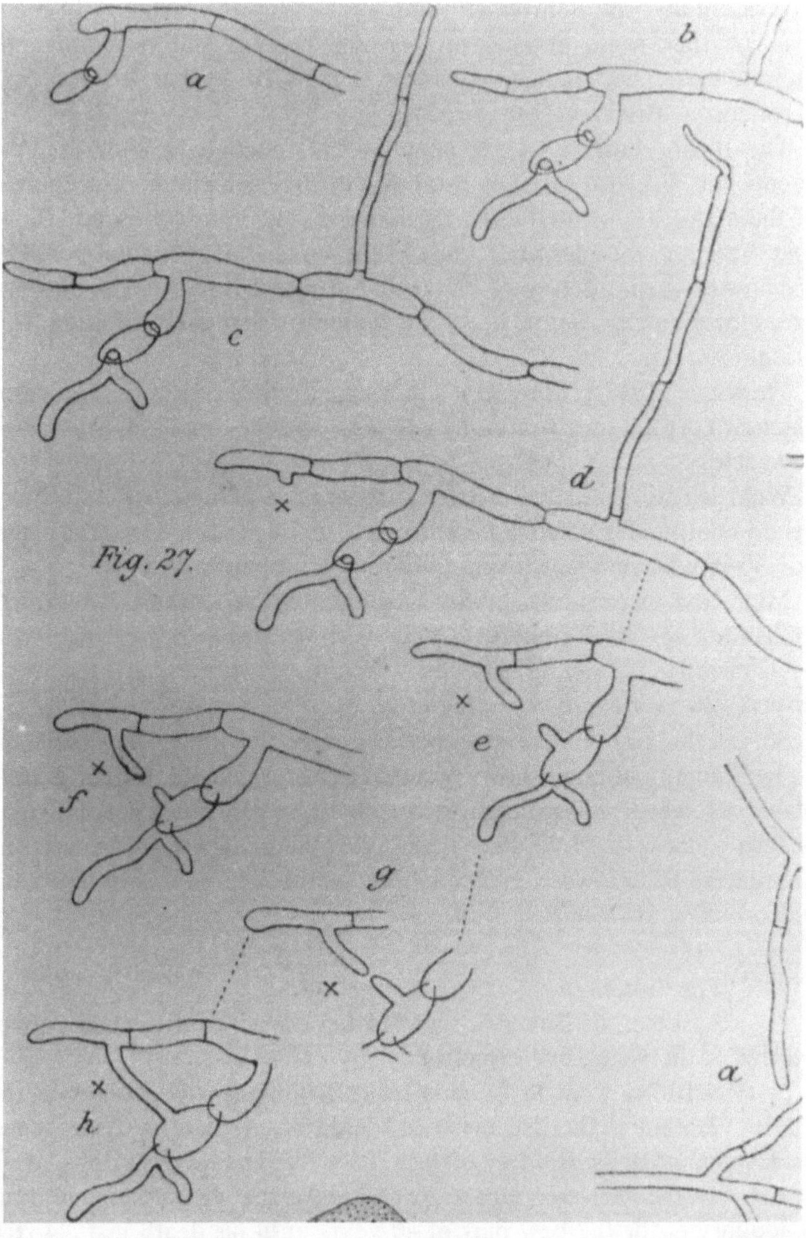

Fig. 1. Stages in formation of a hypha-to-peg fusion in *Botrytis elliptica*. (Drawn by H. Marshall Ward, 1888.)

mycelium is *septate-reticulate.* So, whereas the zygomycete mycelium maps a tree, the higher fungi map a net.

Incidentally, the oomycetes stand apart from this scheme. In many respects they seem anomalous, for instance in their predominantly diploid existence, and in persistent claims of hyphal fusions (e.g. Stephenson, Erwin & Leary, 1974).

The higher fungi then are septate: but functionally they are also coenocytic, because each septum has a central perforation during most of the active life of the hypha. In addition, electron microscopy shows that there are also plasmodesmata in the septa. But the septal pores we are now discussing are much larger than plasmodesmata – ten or twenty times the diameter. Septal pores are evidently completely different from plasmodesmata.

Wahrlich (1893), using the chlor-zinc iodine reagent, discovered abundant cytoplasmic continuity between cells via septal pores in nearly 50 species of higher fungi.

Wahrlich also actually watched granular protoplasm moving from one cell to another in *Eurotium herbariorum*, and he realized that the septal pores serve as channels for migration of protoplasm.

Migration of protoplasm was substantiated by Rothert (1892) with *Sclerotium* sp.; by Reinhardt (1892) with *Sclerotium sclerotiorum*; and by Charlotte Ternetz (1900) with *Ascophanus carneus*. In the zygomycete coenocyte two-way circulation was described by Arthur (1897).

So, at the start of the twentieth century the fungi, especially the higher fungi, had been shown to possess features quite unlike vascular plants. However, these features remained as curiosities and no significance seems to have been accorded them. These scattered and unorganised facts were in need of a synthesiser who arrived in the person of A. H. Reginald Buller.

The Buller era

A sketch of Buller is relevant here in view of the subsequent neglect of the vegetative mycelium.

In 1904 Buller went to Canada from Birmingham University to the Chair of Botany in the University of Manitoba. Arriving in Winnipeg at the Canadian Pacific Railway Station, he walked a few yards down Main Street, saw the McLaren Hotel, turned in for the night and made it his headquarters for the best part of 40 years until his death in 1944 (the hotel had a good billiards table – spore ballistics!). All this is relevant because, when not teaching, the bachelor Buller, unencumbered by

Professor A. H. Reginald Buller, 1874–1944
President of the British Mycological Society, 1914

domestic ties, could devote his time to research like some modern monk, watching a fungus right through the night if occasion required – as it often did. In the favourable environment of his laboratory at the University the functioning of the agaric hymenium was unravelled, as recorded in the volumes of his *Researches on Fungi.*

In 1931 I went to Winnipeg to work under Buller's direction, not in his own laboratory though, but at the Medical College, and on arriving at the CPR Station my first port of call was also the McLaren Hotel to dine with Buller and Bisby.

Buller was excited over the phenomenon of hyphal fusions which in the 1930s was in the mycological air. He was just taking up the line of study which, over the next few years, opened up a new understanding of the vegetative mycelium. My wife and I got to know him quite well: he was at our wedding, and he once dropped in on us unexpectedly for breakfast after an all-night vigil in his laboratory. He was excellent company, had a critical mind with a boyish enthusiasm. His aim was to understand fungi. I was tremendously impressed by Buller; I still am.

I will try briefly to summarise his main contributions on the mycelium as expounded in his *Researches* (Buller, 1931; 1933).

Hyphal fusions

Three types of true fusion were distinguished according to their function: (1) vegetative fusions; (2) sexual fusions; and (3) the parasitic fusions seen when *Chaetocladium* and *Parasitella* attack certain Mucoraceae.

True vegetative hyphal fusions (of the kind watched by Marshall Ward) lead to protoplasmic continuity being established across the bridge, and must be distinguished from mere hyphal contact or hyphal adhesion in which continuity of cell lumen is *not* established. [This distinction is fundamental in mycology.]

Vegetative fusions develop in an older part of the mycelium, converting it into a three-dimensional transport system. The process is self-regulated so that the final number of fusions is limited. (From Buller's illustrations of his *Pleurage curvicola* I estimate that the number of fusions is of the order of 1 per 200 μm length of hypha.)

Buller lists the main functions of hyphal fusions as follows: (1) conduction of food materials; (2) mating between + and – mycelia in a heterothallic species; (3) passage of nuclei during 'diploidization'; (4) reducing the effect of mechanical damage to a hypha by providing an alternative route for movement of protoplasm; and (5) enabling several mycelia of the same species to form a single compound mycelium which acts as a social unit in the production of fruit-bodies and spores from a limited food base.

This last point is illustrated by Buller's diagram of a horse-dung ball colonised by 24 basidiospores of the homothallic *Coprinus sterquilinus*. In competition not one of the 24 mycelia would have enough food to fructify, but their pooled resources after fusion would give a large sporophore – more of this later.

Observations by earlier workers on 'teleomorphosis' and 'zygotropism' were confirmed, though the phenomena remain largely unexplained to this day. Vegetative fusions occur between hyphae of one and the same species, but not between different species. This specificity of fusions was stressed, and the possibility of using the ability of two mycelia to anastomose as a test of conspecificity was explored in early experiments with *Panellus stipticus* (Buller, 1924; and see Davidson, Dowding & Buller, 1932).

By continuous observation with the microscope during all stages of numerous fusions, Buller concluded that *all* vegetative fusions take place between growing hyphal *tips*, albeit the lateral hypha elicited by an approaching hypha may be merely a short 'peg'. Four different kinds of hyphal fusions were distinguished: hypha to hypha; hypha to peg; peg to peg (Fig. 2); and hook to peg in clamp connections.

Protoplasmic streaming

Buller confirmed in general the presence of an open pore in the septum of living hyphae in the higher fungi by observing protoplasm flowing through the septum: in *Sordaria fimicola* he recorded a maximum speed of 6 cm h^{-1}. In *Pyronema* streaming was watched through a series of 161 successive septa, away from hyphae which were being emptied and towards a rapidly growing area of mycelium.

As to the function of septal pores – Buller concluded that perforate septa offer little resistance to cytoplasmic flow, but the pore can be plugged rapidly if a cell is injured and at the same time flow is not interrupted because alternative routes are available through the hyphal network.

Among basidiomycetes, flow was observed in *Rhizoctonia solani*, both through main hyphae and across bridges formed by hyphal fusions. However most hymenomycetes have very clear transparent cytoplasm, so flow could not easily be observed.

But Buller discovered a new phenomenon – migration of nuclei through septal pores. (To anticipate, he saw the structure which Moore & McAlear (1962) thirty years later, using the electron microscope, described as the dolipore septum of hymenomycetes, but Buller concluded that what he saw was an optical illusion because it did not inhibit nuclear migration (more of this later).)

Nuclear migration

Mating tests with *Coprinus lagopus* (– it was actually *C. radiatus*) *in vitro* showed that a fragment of haploid (monokaryotic) mycelium placed at the edge of a large haploid mycelium would rapidly lead to dikaryotisation of the large mycelium. And so, surprisingly, would a fragment of a dikaryotic mycelium used as inoculum. The appropriate nuclei travelled at 1.4 mm h^{-1} into the larger mycelium (or rather their progeny did in a kind of relay race).

The dikaryotisation of a monokaryon by a dikaryon has become known as 'the Buller phenomenon'. Buller concluded that the persistence of conjugate nuclei (instead of fusing to true diploid nuclei) aids dikaryotisation of naturally-occurring haploid mycelia in nature. The

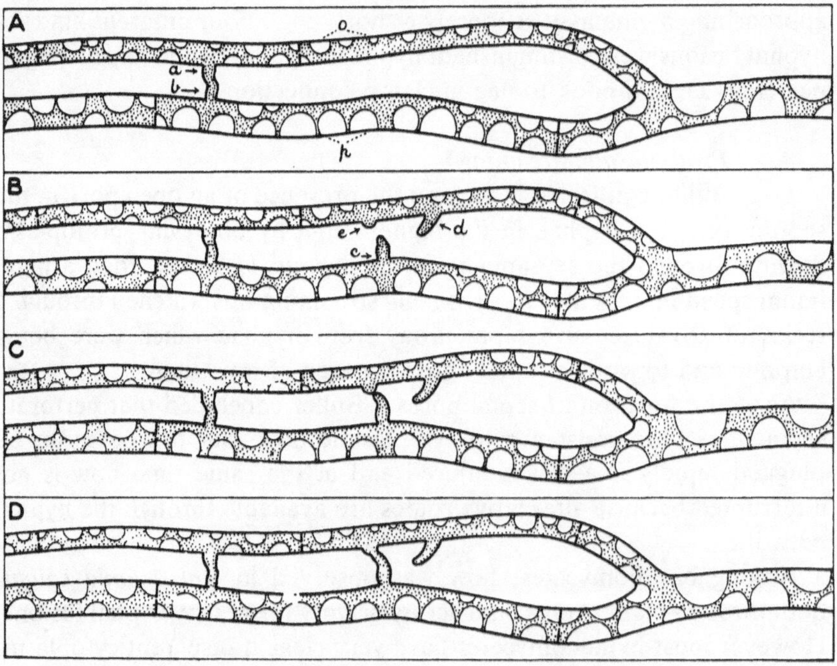

Fig. 2. Stages in formation of a peg-to-peg fusion in *Pyronema*. (Drawn by A. H. R. Buller, 1933, p. 58.)

function of clamp-connections he saw as preserving the correct balance between nuclei and cytoplasm in the terminal cell (never less than two nuclei).

The conclusion was that nuclei can travel through long series of cells, not merely to the next cell as in the sexual processes of the lower fungi and other organisms.

Reactions to Buller's work on the vegetative mycelium

Inevitably, after nearly 50 intervening years of research, Buller's conclusions need some editing. Surprisingly however, while most of his findings have been confirmed, they have never been assimilated into mycological philosophy.

I believe that the main reason is that the facts are disquieting to anyone accustomed to classifying fungi as plants. One can almost hear the traditional botanist despairingly adapt Henry Higgins's comment on Eliza Doolittle in 'My Fair Lady': 'Why *can't* a fungus be more like a plant?'.

Are fungi plants?

The fact is that the life-styles of green vascular plants and of fungi are vastly different. Green plants have an excellent plumbing system for conducting water and solutes through the tissues, but they do not permit living protoplasm with its organelles (sometimes including nuclei) to migrate through up to 160 consecutive cells! Angiosperms resemble fungi most closely in the behaviour of the pollen tube, in other respects their life-styles are poles apart.

In the nineteenth century emphasis was laid on *differences* between species. Today it is customary to emphasise the *unity* of living organisms; but both facets are worth exploring.

Higher plants draw water and inorganic nutrients from the soil, while maintaining their green tops in the air and sunshine where they synthesise elaborate organic compounds. In contrast fungi obtain all their food from the substratum. The mycelium is organised to rootle it out and carry it towards the site where the resting or dispersal organs are forming. All this supports the modern claim for recognising Kingdom Fungi (Whittaker, 1969; etc.). My only reservation is that we still lack comparable information about the algae, where vegetative hyphal fusions undoubtedly occur, as for example in the Laminariaceae (Fritsch, 1945). Perhaps the natural solution will be to recognise Kingdom Thallophyta.

An unconscious cover-up?

So, while one would have expected the new knowledge to be welcomed, instead it was largely ignored, and there has been an unconscious cover-up.

It must be admitted that Buller was somewhat vulnerable, not in his observations but because of his isolation during three quarters of the year in Winnipeg (where he had in fact built up an active centre of mycological research). His *Researches*, published largely at his own expense, are often prolix. And once, when he wrote about a fungus growing on cow dung, he added a photograph of cows in Kildonan Park! It seemed easy to disparage the unwelcome new findings as the product of a boyish enthusiast from the Wild West, and to push the disturbing ideas under the carpet.

To most botanists the new ideas were disturbing indeed. The only parts of his work on the vegetative mycelium referred to in the Royal Society of London's obituary notice are the Buller Phenomenon and his explanation for clamp-connections. (The British Mycological Society did not publish an obituary.)

Only Maurice Langeron in his *Précis de Mycologie* (1945) seems to have grasped the implications of Buller's work, and he drew conclusions more far-reaching than Buller himself ever did. Some years ago I quoted Langeron's brief attempt at a synthesis (Gregory, 1966), which I translate freely as follows.

> 'A fungus is a nucleate cytoplasmic mass which moves in a centrifugal direction, either freely without restraint, or inside tubes which it builds gradually as it moves towards the periphery. At the same time as these tubes are elongating under cytoplasmic pressure, the fungus gradually quits the central part of the thallus, whose elements become vacuolate progressively, then becoming empty and moribund. In this way a fungal thallus, whatever its shape, always comprises a living zone, growing continuously, and a central skeletal zone formed of empty dead tubes. The growth of the thallus is only limited by exhaustion of the cytoplasmic mass, which finishes up by passing into the propagative or reproductive spores' (Langeron, 1945).

Significantly this passage was deleted by the editor of the posthumous second edition of Langeron's *Précis*: part of the unconscious cover-up? The only work I know which gives Buller an impartial treatment is

Burnett's *Fundamentals of Mycology*, and I am pleased to note that the second edition (1976) has *not* been censored!

Observations on living fungi

There was a further reason for neglect of the new ideas, and that lay in technical progress. Buller went on looking down his light microscope at moving, growing fungi, continuing obstinately (but fortunately) into the era when most research had moved over to study dead mycelia, killed and embalmed with elaborate ritual before being stained or cut up with steel, glass or diamond knives. Don't misunderstand me: I am not belittling the brilliant advances gained by the microtome and electron microscope; I am merely stressing what every modern microscopist realises – that one disadvantage of modern techniques is that they not merely cut sections through *tissues*, but they also cut sections through *time*. Unfortunately the success of the modern lethal techniques is so great that information gained from the older techniques in flowing time has been largely forgotten.

For the study of the mycelium, mycology now needs a synthesis between observations on living mycelia and observations on dead sections. First, observations in flowing time must be revived: how else can we interpret EM and SEM pictures unless we have a clear picture of timing of events in the mycelium? Use of a television microscope could help share the tedium of watching. And fortunately development of phase-contrast optics has revived interest in the living mycelium, leading to studies of nuclear migration, for example by Dowding (1958) on *Gelasinospora*, and recently by Niederpruem (1980) on *Schizophyllum*.

Developments since 1950

Although Buller's ideas have not been adopted by general mycology, the phenomenon of hyphal fusions has stimulated two specialised groups: geneticists and virologists. For convenience the virus work will be considered first.

Mycoviruses

Beginning with the identification of mushroom viruses by Hollings (1962), an important development of the last 20 years is the discovery that a large proportion of mycelia of higher fungi (*in vitro* or in the field) are infected with viruses (see reviews by Hollings, 1982; Buck, 1980; Lemke, 1981). Usual methods of experimental inoculation fail with mycoviruses but fusion with another conspecific mycelium may transfer virus infection. On mushroom farms, before spawned beds are covered with casing soil, the mycelium may become infected by fusing with airborne basidiospores acting as virus vectors, the ease of transfer depending on which strains of *Agaricus bisporus* are involved (Schisler, Sinden & Sigel, 1967).

Concept of the mycelium mosaic

The dawn of the antibiotic era in the 1940s with the development of penicillin stimulated an enormous interest in fungal genetics.

Applied workers had been puzzled to account for the variability of their fungi – often in the apparent absence of any sexual process. Köhler (1930), and Brierley (1931) independently suggested that hyphal fusions might facilitate heterokaryosis and so increase variability. Much laboratory experimentation on heterokaryosis followed, especially after Pontecorvo and his colleagues described the 'parasexual cycle' or mitotic recombination (Pontecorvo, 1956), and a new understanding of origins of variability seemed to be opening up.

Perhaps it was Buller's concept of the social unit mycelium and his diagram of *Coprinus sterquilinus* growing on its multiple-inoculated dung ball that sparked off a new concept that all mycelia of a species form a physiological unit.

On tree stumps, Burnett & Partington (1957) found adjacent groups of *Coriolus versicolor* sporophores carrying the same mating-type factors in different combinations. Mycelia derived from these would anastomose freely in culture. One possible interpretation was that in the wild the mycelium 'exists as a single physiological and ecological unit, although genetically a mosaic', thus setting the fungi apart from almost all other organisms in what Buller called their social organisation, and making an individual fungus hard to define (Burnett, 1976).

Raper (1966) and Raper & Flexer (1970) also envisaged mosaicism in *Schizophyllum* and *Coprinus* – 'physiologically unified dikaryons constituted of many accretions of diverse origins. The product is a genetic mosaic, all parts of which may produce fruiting bodies and liberate basidiospores . .. a mosaic population capable of harbouring enormous variability'. It seemed that the concept of an individual could not be applied in the higher fungi – the individual was merged in a cooperative society co-extensive with the species.

However, doubts began to arise when Caten & Jinks (1966) pointed out that in the ascomycetes the induction of heterokaryosis was mainly a laboratory phenomenon, not paralleled in the wild. The role of hyphal fusions began to appear in a new light.

Some complications

The consequences of hyphal fusion proved to be more varied than Buller envisaged. Fusions within one and the same mycelium indeed still appear entirely beneficial. Confrontations between mycelia of *different* species can lead to repulsion or indifference, only rarely to fusions. In some cases approaching hyphae may attract one another, but on making contact their apices remain rounded instead of mutually flattening as in the prelude to true fusion (Köhler, 1930).

Confrontation between different mycelia of the *same* species usually leads to consummation of fusion, but this is often followed by a rejection reaction, with death of groups of cells, as for instance in *Thanatephorus* spp. (Flentje & Stretton, 1964). In Britain, *Aspergillus nidulans* exists in many compatibility groups, between which heterokaryons are not formed (Jinks, Caten, Sinden & Croft, 1966). The latitudinarianism of the 1950s began to wilt.

Concept of the individualistic mycelium

A new phase started a few years ago when Rayner & Todd (1977) and Todd & Rayner (1980) went into the woods with a saw and cut up decaying tree stumps which had multiple natural basidiospore infections. They discovered that hitherto largely unsuspected mycelial-rejection phenomena could operate after successful hyphal fusion to limit the scope of the social unit mycelium concept.

A stump with *Coriolus versicolor* for example showed several decay columns, occupied by mutually antagonistic dikaryons, but whose monokaryotic components were, significantly, fully interfertile. The columns were separated by relatively undecayed dark zone lines. Rayner & Todd conclude that in Nature antagonisms operate to prevent effective vegetative fusion between genetically different mycelia of the same species. They advance the hypothesis that such mechanisms are general in the higher fungi, operating as *vegetative* but not *sexual* mechanisms solely to delimit individuals within freely interbreeding populations. These mechanisms act *only* against genetically distinct mycelia: mycelia of the same clone are accepted. Membership of the co-op is restricted to members of the clone.

If, as seems probable, this situation proves general in the higher fungi, a number of puzzles fall into place and others need re-thinking: the 'barrage' phenomenon studied by Vandendries & Brodie (1933) and Esser (1966); the rejection in *Thanatephorus* described by Flentje & Stretton (1964); zone lines (Campbell, 1933, etc.); and some of the strange features of fairy rings. A corollary of the theory is that in most species the prevention of vegetative out-fusions must somehow be reconciled with the usual requirement for out-breeding.

It seems that such reactions occur not only between established mycelia, but also between a mycelium and a germinating spore. Kemp (1975) studied homing and lethal reactions between oidia and mycelium in *Coprinus*, and suggested that 'oidia can function, not only as male gametes or spermatia, but also as a means of eliminating or reducing the growth of hyphae belonging to closely related competitors'. Niels Fries (1981) hints that basidiospores in *Leccinum* can function in the same way. Biological warfare in the fungi!

When I speculated (Gregory, 1966) that 'besides their function as colonizers, spores may perhaps act as an unreliable airmail service transmitting genes between established mycelia' I did not foresee that the 'mail' delivered would consist largely of letter-bombs and virus packets!

So much for the history of the mycelium. Now for the perspective.

Modes of mycelial functioning

To understand the dynamics of the mycelium it is helpful to use the analogy of a piece of electronic equipment, a 'music centre' for example, which functions in different modes at the touch of a switch. I can perceive nine possible modes – there may well be more. Note that these modes are not necessarily part of a fixed cycle, but depend on the appropriate switch. And note that different parts of a mycelium may be functioning in different modes at one and the same time.

Mode 1: spore germination

Often we know little about what triggers a mode, but this is not the case with the first mode, which has been well studied.

The mycelium is typically initiated at spore germination. Here we merely need to note that in germination the protoplasm (by which term I understand cytoplasm + nucleus) migrates into the germ tube apex leaving a vacuole enclosed by the spore wall; it may even pop back again within the spore wall if conditions become adverse.

Mode 2: divergent growth

During the phase of hyphal construction, in which the available environment is explored, new cell wall is formed at the growing apex of the hypha, and branches are established in a pattern characteristic of the species by initiation of new lateral growing points behind the leader. In this mode young hyphae may approach and grow over each other without initiating a hyphal fusion.

Mode 3: hyphal fusion (anastomosis)

In the higher fungi, but not in the lower fungi, the mode of hyphal fusions establishes a three-dimensional network: this is a phase of middle age. The triggers for this mode are largely speculative: exhaustion of nutrients, accumulation of staling substances perhaps? As already noted, the autonomy of Modes 2 and 3 was noted by Marshall Ward (1888). It was more clearly stated by Buller (1933) who wrote:

> 'In a single mycelium the formation of hyphal fusions sets in in the older parts where the culture medium is becoming exhausted, and in general the condition is promoted by conditions of starvation. . . . The actual formation of hyphal fusions seems to alter the physiological condition in such a way that in the end the vegetative mycelium ceases to produce hyphal fusions'.

It may be noted here that when starting a culture with a dense spore suspension, the resulting germ tubes may enter an anastomosing phase (Mode 3) temporarily, forming a mycelium which then reverts to Mode 2.

In contrast with Modes 1 and 2, the remarkable phenomena of hyphal fusions have been very little studied during the last 50 years. How is the mode triggered? What signal is sent out by the hyphal tip initiating the process, and how is it received so as to evoke the development of a growing apex from one or more points on a neighbouring hypha? What special structures receive the signal? What stimuli orient the two growing points? What signals lead to recognition between mycelia (very complex surely), and lead eventually to complete fusion of both cell wall and cytoplasm – the ultimate criterion of true vegetative hyphal fusions as distinct from contact and adhesion.

Time is now ripe for a fresh enquiry into these mysteries with biochemical and EM techniques. Some answers will follow later in this Symposium.

Mode 4: migration of protoplasm (transmigration)

Hyphal fusions facilitate protoplasmic migration – the mode by which the essential fungus begins to quit the tubes of its mycelium and migrates towards developing sporulating or resting structures – no doubt at times moving with extreme slowness.

Naturally this process can be observed only in living material, and little has been added since Buller's era. But movement does depend on open septal pores, and here electron microscopy has contributed valuable information. Following discovery of the basidiomycete dolipore by Moore & McAlear (1962), it seemed that the parenthosome might prove a barrier to migration of cytoplasm and nuclei. But this cap is now known to have openings, and in some Coprini at least, the cap can disappear in the nuclear migration mode. Bracker & Butler (1964) and Giesy & Day (1965) illustrate nuclei in the act of passing through dolipore septa. Possibly the function of the dolipore is to imprison the nucleus in its cell until the nuclear migration Mode 5 sets in.

Controversy has followed Buller's belief that migration of protoplasm through the mycelium is due to vacuolar pressure from the rear. While some have supported this view, others have produced evidence of suction from in front (Plunkett, 1958); or active creeping could be involved (Isaac, 1964).

What is surprising is that argument about the causes of protoplasmic migration can continue without explicitly acknowledging its function. This is evident from confusion, even 50 years after Buller, between evacuation of a hypha and autolysis. For instance we read statements to the effect that sporulation of some fungus or other is favoured by 'autolysis'. The idea results evidently from ignorance of the phenomenon of migration, the devitalised appearance of the evacuated hypha being incorrectly attributed to autolysis. True autolysis in fungi has been known for a long time, most spectacularly in the deliquescing gills of *Coprinus*.

Radioisotope techniques developed in the 1950s led to much work on translocation in fungi, reviewed by Wilcoxon & Sudia (1968). It is clear to me that in the fungi two distinct phenomena have been confused under the term 'translocation'.

(1) True translocation, movement of water and solutes through the hyphae, as studied for example by Jennings (1982), is a process probably going on continuously through all modes, except in dry, concentrated structures like sclerotia and spores.

(2) Mass migration of living protoplasm along a route in the mycelium

(a phenomenon not met with in higher plants) is the characteristic of this Mode.

Transfer of terms from green plant physiology has sometimes proved inapt for mycology. A new term needs to be added for fungi. I suggest that in addition to 'translocation' we use the term 'transmigration' to denote the distinctive processes by which protoplasm and organelles are carried for long sequences of cells through the mycelium.

In mycological literature it is not always clear whether translocation or transmigration is under discussion. For instance Schütte (1956), when he reported that some species of *Penicillium* and *Aspergillus* did not translocate, seemed mainly to be using the term in the sense of transmigration.

It seems incredible that these moulds should show neither translocation nor transmigration. How else are the aerial phialides supplied from the substratum? There may well be fungi which do not transmigrate, but I interpret Schütte's results as showing that his moulds do not translocate or transmigrate *laterally*. This interpretation would be consistent with Burgess (1960) who found that soil fungi belonging to his 'Penicillium Pattern' sporulated densely on a small nutrient mass, but did not grow out into the surrounding soil.

Transmigration is thus a process compatible with but quite distinct from translocation. In transmigration, movement can be in any direction through the mycelium, it is normally most vigorous towards a reproductive or resting structure which is being furnished, and involves cytoplasm and organelles, at times including nuclei.

Mode 5: nuclear migration

The phenomenon of nuclear migration, first discovered by Buller (1931) and then doubted, has been confirmed by later workers (see Snider, 1963). One trigger is clearly genetic, and the mechanism that suggests itself is traction by protein fibres. During transmigration it seems that nuclei may or may not migrate, but I wish we had more information on this little-explored aspect.

Nuclear migration may well be a common event. Although it is difficult to detect in the absence of genetic effects, there is no reason to suppose that its only function is in dikaryotisation.

Mode 6: mycelium and yeast transition

Some fungi can switch between mycelial and yeast morphology. In *Mucor* the yeast phase is determined by anaerobiosis. Some systemic pathogens of man and animals are mycelial in culture at room temperature but switch to yeast form in the animal body (see Nickerson, 1947). The industrially-important yeasts seem to be permanently fixed in yeast form. Factors controlling the transition are reviewed by Cole & Nazawa (1981).

Mode 7: coherent growth

From spore germination onwards, early growth of the mycelium is potentially three-dimensional. While the hyphae are exploiting the substratum, branches diverge in specifically characteristic patterns. Sooner or later in most higher fungi (but not in lower fungi) the pattern switches from divergent to what I term 'coherent growth', perhaps forming sclerotia, strands or rhizomorphs, or forming sporocarps leading to teleomorph or anamorph reproduction (Modes 8 and 9). Needless to add, the switch is reversible.

Nutritional factors are suggested for the transition from divergent to coherent growth in *Armillaria mellea* by Garrett (1953), in *Serpula lacrimans* by Butler (1958), and in the cultivated mushroom by Chanter & Thornley (1978). But it seems that we still do not understand what factors trigger this extraordinary change from divergent to coherent growth – a crisis in development which is probably universal in the higher fungi. At this point we reach the end of the mycelial modes. There remain two more modes however.

Mode 8: teleomorph reproduction and Mode 9: anamorph reproduction

Factors switching Modes 8 and 9 are reviewed by Müller (1979), but discussion of these modes falls outside the scope of this survey.

Ecology of the mycelium

Viewed at its broadest, the mycelium is seen as an ecological adaptation to a xerophytic environment.

My suggestion (Gregory, 1966) that a fungus resembles a myxomycete, except that its protoplast moves about in a system of tubes, seemed at the time no more than a helpful analogy. But current practice classifies the myxomycetes firmly in the fungi (Martin & Alexopoulos, 1969; Whittaker, 1969; Ainsworth & Bisby, 1971), and therefore the relevance of the myxomycetes may be one of kinship rather than analogy. The unveiled activities of myxomycetes contrast with the more secluded life style of eumycetes. We can equate the myxomycete plasmodium with the eumycete protoplast. Working out correspondences between the functional modes in the two groups is rewarding, but will not be pursued here.

Ecologically the fungi can be arranged in a series showing increasing adaptation to a land environment. (1) The myxomycetes (starting perhaps with such purely aquatic forms as *Labyrinthula*) have plasmodium (or protoplast) more or less naked, apt for ingestion of bacteria in a wet environment, but dangerously vulnerable to drought. (2) The non-lichenised mycelial fungi, with protoplasts moving in tubes, are better adapted to xerophytism, but restricted to a liquid intake. (3) The series culminates in the lichen fungi, some of which must be among the most xerophytic organisms known. Here the mycelium does not explore the substratum, but is differentiated in one direction for anchoring, and in another direction for weaving the fabric of the alga-studded 'solar panels' which cover a substantial part of the Earth's surface.

The mycelial theme

A non-lichenised mycelial fungus is a heterotrophic nucleated cytoplasmic mass which constructs tubular cell walls within which it moves as it feeds off the substratum.

In the lower fungi the mycelium may be coenocytic, branching like a tree but not anastomosed; septa serve to repair wounds or to block off parts of the tubes from which the protoplast has withdrawn.

In the higher fungi the mycelium is semi-cellular, branched, and becomes a three-dimensional transport network through hyphal fusions. Septa remain perforate during active exploration of the substratum, and become closed at need.

When the mycelium switches in whole or in part to coherent growth or to a reproductive mode, the fungus inhabiting the tubes migrates towards the developing sporocarp (or resting body). In sporulation the fungus leaves behind empty tubes, breaking up and launching out in the form of spores to colonise new substrata, forming new mycelia, equipped like the parent mycelium for out-breeding and in-fighting.

On this theme are constructed a myriad variations, but this is the basic life-style of the mycelial fungi.

It is now clear that biologically the vegetative mycelium is the most important part of the fungus.

References

Ainsworth, G. C. & Bisby, G. R. (1971). *Dictionary of the Fungi*, 6th edn. Kew, England: Commonwealth Mycological Institute. 663 pp.

Arthur, J. C. (1897). The movement of protoplasm in coenocytic hyphae. *Annals of Botany*, **2,** 491–507.

Bracker, C. E. & Butler, E. E. (1964). Function of the septal pore apparatus in *Rhizoctonia solani* during protoplasmic streaming. *Journal of Cellular Biology*, **21,** 152–7.

Brierley, W. B. (1931). Biological races in fungi and their significance in evolution. *Annals of Applied Biology*, **18,** 420–34.

Buck, K. W. (1980). Viruses and killer factors of fungi. In *The Eucaryotic Microbial Cell* (ed. G. W. Gooday, D. Lloyd, & A. J. P. Trinci), pp. 329–75. Society for General Microbiology, Symposium No. 30. Cambridge: Cambridge University Press.

Buller, A. H. R. (1924). *Researches on Fungi*, vol. 3. London, UK: Longmans Green.

Buller, A. H. R. (1931). *Researches on Fungi*, vol. 4. London, UK: Longmans Green.

Buller, A. H. R. (1933). *Researches on Fungi*, vol. 5. London, UK: Longmans Green.

Burgess, N. A. (1960). Dynamic equilibrium in the soil. In *The Ecology of Soil Fungi*, ed. D. Parkinson & J. S. Waid, pp. 185–91. Liverpool: Liverpool University Press.

Burnett, J. H. (1976). *Fundamentals of Mycology*, 2nd edn. London: Arnold. 673 pp.

Burnett, J. H. & Partington, M. (1957). Spatial distribution of fungal mating-type factors. *Proceedings of the Royal Physical Society of Edinburgh*, **26,** 61–8.

Butler, G. M. (1958). The development and behaviour of mycelial strands in *Merulius lacrymans* (Wulf.) Fr. *Annals of Botany*, N.S. **22,** 219–36.

Campbell, A. H. (1933). Zone lines in plant tissues. I The black lines formed by *Xylaria polymorpha* (Pers.) Grev. in hardwoods. *Annals of Applied Biology*, **20,** 123–45.

Caten, C. E. & Jinks, J. L. (1966). Heterokaryosis: its significance in wild homothallic ascomycetes and fungi imperfecti. *Transactions of the British Mycological Society*, **49,** 81–93.

Chanter, D. O. & Thornley, J. H. M. (1978). Mycelial growth and the initiation and growth of sporophores in the mushroom crop: a mathematical model. *Journal of General Microbiology*, **106,** 55–65.

Cole, G. T. & Nazawa, Y. (1981). Dimorphism. In *Biology of Conidial Fungi*, vol. 1, ed. G. T. Cole & W. B. Kendrick, chapter 5, pp. 97–133, New York, London: Academic Press.

Davidson, A. M., Dowding, E. S. & Buller, A. H. R. (1932). Hyphal fusions in dermatophytes. *Canadian Journal of Research*, **6,** 1–20.

De Bary, A. (1887). *Comparative Morphology and Biology of the Fungi, Mycetozoa and Bacteria.* Oxford: Clarendon Presses. 525 pp.

Dowding, E. S. (1958). Nuclear streaming in *Gelasinospora. Canadian Journal of Microbiology*, **4,** 295–301.

Esser, K. (1966). Incompatibility. In *The Fungi: an Advanced Treatise*, vol. 2, ed. G. C. Ainsworth & A. S. Sussman, chapter 20. New York & London: Academic Press.

Flentje, N. T. & Stretton, H. M. (1964). Mechanisms of variation in *Thanatephorus cucumeris* and *Thanatephorus praticolus*. *Australian Journal of Biological Science*, **17,** 686–704.

Fries, N. (1981). Recognition reactions between basidiospores and hyphae in *Leccinum*. *Transactions of the British Mycological Society*, **77,** 9–14.

Fritsch, F. E. (1945). *The Structure and Reproduction of the of the Algae*, vol. 2. Cambridge: Cambridge University Press.

Garrett, S. D. (1953). Rhizomorph behaviour in *Armillaria mellea* (Vahl.) Quél. I. Factors controlling rhizomorph initiation by *A. mellea* in pure culture. *Annals of Botany*, N. S. **17,** 63–79.

Giesy, R. M. & Day, P. R. (1965). The septal pores of *Coprinus lagopus* in relation to nuclear migration. *American Journal of Botany*, **52,** 287–93.

Gregory, P. H. (1952). Fungus spores. *Transactions of the British Mycological Society*, **35,** 1–18.

Gregory, P. H. (1966). The fungus spore: what it is and what it does. In *The Fungus Spore*, ed. M. F. Madelin, pp. 1–13. London: Butterworth.

Hollings, M. (1962). Viruses associated with die-back disease of cultivated mushroom. *Nature,* **194,** 962–5.

Hollings, M. (1982). Mycoviruses: viruses that infect fungi. *Advances in Virus Research*, **22,** 1–53.

Isaac, P. K. (1964). Cytoplasmic streaming in filamentous fungi. *Canadian Journal of Botany*, **42,** 787–92.

Jennings, D. H. (1982). The movement of *Serpula lacrimans* from substrate to substrate over nutritionally inert surfaces. In *Decomposer Basidiomycetes*, ed. J. Frankland, J. Hedger & M. J. Swift, chapter 5, pp. 91–108. British Mycological Society, Symposium No. 4. Cambridge: Cambridge University Press.

Jinks, J. L., Caten, C. E., Simchen, G. & Croft, J. H. (1966). Heterokaryon incompatibility and variation in wild populations of *Aspergillus nidulans*. *Heredity*, **21,** 227–39.

Kemp, R. F. O. (1975). Breeding biology of *Coprinus* species in the section *Lantuli*. *Transactions of the British Mycological Society*, **65,** 375–88.

Köhler, E. (1930). Zur Kenntnis der vegetationen Anastomosen der Plize. II Mitteilung. Ein Beitrag zur Frage der spezifischen Pilzwirkungen. *Planta*, **10**, 495–522.

Langeron, M. (1945). *Précis de Mycologie.* Paris: Masson. 675 pp.

Lemke, P. A. (1981). Viruses of conidial fungi. In *Biology of Conidial Fungi*, vol. 2, ed. G. T. Cole & W. Bryce Kendrick, chapter 26, pp. 395–416. New York, London: Academic Press.

Martin, G. W. & Alexopoulos, C. J. (1969). *The Myxomycetes.* Iowa City: University of Iowa Press. 560 pp.

Meyer, A. (1902). Das Plasmaverbindungen und die Fusionen der Pilze den Floridenreihe. *Botanisches Zeitung*, **60,** 139–78.

Moore, R. T. & McAlear, J. H. (1962). Fine structure of mycota. 7, Observations on septa of Ascomycetes and Basidiomycetes. *American Journal of Botany*, **49,** 86–94.

Müller, E. (1979). Factors inducing asexual and sexual sporulation in fungi (mainly Ascomycetes). In *The Whole Fungus: the Sexual and Asexual Synthesis*, ed. W. Bryce Kendrick, chapter 16, pp. 265–78. Ottawa: Ottawa National Museum of Canada.

Nickerson, W. J. (1947). *Biology of Pathogenic Fungi.* Waltham, Mass.: Chronica Botanica. 236 pp.

Niederpruem, D. J. (1980). Direct studies of dikaryotization in *Schizophyllum commune. Archives of Microbiology*, **128,** 162–71.

Plunkett, B. E. (1958). Translocation and pileus formation in *Polyporus brumalis. Annals of Botany*, N. S. **22,** 237–49.

Pontecorvo, G. (1956). The parasexual cycle in fungi. *Annual Review of Microbiology*, **10,** 393–400.

Raper, J. R. (1966). *Genetics of Sexuality in Higher Fungi.* New York: Ronald.

Raper, J. R. & Flexer, A. S. (1970). The road to diploidy with emphasis on a detour. In

Prokaryotic and Eukaryotic Cells, Symposia of the Society for General Microbiology No. 20, pp. 401–32.
Rayner, A. D. M. & Todd, N. K. (1977). Intraspecific antagonism in natural populations of wood-decaying basidiomycetes. *Journal of General Microbiology*, **103,** 85–90.
Rayner, A. D. M. & Todd, N. K. (1979). Population and community structure and dynamics of fungi in decaying wood. *Advances in Botanical Research*, **7,** 333–420.
Reinhardt, M. O. (1892). Das Wachstum der Pilzhyphen. *Jarhbuch für wissenschaftliche Botanik*, **23,** 479–566.
Rothert, W. (1892). Ueber *Sclerotium hydrophilum* Sacc., einen sporenlosen Pilze. *Botanische Zeitung*, **50,** 321–539.
Schisler, L. C., Sinden, J. W. & Sigel, E. M. (1967). Etiology, symptomatology, and epidemiology of a virus disease of cultivated mushrooms. *Phytopathology*, **57,** 519–26.
Schütte, K. H. (1956). Translocation in the fungi. *New Phytologist*, **55,** 164–82.
Snider, P. J. (1963). Genetic evidence for nuclear migration in Basidiomycetes. *Genetics*, **48,** 47–55.
Stephenson, L. W., Erwin, D. C. & Leary, J. V. (1974). Hyphal anastomosis in *Phytophthora capsici*. *Phytopathology*, **64,** 149–50.
Ternetz, C. (1900). Protoplasmabewegung und Fruchtkörperbildung bei *Ascophanus carneus*. *Jahrbuch für wissenschaftliche Botanik*, **35,** 273–312.
Todd, N. K. & Rayner, A. D. M. (1980). Fungal individualism. *Science Progress, Oxford*, **66,** 331–54.
Tulasne, L. R. & Tulasne, C. (1861–1865). *Selecta Fungorum Carpologia*. Paris. (English translation: W. B. Grove, Oxford University Press, 1931.)
Vandendries, R. & Brodie, H. J. (1933). Nouvelles investigations dans la domaine de la sexualité des Basidiomycètes et l'étude expérimentale des barrages sexuels. *La Cellule*, **42,** 165–209.
Wahrlich, W. (1893). Zur Anatomie der Zelle bei Pilzen und Fadenalgen. *Scripta Botanica Horti Universalis Imperialis Petropolitanae*, **4,** 101–55 (and in Russian).
Ward, H. Marshall (1888). On a lily disease. *Annals of Botany*, **2,** 319–82.
Whittaker, R. H. (1969). New concepts of kingdoms of organisms. *Science*, **163,** 150–60.
Wilcoxon, R. D. & Sudia, T. W. (1968). Translocation in fungi. *Botanical Reviews*, **34,** 32–50.
Woronin, M. (1866). Entwicklungsgeschichte des *Ascobolus pulcherrimus*. In *Beiträge zum Morphologie und Physiologie der Pilze II*, ed. A. de Bary, A. & M. Woronin, p. 2.

2

Regulation of hyphal branching and hyphal orientation

A. P. J. TRINCI
Cryptogamic Botany Laboratories, Botany Department, University of Manchester, Manchester, M13 9PL, UK

Bull & Trinci (1977) suggest that the morphology of a vegetative mycelium is determined largely by mechanisms which regulate the polarity and the direction of hyphal growth and the frequency with which hyphae branch. This chapter considers the involvement of the latter two mechanisms in the determination of mycelial morphogenesis. There is little doubt that these mechanisms make a significant contribution to the efficiency with which fungi colonise solid substrata and similar if not identical mechanisms can be identified in mycelia formed by Streptomycetes (bacteria) and *Candida albicans*. A mathematical model has been proposed by Prosser & Trinci (1979) to explain how branching may be regulated in fungi, but less is known about the mechanisms which determine the spatial distribution of hyphae within mycelia.

Spore germination and germ tube growth

Spore 'swelling' is the first morphological stage in germination and during this period spores increase in diameter at a linear rate (Ekundayo & Carlile, 1964; Fletcher, 1969; Gull & Trinci, 1971). However, since spores increase in biomass as well as in volume during spore 'swelling' (Van Etten, Dunkle & Freer, 1977), it is more correct to refer to this period as a phase of spherical growth (Bartnicki-Garcia, 1981). Spherical growth of the spore usually persists for only a few hours, but this phase may be prolonged by incubating spores at supraoptimal temperatures (Anderson & Smith, 1972) or by exogenous cAMP (Paznokas & Sypherd, 1975). However, under normal circumstances spherical or isotropic growth of the spore soon ceases and a germ tube is initiated, that is growth becomes polarised. The site on the spore wall from which the germ tube is produced may be determined by ionic currents generated within the spore; such currents can simultaneously perform work (transport) and provide vectorial information (Harold, 1982). Significantly, De Vries & Wessels (1982) have shown that the point at which hyphae emerge from regenerating protoplasts of *Schizophyllum commune* is influenced by external electrical fields.

Spherical growth of the spore is followed by a phase in which the spore and germ tube increase in length exponentially (Trinci, 1971*a*; Katz, Goldstein & Rosenberger, 1972), and in turn this phase is followed by one in which the extension rate of the germ tube continues to increase but its growth is no longer exponential; exponential growth of the whole germling is, however, maintained by branch formation (see below). In *Aspergillus nidulans* there seems to be a direct relationship between the mean length of the apical compartment of the germ tube and its extension rate (Fiddy & Trinci, 1976*a*); this relationship may reflect the increase in the length of the peripheral growth zone (Trinci, 1971*b*) which must accompany the observed increase in the rate of germ-tube extension. Germ tubes finally attain a linear rate of extension and this rate is characteristic of the strain and cultural conditions; germ tubes of *Aspergillus nidulans* (Fiddy & Trinci, 1976*a*) and *Mucor hiemalis* first attain this rate when they are about 600 and 950 μm long respectively.

In contrast to the behaviour of moulds, germ tubes formed by *Candida albicans* (Fig. 1) and *Streptomyces coelicolor* (Allan & Prosser, 1983) apparently extend at a linear rate from their inception. When *C. albicans* germinates, protoplasm migrates from the yeast cell into the germ tube, and during subsequent growth protoplasm migrates forward

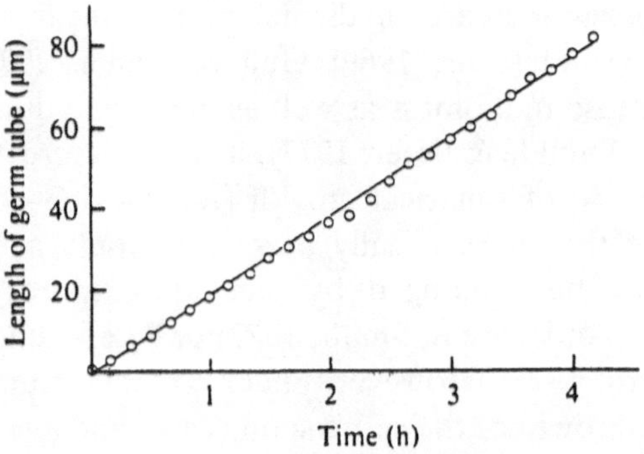

Fig. 1. Growth of a germ tube of *Candida albicans* from a yeast cell on 20% serum agar at 37°C. (From Gow & Gooday, 1982*b*.)

with the extending apex (Gow & Gooday, 1982*a*). A yeast cell of *C. albicans* has a volume (about 95 μm^3) which is more than half that (about 140 μm^3) of the peripheral growth zone of the germ tube (Gow & Gooday, 1982*a*) and therefore the parental cell can make a substantial contribution to the attainment of linear growth by the germ tube. However, a conidium of *Aspergillus nidulans* has a volume (about 90 μm^3) which represents only a very small fraction of the volume (about 4250 μm^3) of the peripheral growth zone of a germ tube extending at a linear rate. It is therefore not surprising that germ tubes of *C. albicans*, unlike those of *A. nidulans*, rapidly attain their maximum rate of extension. The volume measurements for *C. albicans* and *A. nidulans* given above do not take account of the differences which may exist between vacuolation of cytoplasm in the parent cell (yeast cell or spore) and the germ tube, and they therefore provide only first estimates of the biomass actually contributing to growth.

Mycelial growth

The hyphal growth unit

Linear growth of individual hyphae within a mycelium is associated with a continual increase in the number of hyphae, so that growth is proportional to the biomass present, i.e. exponential (Plomley, 1959). Plomley suggested that mycelial growth involved the duplication of a growth unit which consisted of 'a free growing hyphal tip associated with a growing mass of constant size'. He determined the size of this unit by measuring the length of germ tubes when they first branched.

When a fungal mycelium is cultured in the absence of inhibitory substances on a medium containing an excess of all nutrients, its total hyphal length and the number of branches increase exponentially at a specific rate characteristic of the strain and cultural conditions. Growth of *Penicillium chrysogenum* under such conditions (unrestricted conditions, Righelato, 1975) is shown in Fig. 2. If the total hyphal length and the number of branches of the mycelium increase at exactly the same specific rate, the ratio between them will be a constant. This ratio has been called the hyphal growth unit (Caldwell & Trinci, 1973) and growth of a mycelium may thus be considered to involve the duplication of a growth unit which consists of a tip and a certain length of hypha. The length of the hyphal growth unit is species and strain specific (Table 1). If the *mean* diameter of hyphae in the mycelium remains constant as it increases in size, the hyphal growth unit will be of constant volume as well as of constant length. When Caldwell & Trinci (1973) grew

Geotrichum candidum in batch culture they found that the dry weight, total hyphal length and number of hyphal tips of the culture increased exponentially at specific rates of 0.173, 0.187 and 0.198 h^{-1}, respectively. This result suggests that the biomass as well as the length of the growth unit remains constant during mycelial growth and confirms Plomley's (1959) hypothesis.

Mycelia of *Candida albicans* (Fig. 3), *Streptomyces hygroscopicus* (Schuhmann & Bergter, 1976) and *Streptomyces coelicolor* (Allan & Prosser, 1983) have growth kinetics which are similar to those of

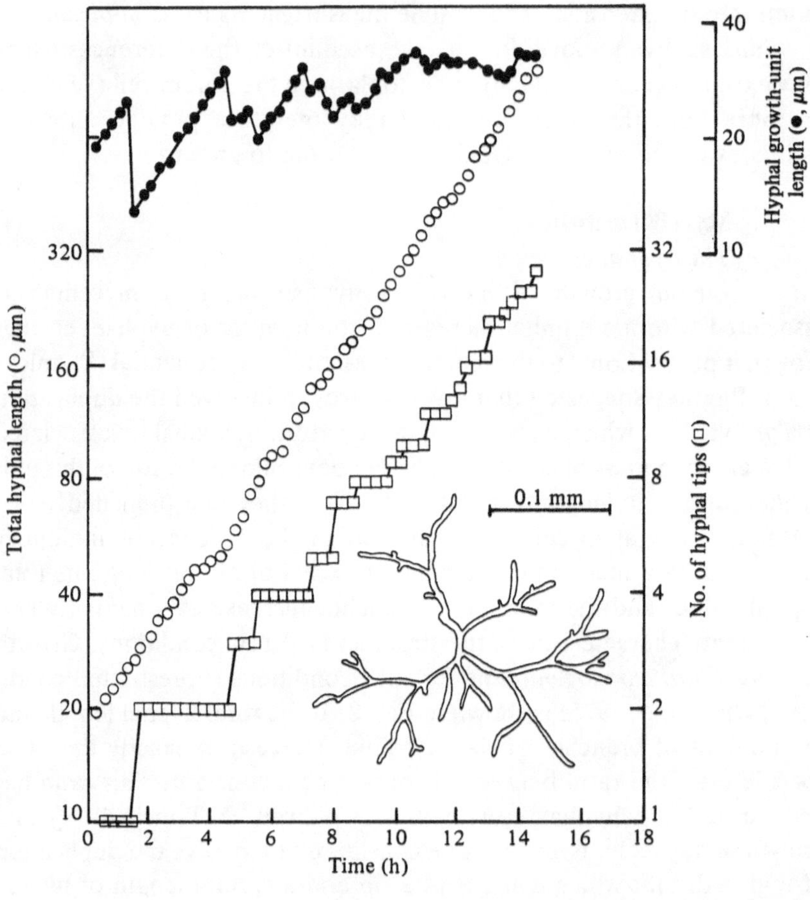

Fig. 2. Growth of a mycelium of *Penicillium chrysogenum* on a glucose-mineral salts medium at 25°C. Number of tips (□), total hyphal length (○) and length of the hyphal growth unit (●). The drawing at the right shows the mycelium at the end of the observation period.

Penicillium chrysogenum except that growth of *Streptomyces hygroscopicus* is apparently biphasic whilst in *C. albicans* there is an initial hyperbolic relationship between log mycelial length and time (Fig. 3). However, in these organisms, as in moulds, mycelial growth eventually involves the duplication of a growth unit consisting of a tip and a certain length of hypha which is species specific (Table 1).

Hyphae of *Candida albicans, Streptomyces coelicolor* and other streptomycetes are appreciably thinner than those of moulds, and therefore the volume of the hyphal growth unit contributing to the extension of the hyphal tip of a mould is much larger than that contributing to the extension of a hypha of *C. albicans* or a streptomycete (Table 1).

It is not known what mechanism regulates hyphal growth unit length, but, as suggested by Plomley (1959) and Katz *et al.* (1972), there may be some relationship between hyphal growth unit length and the maximum rate at which hyphae can extend. This maximum rate may be limited either by the maximum rate at which wall precursors can be transported to the tip or by the maximum rate at which the tip wall can be assembled from these precursors. The observation that some hyphae branch apically (Trinci, 1970; Robinson & Smith, 1980) supports the latter hypothesis since it suggests that the rate of supply of wall precursors to

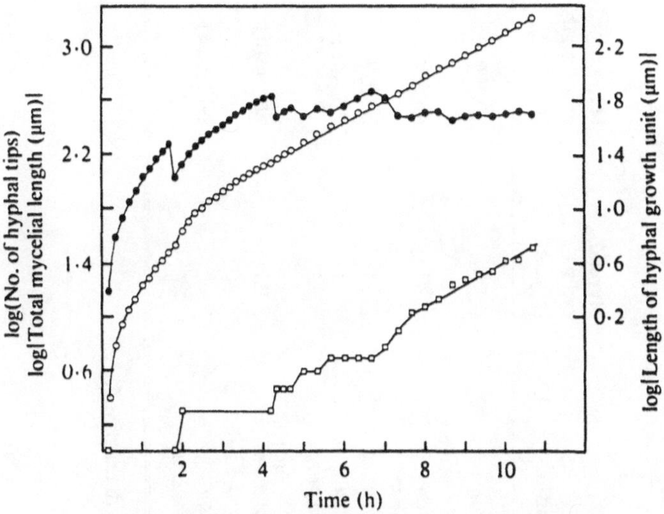

Fig. 3. Growth of a mycelium of *Candida albicans* on 20% serum agar at 37°C. Number of tips (□), total hyphal length (○) and length of the hyphal growth unit (●). (From Gow & Gooday, 1982*b*.)

Table 1. *Hyphal radius, hyphal growth unit length, hyphal growth unit volume and specific growth of micro-organisms grown in batch culture*

Organism	Hyphal radius (r, μm)	Hyphal growth-unit length (G, μm)	Hyphal growth-unit volume (V_g, μm^3)	Specific growth rate of total hyphal length (μ, h^{-1})	Authority
Streptomyces coelicolor	0.51	32 ± 9	26	0.26	Allan & Prosser (1983)
Candida albicans	1.20	48 ± 9	217	0.394	Gow & Gooday (1982*b*)
Neurospora crassa spco 12	2.50	32 ± 9	629	0.203	Trinci (1973*a*, *b*)
Neurospora crassa spco 1	3.32	130 ± 37	4,504	0.322	Trinci (1973*a*, *b*)
Neurospora crassa SYR-17-3A	3.08	402 ± 141	11,986	0.348	Trinci (1973*a*, *b*)
Fusarium avenaceum	*	602 ± 141	*	*	Bull & Trinci (1977)

* not determined.

the tip can, at least under some circumstances, exceed the rate at which precursors can be incorporated into an existing tip wall.

The extension rate of mycelial hyphae

In a fungal mycelium some hyphae will be extending at a linear rate whilst others will be accelerating towards this rate (Trinci, 1974). Thus the actual length of hypha associated with tip extension (its peripheral growth zone) will vary from a minimum value just after branch initiation to a maximum value when the tip is extending at a linear rate. Trinci (1974) determined the mean rate of hyphal extension (E) using the following equation.

$$E = \frac{2(H_t - H_0)}{B_0 + B_t}, \tag{1}$$

where H_0 = total hyphal length of the mycelium at zero time, H_t = total hyphal length 1 h later, B_0 = number of tips at zero time, and B_t = number of tips 1 h later. He found that the mean hyphal extension rate of *Geotrichum candidum* increased until the germling possessed about three tips, but then attained a value ($48 \pm 3\,\mu m\,h^{-1}$) which remained approximately constant as the mycelium increased in size. Steele & Trinci (1975) showed that E was a function of the length of the organism's hyphal growth unit (G) and its specific growth rate (μ). Thus

$$E = G\mu. \tag{2}$$

In contrast to the behaviour of branch hyphae of moulds, branches formed by mycelia of *Streptomyces coelicolor* extend at a linear rate from their inception (Allan & Prosser, 1983). They thus behave like the germ tubes formed from yeast cells of *Candida albicans* (page 24). The volume of the peripheral growth zone of a hypha of *Streptomyces coelicolor* is very small (26 μm) and such hyphae will be expected to attain a linear growth rate when they are still quite short. It is not known whether growth of a branch hypha of a streptomycete, like a branch hypha of a mould (Fiddy & Trinci, 1967*a*, *b*), is initially supported by cytoplasm in the parent hypha.

The relationship shown in equation (2) has also been demonstrated for mycelia of *Streptomyces hygroscopicus* (Schuhmann & Bergter, 1976) and *Candida albicans* (Gow & Gooday, 1982*b*) and thus appears to reflect a fundamental property of fungal and streptomycete mycelia.

Factors which affect branch frequency

Variation in hyphal diameter

When the hyphal growth unit hypothesis was proposed (Caldwell & Trinci, 1973; Trinci, 1974) it was assumed that the mean radius of hyphae in mycelia cultured under unrestricted conditions remained constant. It was therefore assumed that hyphal length was a valid parameter of hyphal volume and possibly also of hyphal biomass. Thus, it was considered that mycelial growth involved the duplication of a growth unit whose length, volume and biomass remain constant. However, as is shown below, some cultural conditions affect hyphal diameter and consequently affect hyphal growth-unit length.

When *Aspergillus nidulans sep A2* is cultured at 15, 20 or 25 °C it forms septate mycelium which is morphologically similar to wild type mycelium (Trinci & Morris, 1979). However, at 37 °C *sep A2* forms non-septate mycelium which grows more slowly than the parental strain; the mutant has a specific growth rate of $0.424\,h^{-1}$ compared with $0.572\,h^{-1}$ for the parental strain. Table 2 compares the hyphal growth unit and hyphal radius of mycelia of *A. nidulans sep A2* grown at permissive (15, 20 and 25 °C) and restrictive (37 °C) temperatures.

Although *Aspergillus nidulans sep A2* has a shorter hyphal growth unit at 37 °C than at lower temperatures, the volume of the hyphal growth unit was not appreciably affected by temperature (Table 2). The decrease in hyphal growth-unit length at 37 °C was thus entirely due to the observed increase in hyphal radius.

If it is assumed that the volume of the hyphal growth unit remains constant during mycelial growth, it follows that relatively small changes in hyphal radius will have appreciable effects on hyphal growth-unit length (G), since, as shown below, G will be inversely related to the square of hyphal radius (r).

$$G = \frac{V_g}{\pi r^2}. \qquad (3)$$

where V_g = hyphal growth unit volume. Robinson & Smith (1979) found that the volume of the hyphal growth unit of *Geotrichum candidum* remained constant when this organism was grown at different dilution rates in glucose-limited chemostat cultures. However, hyphal diameter increased with increase in dilution rate and hyphal growth-unit length decreased with increase in dilution rate (Fig. 4). Thus, under these experimental conditions, G was inversely related to the square of hyphal radius (eq. 3). Robinson & Smith (1979) also found that temperature

Table 2. *Effect of temperature on septation, hyphal radius and hyphal growth-unit length of* Aspergillus nidulans sep A2 (*Data from Trinci & Morris, 1979)*

	Incubation temperature	
Characteristic	15, 20 and 25 °C	37 °C
Hyphal morphology	septate	aseptate
Hypha growth unit volume (V_g, μm^3)	1638	1854
Mean hyphal radius (r, μm)	2.01	2.65
Mean hyphal growth-unit length (G, μm)	129	84

had no appreciable effect on the volume of the hyphal growth unit of *Geotrichum candidum*; hyphal growth unit volumes of about 1500 and 1470 μm^3 were observed at 25 and 30 °C respectively. This result is similar to that obtained with *Aspergillus nidulans sep A2* (Table 3).

In contrast to the results obtained by Robinson & Smith (1979), Fiddy & Trinci (1975) found that when *Geotrichum candidum* was grown in glucose-limited chemostat cultures at dilution rates between 0.07 and 0.33 h^{-1}, specific growth rate had no appreciable effect on hyphal diameter or intercalary compartment length. It may be significant that

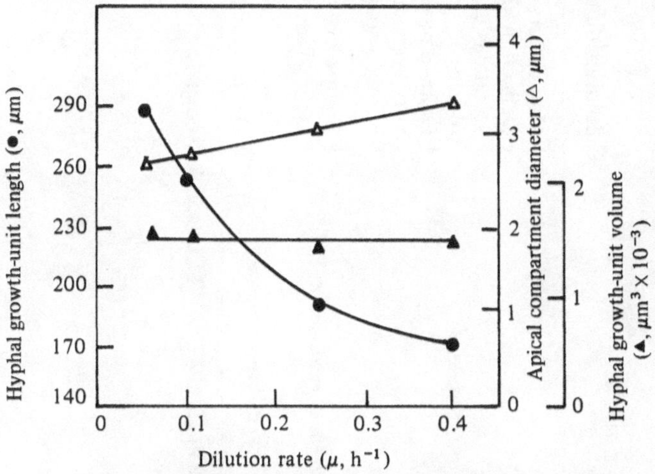

Fig. 4. Effect of dilution rate on hyphal growth unit length (●), hyphal growth-unit volume (▲) and hyphal diameter (△) of *Geotrichum candidum* grown in glucose-limited chemostat culture at 30°C. (Redrawn from Robinson & Smith, 1979.)

Table 3. *Effect of medium composition on specific growth rate, hyphal diameter and hyphal growth-unit length*

Organism	Culture medium	Specific growth rate (μ, h^{-1})	Hyphal diameter (μm)	Hyphal growth-unit length (G, μm)
Aspergillus nidulans[a]	Malt extract	0.36	1.8	~33
	Defined medium with acetate as carbon source	0.14	1.6	~73
	Defined medium with L-tryptophan as nitrogen source	0.11	1.1	~120
Penicillium chrysogenum T14[b]	Complex medium	0.24	—	43
	Defined medium	0.14	—	60
Streptomyces hygroscopicus[c]	Complex medium	0.44	—	34
	Defined medium	0.27	—	13

[a] Katz *et al*. (1972); [b] Morrison & Righelato (1974); [c] Schuhmann & Bergter (1976).

Robinson and Smith's cultures were agitated with an impeller, whereas those of Fiddy and Trinci were stirred magnetically; mycelia grown in cultures stirred with an impeller will be subjected to greater shear stress than mycelia grown in cultures stirred magnetically and shear stress is known to affect hyphal growth-unit length (see below) and hyphal diameter (Ujcova, Fencl, Musilkova & Seichert, 1980). Mycelia grown at various dilution rates in a chemostat will be subjected to the shear stress of the impeller for different periods, the duration of which will be a function of the organism's residence time in the culture vessel.

When Riesenberger & Bergter (1979) grew *Streptomyces hygroscopicus* in glucose-limited chemostat culture they, like Robinson & Smith (1979), observed that hyphal diameter increased with increase in dilution rate and hyphal growth-unit length decreased with increase in dilution rate (between 0.05 and 0.32 h^{-1}). Similarly, Matin & Veldkamp (1978) reported that the width of cell of *Spirillum* and *Pseudomonas* grown in glucose-limited chemostat culture increased with increase in dilution rate. This thinning effect with decrease in dilution rate was particularly pronounced in the *Spirillum* species, whose cell length was relatively constant over a wide range of growth rates.

Variation in medium composition and incubation temperature

The specific growth rate of a batch culture may be varied either by changing the composition of the medium or by varying the incubation temperature. When Katz *et al.* (1972) varied the specific growth rate of *Aspergillus nidulans* by growing it on different media, they found that hyphal diameter increased with increase in specific growth rate and hyphal growth-unit length decreased with increase in specific growth rate (Table 3). This result is similar to that obtained by Robinson & Smith (1979) with *Geotrichum candidum* in glucose-limited chemostat culture (see above) and similar results have been obtained with *Penicillium chrysogenum* and *Streptomyces hygroscopicus* (Table 3).

Trinci (1973*a, b*) found that temperature had no appreciable effect on the length of the hyphal growth unit of *Neurospora crassa spco* 1; the specific growth rate of this organism increased from 0.192 h^{-1} at 20 °C to 0.452 h^{-1} at 37 °C. Thus, temperature affects the rate at which the hyphal growth unit is duplicated, but not its length. In contrast, the length of the hyphal growth unit of the temperature-sensitive mutant, *N. crassa cot 2* varied from 193 ± 55 μm at 15 °C to 38 ± 8 μm at 30 °C (Fig. 5), although the specific growth rate of this organism increased from 0.10 h^{-1} to 0.32 h^{-1} over the same temperature range (Steele & Trinci, 1977*a*). This result can be explained by assuming that the mutation either has a direct effect on the rate of formation of the tip wall or affects hyphal extension indirectly by reducing the rate of transport of wall precursors to the tip.

Growth within Cellophane

When *Phytophthora gonapodyides* grows on the surface of Cellophane laid over nutrient medium, it forms sparsely branched mycelia. However, hyphae occasionally (Fig. 6) penetrate (usually via some weakness in the surface of the Cellophane) and grow within the

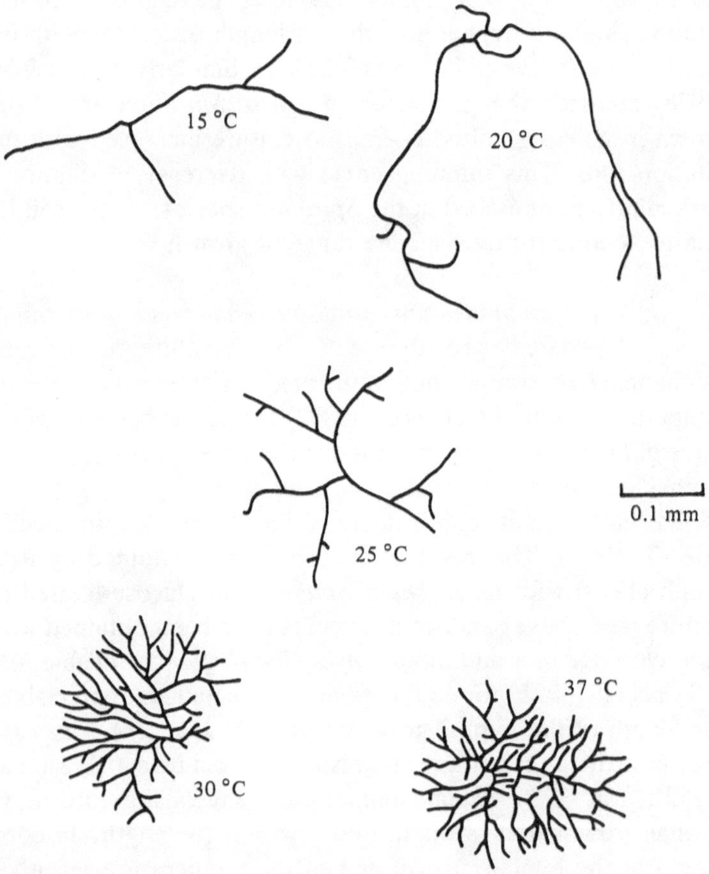

Fig. 5. Mycelia of *Neurospora crassa cot 2* grown on Vogel's medium at various temperatures. (From Steele & Trinci, 1977*a*.)

laminated sheet of Cellophane where they branch profusely and form frond-like mycelia composed of wide hyphae (Fig. 7). The hyphae within the Cellophane contain an abundance of cytoplasmic vesicles and these are similar to those observed at the hyphal tip (Dr Janice Pittis, personal communication). When 'fronded' hyphae emerge from the Cellophane they revert immediately to their normal diameter and branching pattern. No evidence was obtained that *P. gonapodyides* actively degrades the Cellophane and it seems unlikely that the morphological variation induced by growth within Cellophane is caused by nutrient limitation, or lack of oxygen. It is, however, possible that the physical constraint on hyphal extension imposed by growth within Cellophane may have caused an increase in branch frequency (eq. 2, and see below).

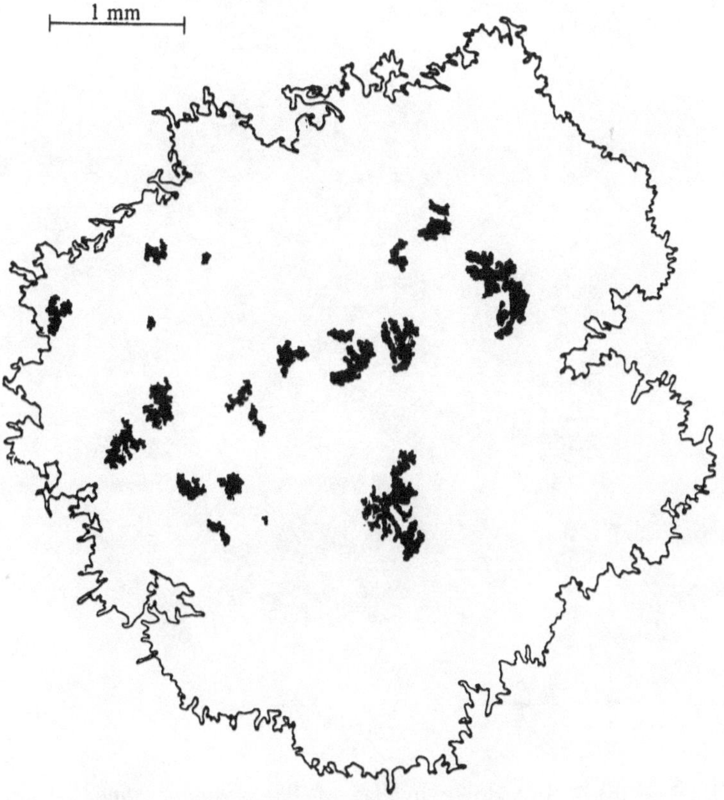

Fig. 6. Location of frond-like hyphae (shaded areas) in a colony of *Phytophthora gonapodyides* grown for 10 days on a medium overlaid with a Cellophane sheet. (Drawing made by Dr Janice Pittis.)

Variation in pH of cultures grown at a constant specific growth rate

A chemostat may be used to keep the specific growth rate (μ) of an organism constant whilst varying other environmental conditions which may affect hyphal extension. Miles & Trinci (1983) showed that

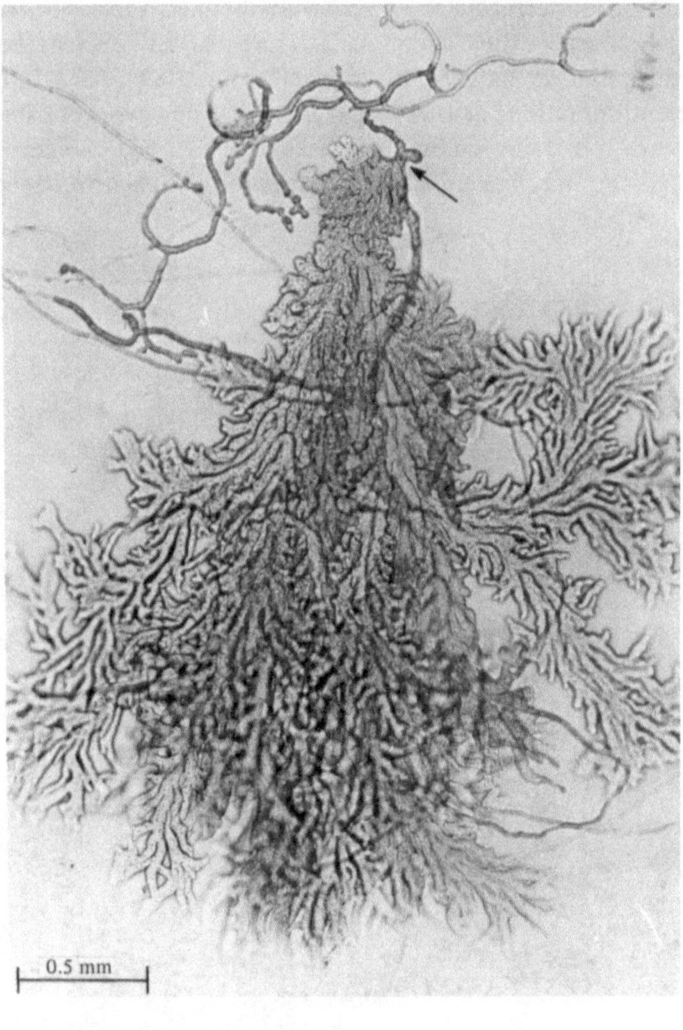

Fig. 7. Frond-like hyphae of *Phytophthora gonapodyides* growing within a sheet of PT 300 Cellophane. The arrow indicates the point at which a normal hypha penetrated the Cellophane. (Photograph provided by Dr Janice Pittis.)

when *Penicillium chrysogenum* was grown at a dilution rate of 0.09 h^{-1} in a glucose-limited chemostat culture, the organism had a significantly longer hyphal growth unit at pH 6.0 than at the other pH values tested (Fig. 8). One explanation of this result is that pH 6.0 is the optimum pH for hyphal extension; if hyphal growth-unit volume and hyphal radius of mycelia grown at a *particular dilution rate* remain constant, the variation in any condition, e.g. pH, which affects hyphal extension will cause a change in hyphal growth-unit length (eq. 2 and see below). Under these circumstances, maximum hyphal growth-unit values will be observed when pH is optimal for hyphal extension; enzymes involved in chitin synthesis are thought to be located in the protoplasmic membrane or the wall itself.

Effect of stirrer speed on mycelial morphology

When *Penicillium chrysogenum* was grown in glucose-limited chemostat culture at a stirrer speed of 1000 rev min^{-1}, Morrison & Righelato (1974) and Miles & Trinci (1983) found that hyphal growth-unit length decreased with increase in dilution rate (Fig. 9); this result is

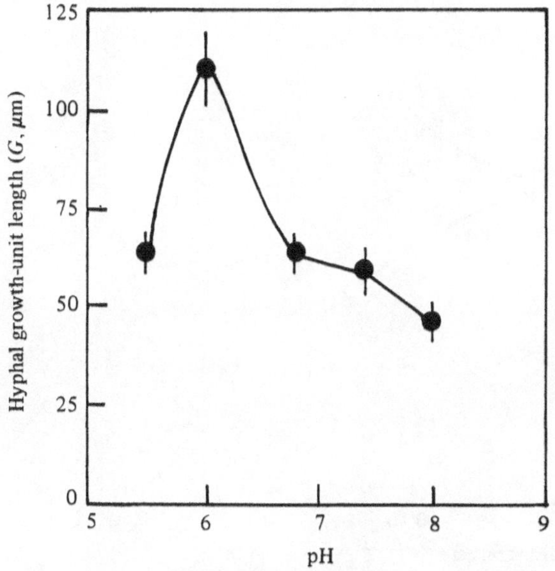

Fig. 8. Effect of pH on the length of the hyphal growth unit of *Penicillium chrysogenum* grown in glucose-limited chemostat culture at 25 °C, a dilution rate of 0.09 h^{-1}, a stirrer speed of 1000 rev min^{-1}, and various pH values. Vertical lines represent 95% confidence limits for the mean. (From Miles & Trinci, 1983.)

similar to that obtained by Robinson & Smith (1979) with *Geotrichum candidum*. However, when *P. chrysogenum* was grown at a stirrer speed of 500 rev min^{-1}, hyphal growth-unit length increased with increase in dilution rate (Fig. 9). Van Suijdam & Metz (1981) have also shown that stirrer speed has a significant effect on mycelial morphology; when *P. chrysogenum* was grown in glucose-limited chemostat culture at a dilution rate of 0.055 h^{-1}, there was an increase in hyphal growth-unit length from 54 μm at an impeller speed of 1000 rev min^{-1} to 89 μm at 450 rev min^{-1}. Thus the shear stress of the impeller has a significant effect on hyphal growth unit length.

Ujcova *et al.* (1980) found that hyphae of *Aspergillus niger* grown in batch culture became thicker with increase in impeller speed, that is, with increase in shear stress. However, it is not known whether the effect of impeller speed on the morphology of mycelia grown in

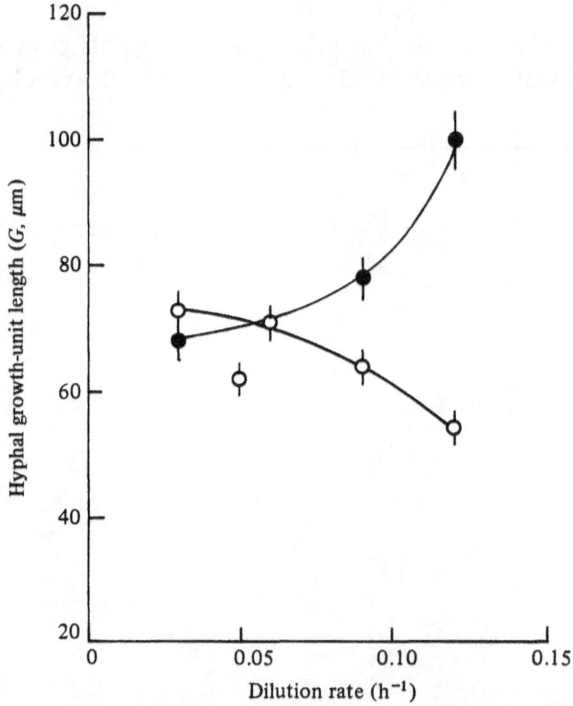

Fig. 9. Effect of dilution rate on the length of the hyphal growth unit of glucose-limited chemostat cultures of *Penicillium chrysogenum* grown at 25 °C, pH 6.8 and an impeller speed of 500 (●) or 1000 (○) rev min^{-1}. Vertical lines represent 95% confidence limits for the mean. (From Miles & Trinci, 1983.)

chemostat culture is due to an effect on hyphal diameter or an effect on hyphal extension.

A general model for mycelial branching

When a mycelium is grown under conditions which support exponential growth, the length, volume and biomass of the hyphal growth unit remain approximately constant. Several workers have now shown that the mean hyphal extension rate (E) of a mycelium grown under such conditions is a function of the length of the organism's hyphal growth unit (G) and its specific growth rate (μ). Equation (2) can be rewritten

$$E = \frac{V_g}{\pi r^2}\,\mu. \qquad (4)$$

This equation shows that variations in hyphal diameter will affect both hyphal extension and hyphal growth-unit length. Further, changes in specific growth rate will only affect hyphal growth unit length if the hyphal extension rate of the mycelium (E) is not linearly related to specific growth rate (μ), i.e. if the ratio E/μ does not remain constant. The observation that the hyphal growth unit of *Neurospora crassa spco 1* was not affected by temperature (Trinci, 1973*a*) suggests that the ratio E/μ is also not affected by temperature and this has been confirmed experimentally (Trinci, 1974). Katz *et al.* (1972) suggested that 'when the apex or apices reach a characteristic maximum rate of extension, which is achieved at different hyphal lengths with different growth rates, a new branch is initiated'. However, in fungi, a germ tube hypha or the branch hyphae of a mycelium initiate new branches before the existing hyphae have attained the organism's characteristic linear growth rate (Trinci, 1974). In addition, when the specific growth rate of *N. crassa spco* 1 is changed by varying the temperature, mycelia attain their characteristic mean rate of hyphal extension at the same hyphal growth-unit length.

Conditions which affect the ratio E/μ will also affect hyphal growth-unit length, i.e. the density of mycelial branching. Such conditions certainly or probably include mutations (the *sep A2* mutant of *Aspergillus nidulans* and the *spco* and *cot 2* mutations of *Neurospora crassa*), growth under confined conditions (growth of *Phytophthora gonapodyides* in Cellophane), impeller speed in fermenters (Fig. 9), paramorphogens such as L-sorbose (Trinci & Collinge, 1973), dilution rate in chemostat culture (Robinson & Smith, 1979, and Fig. 4) and pH changes in chemostat cultures grown at a fixed dilution rate (Miles & Trinci, 1983, and Fig. 8). A change in the ratio E/μ is probably always accompanied by a change in hyphal diameter, e.g. when the *sep A2* mutant of *A. nidulans* is grown at permissive and non-permissive temperatures and when *Geotrichum candidum* is grown at different dilution rates in a chemostat (Fig. 4); but in addition it may also be accompanied by a change in hyphal growth-unit volume, e.g. the *spco* mutations of *N. crassa* (Table 1) and when moulds are treated with paramorphogens such as L-sorbose (Trinci & Collinge, 1973). Thus the present evidence suggests that mycelial branching is determined by mechanisms which regulate hyphal growth-unit volume, hyphal radius and the ratio between the mean extension rate of mycelial hyphae and the organism's specific growth rate (eq. 4).

Septation, branch frequency and branch location

Branch initiation involves some event(s) which results in the transformation of rigid wall into extensible wall; any part of the hyphal wall, including the primary wall of the extension zone and the septum, can potentially form a branch (Trinci & Collinge, 1974; Collinge, Fletcher & Trinci, 1978). The previous part of this chapter has described some factors which affect branch frequency, and in this section the relationship between septation, branch frequency and branch location will be discussed.

When the temperature-sensitive mutant, *Neurospora crassa cot 2* was grown at various temperatures, a direct relationship was observed (Fig. 10) between intercalary compartment length and hyphal growth-unit length (Steele & Trinci, 1977*a*). A similar relationship was observed when *Geotrichum candidum* was grown at different dilution rates in glucose-limited chemostat culture (Robinson & Smith, 1979). In both species there was a ratio of 4 to 4.5 : 1 between hyphal growth-unit length and intercalary compartment length. These and other observations suggest that branch frequency may be regulated by septation. However, this hypothesis cannot explain how branching is regulated in aseptate fungi, and experiments with *Aspergillus nidulans sep A2* have shown that even in septate fungi branching is not regulated by septation. Although an aseptate mycelium of *A. nidulans sep A2* cultured at 37 °C has a hyphal growth unit which is shorter than that of septate mycelia formed at 30 °C and below, the volume of the hyphal growth unit of this strain is not appreciably affected by temperature (Table 1). Thus, the

Table 4. *Location of branches formed by intercalary compartments of* Geotrichum candidum *and* Aspergillus nidulans *grown under unrestricted conditions on solid medium; each compartment usually forms one branch (From Fiddy & Trinci 1976a, b)*

'Regions' of an intercalary compartment	Branches initiated in each 'region' expressed as a % of the total number observed	
	G. candidum	*A. nidulans*
↑ Hyphal tip		
Septum →	71	25
	13	25
	12	17
	2	18
Septum →	2	12

mechanism which regulates branching in *A. nidulans* does not appear to be dependent upon septation.

Although septation apparently does not regulate branch frequency, septa do influence branch location. The relationship between septation and branch initiation has been discussed by Trinci (1979) and Table 4

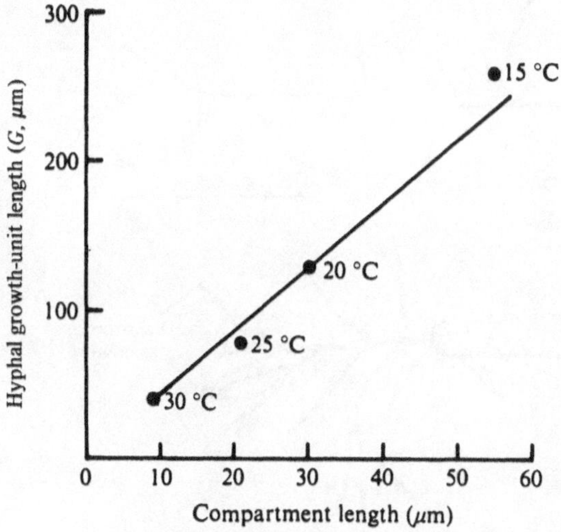

Fig. 10. Relationship between intercalary compartment length and hyphal growth-unit length of mycelia of *Neurospora crassa cot 2* grown on Vogel's medium at the temperatures indicated on the figure. (From Steele & Trinci, 1977*a*.)

shows the location of branches formed by intercalary compartments of hyphae of *Geotrichum candidum* and *Aspergillus nidulans* grown under unrestricted conditions. The higher degree of polarity observed in intercalary compartments of *G. candidum* compared with intercalary compartments of *A. nidulans* may be correlated with the types of septa formed by these two fungi; septa of *G. candidum* contain micropores which are much narrower (8.5 to 9.5 nm in diameter) than the central pores (50 to 500 nm) present in ascomycete septa (Trinci, 1979).

Aseptate mycelia of *A. nidulans sep A2* cultured at 37°C branch apically as well as laterally, but septate mycelia of the parental strain cultured under the same conditions only branch laterally (Fig. 11). Thus, although septation does not regulate branch frequency in *A. nidulans sep A2*, septa do certainly influence branch location.

Apical branching

Hyphae at the margin of colonies of *Geotrichum candidum* branch apically when the initial glucose concentration in the medium exceeds 50 mg l^{-1} (Robinson & Smith, 1980); each leading hypha

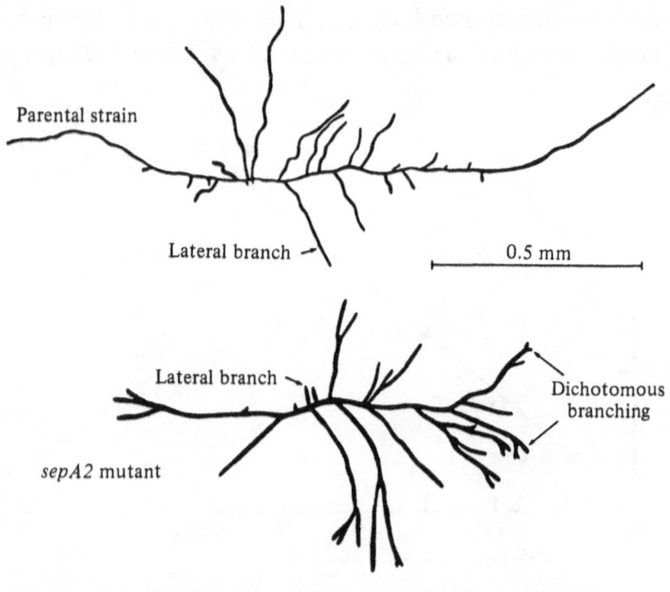

Fig. 11. Branching patterns of an aseptate mycelium of *Aspergillus nidulans sep A2* (lower pattern) and a septate mycelium of the parental strain (upper pattern); both strains were grown in batch culture for 24 h at 37 °C. (From Trinci & Morris, 1979.)

normally produces two branches (Trinci, 1970; Robinson & Smith, 1980). Trinci (1970) showed that apical branching in *Geotrichum lactis* (= *candidum*) was not preceded by any detectable deceleration in the rate of extension of the parent hypha, and on average each apical branch attained the extension rate of its parent hypha within about 16 min of branch initiation (Fig. 12). The observation of Robinson & Smith (1980) that a parent hypha of *G. candidum* and its daughter hyphae had diameters in the ratio 5:4:3 (pairs of apical branches usually differed in diameter) provides the basis of an explanation of how apical branches are able rapidly to attain the same extension rate as the parent hypha (Fig. 2), since the cross-sectional area of the latter (19.64 arbitrary units^2) is exactly equal to the combined cross-sectional areas of the two daughter hyphae (12.57 + 7.07 arbitrary units^2). Thus, the rate of biosynthesis of the parent hypha prior to branching should be sufficient

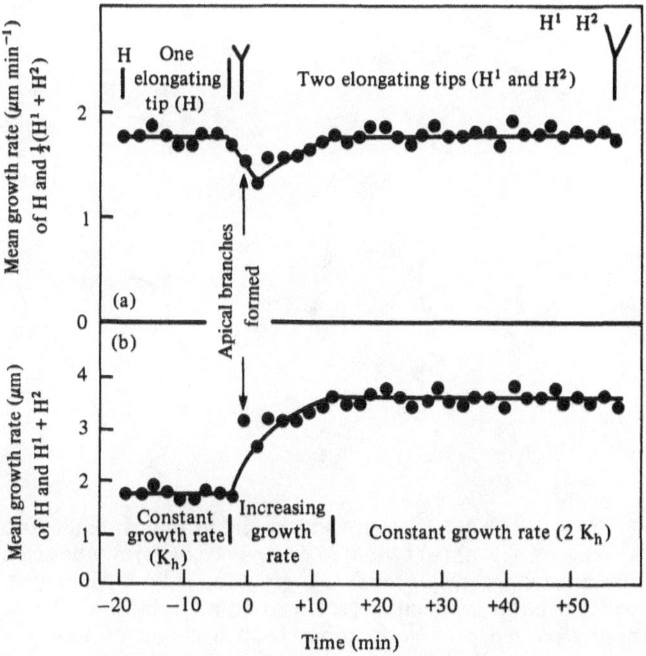

Fig. 12. Kinetics of apical branching of *Geotrichum lactis* (= *candidum*) grown on glucose-mineral salts medium at 25 °C. The results are the mean of 8 hyphae which branched apically. The growth rate of the parent hypha (H) and the mean growth rate of the apical branches 0.5 ($H^1 + H^2$) are plotted in (*a*). The growth rate of the parent hypha and the sum of the growth rates of the apical branches ($H^1 + H^2$) are plotted in (*b*). (From Trinci, 1970.)

to support the growth of two apical branches, each extending at the same rate as the parent hypha. Prosser & Trinci (1979) have formulated a hypothesis to explain how lateral and apical branching may be regulated in fungi.

Regulation of hyphal orientation

Site of germ tube and branch formation

Spores of *Geotrichum candidum* resemble those of many fungi in that their germination is stimulated by contact with one or more other spores (Robinson, 1973*a*). As shown in Fig. 13a and b, the site at which

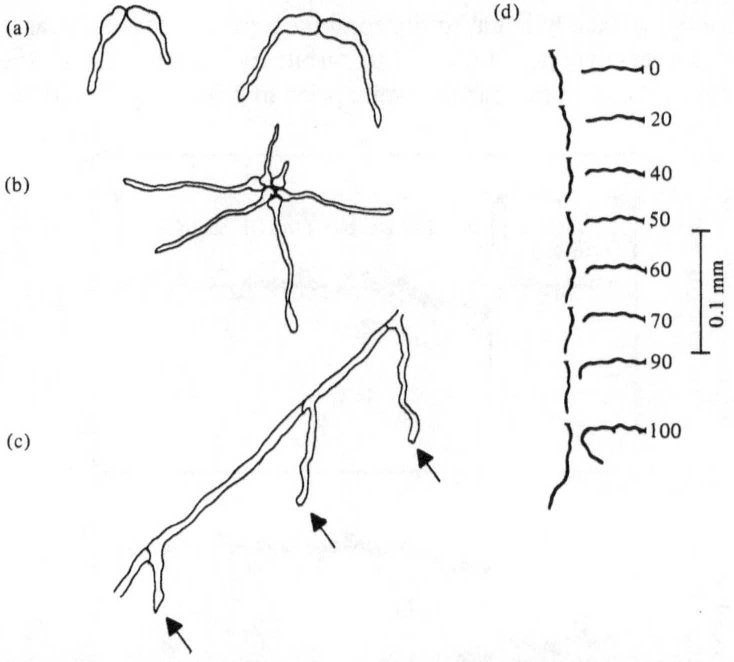

Fig. 13. Oriented germ tube emergence, oriented branch emergence and negative autotropism: (a) Oriented emergence of germ tubes from a pair of spores of *Geotrichum candidum* grown on solid medium in an 'oxygen concentration gradient'. (Redrawn from Robinson, 1973*a*.) (b) Oriented emergence of germ tubes from a clump of spores of *Rhizopus nigricans*. (Redrawn from Stadler, 1952.) (c) Oriented emergence of branches from a hypha of *Geotrichum candidum* grown on solid medium in an 'oxygen concentration gradient'. Each branch (arrowed) has originated on the side of the hypha facing the higher oxygen concentration. (Redrawn from Robinson, 1973*a*.) (d) Negative autotropic response between a pair of germ tubes of *Aspergillus nidulans* grown on solid medium. The figures give times in min. (From Trinci *et al.*, 1979.)

a germ tube is formed may be influenced by the presence of neighbouring spores. When pairs of spores of *G. candidum* germinate they produce germ tubes from the end or corner of the spore which is not in contact with its neighbour (Fig. 13a); single spores form germ tubes from one or both ends. Spore pairs of *Botrytis cinerea* sometimes behave like those of *G. candidum*, but under some circumstances they form germ tubes from the sides of the spores which are in contact with each other. Müller & Jaffe (1965) suggested that *B. cinerea* spores produce a diffusible, high-molecular-weight, labile factor which stimulates germ-tube formation, but Robinson (1973*a*) found no evidence for the production of such a substance by *G. candidum*. However, Robinson did find that when spores of *G. candidum* were subjected to an 'oxygen concentration gradient', they formed germ tubes from the side of the spore facing the higher oxygen concentration. Similarly, when hyphae of *G. candidum* were grown in an 'oxygen concentration gradient', branches were initiated on the side of the hypha facing the higher oxygen concentration (Fig. 13c). Robinson's experiments were made using a plastic coverslip with a central, 0.5-mm-diameter perforation. The coverslip was used to separate a layer of agar medium containing spores from a layer of agar medium alone; provided that the spores were sown at sufficiently high density, an oxygen concentration gradient would be established in the medium in the vicinity of the perforation.

J. I. Prosser (personal communication) found that the branch angle (the angle made by the branch hypha with its parent hypha) in mycelia of *Neurospora crassa* grown from single spores decreased from 90° to 64° as the mycelium increased in size; the branches grew away from the centre of inoculation. Under Prosser's experimental conditions there was a steep decrease in branch angle 22 h after inoculation and this might correspond to the time at which an oxygen concentration gradient was established in the medium.

Chemotropism and autotropism

Observation of a developing mycelium shows that hyphae grow outwards from the inoculum centre (Fig. 13b) and that leading hyphae at the colony margin grow approximately parallel to one another and at approximately the same distance apart. This spatial arrangement is most easily observed in mycelia growing on medium containing a low concentration of nutrients; leading hyphae are further apart on nutrient-'poor' medium than on nutrient-'rich' medium. Clark (1902) suggested that such behaviour could be explained by assuming that germ tubes and branch hyphae react in a negative chemotropic manner to metabolites produced by the fungus. However, results obtained by Robinson (1973*a*) using the perforated-plate technique suggest that germ tubes and mature hyphae exhibit positive chemotropism to oxygen. He (Robinson, 1973*b*) suggests that positive chemotropism results from the formation of a bulge on the side of the tip towards the higher oxygen concentration and believes that positive chemotropism to oxygen can operate over a much larger range (a few millimetres ?) than the perceptive mechanism involved in the promotion of germination between neighbouring spores or the perceptive mechanism involved in determining the site of germ-tube emergence (these mechanisms only operate at distances up to about 50 μm; Robinson, Park & Graham, 1968).

The spacing between hyphae at the colony margin probably involves an 'avoidance' or negative autotropic response between neighbouring hyphae. Such a reaction occurs between germ tubes of *Aspergillus nidulans* (Fig. 13d) and other fungi (Trinci *et al*., 1979), between neighbouring hyphae of *Streptomyces coelicolor* (Hopwood, 1960), but not apparently between hyphae of *Candida albicans* (Gow & Gooday, 1982*b*). In the avoidance reaction illustrated in Fig. 13d, the responding germ tube turned away from a second germ tube which did not deviate from its original direction of growth. The avoidance reaction in *A. nidulans* was initiated when the germ tubes were about 27 μm apart (Trinci, Saunders, Gosrani & Campbell, 1979) and an avoidance reaction was observed in *Mucor hiemalis* when hyphae were 10 to 20 μm apart (Hutchinson, Sharma, Clarke & MacDonald, 1980). Negative autotropism may result from the establishment of an oxygen concentration gradient in the agar medium surrounding growing hyphae, i.e. it may reflect a positive chemotropic response to oxygen. However, no-one has yet shown that an oxygen concentration gradient is established in the medium surrounding hyphae, or produced *direct* evidence that hyphae exhibit positive chemotropism to oxygen.

Chemotropism to nutrients occurs in the Mastigomycotina but not in other fungi; hyphae of *Achlya bisexualis* grow towards sources of casein hydrolysate and amino acids (Musgrove, Ero, Scheffer & Oehlers, 1977), hyphae of *Saprolegnia ferax, Saprolegnia mixta* and *Achlya polyandra* grow towards sources of casein hydrolysate and malt extract (Fischer & Werner, 1955), and rhizoids of *Blastocladiella emersonii* grow towards sources of amino acid mixtures and inorganic phosphate Harold & Harold, 1980). The reason for the absence in 'higher' fungi of a positive chemotropic response to nutrients is not known.

Spiral growth of hyphae

Hyphae do not always grow radially outwards from the point of inoculation but may instead curve in either a left- or right-handed direction, giving the colony a spiral growth pattern which is in either a clockwise (Fig. 14) or anticlockwise direction when viewed from above. 'Spiral' growth of hyphae can be observed in young mycelia and is influenced by agar concentration (Fig. 14); 'spiral' growth may also be observed at the margin of mature colonies (Fig. 15). Of the 157 isolates examined by Madelin, Toomer & Ryan (1978), 8 showed pronounced spiralling in a clockwise direction, 13 showed pronounced spiralling in a counterclockwise direction and 39 showed weak spiralling. Madelin *et al.* (1978) suggested that 'spiral' growth of a hypha is caused by axial rotation of the extension-zone wall, and that this rotation causes the hypha to roll over the surface of the substrate. Observations made by Trinci *et al.* (1979) on the curvature of sporangiophores of *Phycomyces blakesleeanus* and conidiophores of *Aspergillus giganteus* grown in

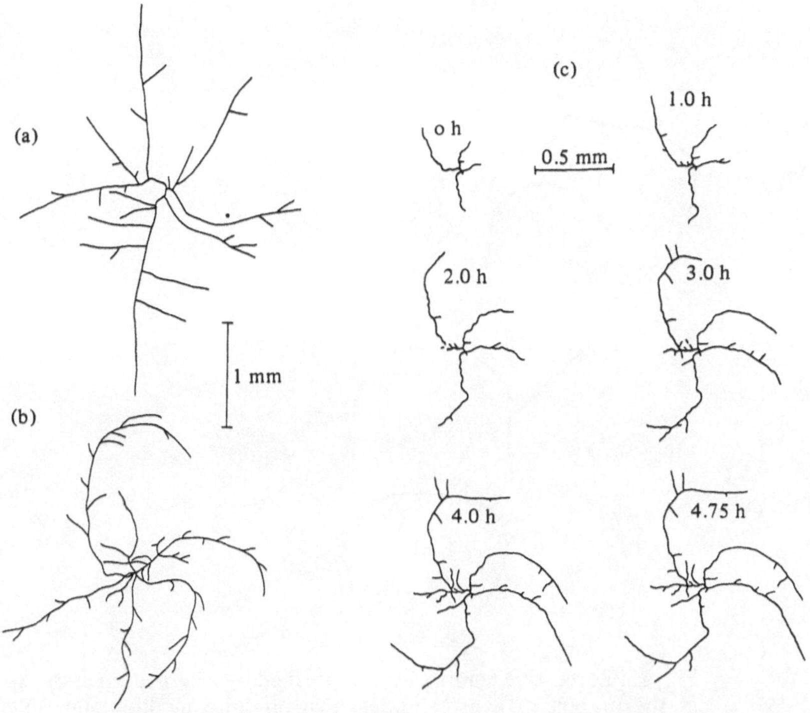

Fig. 14. Mycelia of *Mucor hiemalis* grown for 24 h on media gelled with 0.5% (*a*) and 3% (*b*) agar, and a time-lapse sequence (*c*) of the growth of a mycelium on medium gelled with 4% agar. (From Trinci *et al.*, 1979.)

contact with solid surfaces support this hypothesis; the tips of these aerial hyphae are known to exhibit spiral wall growth. Beever (1980) obtained mutants of *Neurospora crassa* carrying a *coil-1* gene which caused very pronounced clockwise 'spiralling'. The mean radius of 'spiral' growth of hyphae of the parental strain was 1 mm compared with 0.2 mm for the mutant; hyphae of the latter strain frequently formed complete circles! Surprisingly, the linear growth rate of *N. crassa coil-1* in 'race' tubes was the same as that of the parental strain; some observations suggest that 'spiral' growth may be abolished in dense colonies by negative autotropism (Trinci *et al.*, 1979). A mutant of *Aspergillus nidulans* has been obtained by P. Markham (personal communication) whose hyphae coil in the opposite direction to the parental strain. The mechanism of axial rotation of the extension zone wall is not understood but it may be related to the presence of spiral structures in the wall.

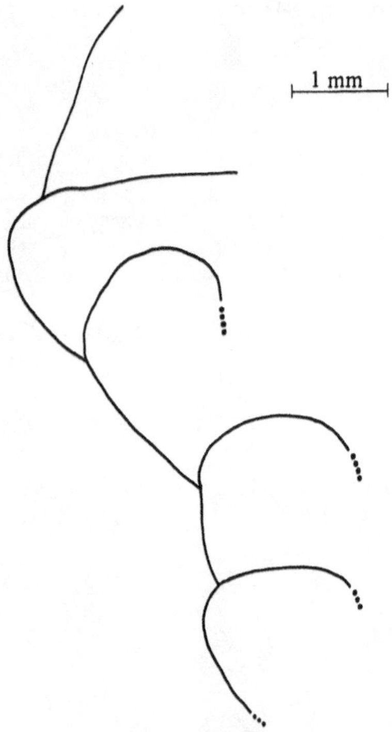

Fig. 15. Successive leading hyphae formed by *Neurospora crassa spco 9* at the margin of a mycelium grown on solid medium; new leading hyphae arise as branches from previous leading hyphae which have curved back towards the centre of the colony. (Drawing made by Dr Graham W. Steele.)

Conclusions

The mechanisms which regulate hyphal branching and the direction of hyphal growth play crucial roles in ensuring that fungi colonise solid substrata efficiently and effectively. In young mycelia, hyphal orientation is regulated by two mechanisms, one of which determines the site of germ-tube emergence whilst the other is responsible for negative autotropic reactions between germ-tube hyphae. The spacing between hyphae at the margin of a mature colony results from the response of the fungus to nutrient concentration and negative autotropism between neighbouring hyphae; negative autotropism is not observed at the centre of mature colonies. Branching in fungi is regulated by a mechanism which is apparently able to monitor biomass. Although the exact nature of this mechanism is not known, Prosser & Trinci (1979) have suggested that it involves the vesicles which are concerned with tip extension. An understanding of the precise mechanism which regulates hyphal branching is clearly desirable because of the profound influence of mycelial morphology on culture rheology (Metz, 1976).

Note added in proof: D. L. Kropf, M. D. A. Lupa, J. H. Caldwell & F. M. Harold (*Science*, **220**, 1385–7, 1983) have shown that in *Achlya bisexualis* an inward-moving electrical current precedes branching and predicts the site of branch emergence. They suggest that the inward current probably acts as an early signal in branch initiation.

References

Allan, E. J. & Prosser, J. I. (1983). Mycelial growth and branching of *Streptomyces coelicolor* on solid medium. *Journal of General Microbiology*, **129**, 2029–36.

Anderson, J. G. & Smith, J. E. (1972). The effects of elevated temperatures on spore swelling and germination in *Aspergillus niger*. *Canadian Journal of Microbiology*, **18**, 289–97.

Bartnicki-Garcia, S. (1981). Cell wall construction during spore germination in Phycomycetes. *The Fungal Spore*, 1981, ed. G. Turian & H. R. Hohl, pp. 533–56. London, UK: Academic Press.

Beever, R. E. (1980). A gene influencing spiral growth of *Neurospora crassa* hyphae. *Experimental Mycology*, **4**, 338–42.

Bull, A. T. & Trinci, A. P. J. (1977). The physiology and metabolic control of fungal growth. *Advances in Microbial Physiology*, **50**, 1–84.

Caldwell, I. Y. & Trinci, A. P. J. (1973). The growth unit of the mould *Geotrichum candidum*. *Archiv für Mikrobiologie*, **88**, 1–10.

Clark, J. F. (1902). On the toxic properties of some copper compounds with special reference to Bordeaux mixture. *Botanical Gazette*, **33**, 26–48.

Collinge, A. J., Fletcher, M. H. & Trinci, A. P. J. (1978). Physiology and cytology of septation and branching in a temperature sensitive colonial mutant (*cot 1*) of *Neurospora crassa*. *Transactions of the British Mycological Society*, **77**, 107–20.

De Vries, S. C. & Wessels, J. G. H. (1982). Polarized outgrowth of hyphae by constant electrical fields during reversion of *Schizophyllum commune* protoplasts. *Experimental Mycology*, **6,** 95–8.

Ekundayo, J. A. & Carlile, M. J. (1964). The germination of sporangiospores of *Rhizopus arrhizus*: spore swelling and germ-tube emergence. *Journal of General Microbiology*, **35,** 261–9.

Fiddy, C. & Trinci, A. P. J. (1975). Kinetics and morphology of glucose-limited cultures of moulds grown in a chemostat and on solid media. *Archives of Microbiology*, **103,** 191–7.

Fiddy, C. & Trinci, A. P. J. (1976*a*). Mitosis, septation, branching and the duplication cycle in *Aspergillus nidulans*. *Journal of General Microbiology*, **97,** 169–84.

Fiddy, C. & Trinci, A. P. J. (1976b). Nuclei, septation, branching and growth of *Geotrichum candidum*. *Journal of General Microbiology,* **97,** 185–92.

Fischer, F. G. & Werner, G. (1955). Eine Analyse des Chemotropismus einiger Pilze, insbesondere der Saprolegniaceen. *Hoppe-Seyler's Zeitschrift für physiologische Chemie, Berlin*, **300,** 211–36.

Fletcher, J. (1969). Morphological and nuclear behaviour of germinating conidia of *Penicillium griseofulvum. Transactions of the British Mycological Society*, **53,** 425–32.

Gow, N. A. R. & Gooday, G. W. (1982*a*). Vacuolation, branch production and linear growth of germ tubes of *Candida albicans. Journal of General Microbiology*, **128,** 2195–8.

Gow, N. A. R. & Gooday, G. W. (1982*b*). Growth kinetics and morphology of colonies of the filamentous form of *Candida albicans. Journal of General Microbiology*, **128,** 2187–94.

Gull, K. & Trinci, A. P. J. (1971). Fine structure of spore germination in *Botrytis cinerea. Journal of General Microbiology*, **68,** 207–20.

Harold, F. M. (1982). Pumps and currents: a biological perspective. *Current Topics in Membranes and Transport*, **16,** 485–516.

Harold, R. L. & Harold, F. M. (1980). Orientated growth of *Blastocladiella emersonii* in gradients of ionophores and inhibitors. *Journal of Bacteriology*, **144,** 1159–67.

Hopwood, D. A. (1960). Phase-contrast observations on *Streptomyces coelicolor. Journal of General Microbiology*, **22,** 295–302.

Hutchinson, S. A., Sharma, P., Clarke, R. R. & MacDonald, I. (1980). The control of hyphal orientation in colonies of *Mucor hiemalis* Wehm. *Transactions of the British Mycological Society*, **75,** 177–91.

Katz, D., Goldstein, D. & Rosenberger, R. F. (1972). Model for branch initiation in *Aspergillus nidulans* based on measurements of growth parameters. *Journal of Bacteriology*, **109,** 1097–1100.

Madelin, M. F., Toomer, D. K. & Ryan, J. (1978). Spiral growth of fungus colonies. *Journal of General Microbiology*, **106,** 73–80.

Matin, A. & Veldkamp, H. (1978). Physiological basis of selective advantage of a *Spirillum* sp. in a carbon-limited environment. *Journal of General Microbiology,* **105,** 187–97.

Metz, B. (1976). From pulp to pellet. Ph.D. Thesis, Technical University of Delft, The Netherlands.

Miles, E. A. & Trinci, A. P. J. (1983). Effect of pH and temperature on the morphology of batch and chemostat cultures of *Penicillium chrysogenum. Transactions of the British Mycological Society*, **81,** 193–200.

Morrison, K. B. & Righelato, R. C. (1974). The relationship between hyphal branching, specific growth rate and colony radial growth rate in *Penicillium chrysogenum. Journal of General Microbiology* **81,** 517–20.

Müller, D. & Jaffe, L. F. (1965). A quantitative study of cellular rheotropism. *Biophysical Journal*, **5**, 317–35.

Musgrave, A., Ero, L., Scheffer, R. & Oehlers, E. (1977). Chemotropism of *Achlya bisexualis* germ hyphae to casein hydrolysate and amino acids. *Journal of General Microbiology*, **101**, 65–70.

Paznokas, J. L. & Sypherd, P. S. (1975). Respiratory capacity, cyclic adenosine 3′,5′-monophosphate, and morphogenesis in *Mucor racemosus*. *Journal of Bacteriology*, **124**, 134–9.

Plomley, N. J. B. (1959). Formation of the colony in the fungus *Chaetomium*. *Australian Journal of Biological Sciences*, **12**, 53–64.

Prosser, J. I. & Trinci, A. P. J. (1979). A model of hyphal growth and branching. *Journal of General Microbiology*, **111**, 153–64.

Riesenberger, D. & Bergter, F. (1979). Dependence of macromolecular composition and morphology of *Streptomyces hygroscopicus* on specific growth rate. *Zeitschrift für allgemeine Mikrobiologie*, **19**, 415–30.

Righelato, R. C. (1975). Growth kinetics of mycelial fungi. In *Filamentous Fungi*, vol. 1, ed. J. E. Smith & D. R. Berry, pp. 77–103. London, UK: Edward Arnold.

Robinson, P. M. (1973*a*). Chemotropism in fungi. *Transactions of the British Mycological Society*, **61**, 303–13.

Robinson, P. M. (1973*b*). Oxygen–positive chemotropic factor for fungi? *New Phytologist*, **72**, 1349–56.

Robinson, P. M., Park, D. & Graham, T. A. (1968). Autotropism in fungal spores. *Journal of Experimental Botany*, **19**, 125–34.

Robinson, P. M. & Smith, J. M. (1979). Development of cells and hyphae of *Geotrichum candidum* in chemostat and batch culture. *Transactions of the British Mycological Society*, **72**, 39–47.

Robinson, P. M. & Smith, J. M. (1980). Apical branch formation and cyclic development in *Geotrichum candidum*. *Transactions of the British Mycological Society*, **75**, 233–8.

Schuhmann, E. & Bergter, F. (1976). Mikroskopische Untersuchungen zur Wachstumskinetik von *Streptomyces hygroscopicus*. *Zeitschrift für allgemeine Mikrobiologie*, **16**, 201–15.

Stadler, D. R. (1952). Chemotropism in *Rhizopus nigricans*: the staling reaction. *Journal of Cellular and Comparative Physiology*, **39**, 449–74.

Steele, G. W. & Trinci, A. P. J. (1975). Morphology and growth kinetics of hyphae of differentiated and undifferentiated mycelia of *Neurospora crassa*. *Journal of General Microbiology*, **91**, 362–8.

Steele, G. W. & Trinci, A. P. J. (1977*a*). Relationship between intercalary compartment length and hyphal growth unit length. *Transactions of the British Mycological Society*, **69**, 156–8.

Steele, G. W. & Trinci, A. P. J. (1977*b*). Effect of temperature and temperature shift on growth and branching of a wild type and a temperature sensitive colonial mutant (*cot 1*) of *Neurospora crassa*. *Archives of Microbiology*, **113**, 43–8.

Trinci, A. P. J. (1970). Kinetics of apical and lateral branching in *Aspergillus nidulans* and *Geotrichum lactis*. *Transactions of the British Mycological Society*, **55**, 17–28.

Trinci, A. P. J. (1971*a*). Exponential growth of the germ tubes of fungal spores. *Journal of General Microbiology* **67**, 345–8.

Trinci, A. P. J. (1971*b*). Influence of the peripheral growth zone on the radial growth rate of fungal colonies. *Journal of General Microbiology*, **67**, 325–44.

Trinci, A. P. J. (1973*a*). The hyphal growth unit of wild type and spreading colonial mutants of *Neurospora crassa*. *Archiv für Mikrobiologie*, **91**, 127–36.

Trinci, A. P. J. (1973*b*). Growth of wild type and spreading colonial mutants of *Neurospora crassa* in batch culture and on agar medium. *Archiv für Mikrobiologie*, **91**, 113–16.

Trinci, A. P. J. (1974). A study of the kinetics of hyphal extension and branch initiation of fungal mycelia. *Journal of General Microbiology*, **81**, 225–36.

Trinci, A. P. J. (1979). The duplication cycle and branching in fungi. In *Fungal Walls and Hyphal Growth*, 1979, ed. J. H. Burnett & A. P. J. Trinci, pp. 319–58. Cambridge, UK: Cambridge University Press.

Trinci, A. P. J. & Collinge, A. J. (1973). Influence of L-sorbose on the growth and morphology of *Neurospora crassa*. *Journal of General Microbiology*, **78**, 179–92.

Trinci, A. P. J. & Collinge, A. J. (1974). Occlusion of the septal pores of damaged hyphae of *Neurospora crassa* by hexagonal crystals. *Protoplasma*, **80**, 56–67.

Trinci, A. P. J. & Morris, N. R. (1979). Morphology and growth of a temperature-sensitive mutant of *Aspergillus nidulans* which forms aseptate mycelia at non-permissive temperatures. *Journal of General Microbiology*, **114**, 53–9.

Trinci, A. P. J., Saunders, P. T., Gosrani, R. & Campbell, K. A. S. (1979). Spiral growth of mycelial and reproductive hyphae. *Transactions of the British Mycological Society* **73**, 283–92.

Ujcova, E., Fencl, Z., Musilkova, M. & Seichert, L. (1980). Dependence of release of nucleotides from fungi on fermenter turbine speed. *Biotechnology and Bioengineering*, **22**, 237–41.

Van Etten, J. L., Dunkle, L. D. & Freer, S. N. (1977). Germination of *Rhizopus stolonifer* sporangiospores. In *Eukaryotic Microbes as Model Developmental Systems*, 1977, ed. D. M. O'Day & P. A. Horgen, pp. 372–4. New York, USA: Marcel Dekker.

Van Suijdam, J. C. & Metz, B. (1981). Influence of engineering variants upon the morphology of filamentous moulds. *Biotechnology and Bioengineering*, **23**, 111–48.

3
Colony ontogeny in basidiomycetes

G. M. BUTLER
Department of Plant Biology, University of Birmingham, Birmingham B15 2TT, UK

The colony of a mycelial fungus consists of divergently-growing, branching and often interconnecting hyphae and its unity is usually recognised by the regularity of the margin and the well-defined spatial arrangement of differentiated structures. Most basidiomycetes form extensive colonies in nature and can be grown readily in culture. The main exception to this generalisation is the Teliomycetes and these specialised plant parasites are not considered here.

Colony organisation is most often studied in individual or clonal mycelia in Petri dishes. The widespread occurrence of vegetative incompatibility amongst basidiomycetes suggests that their colonies in Nature may often be individuals (Rayner & Todd, 1982). The Petri dish culture system provides a model for the development of a colony in and on the surface of one nutrient resource of limited size. There are some fungi, particularly wood-decomposing basidiomycetes, in which the vegetative mycelium is not only assimilative but also a means of spatial spread from one nutrient resource to another. For studying the latter a culture system with a localised nutrient supply is needed (Jennings, 1982; see also Chapters 7, 8, 9, 10).

Two sorts of morphological differentiation can be recognised by comparing hyphae in the outer margin with those within a colony. There may be changes in existing hyphal compartments or new types of hyphal growth may take place leading to the production of, for example, sporulating structures or multihyphal vegetative aggregates. A third form of differentiation occurs in the marginal hyphae as the colony develops so that its recognition requires comparisons over time. There is a juvenile phase before the mature margin of radially oriented hyphae extending at a constant rate is set up (Bull & Trinci, 1977). During this phase there are often changes in morphology as well as in extension rate of the hyphae. Steele & Trinci (1975) have shown that the margin of mature colonies of *Neurospora crassa* consists of wide main hyphae extending at a fast rate with narrower, slower-extending branches, whereas all the hyphae of undifferentiated mycelia, i.e. those growing exponentially from spores, are relatively narrow and extend slowly. The magnitude of this ontogenetic change varies in different fungi. It is large in *N. crassa* whereas in *Aspergillus nidulans* and *Geotrichum lactis* there is little difference between hyphae in the mature margin and in undifferentiated mycelia (Bull & Trinci, 1977).

Changes in the hyphae of the colony margin could be generated in different ways. There could be transformation of individual apices with increasing length of the system. A change of this sort occurs in *G. lactis* where the main hyphae branch dichotomously and each main apex increases in width with increasing length from the last dichotomy. In this example development continues for only a short time and at a critical diameter the tip branches to form two narrower apices (Robinson & Smith, 1980).

Alternatively new hyphal types could be generated which by virtue of their faster extension rate come to occupy a marginal position. This possibility is illustrated by the morphologically distinct unbranched aerial stoloniferous hyphae of *Absidia.* On a relatively concentrated malt agar medium (2%) the extension rate of the colony margin gradually slows down, perhaps because of accumulation of auto-inhibitory substances, and extension of the colony takes place by a series of 'leapfrogs' by the stoloniferous hyphae. These are initiated as branches from the surface hyphae and extend more rapidly through the air until their tips land on the agar ahead of the main colony.

Basidiomycete colonies

Nuclear states of the mycelium

Analysis of colony organisation in basidiomycetes is complicated by the different nuclear states in which the mycelium can exist. Usually the homokaryotic primary mycelium has uninucleate compartments bounded by simple septa and the heterokaryotic secondary mycelium has binucleate compartments with clamp-connections. The latter are not an invariable indicator of dikaryotic mycelium since in addition to species in which they are absent there is a small number of species in which they are formed inconstantly (Boidin, 1971). Primary and secondary mycelia often differ in both hyphal growth form (Buller,

1933) and extension rate (Simchen & Jinks, 1964). Nuclear migration is sometimes restricted (Casselton, 1978; Butler, 1972). Hence changes between homokaryotic and dikaryotic states may be manifested either as sectors or as widespread conversion of the colony margin. Arita (1979) has shown in *Pholiota nameko* the heterokaryotic–dikaryotic margin is replaced during colony ontogeny by homokaryotic segregants. He attributes this replacement to a combination of three unusual features; irregularities in nuclear division leading to formation of homokaryotic tips, more rapid extension of homokaryotic than dikaryotic hyphae, and very slow nuclear migration.

Not all species of basidiomycete fit into the homokaryon–dikaryon scheme. Boidin (1971) and Kuhner (1977) have shown that multinucleate apical compartments occur in a substantial number of species. They did not report outbreeding in holocoenocytic species but there is now evidence for this in *Stereum hirsutum* (Coates, Rayner & Todd, 1981), *S. gausapatum* (Boddy & Rayner, 1982) and *S. rugosum* (Rayner & Turton, 1982). Both monosporous and secondary mycelia form a mixture of simple septa, single clamp-connections and whorled clamp-connections. The multinucleate apical compartments of secondary mycelia in outbreeding holocoenocytic and astatocoenocytic species are probably heterokaryotic, with consequent potential flexibility in nuclear ratio. In astatocoenocytic species the secondary mycelium has two phases, one with multinucleate compartments and simple septa and the other with binucleate, constantly clamped compartments (Kuhner, 1977). In *Phlebia radiata* and *P. rufa* the multinucleate phase is marginal and the clamped phase occurs within the colony and two phases similar in appearance to these but lacking clamp-connections occur in homokaryons (Boddy & Rayner, 1983; Rayner & Webber, Chapter 18, Fig. 6).

Diploidy has been reported in *Armillaria mellea* s.l. and in *Clitocybe tabescens* (Korhonen & Hintikka, 1974; Ullrich & Anderson, 1978; Anderson, 1982). The dikaryon is unstable and is replaced during colony growth by diploid hyphae.

Patterns of change in the colony margin

Despite the pioneering observations of Buller (1933) on colony growth from spores of *Coprinus sterquilinus*, our knowledge of development of hyphae in the colony margin of basidiomycetes is poor. Studies of basidiomycete growth in pure culture often use a large agar inoculum disc which obscures the early stages of colony development. Widespread ontogenetic changes in the margin will be discussed in later sections.

There is evidence that other patterns of marginal change can occur. Coggins, Hornung, Jennings & Veltkamp (1980) describe the initiation of faster-extending hyphae at localised points either within or at the margin of post-acceleration phase colonies of *Serpula lacrimans*. These hyphae spread round the original mycelium. They are stable on subculture and do not appear to be due to genetic change. Their formation is associated with unfavourable growth conditions in the original colony. This 'point growth' is strikingly similar to the behaviour of a sectoring variant of *Schizophyllum commune* analysed by Papazian (1955). In both species the change was more readily ascribed to differentiation than to genetic change. Several features may be important generally in understanding colony ontogeny in basidiomycetes. First, differences during divergent growth may take place at a late stage of colony development; second, such differences may be localised, resulting in directional growth; and third, slow-extending mycelia from which fast-extending forms arise have been isolated directly from nature. The last also occurs in *S. gausapatum* (Boddy & Rayner, 1982).

Early development of small mycelia

Methods of study

Observations of the growth kinetics of small undifferentiated mycelia have clarified ideas about functional interrelationships (Trinci, 1974, 1978). In the dikaryon of constantly clamped basidiomycetes, e.g.

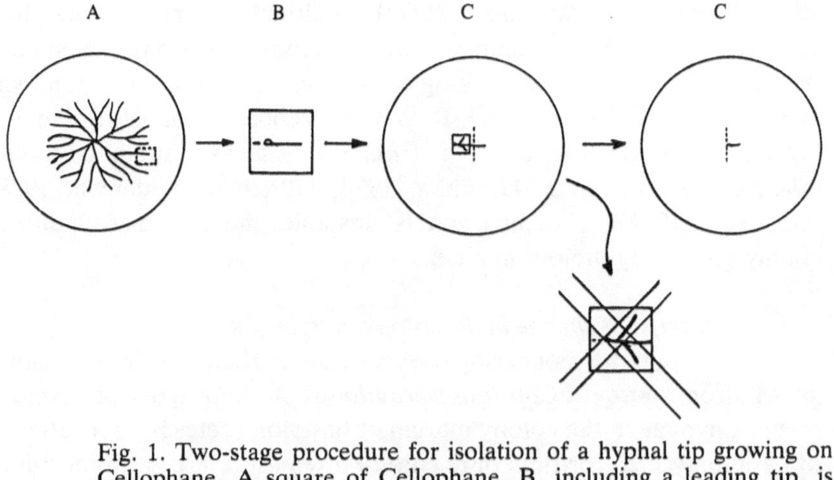

Fig. 1. Two-stage procedure for isolation of a hyphal tip growing on Cellophane. A square of Cellophane, B, including a leading tip, is removed from disc A and after growth onto Cellophane disc C the tip is cut off and B is removed and discarded.

Coprinus cinereus, the formation of clamp-connections would permit the monitoring of nuclear division alongside other growth parameters. However it is difficult to obtain small dikaryotic inocula since only a small proportion of species form asexual dikaryotic spores. Special methods which allow one to obtain small dikaryotic inocula are tip excision (Harder, 1926), maceration (Miles & Raper, 1956) and protoplast formation (Wessels & De Vries, 1973). The method adopted here for kinetic studies on *C. cinereus* was a two-stage excision of whole tip compartments growing on Cellophane (Fig. 1). This method does not involve disturbance of the tip compartment from its position on the Cellophane and hyphae treated in this way most frequently resumed growth within a few minutes from the original position of the apex such that the point of temporary arrest was difficult to distinguish.

Morphological variation

A striking feature of small mycelia produced from isolated tips was the large amount of morphological variation despite the fact that tips were taken from colonies recently derived from a single hyphal tip. Some tips eventually yielded simply-septate homokaryotic colonies even

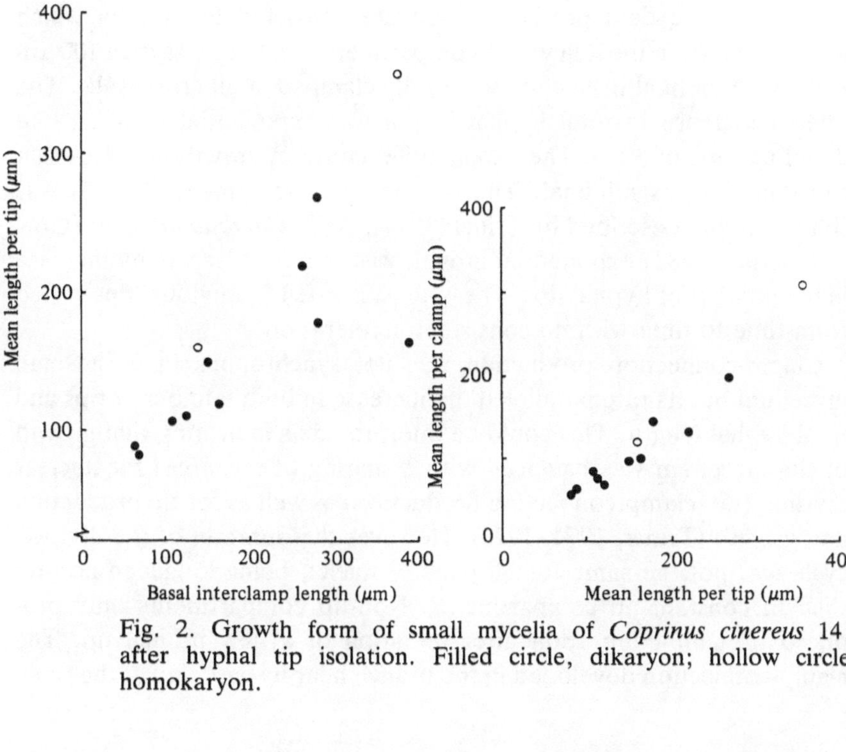

Fig. 2. Growth form of small mycelia of *Coprinus cinereus* 14 h after hyphal tip isolation. Filled circle, dikaryon; hollow circle, homokaryon.

though they formed clamp-connections during early growth. Harder (1927) showed that expression of clamp-connections persisted for a number of nuclear divisions during growth from uninucleate subterminal compartments obtained by operations on clamped hyphal tips. Other components of variation could be segregation of cytoplasmic factors or environmental differences. However compartment length in the original inoculum also seemed to be important. Both mean hyphal length per tip and per clamp-connection increased with increasing length of the basal compartment (Fig. 2). It may be that these features are due to the persistence over the short experimental period of the mode of growth which occurred in that compartment before excision.

Sometimes simply-septate homokaryotic branches were formed in predominantly dikaryotic small mycelia. These were frequently but not invariably from the basal end of the excised compartment, where it is possible that they were due to a nucleus trapped in the unfused clamp arm. Segregation of homokaryotic branches during growth from blended mycelium of *S. commune* is described by Nguyen & Niederpruem (Chapter 4).

Growth kinetics

Kinetic data presented here are restricted to mycelia which were grown from inocula with a compartment length of less than 100 μm and which on final inspection were fully clamped at all cross-walls. The rate of increase in total hyphal length was exponential with a mean doubling time of 3.8 h. The exponential nature of growth indicates that cell damage was minimal. The pattern of development (Fig. 3) was similar to that described by Trinci (1974). As in *Candida albicans* (Gow & Gooday, 1982) exponential growth was achieved largely by increase in the number of hyphal tips. The extension rate of individual tips varied from time to time with no consistent acceleration.

Clamp-connection production was not synchronous in each small mycelium but its rate paralleled the increase in both number of tips and total hyphal length. This could be interpreted as indicating that growth of the mycelium was balanced with a sharing of resources for nuclear division (i.e. clamp-connection production) as well as for tip production and growth (Trinci, 1974, 1978). However the duration of the division cycle was not the same for all pairs of nuclei, being longer in non-tip than in constant-tip compartments. Non-tip compartments only produced a clamp-connection after initiation of a new branch tip. The clamp-connection developed in the branch near its base and at the same

Table 1. *Clamp-connection production in constant-tip and branch-tip compartments of small mycelia of* Coprinus cinereus *growing at 25 °C on a medium containing glucose 1.8 g l^{-1} and ammonium chloride 1.6 g l^{-1}*

	Time from formation of previous clamp-connection (h)	Time from tip initiation (h)	Compartment length after division (μm)	
			Basal	Tip
Constant-tip compartments	2.5	—	38	43
Branch-tip compartments	5.4	2.9	47	41

distance behind the growing apex as in constant-tip compartments. The time interval between branch-tip initiation and clamp-connection formation was similar to the interclamping time interval in constant-tip compartments (Table 1). This suggests that the nuclear division cycle is somehow 'switched on' by the initiation of a new apex.

The switch cannot be absolute since clamp-connections did form exceptionally on the main axis of non-tip and branch-tip compartments

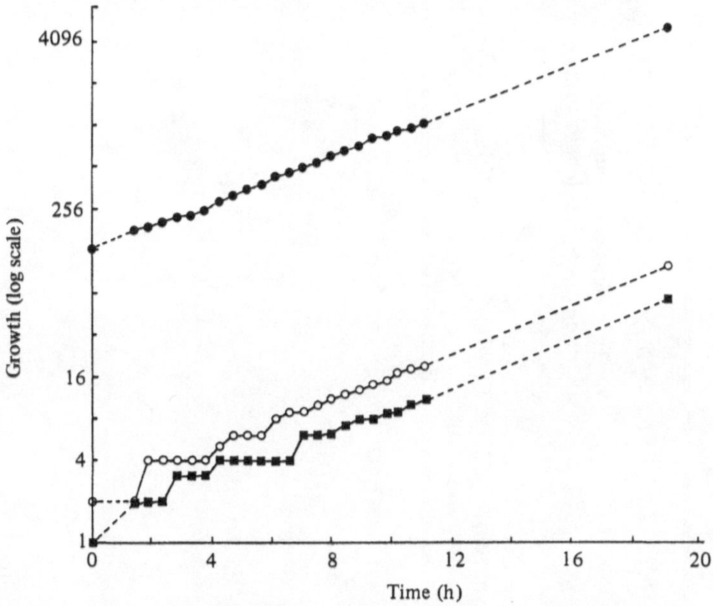

Fig. 3. Growth of a small mycelium of *C. cinereus* on a medium containing glucose ($1.8 g l^{-1}$) and NH_4Cl ($1.6 g l^{-1}$) at 25 °C. Filled circles, total hyphal length (μm); filled squares, total no. of hyphal tips; hollow circles, total no. of clamp-connections.

Table 2. *Hyphal features of small mycelia and mature colony margins in* Coprinus cinereus

		Tip-extension rate (μm h^{-1})	Interclamp length (μm)	Minimum tip-compartment length (μm)	Mean hyphal radius (μm)	Mean tip-compartment volume (μm^3)	Constant-tip compartment generation time (h)
Small mycelia		16	38	43	1.49	432	2.49
Main hyphae of mature margins	Exp. 1	284	269	—	—	—	0.95
Main hyphae of mature margins	Exp. 2	—	259	140	2.95	7354	—

Table 3. *Growth features in a developing colony of* Coprinus disseminatus

Feature	Time (h) 66	77	89	98	113	138	161	185	232
Mean colony diameter (mm)	0.97	1.55	2.72	3.48	5.35	9.50	—	—	—
Mean leader extension rate [a] ($\mu m\ h^{-1}$)	42	46	59	71	71	120	174	189	171
Mean angle between primary branch and parent axis	82	70	59	53	58	49	40	35	33
Mean primary-branch extension rate [b] ($\mu m\ h^{-1}$)	4	21	14	23	25	50	111	107	111
Mean relative primary-branch extension rate (%) [c]	10	46	24	32	35	42	64	57	65
Presence of aerial hyphae at margin	–	–	–	–	–	–	–	+	+
Clamp-connections frequent on marginal hyphae	–	–	–	–	–	–	–	–	+

Growth took place on Cellophane over D/5 medium (glucose $2\,g\,l^{-1}$, sodium nitrate $0.24\,g\,l^{-1}$) and growth measurements were made over a 1-h interval at each time of assessment.

[a] A leader hypha is defined as the fastest-extending hypha in each 1000 μm of the colony circumference.

[b] Primary branches arising within 1000 μm of the parent tips.

[c] Relative primary-branch extension rate (%)

$$= \frac{\text{extension rate of longest primary branch per node}}{\text{extension rate of parent hypha}} \times 100.$$

in *C. cinereus* and such clamp-connections have been observed regularly in *S. commune* (Niederpruem, Jersild & Lane, 1971). In the former species these exceptional clamp-connections occurred in compartments in which branch initiation was delayed or did not take place. The crucial question may be whether the duration of the division cycle is shortened by the presence of a growing tip. Raudaskoski & Kaukonen (1978) have suggested that the S phase is activated by DNA synthesis-promoting factors formed at the apex. There is as yet no information about the duration of the phases in the DNA cycle in different sorts of compartment.

Ontogeny of the margin

Maximum hyphal extension rate

In both *C. cinereus* (Table 2) and *C. disseminatus* (Table 3) there is a large difference in extension rate between hyphae of small mycelia and main hyphae of the mature colony margin. If it is assumed that the mycelial length-doubling time of exponentially growing small mycelia is a satisfactory measure of the specific growth rate (α) of the cytoplasm in the main marginal hyphae, a theoretical peripheral growth zone (w_t) required to maintain by synthesis the observed extension rate (K_r) of the main hyphae can be calculated using the formula $w_t = K_r/\alpha$ (Trinci, 1971). These calculated lengths, 1556 μm for *C. cinereus* and 2522 μm for *C. disseminatus* are considerably longer than their apical compartment lengths (270 and 431 μm, respectively). On this basis it seems that main-tip extension rate in the mature margin must rely on parts of the hypha behind the apical compartment, either by synthesis or relocation of protoplasm (Zalokar, 1959).

Attainment of maximum hyphal extension rate in *C. disseminatus* when growing on a dilute nutrient medium takes a long time, 6–7 days in Table 3. This colony was about 7 mm in radius at the time of maturation, considerably longer than the theoretical length required to maintain the main tips by synthesis. Although this does not provide insight into mechanisms it does indicate interdependence over a large volume of mycelium for the initiation of fast-extending tips.

Rate of tip compartment division

Features of the mature margin. In attempting to understand the mechanisms underlying rates of division in basidiomycete hyphae three features of the mature margin need to be taken into account. First the rate of nuclear division in tip compartments of both basidiomycetes and other mycelial fungi is high (Table 4), of the same order of magnitude as minimum doubling times of unicellular eukaryotes growing at the same temperature. The media used for such measurements have not been selected for maximum specific growth rate. Second, although primary branch hyphae extend more slowly and are narrower than their parent hyphae, the rate of production of clamp-connections is more or less equal in the tips of the two sorts of hyphae. Nuclear division in tip compartments of the same branching system is not synchronous and division in intervening subterminal compartments frequently ceases after one further division. The third feature is the differential effect of the inhibitor cycloheximide on rates of extension growth and tip-compartment division. Over a range of concentrations of cycloheximide which halved the extension rate, the rate of tip compartment division was virtually unchanged in *C. disseminatus* (Butler, 1982). Under these conditions, presumably of reduced protein synthesis, the cytoplasmic

Table 4. *Mature tip compartment generation times in different fungi*

	Generation time (h)	Temperature (°C)
Dikaryons		
Coprinus cinereus	1.0	25
C. disseminatus	1.3	25
Schizophyllum commune (Niederpruem, Jersild & Lane, 1971)	1.2	25
Sistotrema brinkmannii	1.3	25
Coriolus versicolor (Girbardt, 1968)	2.1	24
Others		
Alternaria solani (King & Alexander, 1969)	0.95	25
Aspergillus nidulans (Fiddy & Trinci, 1976)	2.1	25
Polyporus arcularius (Valla, 1973)	2.1	26–27

volume at division of the tip compartment was also reduced. These features are difficult to reconcile with the concept that the tip compartment behaves as a duplicating entity (Trinci, 1979). Trinci (1978) has suggested that cytoplasm synthesized within intercalary compartments may be transported to the tip compartment.

Comparison between small mycelia and mature margins. The inter-clamping time interval in tip compartments was significantly shorter in main hyphae of the mature margin than in exponentially growing small mycelia (Table 2). The mean tip-compartment volume was also larger so that division did not take place at a constant cytoplasmic volume per compartment. An alternative hypothesis is that the rate of nuclear division is related to the total amount of synthesis of a division factor by different volumes of cytoplasm per nuclear pair. It is not at present possible to assess whether the volume of cytoplasm within the apical compartment is regulating the rate of division. Information about the duration of different phases in the division cycle is needed.

Expression of clamp-connections

Expression in C. disseminatus. This fungus is a heterocytic heterothallic species in which homokaryons lack clamp-connections and the secondary mycelium has predominantly binucleate but sometimes trinucleate compartments. Clamp-connections occur inconstantly in the secondary mycelium and there is a consistent association between expression of

Table 5. *Growth thresholds for clamp-connection formation in* Coprinus disseminatus

Condition	Threshold[a] colony radius (mm)	Threshold colony radial growth rate (μm h^{-1})
Small mycelia		
2% malt	7.5	190 (50% leaders)
D/5[b] + Cellophane	6.8	170
D/5 + Cellophane	6.5	179
Mature colonies		
Temperature		152
Cycloheximide		193
p-fluoro-DL-β-phenylalanine		150
Emetine		210
Sodium taurocholate		165
Copper sulphate		195
2-Deoxy-D-glucose		115
Sucrose		190
Potassium chloride		175

[a] Threshold: at least 1 clamp-connection visible in 50% standard 2.69-mm^2 microscope fields of view.
[b] D/5: medium containing glucose 2 g l^{-1} and sodium nitrate 0.24 g l^{-1}. Cultures were grown at 25 °C except in temperature-variation experiments.

clamp-connections and fast extension rate of wide hyphae. Thus although main hyphae in the mature colony margin are constantly clamped, primary branches are at first simply septate and begin to form clamp-connections later, when they extend more rapidly and have wider tips (Butler, 1972). When colony growth was initiated from a small heterokaryotic inoculum there was an ontogenetic delay before full expression of clamp-connections in main hyphae extending at their maximum extension rate (Table 5). The colony radius at full expression of clamp-connections was between 6 and 12 mm. When the radial extension rate of the mature margin was altered using a range of different environmental changes, clamp-connections were consistently expressed in growth at or above a characteristic threshold extension rate. This was similar to the ontogenetic threshold extension rate (Table 5). Clamp-connections were restricted to wide hyphae and the change was reversible on transfer between permissive and non-permissive conditions (Butler, 1982). Formation of simple septa in hyphae just below the clamping threshold in all of these situations was associated with synchronous division of two or occasionally three nuclei and tip isolation showed that they were almost always heterokaryotic for mating type.

The ontogenetic delay seemed to depend on growth from a sufficiently small inoculum. It was observed during growth from a 1.2-mm^3 but not from a 15-mm^3 inoculum disc. In excised hyphal tips a sequence from true clamp-connection formation through pseudoclamp formation to simple septum formation occurred in the main axis, the change being immediate in growth from isolated apical compartments. A transient pseudoclamp phase also occurred at the clamping threshold imposed by both colony size and environmental change. Pseudoclamps complicate subsequent development since they result in homokaryotic subterminal compartments.

Formation of clamp-connections in *C. disseminatus* is thus a feature of differentiated main hyphae in the mature colony margin. This situation can be compared with that in other species in which an ontogenetic change in expression of clamp-connections without change in the kinds of nuclei present in the mycelium has been reported.

Expression in species with binucleate compartments. S. commune has constantly clamped binucleate compartments in the secondary mycelium. However clamp-connections are not expressed immediately during regeneration from dikaryotic protoplasts and the duration of the ontogenetic delay seemed to be related to the past history of cohabitation of the pair of nuclei (Raper, 1978). In dikaryotic protoplasts derived from dikaryotic hyphae there was a brief transient phase of pseudoclamp formation before the full expression of clamp-connections, suggesting that this element was a cytoplasmic factor (Wessels, Hoeksema & Stemerding, 1976). In protoplasts synthesised by fusion of homokaryotic protoplasts there was a longer ontogenetic delay, perhaps involving nuclear interaction.

In two clamped species of *Pholiota* (*P. adiposa* and *P. nameko*) heterokaryotic dikaryotic asexual spores form simple septa for the first few synchronous divisions after germination (Arita, 1979). Here the pair of nuclei had co-existed in an inactive state in the spore.

Dedikaryotisation of secondary mycelium has been induced in a number of heterothallic clamped species by means of various growth-retarding chemicals (Miles & Raper, 1956; Kerruish & Da Costa, 1963). In some of these species this change is associated with pseudoclamp formation (Miles & Raper, 1956).

Clamp-connection expression is delayed until the colony is a characteristic and large size in *C. sterquilinus* (Harder, 1926; Buller, 1933) and in homothallic strains of *Sistotrema brinkmannii* (Ullrich & Raper, 1975). These are constitutively homothallic and binucleate compartments are constantly clamped. The ontogenetic delay is associated with establishment of binucleate compartments. Once established the clamped state in *S. brinkmannii* is not regularly reversed by isolation of tip compartments.

Expression in species with multinucleate compartments. The system in holocoenocytic species has most in common with that in *C. disseminatus*. The mature colony margin contains different sorts of hyphae, with simple septa, single or whorled clamp-connections. There is an ontogenetic delay in expression of whorled clamp-connections (Gäumann, 1952). In *S. gausapatum* and *S. hirsutum* whorled clamp-connections are associated with wide hyphae extending at a fast rate (Boddy & Rayner, 1982; M. N. Turton via A. D. M. Rayner, personal communication). In the former, clamp expression is reversed when growth is inhibited by a nearby slow-growing isolate. However unlike the situation in *C. disseminatus* clamp-connections are formed on both monosporous and secondary mycelia.

Two themes may be extracted from these diverse ontogenetic changes. First, in both constantly and inconstantly clamped species there is evidence that pseudoclamp formation is associated with slow extension rate. Second, clamp-connection expression in the margin of inconstantly clamped species is a manifestation of the differentiation of wide, fast-extending hyphae. The evidence from experiments on reversibility of expression indicates a close link between the mechanisms controlling extension rate and clamp-connection formation.

Density of the margin

The mature colony margin is usually described as having not only a constant extension rate but also a constant density of radially oriented hyphae (Bull & Trinci, 1977). The experimental basis for this view is slender. When colonies of *C. cinereus* and *C. disseminatus* are grown from small inocula there is a visually-dramatic ontogenetic change from a margin of predominantly surface and submerged hyphae to a fluffy margin containing many aerial hyphae. Some information on the time sequence of different ontogenetic changes is provided by *C. disseminatus* growing on Cellophane from a small inoculum (Table 3). As the young colony developed, the angle of branch growth gradually diminished. One other feature of branch growth stood out as showing a marked increase during colony maturation. As the extension rate of the main hyphae increased so did the mean extension rate of primary branch hyphae within 1 mm of the main tips. Main tips and their primary branches attained their maximum extension rates concomitantly. Moreover the *relative* extension rate of primary branches, expressed as a percentage of the extension rate of parent hyphae, only reached its maximum, about 62%, in this late stage of maturation. The fluffy margin with aerial hyphae was first evident after the main hyphae had attained their maximum extension rate.

Hutchinson, Sharma, Clarke & Macdonald (1980) show how the generation of a circular colony shape can be simulated using known branching parameters of *Mucor hiemalis* without the need to introduce interhyphal-orienting factors or self-regulatory co-ordination of parent and branch growth. However *M. hiemalis* lacks a hierarchical branching system whereas basidiomycetes usually show a clear distinction between main and branch hyphae. In *C. disseminatus* the increase in relative primary-branch growth rate occurred when the margin was becoming denser and thus when growth-reducing changes in the medium associated with uptake of nutrients and release of metabolites were likely to be increasing. This supports the view that there is internal competition for resources within a considerable length of the expanding mycelium. This situation could perhaps be accommodated in the model of Prosser & Trinci (1979) by adjusting the proportion of vesicles allowed to flow between adjacent hyphal compartments.

Since it appears that expression of clamp-connections in *C. disseminatus* occurs only in hyphae extending at or above a critical extension rate, earlier data (Butler, 1968) on the frequency of clamp-connections on primary branch hyphae provide insight as to how margin density is maintained as the colony expands. Of those primary branches which continue to grow an increasing proportion form clamp-connections at successive nodes from the branch base. It appears that on nutrient-rich media the majority of these primary branches are gradually increasing their rate of growth.

This pattern of branch growth could help to explain the maintenance of margin density and the extensive aerial growth at the margin. Branches would still lag behind their parent hyphae if the latter were radially oriented in the margin and continued to grow at a constant rate. Neither of these conditions has been substantiated. Hyphal orientation is not strictly radial but is at various acute angles to the colony radius, with leading tips tending to meet and cross indiscriminately. There are no long-term population studies on the survival rate of main tips in the margin. Short-term observations show that growth is often monopodial but in *S. lacrimans* some wide main hyphae grow ahead of the majority and these have limited growth, the tip being replaced by a primary branch from the subterminal compartment (Hornung & Jennings, 1981).

Although hyphae on the surface of the medium are easier to observe, the colony margin in many basidiomycetes consists largely of aerial hyphae. Aerial hyphae must be using resources from the substratum hyphal system and their rate of extension is likely to be faster than that of surface hyphae under conditions in which non-volatile auto-inhibitory substances accumulate in the medium. This factor cannot be involved in margin organisation in those basidiomycetes in which the margin is largely submerged, e.g. *P. radiata* and *P. rufa* (Boddy, Chapter 12; Rayner & Webber, Chapter 18).

Conclusions

The high differentiative capacity of basidiomycete mycelia is shown not only by the range of differentiated hyphae and multihyphal structures formed within colonies (Nobles, 1971; Watkinson, 1979) but also by the characters of divergently-growing hyphae. For species with an advancing margin of aerial hyphae a theme is emerging of increase during ontogeny not only in extension rate but also in hyphal diameter, rate of tip-compartment division, branching capacity and, in certain species, frequency of clamp-connections. Maximum expression of these changes occurs only in colonies of substantial size and this poses problems of physiological mechanisms. What length of system is required to maintain the changed forms of growth is not clear although the data suggest that a hyphal length greater than that of the apical compartment is involved.

The association between morphological features and extension rate does not seem to occur in the change to fast-extending hyphae in 'point growth' of *S. lacrimans* (Coggins *et al.*, 1980). 'Point growth' and its derivation from slow-growing forms illustrates the increased awareness of the mycelial capacities of basidiomycetes which is coming from studies of the behaviour of mycelia in nature and in isolates obtained directly from nature. The ecological significance of these capacities is discussed in other chapters (Boddy, Chapter 12; Thompson, Chapter 9).

Many of these changes during colony ontogeny seem not to be

attributable to differences in nuclear state of the mycelium. It should now be possible to choose for further study species which combine important mycelial characters with those needed for genetic analysis.

References

Anderson, J. B. (1982). Bifactorial heterothallism and vegetative diploidy in *Clitocybe tabescens. Mycologia*, **74**, 911–16.

Arita, I. (1979). Cytological studies on *Pholiota. Reports of the Tottori Mycological Institute*, **17**, 1–118.

Boddy, L. & Rayner, A. D. M. (1982). Population structure, inter-mycelial interactions and infection biology of *Stereum gausapatum. Transactions of the British Mycological Society*, **78**, 337–51.

Boddy, L. & Rayner, A. D. M. (1983). Mycelial interactions, morphogenesis and ecology of *Phlebia radiata* and *P. rufa* from oak. *Transactions of the British Mycological Society*, **80**, 437–48.

Boidin, J. (1971). Nuclear behaviour in the mycelium and the evolution of the basidiomycetes. In *Evolution in the Higher Basidiomycetes*, ed. R. H. Petersen, pp. 129–48. Knoxville: The University of Tennessee Press.

Bull, A. T. & Trinci, A. P. J. (1977). The physiology and metabolic control of fungal growth. *Advances in Microbial Physiology*, **15**, 1–84.

Buller, A. H. R. (1933). *Researches on Fungi*, vol. 5. London: Longmans Green.

Butler, G. M. (1968). Environmentally induced changes in clamp connection incidence in hyphal branching systems of *Coprinus disseminatus. Annals of Botany*, **32**, 847–62.

Butler, G. M. (1972). Nuclear and non-nuclear factors influencing clamp connection formation in *Coprinus disseminatus. Annals of Botany*, **36**, 263–79.

Butler, G. M. (1982). Effect of growth-retarding environmental factors on growth kinetics and clamp connection occurrence in dikaryon of *Coprinus disseminatus. Transactions of the British Mycological Society*, **77**, 593–603.

Casselton, L. A. (1978). Dikaryon formation in higher basidiomycetes. In *The Filamentous Fungi*, vol. 3, *Developmental Mycology*, ed. J. E. Smith & D. R. Berry, pp. 275–97. London: Arnold.

Coates, D., Rayner, A. D. M. & Todd, N. K. (1981). Mating behaviour, mycelial antagonism and the establishment of individuals in *Stereum hirsutum. Transactions of the British Mycological Society*, **76**, 41–51.

Coggins, C. R., Hornung, U., Jennings, D. H. & Veltkamp, C. J. (1980). The phenomenon of 'point growth' and its relation to flushing and strand formation in mycelium of *Serpula lacrimans. Transactions of the British Mycological Society*, **75**, 69–76.

Fiddy, C. & Trinci, A. P. J. (1976). Mitosis, septation, branching and the duplication cycle in *Aspergillus nidulans. Journal of General Microbiology*, **97**, 169–84.

Gäumann, E. A. (1952). *The Fungi*, translated F. L. Wynd. New York: Hafner Publishing Company.

Girbardt, M. (1968). Ultrastructure and dynamics of the moving nucleus. In *Aspects of Cell Motility* (*22nd Symposium of the Society for Experimental Biology*), ed. P. L. Miller, pp. 249–59. Cambridge: Cambridge University Press.

Gow, N. A. R. & Gooday, G. W. (1982). Growth kinetics and morphology of colonies of the filamentous form of *Candida albicans. Journal of General Microbiology*, **128**, 2187–94.

Harder, R. (1926). Mikrochirurgische Untersuchungen uber die geschlechtliche Tendez der Paarkerne des homothallischen *Coprinus sterquilinus* Fries. *Planta*, **2,** 446–53.

Harder, R. (1927). Zur Frage nach der Rolle von Kern und Protoplasma im Zellgeschehen und bei der Ubertragung von Eigenschaften. *Zeitschrift für Botanik*, **19,** 337–407.

Hornung, U. & Jennings, D. H. (1981). Light and electron microscopical observations of surface mycelium of *Serpula lacrimans*: stages of growth and hyphal nomenclature. *Nova Hedwigia*, **34,** 101–26.

Hutchinson, S. A., Sharma, P., Clarke, K. R. & Macdonald, I. (1980). Control of hyphal orientation in colonies of *Mucor hiemalis. Transactions of the British Mycological Society*, **75,** 177–91.

Jennings, D. H. (1982). The movement of *Serpula lacrimans* from substrate to substrate over nutritionally inert surfaces. In *Decomposer Basidiomycetes: Their Biology and Ecology*, ed. J. C. Frankland, J. N. Hedger & M. J. Swift, pp. 91–108. Cambridge: Cambridge University Press.

Kerruish, R. M. & Da Costa, E. W. B. (1963). Monocaryotization of cultures of *Lenzites trabea* (Pers.) Fr. and other wood-destroying basidiomycetes by chemical agents. *Annals of Botany*, **27,** 653–69.

King, S. B. & Alexander, L. J. (1969). Nuclear behaviour, septation and hyphal growth of *Alternaria solani. American Journal of Botany*, **56,** 249–53.

Korhonen, K. & Hintikka, V. (1974). Cytological evidence for somatic diploidization in dikaryotic cells of *Armillariella mellea. Archives of Microbiology*, **95,** 187–92.

Kuhner, R. (1977). Variation in nuclear behaviour in the homobasidiomycetes. *Transactions of the British Mycological Society*, **68,** 1–16.

Miles, P. G. & Raper, J. R. (1956). Recovery of the component strains from dikaryotic mycelia. *Mycologia*, **48,** 484–94.

Niederpruem, D. J., Jersild, R. A. & Lane, P. L. (1971). Direct microscopic studies of clamp connection formation in growing hyphae of *Schizophyllum commune. I.* The dikaryon. *Archiv für Mikrobiologie*, **78,** 268–80.

Nobles, M. K. (1971). Cultural characters as a guide to the taxonomy of the Polyporaceae. In *Evolution in the Higher Basidiomycetes*, ed. R. H. Petersen, pp. 169–92. Knoxville: University of Tennessee Press.

Papazian, H. P. (1955). Sectoring variants in *Schizophyllum. American Journal of Botany*, **42,** 394–400.

Prosser, J. I. & Trinci, A. P. J. (1979). A model for hyphal growth and branching. *Journal of General Microbiology*, **111,** 153–64.

Raper, C. A. (1978). Control of development by the incompatibility system in basidiomycetes. In *Genetics and Morphogenesis in the Basidiomycetes*, ed. M. N. Schwalb & P. G. Miles, pp. 3–29. London: Academic Press.

Raudaskoski, M. & Kaukonen, P. (1978). Effects of hydroxyurea on development of uninucleate apical cells of *Schizophyllum commune. Experimental Mycology*, **2,** 239–44.

Rayner, A. D. M. & Todd, N. K. (1982). Population structure in wood-decomposing basidiomycetes. In *Decomposer Basidiomycetes: Their Biology and Ecology*, ed. J. C. Frankland, J. N. Hedger & M. J. Swift, pp. 109–28. Cambridge: Cambridge University Press.

Rayner, A. D. M. & Turton, M. N. (1982). Mycelial interactions and population structure in the genus *Stereum*: *S. rugosum, S. sanguinolentum* and *S. rameale. Transactions of the British Mycological Society*, **78,** 483–93.

Robinson, P. M. & Smith, J. M. (1980). Apical branch formation and cyclic development in *Geotrichum candidum. Transactions of the British Mycological Society*, **75,** 223–8.

Simchen, G. & Jinks, J. L. (1964). The determination of dikaryotic growth rate in the basidiomycete *Schizophyllum commune*: a biometrical analysis. *Heredity*, **19**, 629–49.

Steele, G. C. & Trinci, A. P. J. (1975). Morphology and growth kinetics of hyphae of differentiated and undifferentiated mycelia of *Neurospora crassa*. *Journal of General Microbiology*, **91**, 362–8.

Trinci, A. P. J. (1971). Influence of the peripheral growth zone on the radial growth rate of fungal colonies. *Journal of General Microbiology*, **67**, 325–44.

Trinci, A. P. J. (1974). A study of the kinetics of hyphal extension and branch initiation of fungal mycelia. *Journal of General Microbiology*, **81**, 225–36.

Trinci, A. P. J. (1978). The duplication cycle and vegetative development in moulds. In *The Filamentous Fungi*, vol. 3, *Developmental Mycology*, ed. J. E. Smith & D. R. Berry, pp. 132–63. London: Edward Arnold.

Trinci, A. P. J. (1979). The duplication cycle and branching in fungi. In *Fungal Walls and Hyphal Growth*, ed. J. H. Burnett & A. P. J. Trinci, pp. 319–58. Cambridge: Cambridge University Press.

Ullrich, R. C. & Anderson, J. B. (1978). Sex and diploidy in *Armillaria mellea*. *Experimental Mycology*, **2**, 119–29.

Ullrich, R. C. & Raper, J. R. (1975). Primary homothallism – relation to heterothallism in the regulation of sexual morphogenesis in *Sistotrema*. *Genetics*, **80**, 311–21.

Valla, G. (1973). Divisions nucléaires, septation et ramification chez l'haplonte de *Polyporus arcularius* (Batsch) ex Fr. *Le Naturaliste Canadien*, **100**, 479–92.

Watkinson, S. C. (1979). Growth of rhizomorphs, mycelial strands, coremia and sclerotia. In *Fungal Walls and Hyphal Growth*, ed. J. H. Burnett & A. P. J. Trinci, pp. 93–113. Cambridge: Cambridge University Press.

Wessels, J. G. H. & De Vries, O. M. H. (1973). Wall structure, wall degradation, protoplast liberation and wall regeneration in *Schizophyllum commune*. In *Yeast, Mould and Plant Protoplasts*, ed. J. R. Villanueva, I. Garćia Acha, S. Gascón & F. Uruburu, pp. 295–306. London: Academic Press.

Wessels, J. G. H., Hoeksema, H. L. & Stemerding, D. (1976). Reversion of protoplasts from dikaryotic mycelium of *Schizophyllum commune*. *Protoplasma*, **89**, 317–21.

Zalokar, M. (1959). Growth and differentiation of *Neurospora* hypha. *American Journal of Botany*, **46**, 602–10.

4

Hyphal interactions in *Schizophyllum commune*: the di-mon mating

TAN T. NGUYEN and
DONALD J. NIEDERPRUEM
Department of Microbiology and Immunology, Indiana University School of Medicine, 635 Barnhill Drive, Indianapolis, Indiana 46223, USA

Despite genetic advances over the years regarding the roles of A and B incompatibility loci in governing heterokaryosis in *Schizophyllum commune* (cf. Raper, 1966; Casselton, 1978), very little is actually known about the primary events which occur during somatogamy, that is hyphal fusion and the initial entry of donor nuclei into recipient hyphae of this fungus. While these hyphal interactions are central to heterokaryosis, the relative infrequency with which they can be observed in culture has limited direct microscopic studies of mating behaviour *in vivo*. Thus, whereas the occurrence of hyphal fusions and subsequent *events* in the eventual formation of the stable dikaryon of *S. commune* have been reported previously, the initial mating events were not discerned (Niederpruem, 1980*a*, *b*). These difficulties are further compounded by the fact that *vegetative* hyphal *fusions* within the same monokaryon must clearly be delineated from *sexual fusions* between different compatible monokaryons, which is all the more difficult since the opposing monokaryotic hyphae are morphologically indistinguishable. However, in *S. commune*, the dikaryon is not only binucleate but also bears conspicuous clamp-connections while the monokaryon is uninucleate and without clamp-connections. We therefore decided to circumvent difficulties of identifying sexual fusions by examining live cellular behaviour during mating of dikaryotic with monokaryotic hyphae (di-mon mating) of *S. commune*.

Retrospective

The Buller phenomenon

. . . much of human inquiry is
little more than the elaborate
elucidation of the obvious.

Raper and Flexer (1970)

Any cursory review of the literature on experimental mycology would reveal that Buller (1931) told us all (well, almost all) there was to know about hyphal fusions and nuclear exchange in Basidiomycotina especially in species of *Coprinus*. The phenomenon of vegetative hyphal fusions was observed and discussed: 'The hyphal fusions which have just been described have nothing to do with sex . . . the hyphal bridges come into play, and the liquid contents of the mycelial hyphae flow in all directions toward the fruit-body, thus enabling it to complete its development' – stated more simply, all nuclear exchanges involve hyphal fusions, but not all hyphal fusions involve nuclear exchange. Commenting further on the significance of vegetative hyphal fusions, Buller postulated that 'it is probable that the monosporous mycelia join together to form a single compound net-mycelium which acts as a unit in the formation of one or more fruit-bodies'. This concept, at least as applied to genetically different entities, is now under vigorous challenge (see Aylmore & Todd: Chapter 5; Rayner *et al*: Chapter 23).

Returning to the problem of sexual hyphal fusions, Buller (1931) delineated the sequence of events which was believed to occur in the fungus he called *Coprinus lagopus* (it was actually *C. radiatus*) when two haploid mycelia, that is, monokaryons of opposite sex, were grown near one another. The hypothesis was offered that chemotropism was initially involved and, using his own observations in conjunction with those of Lehfeldt (1923) on mating in *Typhula erythropus*, the following sequence of events was proposed. Directed hyphal growth towards the recipient hypha is followed by fusion, donor nucleus division in the fusion bridge, one progeny nucleus entering the recipient thus dikaryotizing it, and septum formation in the fusion bridge near where the prior mitosis had occurred. The donor nucleus in the recipient now divides and septal erosion occurs, thus enabling one daughter nucleus to migrate to the adjacent cell (Fig. 1). Note in this particular sequence that the emergence of a dikaryon bearing clamp-connections was not featured. However, at the end of the same volume, Buller (1931) made a second attempt to visualise dikaryotisation in *C. lagopus*, this time noting directed growth of the donor hypha towards the recipient, fusion, the donor nucleus entering the recipient, conjugate nuclear division and the emergence of the first clamp-connection in the recipient compartment with no septal breakdown or nuclear migration mentioned in the overall process.

More relevant to our current studies in *S. commune*, Buller had also examined di-mon pairings of *C. lagopus*. His final conclusion was that matings between dikaryons and monokaryons occurred in essentially the same manner as those between monokaryons (Buller, 1930, 1931). His summary of the supposed sequence of events in the di-mon mating is shown in Fig. 2. Similarities included directed growth (in this case, the

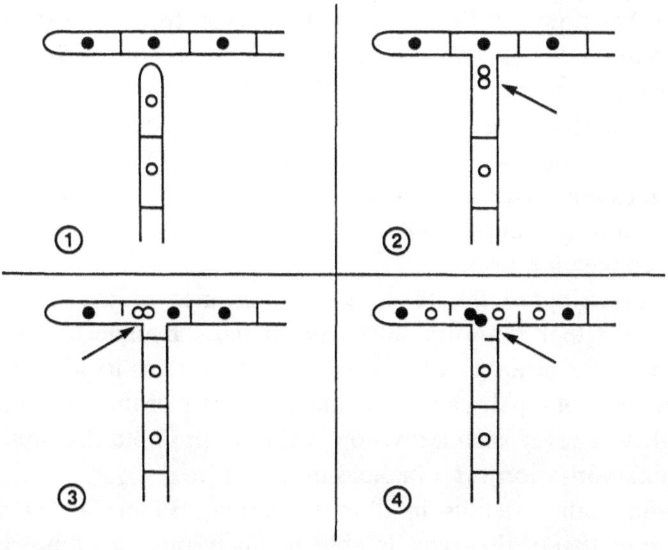

Fig. 1. Mon-mon mating of *Coprinus lagopus* (after Buller, 1931).

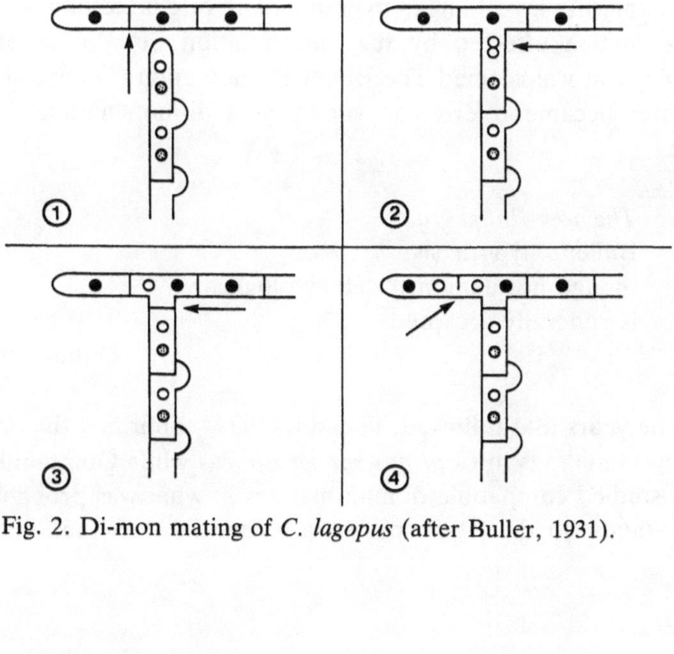

Fig. 2. Di-mon mating of *C. lagopus* (after Buller, 1931).

dikaryon, towards the monokaryon), hyphal fusion, division of one donor nucleus from the dikaryon now in the fusion bridge, entry of one daughter nucleus into the recipient monokaryon and again division of the donor nucleus while in the recipient cell. One of the progeny enters the adjoining compartment *via* septal breakdown. Note also that a septum has now cut off the original dikaryon from the original monokaryon in the fusion bridge itself. Further mitosis of the donor nuclei in the recipient compartments, septal breakdown and nuclear migration occur such that four originally monokaryotic compartments are now dikaryons yet devoid of clamp-connections. Finally, one primary branch undergoes conjugate nuclear division with the emergence of the first true clamp-connection and then the dikaryotic hyphal apex forms a clamp-connection also.

Buller attributed the functional significance of the occurrence of conjugate rather than diploid nuclei in these hyphae to the fact that 'each member of a pair of conjugate nuclei retains its identity, so that one member of a pair can divide independently of the other member of the pair whenever such a division is able to promote the diploidisation (i.e. dikaryotisation) of a haploid mycelium'.

Among other details in di-mon mating, Buller (1931) noted the following: that a dikaryon is able to dikaryotise a monokaryon if it comes in contact with the latter at any point; older mycelia effect dikaryotisation or are dikaryotised more slowly than younger mycelia; dikaryotisation of a monokaryotic mycelium can be effected as rapidly by a dikaryotic mycelium as by a monokaryotic mycelium.

The mating effected by the confrontation of a dikaryon with a monokaryon was termed The Buller Phenomenon (Quintanilha, 1937) but later became referred to simply as a di-mon mating (Papazian, 1950).

The post-Buller era

Buller's hypothesis,
though not demonstrated cytologically,
is generally accepted.

Quintanilha (1939)

In the years that followed, Dickson (1934) confirmed the occurrence of di-mon matings in *Coprinus sphaerosporus* while Quintanilha (1933, 1934) studied compatible di-mon matings in what was probably *Coprinus radiatus.* In the latter species, a bifactorial incompatibility system governs sexual reproduction such that the following di-mon mating could be examined:

(A1B1 + A2B2) × (A3B3)

Although the resultant new dikaryons, (A1B1 + A3B3) and (A2B2 + A3B3), might be expected to occur with equal frequency, disparate recovery of one dikaryon ensued over the other, indicating non-random selection from the two nuclear types in the parent dikaryon. The particular phenomenon of internuclear selection in di-mon matings has also been documented in *Coprinus macrorhizus* f. *microsporus* and *Psilocybe coprophila* (Kimura, 1958), *Pleurotus ostreatus* (Terekawa, 1957), *Coprinus radiatus* (Prevost, 1962), and *Schizophyllum commune* (Crowe, 1963; Ellingboe, 1964; see also Rayner *et al*: Chapter 23). Proposed mechanisms have implicated the A factor, the B factor, multiple alleles independent of the A and B factors, cytoplasmic factors, earlier migration of the favoured nuclear type and other, as yet unknown factors. Raper aptly summarised it when he stated that 'even the simplest question that can be asked about the Buller Phenomenon can elicit only a complex answer' (1966). Some more recent observations and proposals will, however, be discussed by Rayner *et al.* (Chapter 23).

Live cellular behaviour and ultrastructure in mating

Before one can ask *how*
something happens, one must
first know *what* has happened

Anonymous

Much of the early work on mating and nuclear migration in Basidiomycetes in general and *Schizophyllum* in particular, depended upon evaluation of the appearance of clamp-connections on mated hyphae or, even more distal to the initial mating, the formation of basidiocarps. As geneticists became more involved with the control of heterokaryosis, isolation of basidiospores and determination of mating types became powerful adjuncts to analyse various mating situations. Beyond this, several imaginative techniques using auxotrophic recipient strains of *S. commune* were developed to follow nuclear migration indirectly. The auxotroph was grown on complete medium and then a compatible prototrophic strain implanted upon it; mycelial plugs were then removed sequentially from the recipient and implanted on minimal medium; growth here indicated the presence of the prototroph and hence the extent of nuclear migration (Snider & Raper, 1958). While intrusive growth of the prototroph could not be discounted entirely, much valuable information was deduced from these early works (cf. Casselton, 1978).

Unfortunately, none of these studies clarified the nature of live, initial events which are the preamble to heterokaryosis.

However, a few reports of attempts to delineate live mating behaviour have been provided. These include an elegant study on *Coriolus versicolor* (Lange, 1966), a brief account of *Schizophyllum commune* and *Coprinus 'lagopus'* (Sicari & Ellingboe, 1967), and a description of mating reactions in *Clitocybe truncicola* (Bistis, 1970). Extensive attempts to identify the primary mating events have also been made for *S. commune* (Parag, 1970; Raudaskoski, 1973; Niederpruem, 1980*a*, *b*) and *Coprinus disseminatus* (Butler, 1972). While these diverse matings show some common characteristics, other features may be unique to the particular fungus in question. Furthermore so few live mating situations have been documented *from their inception* that it is very difficult to obtain an overview.

There is a consensus, although intuitive in most cases, that *chemotropism* is actually involved in hyphal interactions of Basidiomycetes. This is apparently based on the fact that opposed hyphae show growth curvatures towards one another. In certain mating situations, moreover, basidiospores or oidia seem to attract hyphae, as reported in *Clitocybe truncicola* by Bistis (1970), *Coprinus* spp. by Kemp (1975, 1980) and *Leccinum* spp. by Fries (1981). Environmental and genetic control of hyphal fusions has been investigated in *S. commune* (Ahmad & Miles, 1970*a, b*) and *Coprinus lagopus sensu* Buller (Smythe, 1973). In the case of *S. commune* (Ahmad & Miles, 1970*a*), environmental conditions including temperature, pH and osmotic pressure apparently were without effect on hyphal fusions (mon × mon) as was the concentration, above 0.5%, of carbohydrate in the culture medium. However, lack of carbohydrate coincided with increased fusions. Furthermore, regarding chemotaxis in *S. commune*, there was 'attracted growth' over a distance of 10 μm, which was only discerned between compatible mating types (mon × mon). Membrane studies indicated that a substance released into the medium was responsible for much higher fusion frequencies (Ahmad & Miles, 1970*b*). Finally, it was concluded that high fusion frequency is associated with heterozygosity at the A incompatibility locus in *S. commune* (Ahmad & Miles, 1970*b*). Regarding this particular work, it is important to note that increased hyphal fusions seen in the absence of exogenous carbohydrate may simply reflect enhancement of vegetative fusions. Moreover, Smythe (1973) noted that only a minor role can be attributed to the incompatibility factors in regulating hyphal fusion in mon × mon matings of *Coprinus lagopus sensu* Buller, and that there was no evidence for attracted growth. The hyphal fusion types were mainly tip-to-peg, peg-to-peg or tip-to-tip while the remainder were tip-to-side, appressions or hyphal cross-overs; importantly, hyphal fusions here *were* temperature-dependent with a high Q_{10}. Noteworthy here was the fact that all studies of fusions were performed on a rich culture medium (i.e. the complete medium of Fries). Sicari & Ellingboe (1967) noted that the degree of sexual incompatibility in various matings of *S. commune* or *C. 'lagopus'* correlated with the frequency of hyaline hyphal tips adjacent to hyphal fusion; hyaline hyphal tips reflected apparent lysis of the cytoplasm although no cytoplasmic extrusion or ruptured cell walls were evident. This interesting phenomenon has not, to our knowledge, been verified by other workers or pursued further.

Certainly one of the more extensive photomicroscopic studies of mating in *S. commune* was that of Parag (1970). Among his important observations were that fusions were *not* restricted to pairs of compatible hyphae and that certain hyphae twined tightly around other 'target' hyphae. We interpret these findings as being important in vegetative fusions (see later). In attempts to study sexual hyphal fusions, Parag (1970) stated that nuclear migration was *never* caught in action. In certain hyphae, many cross-walls were destroyed and the number of nuclei varied from none to six. Moreover, presumably as an aftermath of fusion and nuclear migration in the establishment of the dikaryon, clamp-connections were oriented rearward *or* forward.

In the same year, Bistis (1970) observed the fusion of oidia with hyphae of *Clitocybe truncicola* and documented the events of one sequence as follows: within 8 h, the septa were disrupted sequentially and nuclei migrated through the disrupted pores. Nuclear division accompanied migration here. He also noted that secondary septa arose in the absence of mitosis (66 h later). In a second documented sequence for *C. truncicola,* Bistis (1970) noted that oidia attracted hyphae and fused with them, septa became distorted, disintegrated and the lateral wall near them bulged out. Vacuolation ensued, *nuclei disappeared* and many septa were laid down in the absence of mitosis in now nearly empty homokaryotic hyphae. Migrant nuclei were associated with resident nuclei and the first conjugate nuclear divisions were abnormal because reversed clamp-connections occurred which lacked septa and nuclear distributions were uneven. Finally, later divisions appeared normal as the stable dikaryon occurred in secondary and tertiary hyphal branches. Unfortunately, the time points utilised for these direct observations were so widely separated that identification of the actual sequence of events is impossible.

Post-mortem electron microscopy of hyphal matings in *Coprinus* and *Schizophyllum* has revealed some additional cytological features not resolved by ordinary light microscopy. The deceptively simple septa revealed by light microscopy are in reality complex pores (Girbardt, 1958) which could provide physical barriers to nuclear migration after hyphal fusion. Moore & McAlear (1962) showed that diverse basidiomycetous fungi are endowed with these complex septa and coined the terms *dolipore septa* and for their adjoining parenthesis-like membranes, *parenthesomes.* Thus, migrant nuclei must not only breach the dolipore but also the parenthesome. However, the latter *is* porous, and at least in *Rhizoctonia solani*, it has been suggested that migrant nuclei are carried by cytoplasmic streaming and traverse intact septal pores (Bracker & Butler, 1964). However, while cytoplasmic streaming and mitochondria traversing intact septa have been observed, no nuclei were ever seen migrating through intact septa in live hyphae of *R. solani* (C. E. Bracker, personal communication). Subsequent electron-microscope studies with other basidiomycetes, including *Coprinus cinereus* (Giesy & Day, 1965), the common A heterokaryon of *Schizophyllum commune* (Jersild, Mishkin & Niederpruem, 1967), and B-mutants of *S. commune* (Koltin & Flexer, 1969; Niederpruem, 1971), all indicate that septal breakdown is associated with nucleus migration. In retrospect, both Lehfeldt (1923) and Buller (1931) evoked septal disorganisation for the facilitation of nuclear migration in matings of *Typhula erythropus* and *Coprinus 'lagopus'*, respectively. Also studies of live hyphal behaviour during mating in *Coriolus* and *Clitocybe* indicated septal disruption (Lange, 1966; Bistis, 1970). The biochemical basis of septal disruption has been elucidated only in the case of *S. commune* (Wessels & Niederpruem, 1967; Wessels, 1969; Janszen & Wessels, 1970), where the activity of a cell-wall beta-glucanase (R-glucanase) whose activity is apparently governed by the B-incompatibility appears to be responsible.

Later electron-microscopic studies of dikaryotic hyphae in *S. commune* revealed microtubules and microfilaments (Raudaskoski, 1973). In subsequent work dealing with a unilateral mating in *S. commune*, migration hyphae were designated as those bearing dissolved septa and containing several nuclei per cell while pseudoclamps and true clamp-connections were also evident (Raudaskoski, 1973). Moreover, because microtubules were more common within migration hyphae, and, in a few instances, attached to the nuclear envelope, it is appealing to think that these cytoplasmic entities play some as yet undisclosed role in nuclear migration. Alternatively, they may have been involved in nuclear division. Girbardt (1968) first drew attention to the role of microtubules in directing nuclear movements in vegetative dikaryotic hyphal apices of *Coriolus versicolor*.

Mayfield (1974) likewise autopsied moribund hyphae of *S. commune* by electron microscopy. Three distinctive features were noted: multi-vesicular bodies associated with dissolving septa, vacuoles associated with septal disruption, and the occurrence of intra-hyphal hyphae. It was even speculated that the latter could be due to dikaryon development within an old homokaryotic hypha and the significance of this speculation will become apparent later.

Our rationale for presenting the work of Lange (1966) as a benediction for the foregoing is because we believe that it represents a definitive account of mating and nuclear migration in a basidiomycete. Prior work dealing with either stained material or fine structure in our opinion, only marked 'where the vital tides have been'. The same admonition applies to scoring mating *via* the emergence of clamp-connections or basidiocarps.

The approach of Lange (1966), and our intent in the current work, was to document live cellular behaviour from the inception of hyphal mating and nuclear exchange to the establishment of the first stable dikaryotic cell in the same recipient hyphae. Five post-hyphal fusion events in mon × mon matings were documented with time including nuclear migration, mitosis of donor and/or recipient nuclei, dissolution of septa and the sequence of their disruption as well as the direction of nuclear migration. Septal dissolution occurred slowly initially (2.5 h), but adjacent septa dissolved in shorter times thereafter. The disruption of these septa could occur apically or basipetally; if dissolution was towards the hyphal apex, the nuclei migrated apically; if dissolution was of basipetal septa, the nuclei migrated away from the hyphal apex. This thorough documentation of the initial live events in mating is exemplified by the statement that 'Die sicheriste Nachweis für eine Wanderung der Kerne durch die unveranderten Querwande ist zweifellos die direkte mikroskopische Lebendbeobachtung' (Lange, 1966).

The present study

Methodology

In the experiments reported here, fully compatible matings between dikaryons and monokaryons were studied in various combinations. *Schizophyllum commune* prototrophic strains 699 A41B41, 845 A51B51, 2145 A43B26 coh l$^-$, 2151 A41B41 coh l$^-$, 88 A93B93, and 4–40 A43B43 were employed in these matings. Dikaryotic mycelia were obtained from crosses involving 699 × 845, 88 × 40 and 2145 × 2151, respectively. The cultures were maintained at 25 °C and periodically subcultured on C + Y agar medium (Niederpruem, 1980*a*). To effect the di-mon encounter, pairs of plug inocula of the preformed dikaryotic and the monokaryotic mycelia were positioned in proximity with each other on thin films of C + Y medium supplemented with 18% gelatin in lieu of agar. Subsequently, mating behaviour was observed under phase-contrast microscopy in areas where hyphae from the paired mycelia interdigitated sufficiently to establish contact. These studies were normally conducted after 24–36 h of incubation to avoid obtrusive growth of aerial hyphae which interfered with the observations.

We attempted to mix *macerated* dikaryotic and monokaryotic mycelia (Waring blender) to enhance the frequency of di-mon hyphal contacts, and thus the hyphal anastomoses. However, the macerated dikaryon alone appeared unable to stabilise morphologically when allowed to grow in microculture; hence, no further appraisal of di-mon matings using this process was conducted. Figure 3 illustrates a typical pattern of irregular hyphal growth emerging from a fragmented dikaryon which was comprised of anucleated, multinucleated, dikaryotic and monokaryotic compartments, and some with pseudoclamp-connections.

Results and conclusions

Hyphal fusions. Examination of the confrontation zones revealed ample evidence for vegetative hyphal fusions between hyphae of the same dikaryotic or monokaryotic mycelium and, to a lesser extent, sexual anastomoses between hyphae which originated from di-mon matings. Various features common to both vegetative and sexual fusions were observed, and representative forms are shown in Fig. 4. Anastomosis occurred either when hyphae were in close proximity or touching (contact fusion), or over distances of more than 20 μm (distance fusion).

It was not uncommon to observe a single vegetative hypha exhibiting both contact and distance fusions. During distance fusion, the anastomatic hyphal bridges did not grow directly across to the other hyphae but curved or turned at various angles, and the hyphal bridge frequently bypassed nearby hyphae of the opposing mycelium to reach one farther away. A common pattern in contact fusion involved a lateral peg which, in some instances, originated from the clamp-connection and 'homed' towards the adjacent hypha to establish anastomosis. Figure 5 shows a pair of di-mon hyphae that grew alongside each other at a distance of

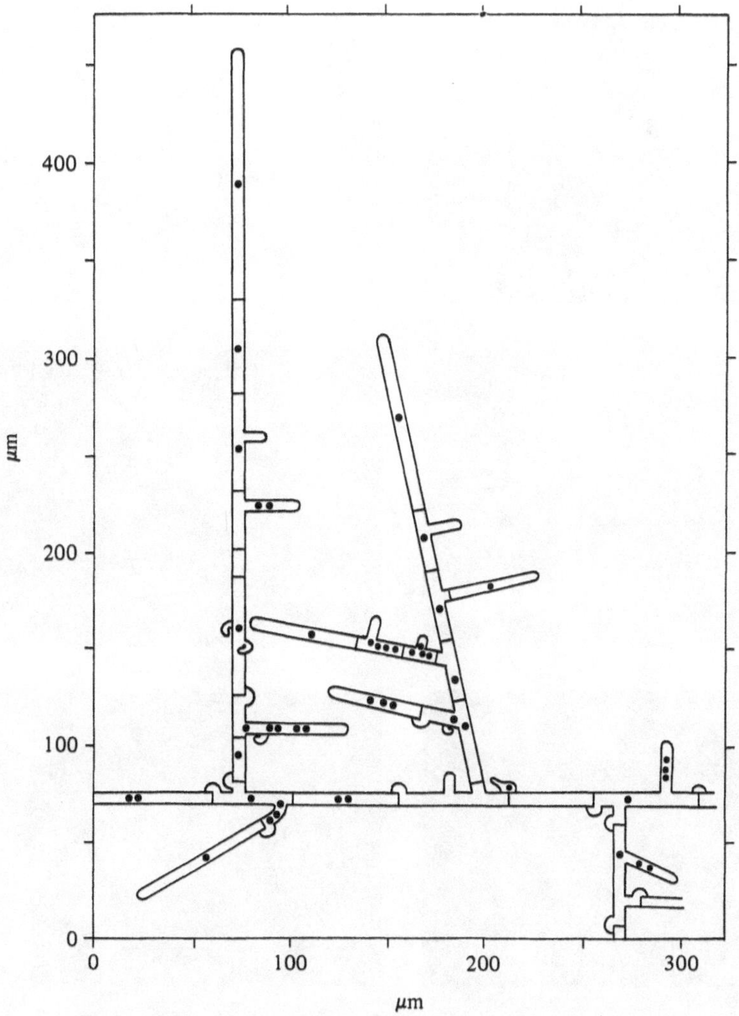

Fig. 3. Aberrant hyphal regeneration patterns from a macerated dikaryon, 699 × 845. Note the irregular presence of dikaryotic, monokaryotic and multinucleated cells.

more than 20 μm apart. Lateral bending in the direction of apical growth brought apices in close proximity where the dikaryotic hyphal tip grew directly toward the partner to fuse with it. The visible swellings at the termini of the anastomosed hyphae may have been the result of a localised production of cell wall-softening enzymes to complete the fusion process.

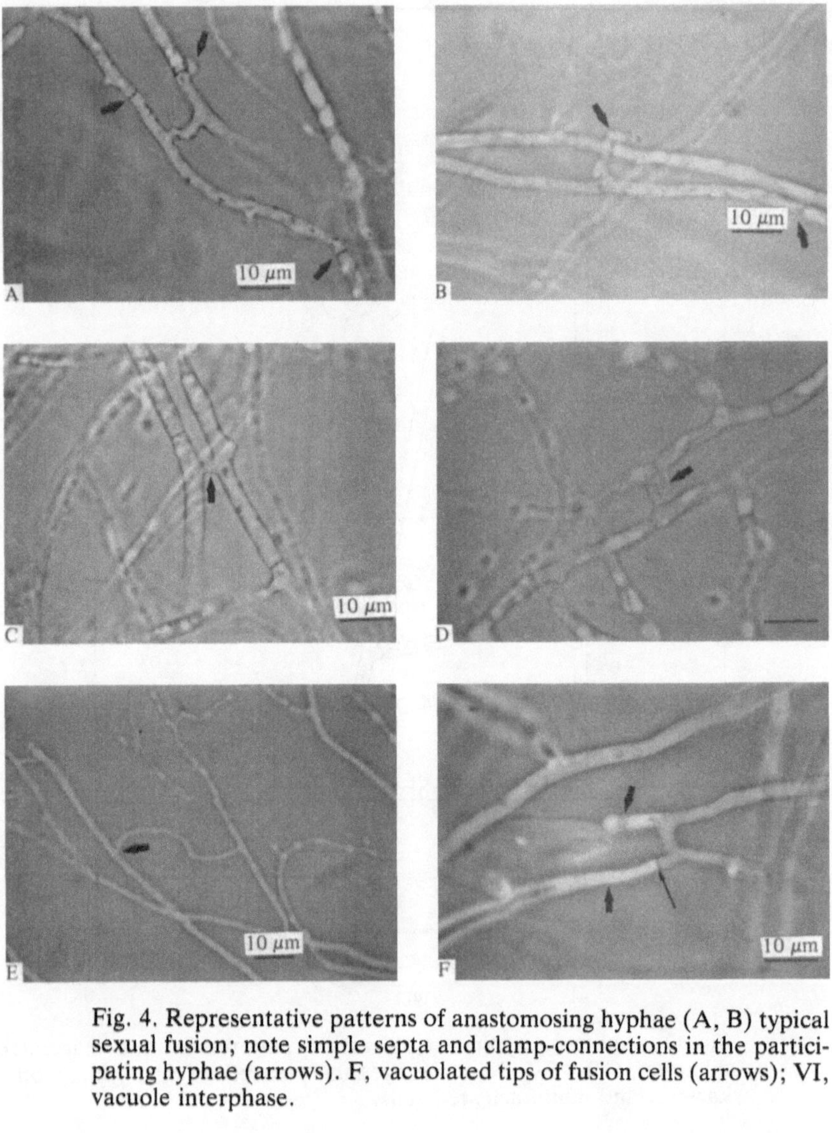

Fig. 4. Representative patterns of anastomosing hyphae (A, B) typical sexual fusion; note simple septa and clamp-connections in the participating hyphae (arrows). F, vacuolated tips of fusion cells (arrows); VI, vacuole interphase.

During dikaryosis, the dikaryotised hyphae that originated from the mated monokaryon were also observed to anastomose frequently with the nearby hyphae (Fig. 8). These bridges, in addition to the existing networks of hyphal channels via vegetative fusions in the original monokaryon, appeared to facilitate the process of nuclear migration.

Vacuolated hyphal tips at the sites of apparent sexual hyphal fusions (Sicari & Ellingboe, 1967) and the partial lysis of cells by vacuoles have been reported (Niederpruem, 1978). Occasional lysed tips were observed in anastomotic hyphae (Fig. 4F), and the presence of these lysed tips did not seem to correlate with any particular mating types or mating combinations as has been reported in the literature (Sicari & Ellingboe, 1967). Our observation indicated that cellular stasis of fusion cells often resulted in a high degree of vacuolisation which subsequently led to lysis of parts or the entire cells. The interface between the vacuole and the cytoplasm may easily be mistaken for a simple septum due to their similar appearances (Fig. 4F).

Nuclear exchange: vegetative fusions. Nuclear interaction patterns subsequent to vegetative hyphal fusions in dikaryons and monokaryons are reported by Aylmore & Todd in *Coriolus versicolor* (Chapter 5). Limited observations to date in *S. commune* revealed some evidence for nuclear interactions in the vegetative anastomotic hyphae in the monokaryotic and dikaryotic mycelial boundaries which took place prior to or simultaneously with sexual di-mon fusions.

Similar to their observations, vegetative fusion of dikaryotic hyphae established a transient tetranucleated state which resulted in sequential

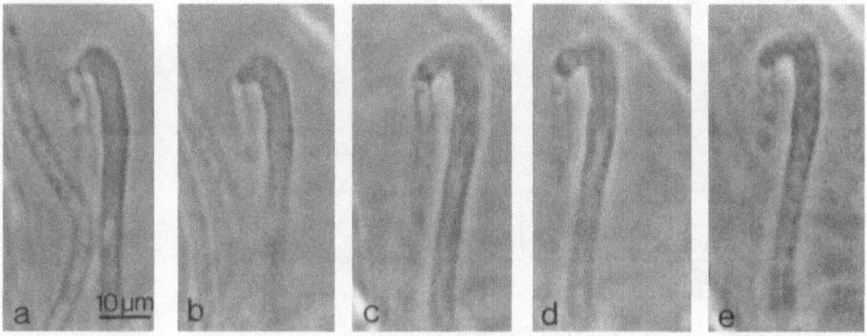

Fig. 5. Directed growth of apical hyphae participating in anastomosis. (a) 0 min; (b) 10 min; (c) 25 min; (d) 35 min; (e) cytoplasmic continuity was evident at this point (45 min).

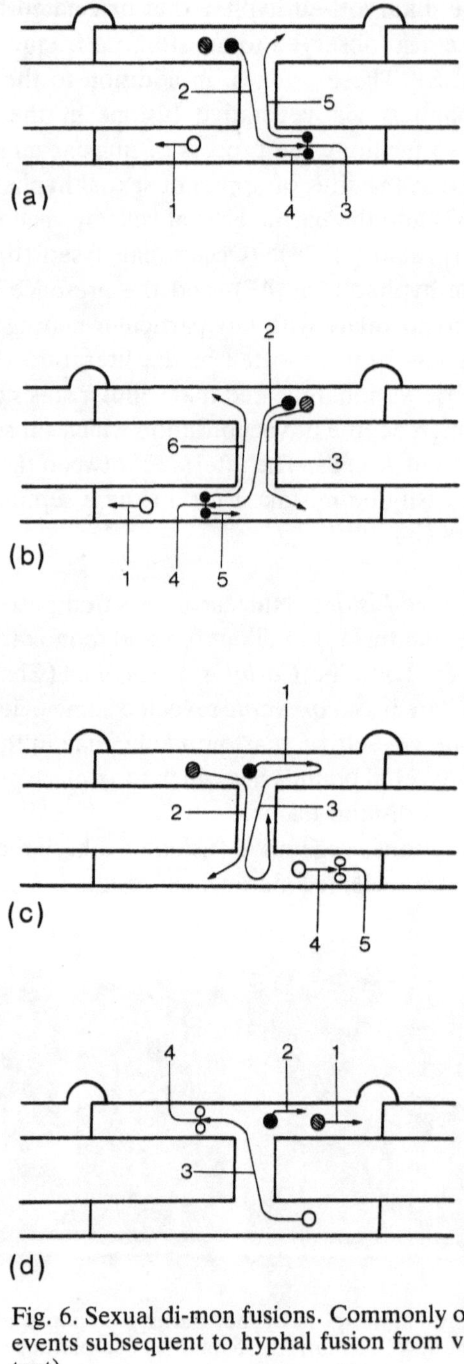

Fig. 6. Sexual di-mon fusions. Commonly observed patterns of nuclear events subsequent to hyphal fusion from various di-mon matings (see text).

disappearance of the conjugate pair of the resident nuclei (or recipient nuclei, Chapter 5) although no conjugate division of the migrant nuclear pairs (donor nuclei, Chapter 5) has yet been observed in the fusion cells. In contrast, erratic nuclear interactions were observed in vegetative fusion cells following anastomoses between hyphae of the same monokaryon. The binucleated interactions appeared to induce, in different cases, disappearance of either the resident nucleus or of one mitotic progeny originated from the mitosis of the migrant nucleus near the anastomosis site. This phenomenon of nuclear interactions has not yet been perceived during sexual di-mon fusions in *S. commune*.

Nuclear exchange: sexual di-mon fusions. A wide variety of capricious nuclear exchange patterns were observed, as ambiguous as the corresponding findings via genetic analyses. The apparent absence of regularity of nuclear mobility reflected in the diverse observations documented in Fig. 6a–d, serves to emphasise the dynamic nature of nuclear interactions in di-mon matings.

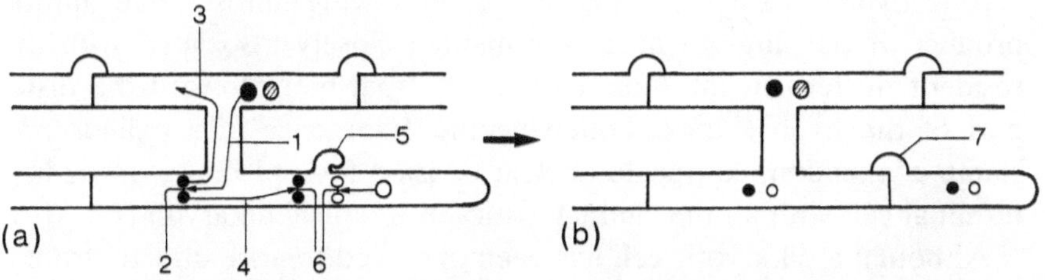

Fig. 7. Complete sequence from a di-mon mating to the emergence of the dikaryon. Note reverse clamp-connection at hyphal apex. Filled and hatched circles, nuclei from the dikaryon; hollow circle, nucleus of the monokaryon.

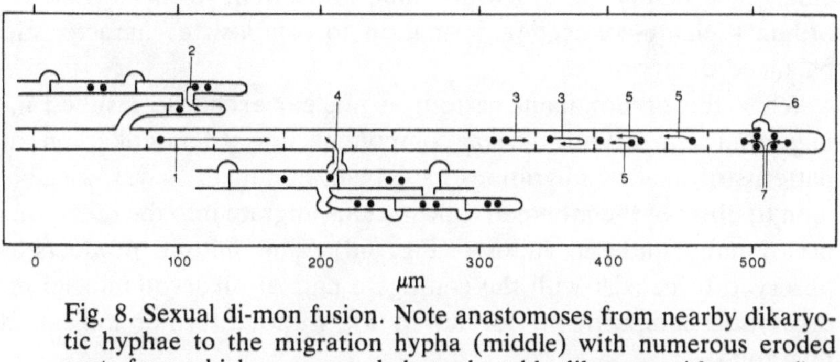

Fig. 8. Sexual di-mon fusion. Note anastomoses from nearby dikaryotic hyphae to the migration hypha (middle) with numerous eroded septa from which emerges a balanced, stable dikaryon with a true (i.e. rearward) clamp-connection emerged at the hyphal apex.

The most common nuclear migratory pathways following anastomosis were those illustrated in Fig. 6a, b, c. In these cases, nuclei of dikaryotic components migrated to the monokaryon although one migrant nucleus remained close to or subsequently returned to the fusion bridge and back to the dikaryotic compartment. Preferential entrance of migrant nuclei appeared independent of their locations distal or proximal to the fusion bridge. While in many cases, the proximal nucleus located nearest the fusion bridge entered it first (Fig. 6a, b; Fig 7a), note that in Fig. 6c, the originally proximal nucleus moved away from the fusion bridge while the more distal one moved forward to migrate into it before the other. Thus the order of entry may be somewhat fortuitous.

Events following initial nuclear migration in the trinucleate fusion cells were also rather unpredictable. These nuclear interactions frequently involved the mitosis of one migrant nucleus; thereafter, one mitotic progeny relocated back to the original dikaryotic boundary. One complete sequence of di-mon mating from fusion and nuclear exchange to the establishment of the first dikaryotic cell in the original monokaryotic hypha is shown in Fig. 7a, b. Following mitosis, one mitotic product of the migrant nucleus remained closely associated with the resident nucleus in the monokaryotic compartment toward the distal part of the hyphal apex. Following the emergence of a palindromic clamp-connection, conjugate nuclear division took place, resulting in a terminal cell with all the characteristics of a typical dikaryon (Fig. 7b).

Although a dikaryotic cell has been observed to arise directly from a fusion cell, the establishment of a balanced dikaryon frequently took place after extensive nuclear migration, as evidenced by numerous eroded septa in the migration hyphae (Fig. 8). Consequently, from the hyphal apex which somehow avoided superfluous nuclear aggregation emerged a dikaryotic cell which then underwent further division with ordinary clamp-connection formation to establish a characteristically balanced dikaryon.

While the predominant pattern of nuclear exchange resulted in the migration of dikaryotic nuclear components into the monokaryon, other patterns of nuclear migrations have been obtained. It was not uncommon to observe the monokaryotic nucleus migrate into the dikaryon and occasionally undergo mitosis (Fig. 6d). The mitotic products were observed to coexist with the conjugate pair of dikaryotic nuclei in the dikaryotic compartment for up to 3 h without further indication of migrating.

Nuclear migration. Dikaryosis proceeded with dikaryotising nuclei in passage through the established hyphal networks of the monokaryotic mycelium. This phenomenon of nuclear migration bears many morphological similarities to the patterns established by two fully compatible mon-mon matings which have been extensively described (Niederpruem, 1980*a, b*) and need not be mentioned at length here. Anucleated compartments adjoined by partially dissolved septa comprised most of the main migration hyphal axis, whilst a few compartments were filled with migrant nuclei which usually numbered from five to ten or more. None of these nuclei showed extensive intercellular migratory activity even after prolonged observation. Localized intercellular nuclear migrations were discerned within portions of migration hyphae bounded by two to three intercalary compartments. Frequently, these nuclei moved in close association, initiated clamp-connections and replicated in synchrony. The heterokaryotic nature of these nuclear components often resulted in pseudoclamp-connections reminiscent of common B heterokaryons. The mitotic progeny then migrated bidirectionally to occupy adjacent empty hyphal compartments. True dikaryotic cells were occasionally found intercalated between other anucleated or multinucleated compartments along migration hyphae and these often propagated lateral branchings of a dikaryotic nature which subsequently were found to anastomose with other monokaryotic or other migration hyphae in the vicinity (Fig. 9a,b).

General apical migration of nuclei proceeded rapidly toward the peripheral hyphae where nuclear aggregates were most commonly encountered. The eventual emergence of balanced dikaryons at the peripheral hyphal apices followed complex processes of nuclear reduction which involved unequal nuclear redistribution, and, more commonly, nuclear entrapment in clamp-connections. Figure 10 illustrates a means by which unequal segregation of nuclear progeny subsequent to conjugate nuclear division favoured a dikaryotic nuclear distribution in apical cells whereas penultimate cells contained the redundancies. Failure of clamp-connections to fuse with penultimate cells and which frequently served to entrap at least one nucleus occurred throughout the dikaryotised hyphae, as was also seen in the matings between two compatible monokaryotic strains of *S. commune* (Niederpruem, 1980*b*). Occasionally observed in multi-nucleated hyphal apices was the sequential initiation of two clamp-connections in the same vicinity prior to conjugate nuclear divisions (Fig. 11). Subsequently, the second

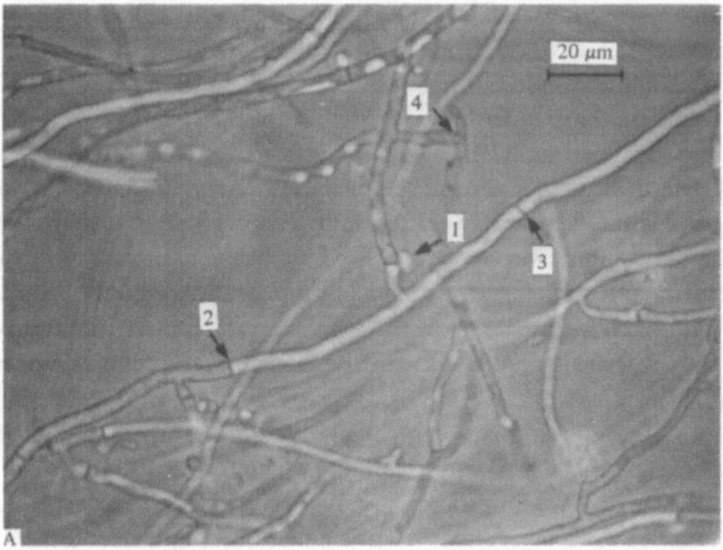

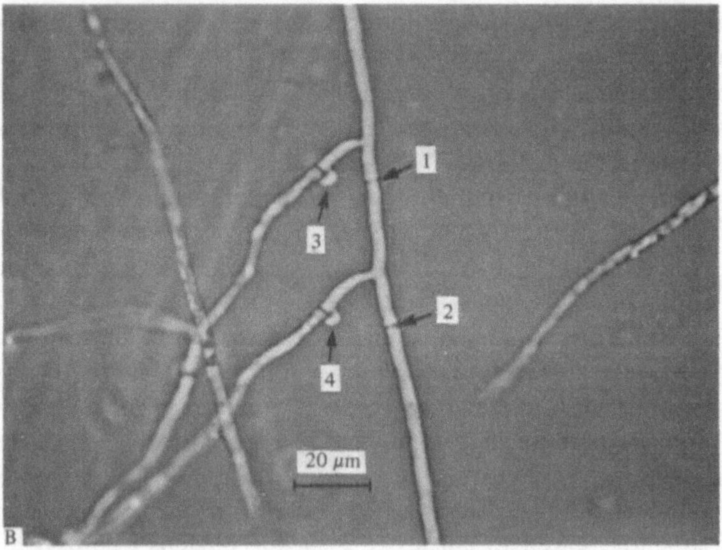

Fig. 9. A, unstable dikaryotised hypha. Note the characteristic morphology of dikaryon with clamp-connections (1) and monokaryon with simple septum (2) and eroded septum (3, 4) and anastomosis (4). B, migration hypha with eroded septa (1, 2) and lateral branching of dikaryotic cells bearing clamp-connections (3, 4).

clamp-connection anastomosed with the penultimate cell whereas the one initiated earlier became a pseudoclamp within which were entrapped two nuclei.

During the unstable dikaryotic phase, nuclear conditions characteristic of both dikaryon and monokaryon were observed within the same hypha during growth (Fig. 12). The patterns of nuclear divisions and subsequent relocations of mitotic progeny ultimately appeared to reduce the number of superfluous nuclei, stabilising the number of nuclei at the hyphal apex. Thus, through the reduction of excess nuclear content via at least these observed patterns did the hyphal apex change from the multikaryotic nature to eventually establish its dikaryotic identity. After two to three days, the entire peripheral mycelium originally derived from the monokaryon was dikaryotised and consisted completely of balanced dikaryotic hyphae.

Hyphal interaction: self parasitism? Frequently observed in di-mon matings involving dikaryons from the cross 699 × 845, 88 × 4–40, or monokaryons from strains 2145 and 2151 at the junction where interdigitating hyphae from both mycelia came in close contact were

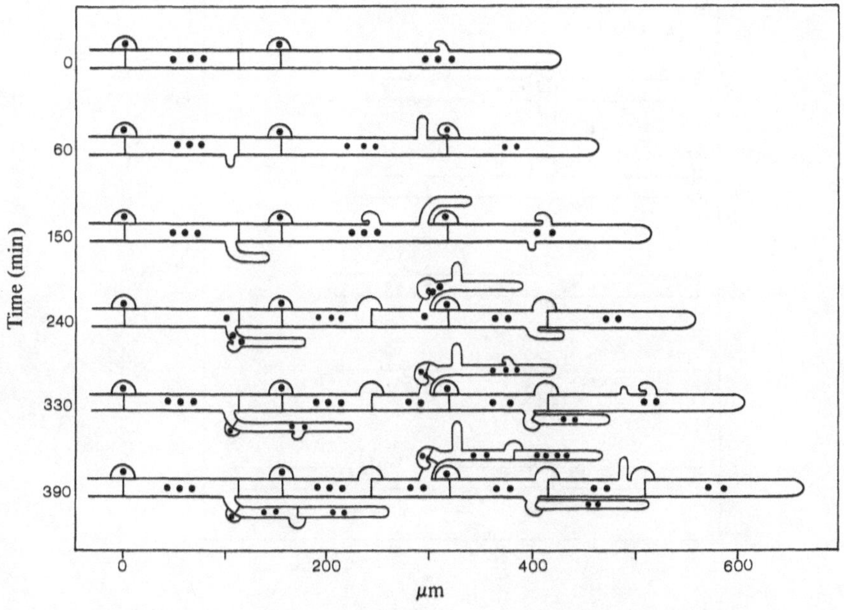

Fig. 10. Nuclear reduction in a multinucleated hyphal apex. Note nuclei entrapped inside clamp-connections which failed to fuse with penultimate cells.

conspicuous thigmotrophic hyphal interactions which bear a resemblance to the coiling reaction characteristic of some types of mycoparasitism. None to date have been observed to occur between hyphae of the same mycelial origin. Although mycoparasitic behaviour of *S. commune* towards a variety of fungi has been demonstrated by Tzean & Estey (1978), the evidence presented here shows that species-specific hyphal interactions mimicking parasitism also occurred within several strains of *S. commune* tested.

These self-parasitic interactions commenced somewhat early in the di-mon confrontation, prior to any indication of sexual mating events (18–24 h of incubation). One frequently observed interaction involved the production of profuse whorls of short branching hyphae which progressively entwined certain advancing hyphae from the opposing mycelium although other opposing hyphae within close proximity were not affected in this way (Fig. 13). The zone of entrapment may extend to at least two or three compartments along the encircled hyphae. In time,

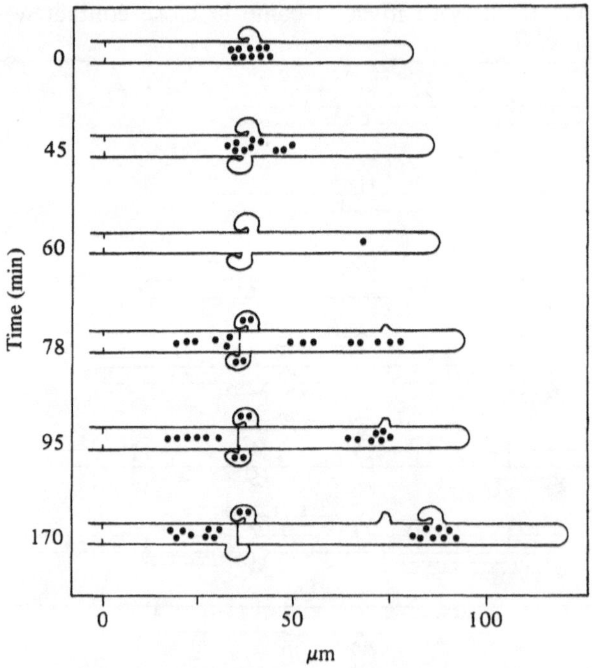

Fig. 11. Double clamp-connections in a multinucleated hyphal apex. Note the entrapment of two nuclear components in pseudoclamp-connection.

conspicuous thickenings of the hyphal wall were deposited along the partially disintegrated hyphae (Fig. 13c). The apical hyphal compartments, in many instances, escaped and continued to grow with normal branching, whereas most entwined vegetative hyphae gradually showed a senescent appearance in the complete absence of anastomoses between these interacting hyphae.

Other concurrent interactions resulted in hyphal invasion (Fig. 14). A wall-to-wall interface was established at the contact surface between the

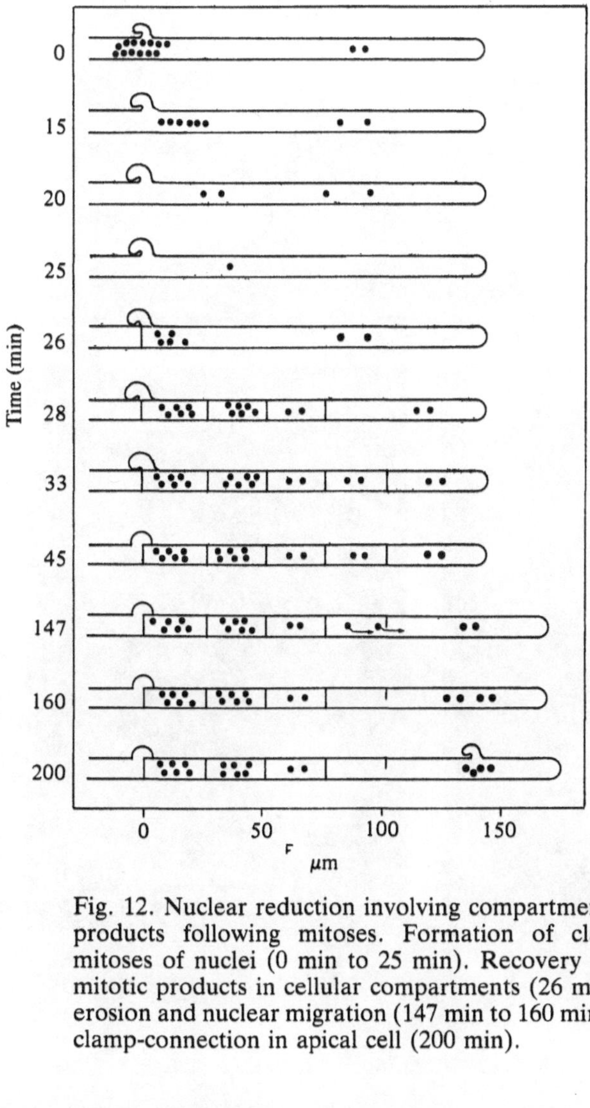

Fig. 12. Nuclear reduction involving compartmentalisation of nuclear products following mitoses. Formation of clamp-connection and mitoses of nuclei (0 min to 25 min). Recovery and redistribution of mitotic products in cellular compartments (26 min to 45 min). Septal erosion and nuclear migration (147 min to 160 min). Initiation of a new clamp-connection in apical cell (200 min).

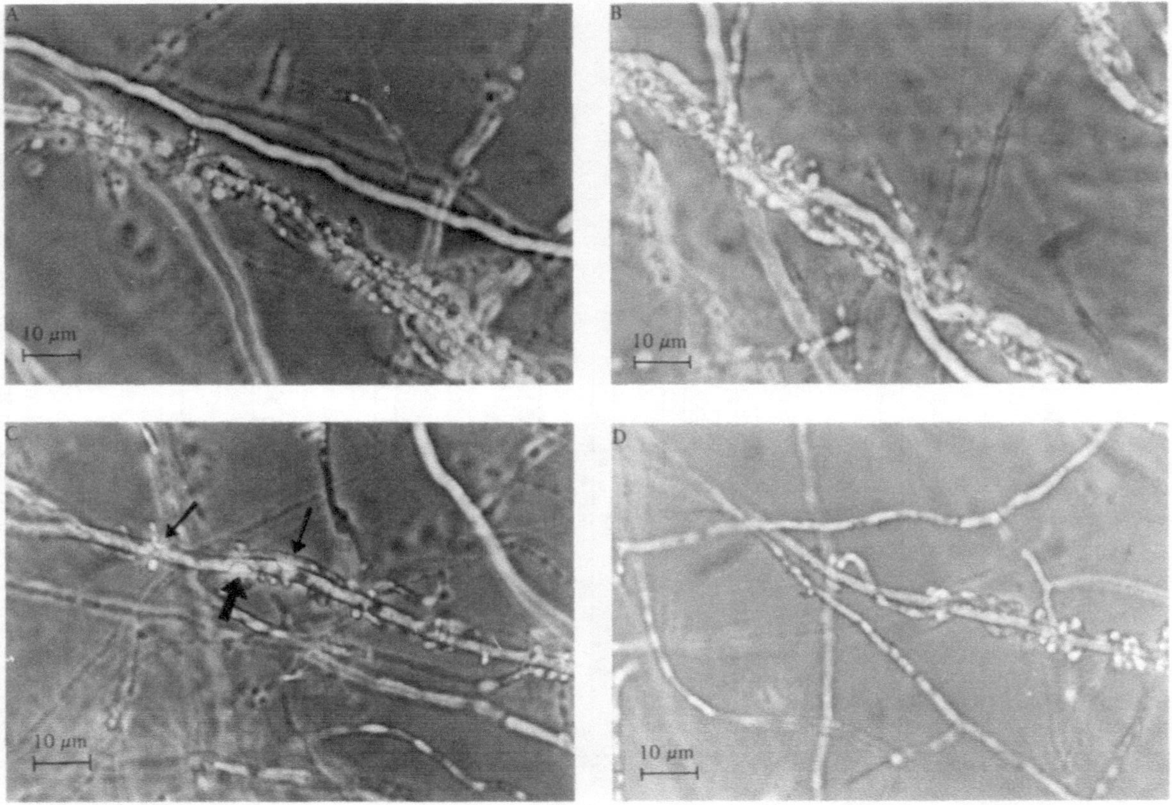

Fig. 13. Entwining interactions between hyphae from dikaryotic and monokaryotic mycelia. Note thickenings of the hyphal wall (arrows, in C).

invading hyphal apex and the vegetative cell. Subsequently, in lieu of fusion, the contact site invaginated due to the elongation of the penetrating hyphal tip. An apparent collar-like wall apposition formed around the apical extremity of the penetrating hypha and appeared to aid in anchoring it until the hyphal apex was encased. Multiple penetrations by invasive hyphal pegs tended to take place close to the adjoining areas between cells (Fig. 14c) but intrahyphal hyphae have not been observed to perforate septa. Once inside, the hyphal tip

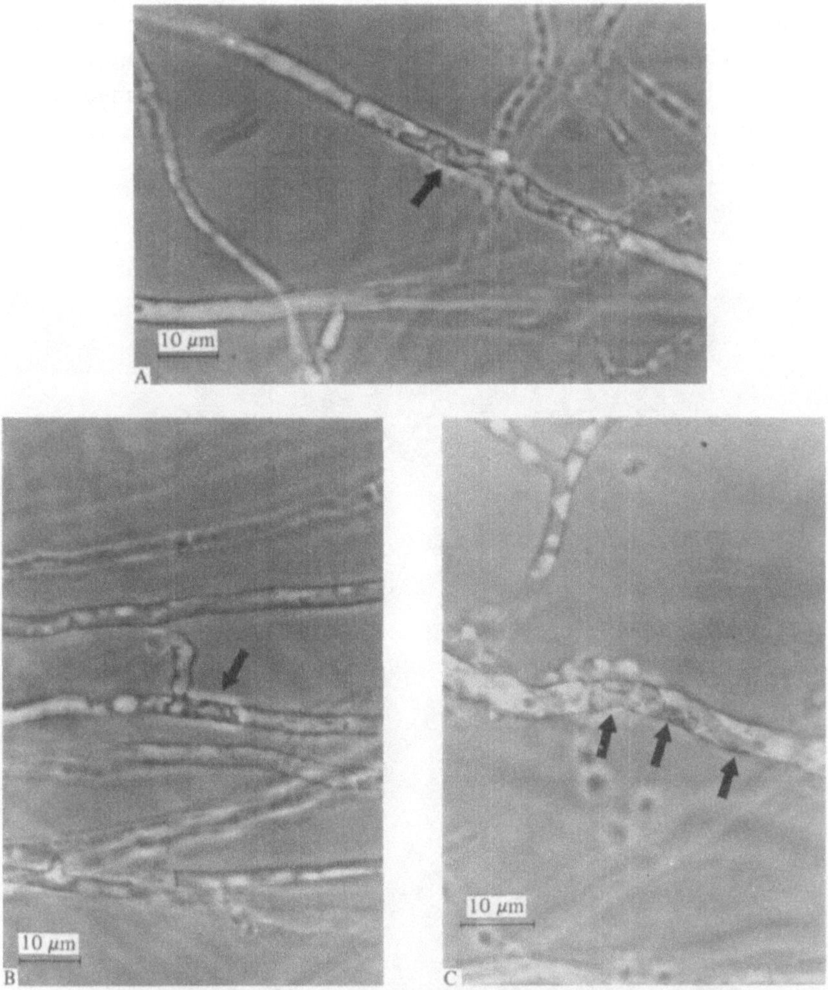

Fig. 14. Hyphal invasions and intrahyphal hyphae. (A, B) Note the characteristic convoluted growth of intrahyphal hypha; (C) multiple intrahyphal hyphae within a single vegetative hypha (arrows).

continued to elongate and grow in close confines within the invaded cell, and much of its surface was tightly appressed to the encasing walls (Fig. 14a,b). Frequently, the intrahyphal growth of invasive hyphae was characterised by a high degree of convolution. Such entrapment of intrahyphal growth was not necessarily permanent; in some instances, after transient intrahyphal development, some invasive hyphae exited apically or via lateral branching (Fig. 15a, schematic diagram in b).

Apparently, this mode of interaction did not always produce an obvious lethal effect on the invaded hyphal cells. Among those which survived the penetration, normal divisions often occurred and new septa arose which may serve to contain the invasion process. Cellular integrity was fully conserved and in no case did a nuclear exchange occur between interacting hyphae. While the significance of these intrahyphal

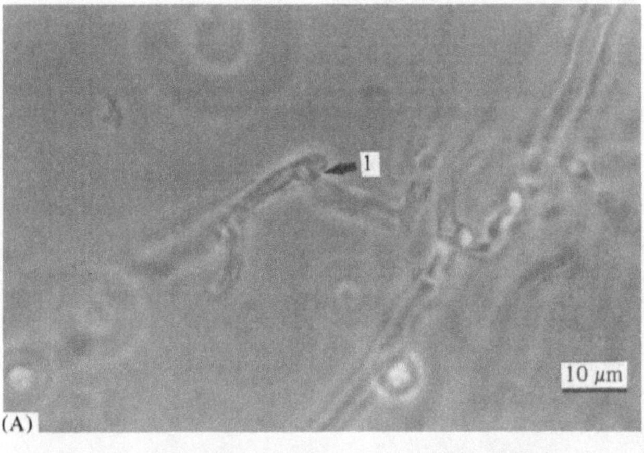

(A)

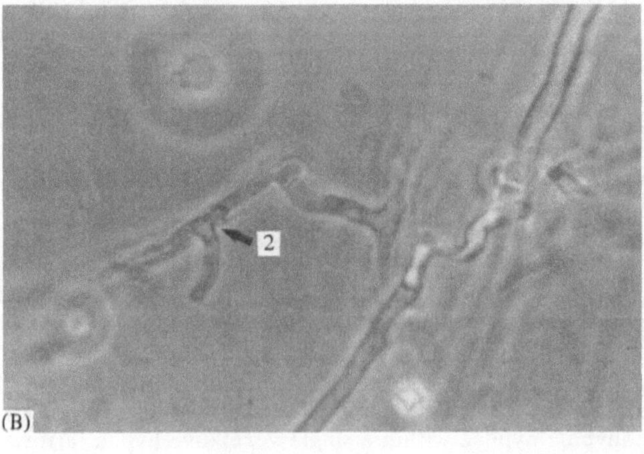

(B)

hyphae remains unclear in *S. commune*, Tewari (1983) noted the presence of intrahyphal hyphae during sexual morphogenesis in *Petromyces aliaceres*.

Vacuolar involvement in nuclear migration. An isolated example of a possible dynamic involvement of vacuoles in altering the structural integrity of a septum, through which nuclear migration subsequently occurred, is shown in Fig. 16. Prior to initiation of vacuolar movement in the hypha, an increase in localised organelle and nucleus motility in the cytoplasm of the cell on the left of the septum was observed. Within minutes, several successive penetrations of vacuoles occurred unidirectionally from left to right (corresponding to the forward direction toward the hyphal apex). Figure 16a–d illustrates the behaviour of the

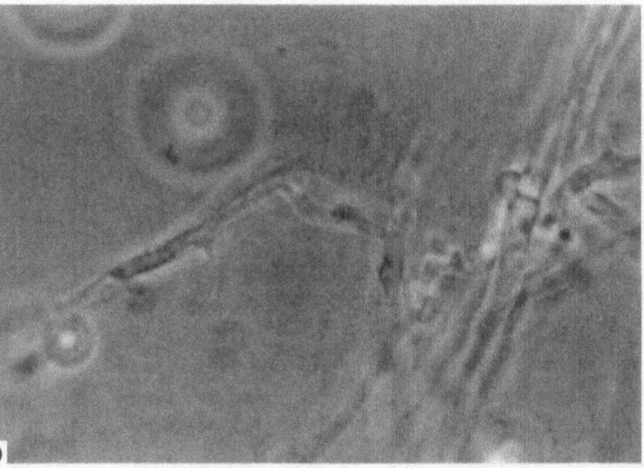

(C)

(D)

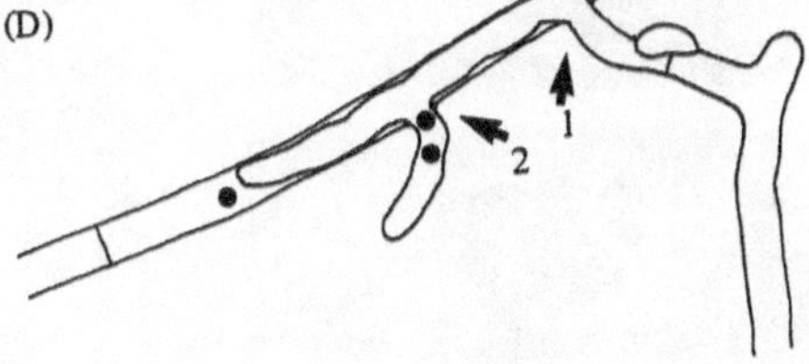

Fig. 15. A–C, hyphal invasion (1) and lateral exit (2) of intrahyphal hypha (arrows). D, schematic diagram.

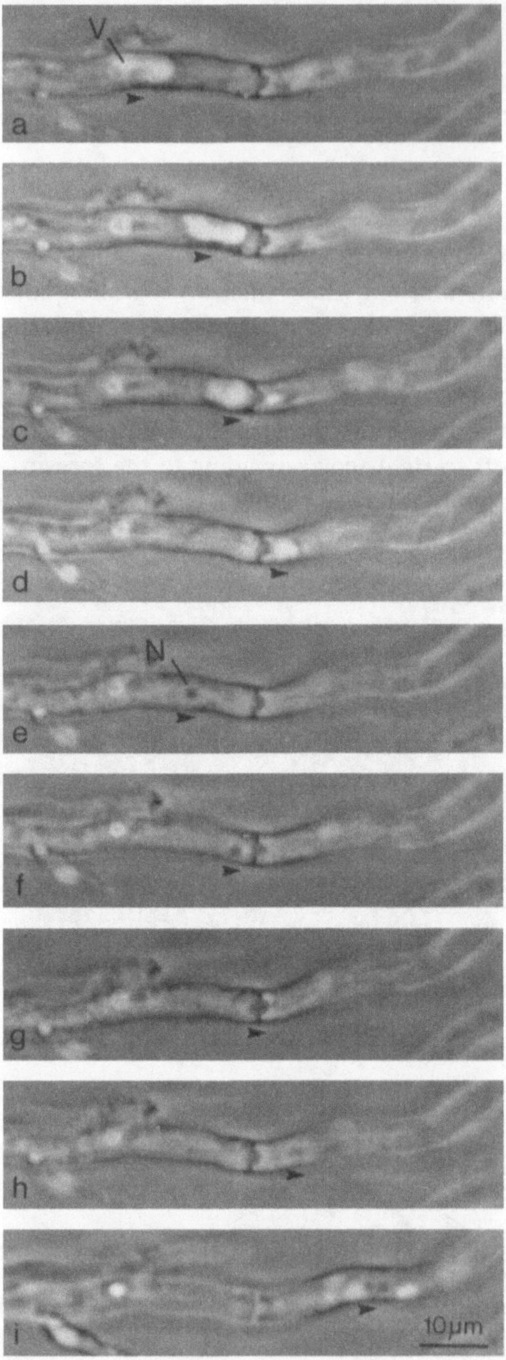

Fig. 16. Vacuole involvement in nuclear migration. Arrow indicates direction of vacuolar (a – d) and nuclear movements (e – i). N indicates nucleus and V, vacuole.

second vacuole to penetrate the septum. This resulted in a visibly-altered septal structure showing a distinctive convex projection on the right side. The ensuing migrant nucleus immediately traversed the orifice of the altered structure into the adjoining cell and came to a rest shortly afterward (Fig. 16e–g). Fig. 16g shows the entry of the migrating nucleus after passing through the altered septum which was then recovered in the cytoplasm of the invaded cell, juxtaposed to the resident nucleus (Fig. 16i). The convex septum gradually returned to an almost normal appearance with time until it could hardly be distinguished from any other septum.

Subsequent events are summarised diagramatically in Fig. 17. Mitosis of the migrant and the resident nuclei (Fig. 17D,E,F) resulted in establishment of an unbalanced distribution of nuclei in the daughter cells (Fig. 17G).

Except for its rarity, the occurrence of this phenomenon raises the possibility of a different mechanism for facilitating nuclear migration through septa. Although isolated fungal vacuoles have been shown to contain high levels of several lysosomal enzymes in *Neurospora* (Martinoia *et al.*, 1979; Matile, 1971) and in yeast cells (Matile &

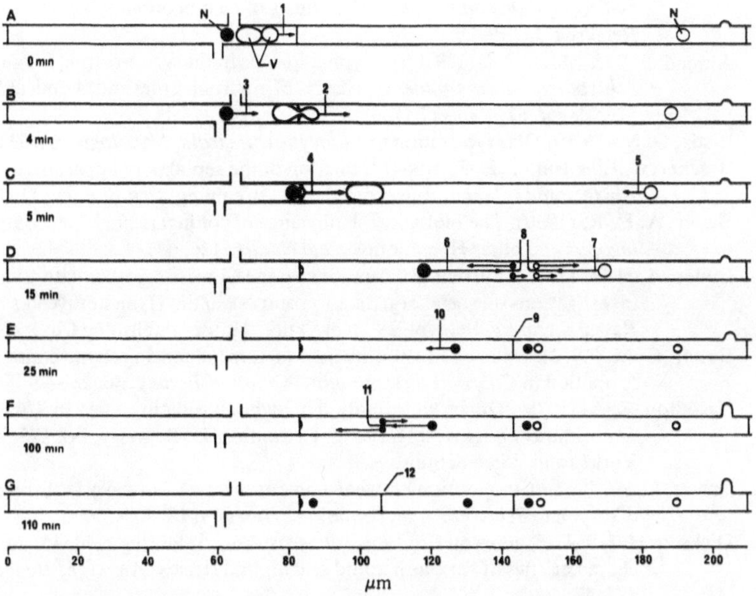

Fig. 17. Schematic diagram of events prior and subsequent to vacuole-facilitated nuclear migration.

Wiemken, 1967), cellular compartmentalisation for either cell wall-degrading enzymes or parenthesome alterations has not yet been demonstrated in vacuoles of *S. commune*. Vacuoles, however, have been observed to contribute a role in physically altering a septum to facilitate nuclear migration in the present study. We have, in other instances, observed an eroded septum immediately adjoining another septum with similar structural appearance to the convex septum shown above, in the same hypha in which nuclear migration had evidently taken place.

In the foregoing account of di-mon matings, we have attempted to trace the initial events in live hyphal interactions of *S. commune*, bearing in mind the following:

> It is the customary fate of new truths to begin
> as heresies and to end as superstitions.
>
> Thomas Henry Huxley

References

Ahmad, S. S. & Miles, P. G. (1970*a*). Hyphal fusions in the wood-rotting fungus *Schizophyllum commune*. I. The effects of incompatibility factors. *Genetical Research*, **15**, 19–28.

Ahmad, S. S. & Miles, P. G. (1970*b*). Hyphal fusions in the wood-rotting fungus *Schizophyllum commune*. II. Effects of environmental and chemical factors. *Mycologia*, **62**, 1008–17.

Bistis, G. N. (1970). Dikaryotization in *Clitocybe truncicola*. *Mycologia*, **62**, 911–24.

Bracker, C. E. & Butler, E. E. (1964). Function of the septal pore apparatus in *Rhizoctonia solani* during protoplasmic streaming. *Cell Biology*, **21**, 152–7.

Buller, A. H. R. (1930). The biological significance of conjugate nuclei in *Coprinus lagopus* and other Hymenomycetes. *Nature*, **126**, 686–9.

Buller, A. H. R. (1931). Further observations on the *Coprini* together with some investigations on social organization and sex in the Hymenomycetes. In *Researches on Fungi*, vol. IV. New York: Hafner Publishing Co.

Butler, G. M. (1972). Nuclear and non-nuclear factors influencing clamp connection formation in *Coprinus disseminatus*. *Annals of Botany*, **36**, 263–79.

Casselton, L. A. (1978). Dikaryon formation in higher basidiomycetes. In *The Filamentous Fungi*, vol. III, ed. J. E. Smith & D. R. Berry, pp. 275–97. New York: John Wiley & Sons.

Crowe, L. K. (1963). Competition between compatible nuclei in the establishment of a dikaryon in *Schizophyllum commune*. *Heredity*, **18**, 525–33.

Dickson, H. (1934). Studies on *Coprinus sphaerosporus*. I. Pairing behavior and the characteristics of various haploid and diploid strains. *Annals of Botany*, **48**, 527–47.

Ellingboe, A. H. (1964). Nuclear migration in dikaryotic-homokaryotic matings in *Schizophyllum commune*. *American Journal of Botany*, **51**, 133–9.

Fries, N. (1981). Recognition reactions between basidiospores and hyphae in *Leccinum*. *Transactions of the British Mycological Society*, **77**, 15–20.

Giesy, R. M. & Day, P. R. (1965). The septal pores of *Coprinus lagopus* in relation to nuclear migration. *American Journal of Botany*, **52**, 287–94.

Girbardt, M. (1958). Uber die Substruktur von *Polystictus versicolor* L. *Archiv für Mikrobiologie*, **28**, 255–69.

Girbardt, M. (1968). Ultrastructure and dynamics of the moving nucleus. In *XXIInd Symposium of the Society for Experimental Biology, Oxford*. Cambridge: Cambridge University Press, pp. 249–59.

Janszen, F. H. A. & Wessels, J. G. H. (1970). Enzymic dissolution of hyphal septa in a Basidiomycete. *Antonie Van Leeuwenhoek Journal of Microbiology and Serology*, **36**, 255–7.

Jersild, R., Mishkin, S. & Niederpruem, D. J. (1967). Origin and ultrastructure of complex septa in *Schizophyllum commune* development. *Archiv für Mikrobiologie*, **57**, 20–32.

Kemp, R. F. O. (1975). Breeding biology of *Coprinus* species in the section *lanatuli*. *Transactions of the British Mycological Society*, **65**, 375–88.

Kemp, R. F. O. (1980). Production of oidia by dikaryons of *Flammulina velutipes*. *Transactions of the British Mycological Society*, **74**, 557–60.

Kimura, K. (1958). Diploidization in the Hymenomycetes. II. Nuclear behavior in the Buller Phenomenon. *Biological Journal of Okayama University*, **4**, 1–59.

Koltin, Y. & Flexer, A. S. (1969). Alterations of nuclear migration in B-mutant strains of *Schizophyllum commune*. *Journal of Cell Science*, **4**, 739–49.

Lange, I. (1966). Das Bewegungsverhalten der Kerne in fusionierten Zellen von *Polystictus versicolor* (L.). *Flora Abteilung A Physiologie und Biochemie* (Jena), **156**, 487–97.

Lehfeldt, W. (1923). Uber die Enstehung des Paarkernmycels bei heterothallischen Basidiomyceten. *Hedwigia*, **64**, 30–51.

Martinoia, E., Heck, V., Boller, T. H., Wiemken, A. & Matile, P. H. (1979). Some properties of vacuoles isolated from *Neurospora crassa* slime variant. *Archives of Microbiology*, **120**, 31–4.

Matile, P. H., (1971). Vacuoles, lysosomes of *Neurospora*. *Cytobiologie*, **3**, 324–30.

Matile, P. H. & Wiemken, A., (1967). The vacuole as the lysosome of the yeast cell *Archiv für Mikrobiologie*, **56**, 148–55.

Mayfield, J. E. (1974). Septal involvement in nuclear migration in *Schizophyllum commune*. *Archives of Microbiology*, **95**, 115–24.

Moore, R. T. & McAlear, J. H. (1962). Fine structure of mycota. 7. Observations on septa of Ascomycetes and Basidiomycetes. *American Journal of Botany*, **49**, 86–94.

Niederpruem, D. J. (1971). Kinetic studies of septum synthesis, erosion and nuclear migration in a growing B-mutant of *Schizophyllum commune*. *Archiv für Mikrobiologie*, **75**, 189–96.

Niederpruem, D. J. (1978). Morphogenetic processes in *Schizophyllum* and *Coprinus*. In *Genetics and Morphogenesis in the Basidiomycetes*. New York: Academic Press.

Niederpruem, D. J. (1980*a*). Direct studies of dikaryotization in *Schizophyllum commune*. I. Live inter-cellular nuclear migration patterns. *Archives of Microbiology*, **128**, 162–71.

Niederpruem, D. J. (1980*b*). Direct studies of dikaryotization in *Schizophyllum commune*. II. Behavior and fate of multikaryotic hyphae. *Archives of Microbiology*, **128**, 172–8.

Papazian, H. P. (1950). Physiology of the incompatibility factors in *Schizophyllum commune*. *Botanical Gazette*, **112**, 143–63.

Parag, Y. (1970). Genetics of tetrapolar sexuality in higher fungi: the B-factor, common B heterokaryosis and parasexuality. *USDA Project No. FG-Is-228*. Israel: The Hebrew University of Jerusalem.

Prevost, G. (1962). *Etude génétique d'un Basidiomycète: Coprinus radiatus*. Thesis, Université de Paris, Paris.

Quintanilha, A. (1933). Le problème de la sexualité chez les champignons. *Boletim da Sociedade Broteriana*, **8,** 1–99.

Quintanilha, A. (1937). Contribution a l'étude génétique du phénomène de Buller. *Compte Rendu Hebdomadaire des Séances de l'Académie des Sciences*, **205,** 745–7.

Quintanilha, A. (1939). Etude génétique du phénomène de Buller. *Boletim da Sociedade Broteriana*, **13,** 425–86.

Raper, J. R. (1966). *Genetics of Sexuality in Higher Fungi*. New York: The Ronald Press Company.

Raper, J. R. (1970). The road to diploidy with emphasis on a detour. In *Prokaryotic and Eukaryotic Cells, Symposia of the Society for General Microbiology*, **20,** 401–32.

Raudaskoski, M. (1970). Occurrence of microtubules and microfilaments, and origin of septa in dikaryotic hyphae of *Schizophyllum commune. Protoplasma,* **79,** 415–22.

Raudaskoski, M. (1973). Light and electron microscope study of unilateral mating between a secondary mutant and a wild-type strain of *Schizophyllum commune*. *Protoplasma*, **76,** 35–48.

Sicari, L. M. & Ellingboe, A. H. (1967). Microscopic observations of initial interactions in various matings of *Schizophyllum commune* and of *Coprinus lagopus*. *American Journal of Botany*, **54,** 437–9.

Smythe, R. (1973). Hyphal fusions in the Basidiomycete *Coprinus lagopus sensu* Buller. I. Some effects of incompatibility factors. *Heredity*, **31,** 108–11.

Snider, P. J. & Raper, J. R. (1958). Nuclear migration in the Basidiomycete *Schizophyllum commune*. *American Journal of Botany*, **45,** 538–46.

Terekawa, H. (1957). The nuclear behavior and the morphogenesis in *Pleurotus ostreatus*. *Scientific Papers of the College of General Education of the University of Tokyo*, **7,** 61–88.

Tewari, J. P. (1983). Stromatic cell autolysis in *Petromyces alliaceus* during ascocarp formation. *Transactions of the British Mycological Society*, **80,** 127–30.

Tzean, S. S. & Estey, R. H. (1978). *Schizophyllum commune* Fr. as a destructive mycoparasite. *Canadian Journal of Microbiology*, **24,** 780–4.

Wessels, J. G. H. (1969). A β-1,6-glucan glucanohydrolase involved in hydrolysis of cell-wall glucan in *Schizophyllum commune, Biochimica et Biophysica Acta*, **178,** 191–3.

Wessels, J. G. H. & Niederpruem, D. J. (1967). Role of a cell wall glucan-degrading enzyme in mating *Schizophyllum commune*. *Journal of Bacteriology*, **94,** 1594–1602.

Wiemken, A., Matile, P. & Moor, H., (1970). Vacuolar dynamics in synchronously budding yeast. *Archiv für Mikrobiologie*, **70,** 89–103.

5

Hyphal fusion in *Coriolus versicolor*

R. C. AYLMORE and N. K. TODD

Department of Biological Sciences, Washington Singer Laboratories, University of Exeter, Exeter, EX4 4PS, UK

Background

The fact that somatic hyphae of higher fungi can fuse together readily, creates, as has already been stressed by Gregory (Chapter 1), interesting possibilities in terms of the interactions that can occur both within and between mycelia. Further, as indicated by Nguyen & Niederpruem (Chapter 4), who also review much of the literature presently available concerning the phenomenon, in considering the consequences of hyphal fusion, a distinction must be made between those types which ultimately result in sexual reproduction, as between compatible homokaryons of heterothallic Basidiomycotina, and those which result purely in vegetative continuity between or within individuals. To this end, the outcome of fusions and hence their biological consequences, is crucially mediated via the sexual and somatic incompatibility systems acting either independently, or, as is discussed in detail by Rayner *et al.* (Chapter 23), in combination.

Whilst much progress has been made in understanding the genetic basis of sexual and, to a lesser extent, somatic incompatibility systems, their mechanisms, which appear to involve a post-fusion cellular recognition event at or near the site of interaction, remain obscure. This must be because, as so often happens in biology, details of the fundamental process, in this case hyphal fusion, have been overlooked in favour of its more interesting consequences. Our aim therefore is to redress the balance by describing some of our own direct observations on the basic cytology of hyphal fusion and immediate post-fusion events. Unlike Nguyen and Niederpruem, who stress those fusions which are sexual in consequence, and which must occur only rarely during the life of a mycelium, our emphasis will be on the everyday vegetative fusions which govern its internal affairs and relationships with its neighbours.

Even though hyphal fusion was reported over a century ago by the brothers Tulasne (Tulasne & Tulasne, 1863, cited in Buller, 1931) and a major contribution made by Buller (1931, 1933) in the thirties, what work does exist is mostly associated with taxonomic studies (Robak, 1942; Cabral, 1951; Brodie, 1955) or is on various factors affecting the process (Bourchier, 1957; Ahmad & Miles, 1970*a,b*; Smythe, 1973). There are, however, several notable exceptions including the work on hyphal fusion in relation to sexual (Lehfeldt, 1923; Noble, 1937; Lange, 1966; Bistis, 1970; Watrud & Ellingboe, 1973) and vegetative (Garnjobst & Wilson, 1956; Flentje & Stretton, 1964; Parmeter, Sherwood & Platt, 1969) incompatibility mechanisms. Despite these, it is still fair to say that the cytological details of the process are despairingly few and, as Burnett (1976) states, virtually nothing is known at the ultrastructural level. Our interest arose from work on intraspecific antagonism (a post-fusion somatic incompatibility mechanism) in the fungus *Coriolus versicolor.* At the macroscopic level, the antagonistic reaction between isolates paired in culture typically takes the form of a narrow pigmented zone of relatively sparse mycelium (Fig. 4a) and a microscopic examination of the zone reveals what have been called spindle cells (Rayner & Todd, 1979; Todd & Rayner, 1980; Fig. 4b). For the reasons outlined above, it was felt that the cytology of pre- and post-fusion events occurring in this fungus would be particularly informative in trying to understand the mechanism of this interaction.

To achieve this, a technique was developed which combined both phase and electron microscopy. Mycelia were grown on membranes under low nutrient conditions in a vented microculture chamber that allowed continuous observation of living hyphae over a period of several days. In order to examine hyphal fusion at the ultrastructural level, procedures were devised that allow processing and sectioning for electron microscopy of specific hyphae examined in their living state moments before fixation. Use of this approach made it possible to undertake time-course studies of fusing hyphae as well as providing for the often-useful comparison between the living and fixed states (for example, see Figs. 2, 3). The work has yielded some surprising observations, particularly concerning the behaviour of nuclei which show novel patterns of intracellular migration, disintegration and division.

The fusion process

Tip-to-side fusion

From work with a number of species, Buller (1933) generalised that all hyphal fusions are of the tip-to-tip type, occurring exclusively between the apical regions of hyphae and invariably involving pre-contact stimuli which bring extending apices together. Such patterns of fusion and the accompanying 'teleomorphosis' (the situation in which one hypha, acting at a distance, stimulates another to produce an opposing peg), and 'zygotropism' (the subsequent growth of the two apices towards each other) have become widely accepted, although little is known of the mechanism responsible for such behaviour (for reviews and discussion see Gooday (1975) and Burnett (1976)). There have, however, been several reports that hyphal fusions may be of the tip-to-side type, involving a single growing apex and a morphologically normal lateral wall. Indeed Buller himself (Buller, 1933) revoked his earlier claims of tip-to-side fusions (Buller, 1931) and disputed similar observations by Bensuade (1918) and others. More recent reports of tip-to-side fusions have been made for *Schizophyllum commune* (Ahmad & Miles, 1970*a,b*; Raudaskoski, 1973) and *Coprinus cinereus* (Smythe, 1973). Watkinson (1978) specifically demonstrated tip-to-side fusion in young mycelia of *Penicillium claviforme*. In *C. versicolor*, our observations confirm that under the experimental conditions used, as many as 80% of fusions may be of the tip-to-side type (Fig. 1). Many appeared to result from random apex–lateral wall contact with neither of the participating hyphae showing overt signs of any pre-contact response. Occasionally, fusion was preceded by an unmistakable homing of an extending apex towards a lateral wall but in such instances telemorphotic peg induction failed and the fusion remained tip-to-side in nature (Fig. 1c,d).

Only one in several hundred tip-to-side encounters actually resulted in successful fusion. In most instances contact resulted in a brief period in which growth ceased and the apex flattened against the side wall, followed by resumption of growth and deflection of the apex either over or along the obstructing hypha. Bifurcation of the apex was regularly observed (Fig. 1a) though often only one of the resulting growing points would maintain extension, the other ceasing growth, sometimes becoming totally vacuolated. Separate encounters involving the same hyphal compartment often behaved differently (see Fig. 1a,b).

In encounters that eventually fused, the initial flattening of the apex was followed by tight appression of the walls rather than by resumed growth. Within several minutes a slight swelling often formed in the lateral wall at the point of appression, probably as a result of incipient wall softening. Under phase optics, fusion appeared to involve formation of a single opening which expanded until all remnants of the cross-wall disappeared. The whole process was usually complete within 25–35 min, the initial opening forming after 15–20 min. Mitochondria traversed the fusion soon after cytoplasmic continuity had been established and bidirectional movement of granules between the fused cells quickly occurred. Bulk flow of cytoplasm between cells was seldom observed.

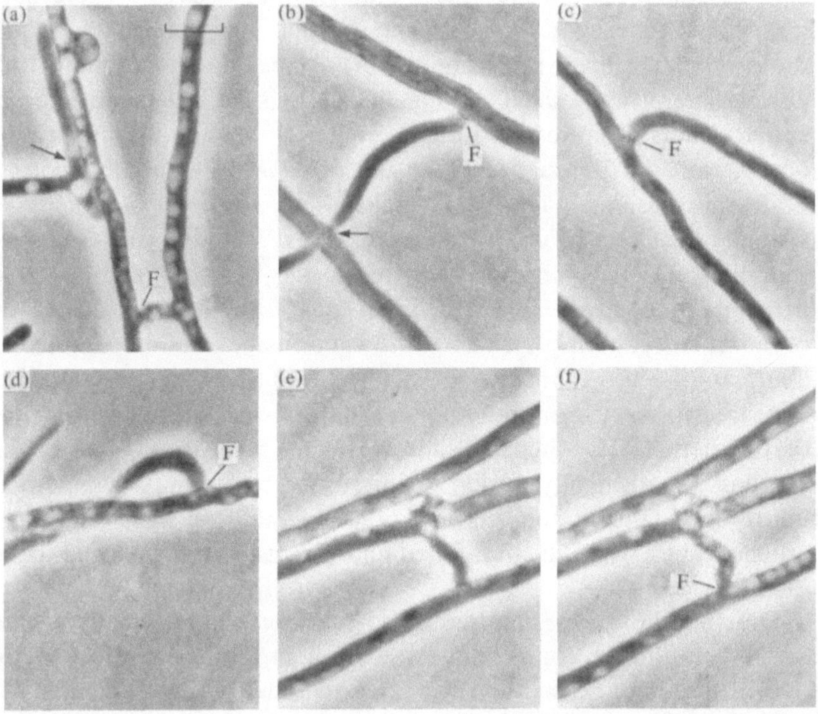

Fig. 1. Micrographs showing tip-to-side fusions between dikaryotic hyphae. (a) Of two apex-side wall contacts forming in the same compartment one has successfully fused (F) while the other (arrowed) resulted in bifurcation and continued growth. Scale bar, 10 μm. (b) Fusion forming at F; note that the previous contact (arrowed) failed to fuse. (c), (d) Apices showing attracted growth before contact and eventual fusion (F). (e), (f) Appression of a growing apex against a morphologically normal lateral wall results in fusion (F). Magnification in (b)–(f) as (a).

Ultrastructure of fusion

From the electron micrographs of hyphae in various stages of fusion (Figs. 2, 3) three basic features were evident. First, the tip-to-side nature of fusion was confirmed, and it was clear that the cytoplasm underlying the lateral wall does not form a nascent growing point (Fig. 3a). Rather than irregular breakdown, perforation and general lysis of wall material, fusion involves the formation of a single opening (Fig. 2f) which undergoes regular and symmetrical enlargement to the full hyphal diameter (Fig. 3a,e). This is accomplished by localised lysis of wall material comprising the inner rim of the remaining cross-wall (Fig. 2a). It is worth noting that fusion of Zygomycete gametangia involves the development at the centre of the fusion wall of openings which enlarge until the whole area is dissolved (Hawker & Gooday, 1969; Hawker & Beckett, 1971).

With the wall-lytic capacity presumed for some of them (Bartnicki-Garcia, 1973), the fate of the apical vesicles and their possible involvement in fusion provides the second point of interest. Figures 2b,c show a potentially fusing contact fixed early in the process in which vesicles are closely associated with the site of wall dissolution. This pattern, however, was not always evident. Whilst single vesicles and multivesicular bodies were usually present in the cytoplasm, there was generally little evidence to suggest that they accumulated preferentially at the site of wall lysis or that they coalesced with the plasmalemma at that point. In the fusion shown in Figs. 2e,f, vesicles, probably derived from the apical apparatus, appear to have become disaggregated soon after contact. Although much biochemical and ultrastructural evidence supports the concept of vesicle-mediated wall lysis during morphogenesis (see Burnett & Trinci, 1979), we feel as yet this study failed to confirm convincingly a role for vesicles in hyphal fusion. Since other studies on various fusion events in fungi suggest that they are involved, the situation remains to be clarified (Hawker & Gooday, 1969; Hawker & Beckett, 1971; Harvey, 1975; Rijkenberg & Truter, 1975; van der Valk & Marchant, 1978).

The third feature was the presence of distinctive sac-shaped organelles regularly and closely associated with sites of wall dissolution. These were present before the opening of the fusion pore, apparently arising from a fold of double membrane forming from invagination of the plasmalemma (Fig. 2b). Several could be present, sometimes forming on both sides of the appressed walls (Fig. 2e). The infolded regularly-spaced double membrane was invariably associated with

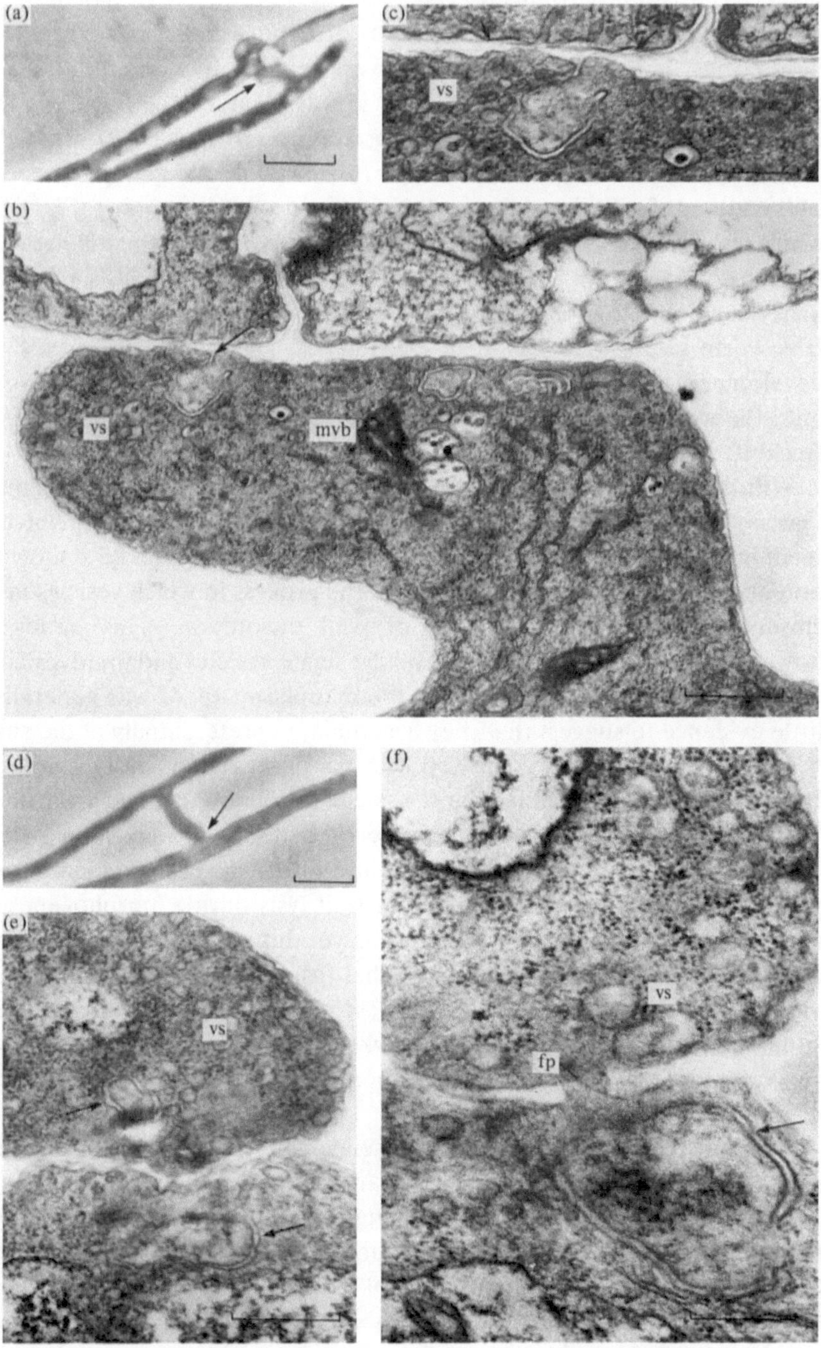
(a)
(c)
vs
(b)
vs
mvb
(d)
(f)
vs
fp
(e)
vs

regions of amorphous electron-translucent material devoid of ribosomes and distinct from surrounding cytoplasm (Fig. 2c). The significance of these 'fusion-associated organelles' is not clear, but they may represent sites at which partially solubilised wall material is taken into cells during the fusion process.

Post-fusion events

Intraspecific antagonism: fusion of dikaryotic hyphae

As mentioned earlier, our interest in hyphal fusion arose from work on intraspecific antagonism in *C. versicolor.* At the microscopic level, we noted the presence of spindle-shaped cells in the interaction region about 48 h after merging of mycelia and these most probably originated from points of fusion between genetically different hyphae. Electron microscopy of such cells (Fig. 4c) revealed that they contain densely packed non-vacuolated cytoplasm rather than the vacuolated diffuse contents present in the surrounding hyphae, including those of adjacent compartments. Spindle-cell cytoplasm characteristically contains numerous small vesicular structures, having electron-opaque contents, the function of which is unknown. There was some evidence that septa delimiting these cells were plugged. Whilst the significance of these swollen cells remains unclear, it is certain that they are found only in the interaction zone between antagonistic strains. Although they probably do reflect the action of somatic incompatibility they do not appear to be the centres of lysis as originally suspected (Rayner & Todd, 1979).

In the hope of discovering more about somatic incompatibility in *C. versicolor*, single fusions between genetically different ('non-self')

Fig. 2. The ultrastructural features of fusion between dikaryotic hyphae. (a) Photomicrograph showing hyphae in the early stages of fusion after close appression of the cell walls (arrowed). Scale bar, 10 μm. (b) A near-median section through the point of contact shown in (a). Vesicles (vs) and multivesicular bodies (mvb) are evident. The plasmalemma is invaginated at the site of wall dissolution (arrowed). Scale bar, 0.5 μm. (c) An enlargement from (b). Note the invaginated fold of membrane encircling amorphous electron-translucent material. The cell walls are beginning to dissolve (arrowed). Several vesicles (vs) are present in this region. Scale bar, 0.25 μm. (d) Photomicrograph showing a tip-to-side contact (arrowed). Scale bar, 10 μm. (e), (f) Near-median sections through the point of contact shown in (d). Note the vesicles (vs) and the large double-membrane-bound organelles (arrowed). A single fusion pore (fp) has formed in the region of wall appression. Scale bars, 0.5 μm.

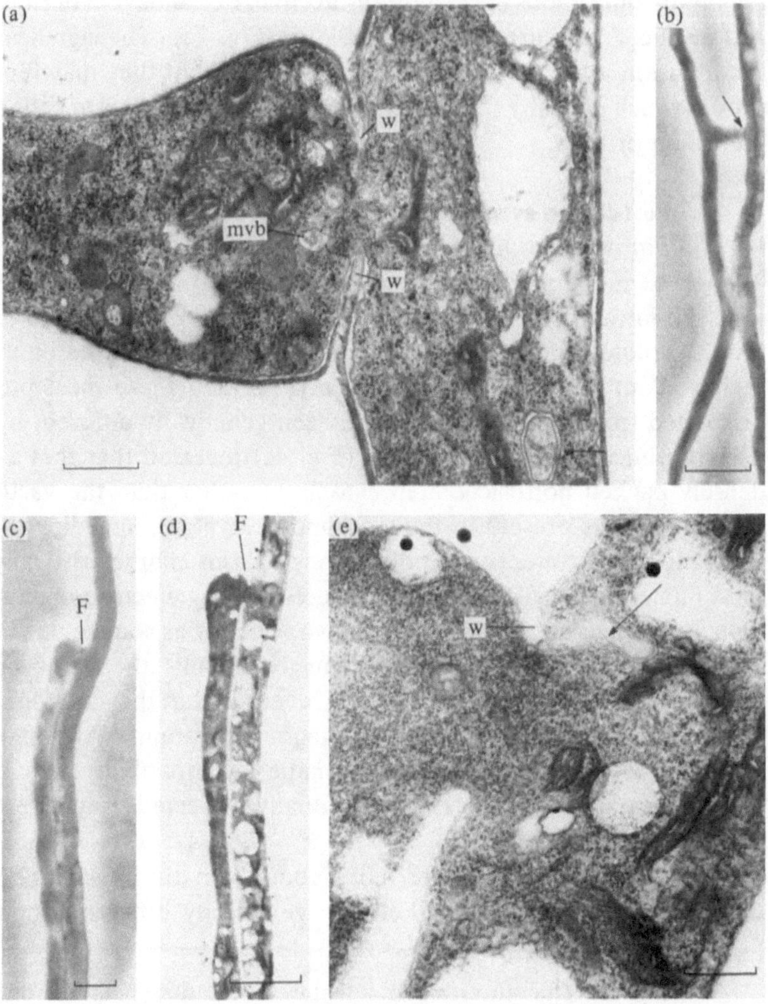

Fig. 3. The ultrastructural features of fusion between dikaryotic hyphae. (a), (b) a tip-to-side contact (arrowed in photomicrograph b) fixed immediately after visible formation of the fusion pore. Note the single opening and moulding of the lateral wall to the flattened apex. The cell walls merge (w) at the region of pore expansion. A double-membrane-bound structure (arrowed) is present in the cytoplasm and multivesicular bodies (mvb) occur close to the site of wall dissolution. Scale bars: (a), 0.5 μm; (b), 10 μm. (c)–(e) Corresponding light (c) and electron micrographs (d), (e) showing the later stages of fusion (F). The fusion pore was enlarged with little of the cross-wall (w) remaining. Amorphous material surrounded by a double membrane (arrowed) occurs in this region. Scale bars: (c), (d), 5 μm; (e), 0.5 μm.

dikaryotic hyphae were examined over periods up to several days. As controls, 'self' fusions between genetically identical hyphae were studied similarly. The results of this work were surprising, not least in the novel behaviour exhibited by nuclei following fusion, but further in

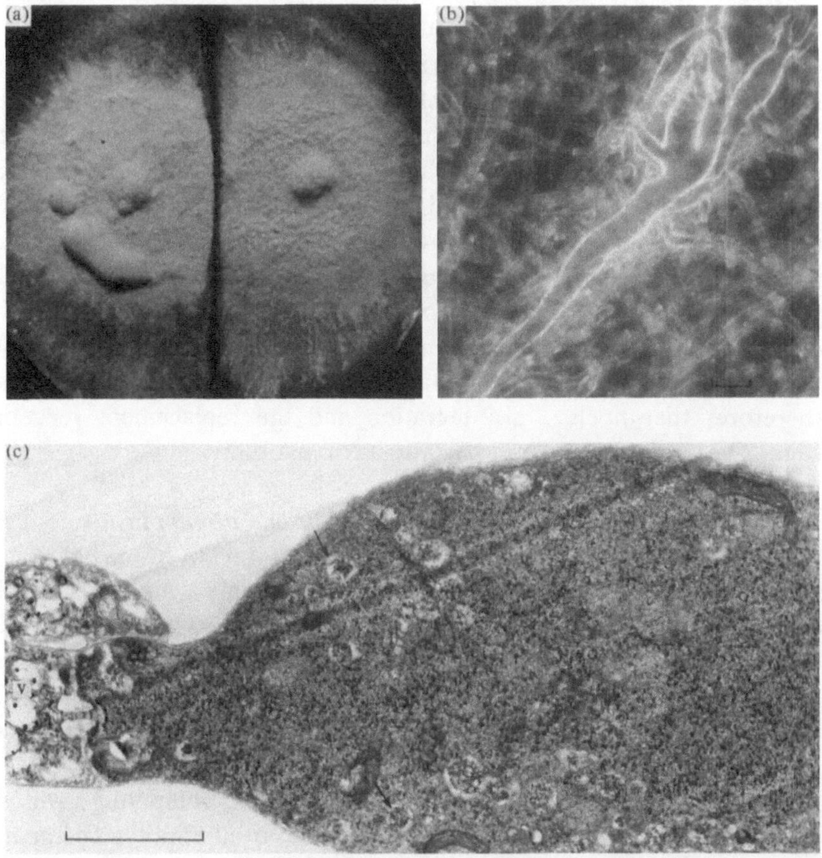

Fig. 4. Intraspecific antagonism between dikaryotic mycelia. (a) Photograph of the interaction in plate culture. (b) Photomicrograph of a spindle cell present in the interaction zone, photographed under dark-field optics. Note the enlargement of the hyphae and highly refractile walls. Scale bar, 10 μm. (c) Electron micrograph showing a near-median section through the swollen portion of a spindle cell. Note the densely packed relatively non-vacuolated cytoplasm containing small vesicles full of electron-opaque material (arrowed). By comparison the cytoplasm of the adjacent compartment is highly vacuolated (v). The dolipore septum delimiting the spindle cell appears plugged with densely staining material. Scale bar, 2 μm.

that a consistent pattern of behaviour occurred, which was identical in both situations (Fig. 5). The participating cells displayed a donor–recipient relationship in which both nuclei of the recipient cell degenerated and were replaced by a normal conjugate division of the donor pair, the original tetranucleate fusion cell being converted into two binucleate compartments which remained separated by a normal clamp-connection that could grow out in either direction (see Fig. 8a–e).

Obviously, in 'self' fusions each set of nuclei is replaced by an identical one and only in 'non-self' situations are the nuclei different. After completion of the 'nuclear-replacement reaction' either compartment could initiate lateral branches and undergo further mitotic division, with each appearing to retain normal viability. Hyphae of a given mycelium could behave either as donors or as recipients on encountering other dikaryons, with the outcome depending solely on which provided the apex in the tip-to-side fusion. There was never any evidence that fusion resulted in karyogamy or disturbance of the dikaryotic state and septal dissolution was never triggered. It seems, therefore, that nuclear disintegration and the replacement reaction cannot be ascribed a role in somatic incompatibility.

Sexual incompatibility: monokaryotic and di-mon fusion

The behaviour of nuclei was also examined in fusions between monokaryons and in di-mon interactions. Our primary aim was to look at the influence of sexual incompatibility on these events. However, it should be mentioned that since somatic incompatibility also seems to operate at the monokaryon level, in somatogamous species such as *C. versicolor* this may have a bearing on mating competence (see Rayner *et al.*, Chapter 23).

In studies of monokaryons, 'self' fusions were compared to those forming between genetically different hyphae, of mating-type compatible isolates. In 'self' fusions, the nuclei were found to undergo a replacement reaction exactly comparable to that described for dikaryons (see Fig. 6). The single nucleus present in the recipient compartment degenerates and is replaced by a normal septum-forming division of the donor nucleus, usually at a point close to the site of fusion (see Fig. 8f). Thus two uninucleate compartments separated by a simple septum are formed from the original binucleate fusion segment, both resulting compartments showing further growth and mitotic division.

In comparison with self fusions, those between sexually compatible monokaryons were rare and only four were recorded in any detail. Two

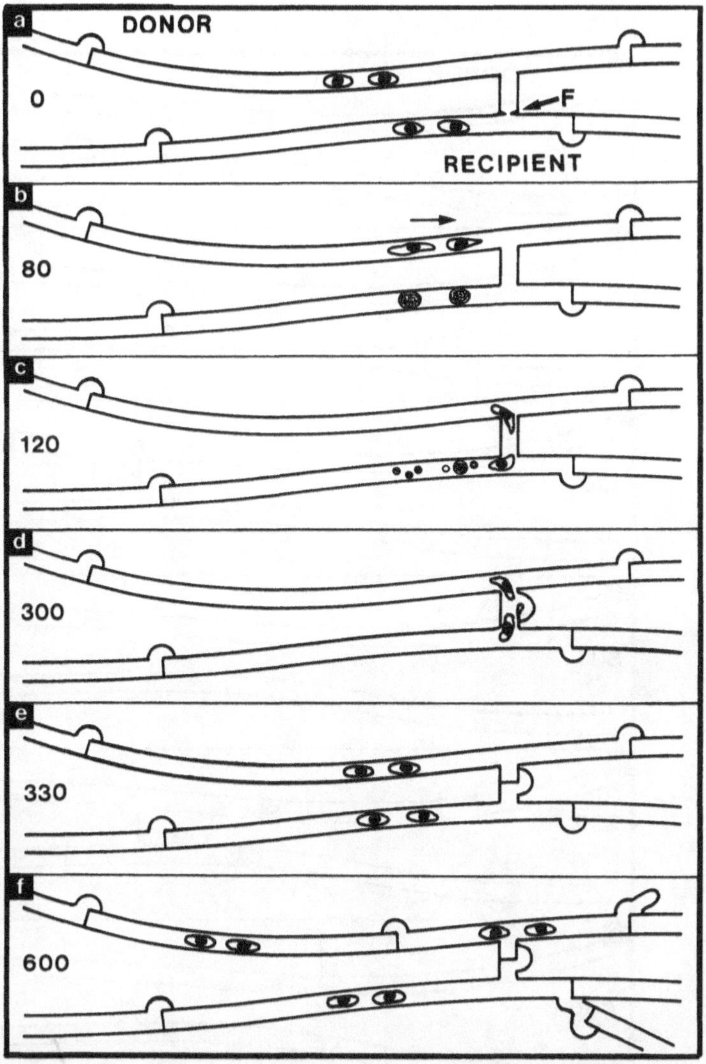

Fig. 5. The nuclear replacement reaction following fusion of dikaryotic hyphae. The numbers indicate minutes after fusion. (a) Tip-to-side fusion (F) between donor and recipient leads to a transient tetranucleate compartment. (b) Both donor cell nuclei migrate towards the fusion site while the stationary recipient pair round up and begin to degenerate. (c) The donor cell nuclei approach the fusion and often enter the recipient compartment. The degenerating recipient nuclei fragment and eventually fade from view. (d) The donor nuclei enter a normal conjugate division with the resulting intercalary clamp-connection growing out in either orientation. (e) The fused cells are converted into two binucleate compartments each possessing nuclei derived from the donor. (f) Both compartments show further growth and mitotic division.

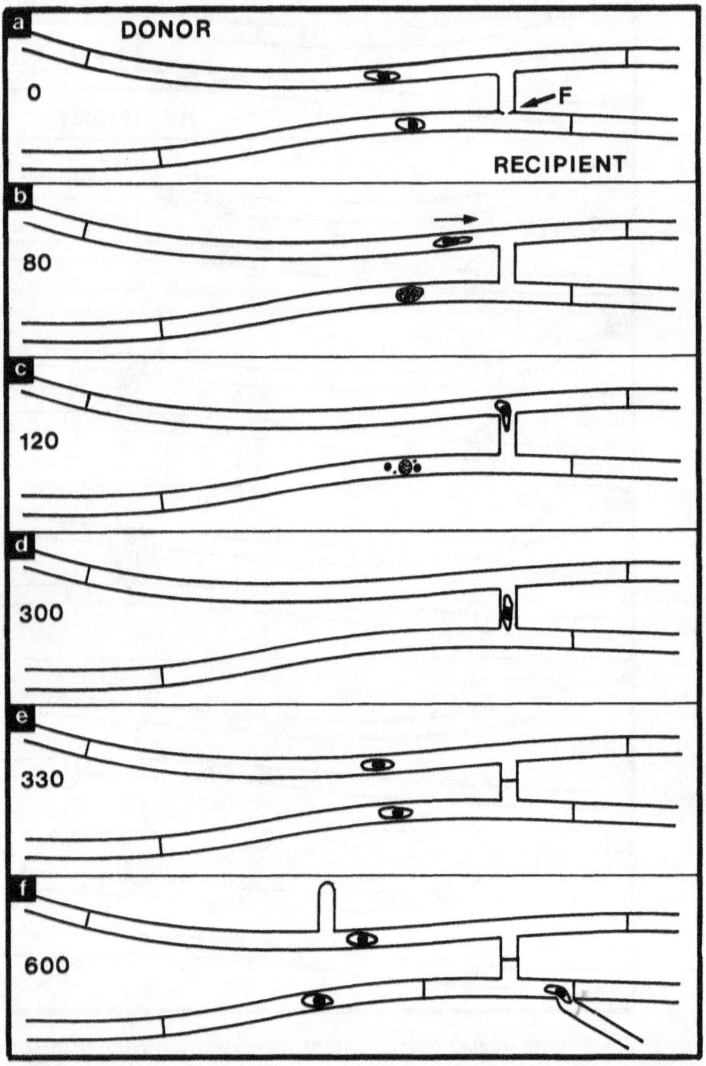

Fig. 6. The nuclear replacement reaction following fusion of monokaryotic hyphae. The numbers indicate minutes after fusion. (a) Tip-to-side fusion (F) between donor and recipient leads to a transient binucleate compartment. (b) While the donor cell nucleus migrates towards the fusion, that of the recipient remains stationary, rounds up and begins to degenerate. (c) The recipient nucleus fragments and fades from view. (d) The donor nucleus undergoes a normal septum-forming mitotic division. (e) The fused cells are converted into two uninucleate compartments each possessing nuclei derived from the donor. (f) Both compartments show further growth and mitotic division.

of these behaved as described above (as Fig. 6), although there were signs of unusual persistence of the recipient compartment nucleus and delay (indeed in one case, failure) of the first mitotic division of the donor nucleus. Further, cellular compartmentalisation was strictly maintained with no evidence, for up to 6 h after fusion, that septal dissolution and accompanying nuclear migration were initiated. The other two fusions behaved totally differently. Rather than degenerating, the nucleus present in the 'recipient' compartment remained and, after limited intracellular movement, became associated with the nucleus from the donor, at a point close to the anastomosis site. Two hours after fusion both nuclei became attenuated and appeared to divide without the formation of septa. The cytological events then became complicated and difficult to follow. However, it appeared that the apical-end septa in both cells then dissolved and the nuclei migrated rapidly into adjacent compartments. Within 50 min the septa of adjacent compartments had also become eroded and migration through the hyphae continued. Incomplete septa were then produced in these cells as the reaction spread along the hyphae. During migration, nuclei moved with considerable speed and attained rates of 25–35 μm min^{-1}. This contrasts with the slower migration (1–2.5 μm min^{-1}) observed in the nuclear movements during replacement reactions. Although observations were made for up to 30 h, no clamp or pseudoclamp structures were found anywhere on the fused hyphae. These events are represented in Fig. 7.

Working on the initial events of mating in *C. versicolor*, Lange (1966) stressed the complexity of the situation, and described the division and migration through dissolved septa of nuclei in the fusion compartment in patterns vaguely similar to those found here. Our observations stop at this point but they seem to dovetail well with the elegant cytology of Niederpruem (1980*a, b*) and Nguyen & Niederpruem (Chapter 4) which describes the dikaryotisation process in *S. commune* after the initial fusion events.

Finally, turning to di-mon fusions, combinations of mycelia were selected in which all nuclei were sexually compatible and dikaryotised readily. Virtually all the fusions studied involved the apex of a dikaryon fusing into the lateral wall of a monokaryon. A standard replacement reaction occurred in all cases with the recipient (monokaryotic) cell nucleus degenerating followed by the division of the conjugate pair present in the donor dikaryon. This resulted, in essence, in the direct dikaryotisation of the single monokaryotic compartment. Any further growth was always dikaryotic, involving perfectly normal conjugate

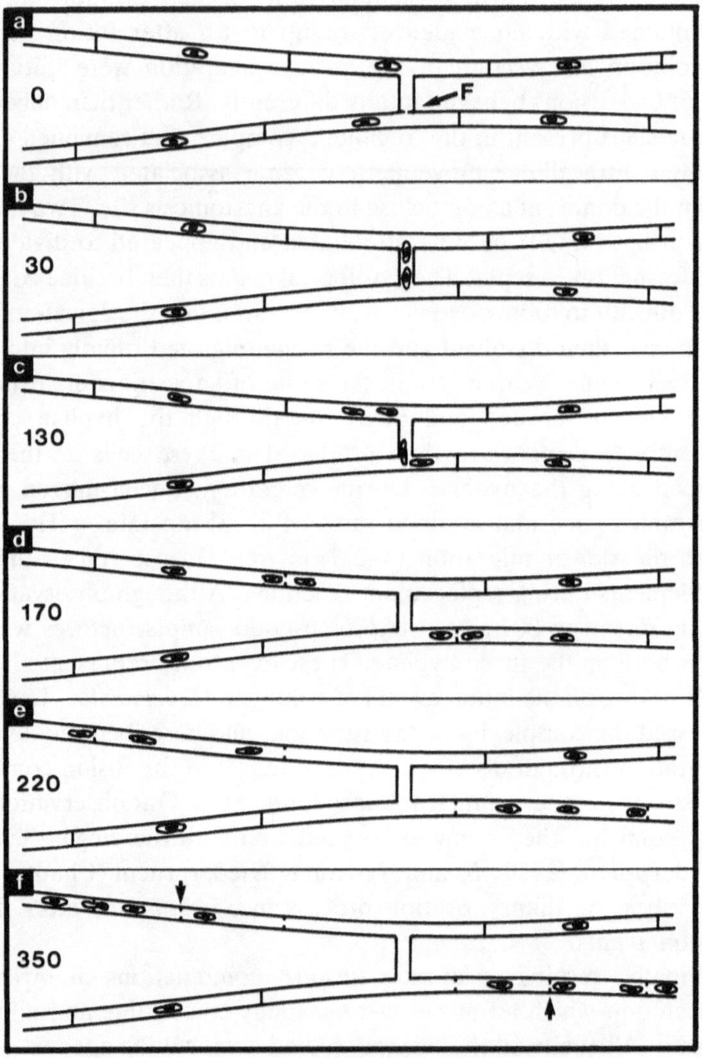

Fig. 7. A simplified representation of events following fusion between compatible monokaryons. The numbers indicate minutes after fusion. (a) A tip-to-side fusion forms at F. (b) Both nuclei associate close to the site of fusion. (c) The nuclei undergo mitotic division without the formation of septa. (d) Septa become eroded and nuclear migration begins. (e) Septa of adjacent cells dissolve and migration continues. (f) Further mitoses occur and incomplete septa are synthesised (arrowed).

division of the two nuclei. Although all nuclei were potentially compatible, it was not possible to find any evidence that septal dissolution and nuclear migration ever occurred in these fusion compartments.

Nuclear degeneration

Whilst the patterns of nuclear replacement observed in this study have not been documented in any detail elsewhere, there are reports that suggest similar behaviour in other basidiomycetes. Bensuade (1918) recorded the lysis of nuclei following both dikaryotic and di-mon fusions in *Coprinus cinereus* (cited in Noble, 1937; Papazian, 1958) and also noted the formation of clamps in association with hyphal fusions (cited in Buller, 1933). Working with *Typhula trifolii*, Noble (1937) observed that, *within* dikaryotic mycelia, 'after anastomosis is complete it appears that the nuclei of one of the two cells thus connected pass over to the other cell' and also 'the nuclei of the cell which is thus invaded gradually disintegrate, sometimes even before the invading nuclei have entered'. Similarly during di-mon fusions in this species, Noble implied the degeneration of the monokaryotic nucleus on entry into the cell of the conjugate pair. More recently, following discussion with us, Nguygen & Niederpruem (Chapter 4) found the degeneration of nuclei in association with vegetative fusions in *S. commune*, although the replacement reaction as seen in *C. versicolor* apparently failed to occur.

There are several other examples in the filamentous fungi in which different nuclei, sharing a common cytoplasmic system, show selective degeneration. Whilst in many instances this might be associated with mitotic or meiotic division (for examples see Olive, 1953) it is also found in connection with various forms of plasmogamy. Examples of the latter include gametangial fusion in certain Zygomycetes (see Webster, 1980), trichogyne–antheridial fusion in *Pyronema confluens* (Moreau & Moreau, 1930, cited in Olive, 1953), and similarly sexual fusion in *Podospora anserina* (Esser, 1965). In basidiomycetes, nuclei may also degenerate during the dikaryotisation process in cells remote from the original fusion (see observations of Lehfeldt, 1923, cited in Noble, 1937; Noble, 1937; Bistis, 1970).

From our work on degeneration, the nuclei appear as enlarged spherical bodies, containing diffuse nucleoli, which fragment and eventually fade from view. Electron micrographs of this process (Figs. 8i, 9a,b) revealed irregular nuclear profiles and nucleoplasm with densely-staining regions, suggesting uneven condensation of chromatin. Also,

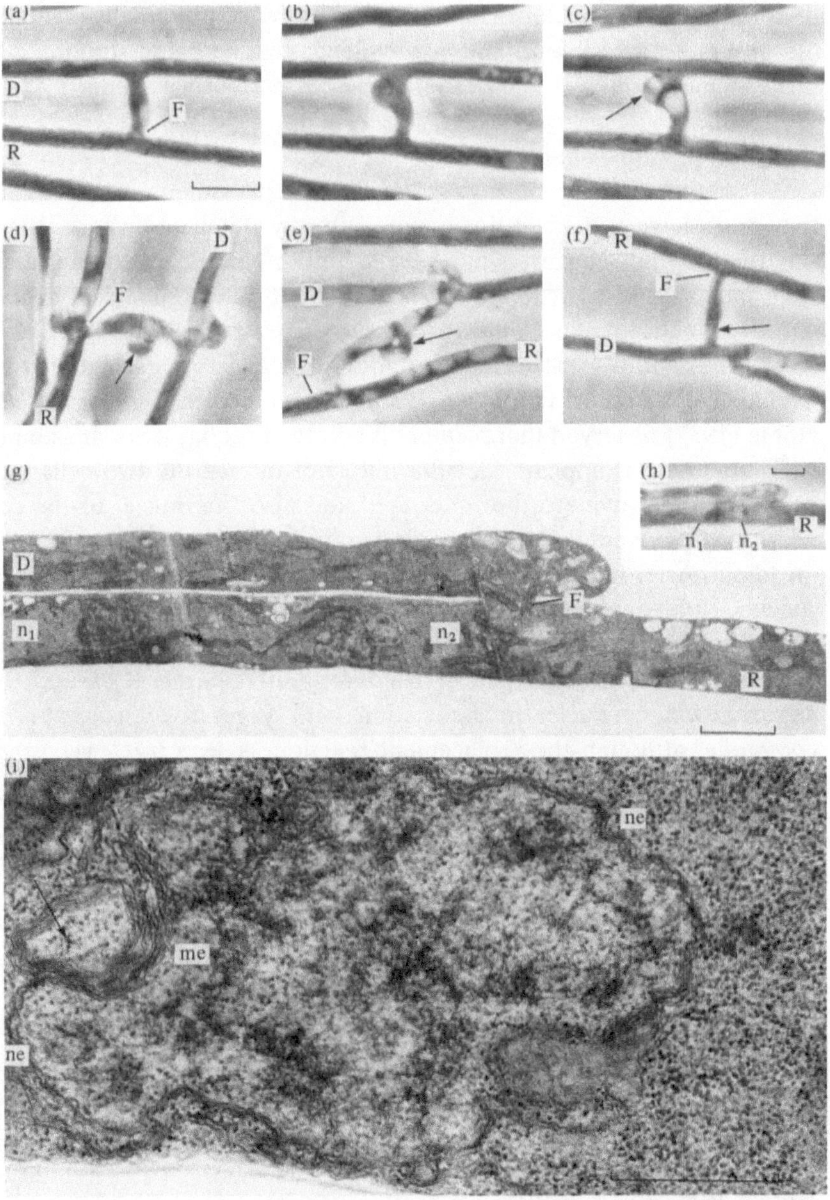

Fig. 8. Micrographs of hyphae after 'self' fusion. D, donor; R, recipient. (a)–(c) Time course of a fusion (F) between dikaryons showing intercalary clamp-connection formation. Scale bar on (a), 10 μm. (d), (e) Examples of clamp-connection formation (arrowed) after fusion (F) between dikaryons. Magnification as in (a). (f) Fusion (F) between monokaryons showing septum on bridge (arrowed). Magnification as

several layers of membrane were found in close association with the nuclear envelope (Fig. 9a). In more advanced stages (Fig. 9b), the nucleoplasm loses its integrity and shows signs of cleavage. By way of comparison, Fig. 9c is of a donor nucleus fixed during the process of migration some 30 min after fusion. Notice its typical structure, with well-defined nuclear envelope, homogeneous nucleoplasm and single nucleolus, and remember that the recipient nuclei in the same fusion have already started to degenerate.

These ultrastructural features of the degeneration process give little clue as to its mechanism. However, it is of interest that there seem to be vague parallels with other examples of selective nuclear breakdown in fungi. For example, the autolysis of nuclei in species of *Saprolegnia* involves uneven condensation of nucleoplasm (Beakes & Gay, 1977; Beakes, 1980). Also, encirclement by membrane is associated with the process of nuclear degeneration in *Phytophthora palmivora* (Hemmes & Hohl, 1973).

Nuclear replacement

Whilst nuclear replacement is a fact of life for these fungi, how and why it occurs is a matter for speculation. At present, there is no evidence that the roles of the participant compartments in a fusion are predetermined by their growth rate or stage in the cell cycle. Rather, each apparently has the potential to behave as either a donor or a recipient with this being solely determined by the part played in the fusion act. Since nuclei are generally absent from hyphal apices (Zalokar, 1959) it may be that conditions in this region are deleterious to their functioning and it is the introduction of apical cytoplasm into recipient cells that causes the degeneration of any nuclei present.

Why the replacement reaction occurs is still unclear, although it seems to be a mechanism for preventing the formation of cells with irregular numbers of nuclei. Indeed, Papazian (1958) in reviewing

---

Caption for Fig. 8 (*cont.*).
in (a). (g), (h) Longitudinal section and corresponding photomicrograph (inset) showing fusion (F) between dikaryotic hyphae. Two nuclei (n1 and n2) are present in the recipient cell (R). Scale bars: (g), 2 μm; (h), 5 μm. (i) Enlargement of nucleus n2. Note the irregular profile, uneven condensation of nucleoplasm and absence of a typical nucleolar region. A fragment, or sectioned lobe, of nuclear material (arrowed) is present close to the surface of the nucleus. The nuclear envelope (ne) in this region is associated with several layers of membrane (me). Scale bar, 0.5 μm.

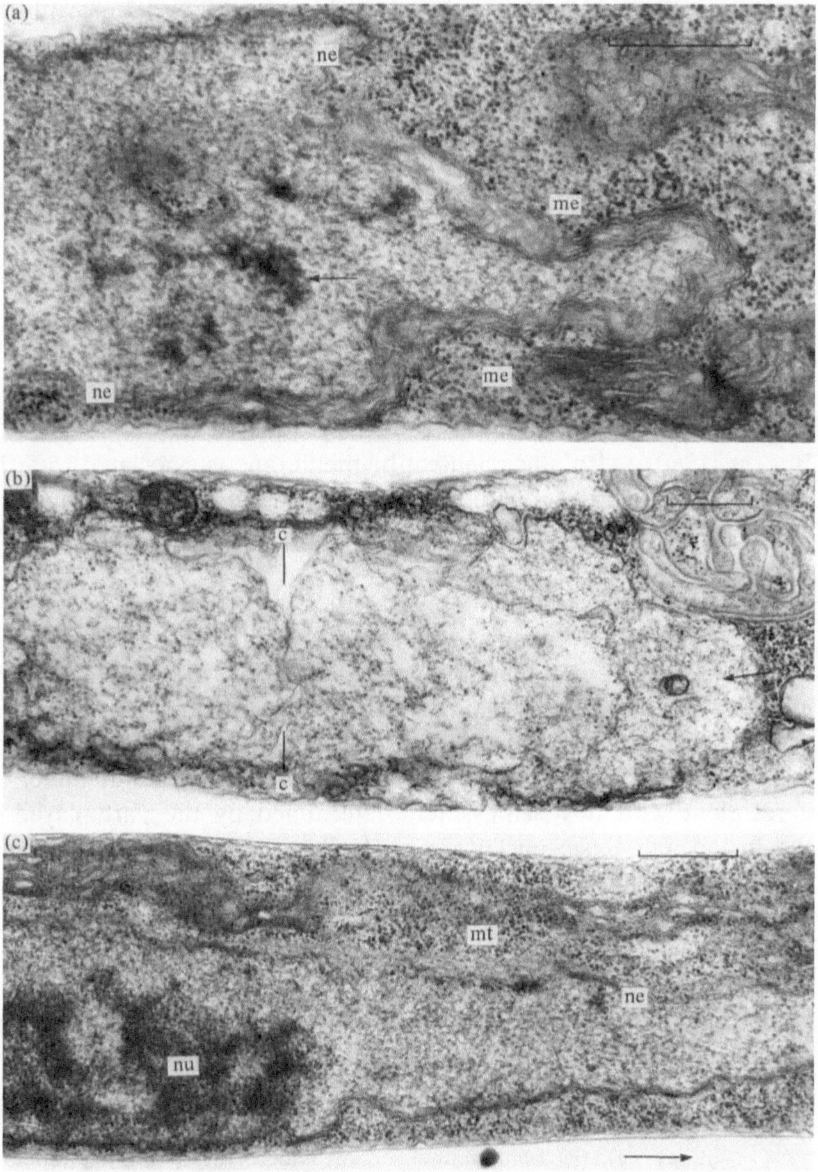

Fig. 9. Electron micrographs of nuclei in the fusion segment. (a) A recipient nucleus fixed during the process of degeneration. Several parallel layers of membranes (me) are closely associated with the nuclear envelope (ne). Note the irregular profile and presence of densely-staining regions (arrowed) dispersed within the nucleoplasm. The surrounding cytoplasm shows no signs of lysis or vacuolation. Scale bar, 0.5 μm. (b) A recipient nucleus fixed at a later stage. This

Bensuade's original reports of fusion-induced nuclear lysis, arrived at a similar conclusion. That the presence of extra nuclei in cells is apparently so deleterious might suggest a considerable degree of compartmental autonomy in these fungi and imply that the hyphae may not be as functionally coenocytic as is often supposed. This highlights the importance of septation and consequently the key role of the dolipore in modifying communication between compartments. Extending this further, it may be that nuclear replacement has some relevance to the phenomenon of somatic recombination in these fungi. Movement of donor nuclei into cytoplasm containing fragmented recipient nuclei might facilitate uptake and incorporation of DNA and hence lead to recombination in a manner similar to specific factor transfer postulated for *Schizophyllum commune* (Ellingboe & Raper, 1962; Ellingboe, 1963). Finally, in cases where aberrant division leads to formation of atypical cells, fusion and the ensuing behaviour of nuclei might serve as a means of cellular correction, assuming that such cells have reduced viability and hence are more likely to act as recipients than donors.

Conclusions

In this chapter we have provided a comprehensive description and the first ultrastructural details of events occurring during and after hyphal fusion in a filamentous fungus. We have established that hyphal fusion involves the expansion of a single pore by carefully controlled lysis of existing wall material. Further we discovered a surprisingly elaborate and highly regulated series of events that occur immediately after fusion.

It is quite clear that our observations on post-fusion events are apparently at variance with information from cultural studies of somatic and sexual interactions. The action of somatic incompatibility in several Ascomycotina (Garnjobst & Wilson, 1956) and in Basidiomycotina (Flentje & Stretton, 1964; Rayner *et al.*, Chapter 23) is usually associated with cytoplasmic lysis within the fused hyphae. In other

---

Caption for Fig. 9 (*cont.*).
structure, barely visible under phase optics, appears to be cleaving (c). The nucleoplasm is diffuse and a cap-like structure (arrowed) is present at one end. Scale bar, 0·5 μm. (c) A donor nucleus fixed during migration towards the point of fusion (arrowed). Note the regular profile which tapers in the direction of movement. The nucleoplasm is homogeneous and contains a typical nucleolar region (nu). Several longitudinally-oriented microtubules (mt) are associated with the nuclear envelope (ne). Scale bar, 0·5 μm.

examples, a reduction in viability of the fused cells is presumed to occur (see Caten & Jinks, 1966). However, in *C. versicolor*, to our surprise, observation of 'non-self' dikaryotic fusions for up to four days failed to detect any cellular expression of an incompatible reaction. Clearly the mechanism of somatic incompatibility which is so obvious in this fungus at the macroscopic level on plates and in Nature remains to be clarified at the cellular level.

As far as sexual interactions are concerned, it is obvious from the events which occur after monokaryotic and di-mon fusion, that not all potentially compatible anastomoses are successful in initiating dikaryotisation. Interestingly, from genetically based studies on monokaryons of *S. commune*, Snider (1968) and Snider & Raper (1958) also obtained evidence that the number of initiations of nuclear migration is low compared to the number of fusions that might be expected. It may be that the switching on of the morphogenetic sequence leading to dikaryotisation takes place only when degeneration fails. There are clearly two areas of fungal decision-making: whether to undergo the nuclear replacement reaction and maintain the *status quo* or for the nuclei to co-exist and trigger the compatible response. On the other hand, triggering may not be totally excluded but merely delayed in instances were nuclei are replaced. This is feasible because adjacent compatible cells joined by fusion could still exchange cytoplasmic signals since there is presumed humoral continuity through dolipores. The nuclear-replacement reaction found in di-mon fusions is, at present, difficult to reconcile with the results from cultural studies which showed nuclear selection and somatic recombination (see Raper, 1966) during such pairings, and with our own work demonstrating 'track-formation' in *C. versicolor* (Todd & Rayner, 1978). Obviously, in this case, our observations need extending as this is not the whole story.

To conclude, the events accompanying hyphal fusion appear to be more complex than was previously imagined. It should be remembered that the observations reported here were made under a particular set of cultural conditions and due to the difficulty of such work, only a limited number of fusions were followed in their entirety. However, we feel sure that our approach will be helpful in unravelling further the intricacies of the process and thereby provide a valuable complement to existing information.

We thank the Science and Engineering Research Council for financial support.

References

Ahmad, S. S. & Miles, P. G. (1970*a*). Hyphal fusions in the wood-rotting fungus *Schizophyllum commune*. 1. The effects of incompatibility factors. *Genetical Research (Cambridge)*, **15**, 19–28.

Ahmad, S. S. & Miles, P. G. (1970*b*). Hyphal fusion in *Schizophyllum commune*. 2. Effects of environmental and chemical factors. *Mycologia*, **62**, 1008–17.

Bartnicki-Garcia, S. (1973). Fundamental aspects of hyphal morphogenesis. In *Microbial Differentiation*, 23rd Symposium of the Society for General Microbiology, ed. J. O. Ashworth & J. E. Smith, pp. 245–67. Cambridge, UK: Cambridge University Press.

Beakes, G. W. (1980). Electron microscopic study of oospore maturation and germination in an emasculate isolate of *Saprolegnia ferax*. 4. Nuclear cytology. *Canadian Journal of Botany*, **58**, 228–40.

Beakes, G. W. & Gay, J. L. (1977). Gametangial nuclear division and fertilization in *Saprolegnia furcata* as observed by light and electron microscopy. *Transactions of the British Mycological Society*, **69**, 459–71.

Bensuade, M. (1918). Récherches sur le cycle évolutif et la sexualité chez les Basidiomycètes. Thesis, Neumours.

Bistis, G. N. (1970). Dikaryotisation in *Clitocybe truncicola. Mycologia*, **62**, 911–24.

Bourchier, R. J. (1957). Variation in cultural conditions and its effect on hyphal fusion in *Corticium vellereum. Mycologia*, **49**, 20–8.

Brodie, H. J. (1955). Morphology and culture characteristics of a highly aberrant *Cyathus. American Journal of Botany*, **42**, 168–76.

Buller, A. H. R. (1931). *Researches on Fungi, Vol. IV*. London: Longmans Green.

Buller, A. H. R. (1933). *Researches on Fungi, Vol. V*. London: Longmans Green.

Burnett, J. H. (1976). *Fundamentals of Mycology*, 2nd edn. London: Edward Arnold.

Burnett, J. H. & Trinci, A. P. J., eds (1979). *Fungal Walls and Hyphal Growth*, Symposium of the British Mycological Society. Cambridge, UK: Cambridge University Press.

Cabral, R. V. de G. (1951). Anastomoses miceliais. *Boletim da Sociedade Broteriana (Ser. 2)*, **25**, 291–362.

Caten, C. E. & Jinks, J. L. (1966). Heterokaryosis: its significance in wild homothallic ascomycetes and fungi imperfecti. *Transactions of the British Mycological Society*, **49**, 81–93.

Ellingboe, A. H. (1963). Illegitimacy and specific factor transfer in *Schizophyllum commune. Proceedings of the National Academy of Sciences of the USA*, **49**, 286–92.

Ellingboe, A. H. & Raper, J. R. (1962). Somatic recombination in *Schizophyllum commune. Genetics*, **47**, 85–98.

Esser, K. (1965). Heterogenic incompatibility. *Incompatibility in Fungi*, ed. K. Esser & J. R. Raper. Berlin: Springer.

Flentje, N. T. & Stretton, H. M. (1964). Mechanisms of variation in *Thanatephorus cucumeris* and *T. praticolus. Australian Journal of Biological Sciences*, **17**, 686–704.

Garnjobst, L. & Wilson, J. F. (1956). Heterocaryosis and protoplasmic incompatibility in *Neurospora crassa. Proceedings of the National Academy of Sciences of the USA*, **42**, 613–18.

Gooday, G. W. (1975). Chemotaxis and chemotropism in fungi and algae. In *Primitive Sensory and Communication Systems*, ed. M. J. Carlile, pp. 155–204. London: Academic Press.

Harvey, I. C. (1975). Development and germination of chlamydospores in *Pleiochaeta setosa*. *Transactions of the British Mycological Society*, **64**, 489–95.

Hawker, L. E. & Beckett, A. (1971). Fine structure and development of the zygospore of *Rhizopus sexualis* (Smith) Callen. *Philosophical Transactions of the Royal Society of London*, **B263**, 71–100.

Hawker, K. E. & Gooday, M. A. (1969). Fusion, subsequent swelling and final dissolution of the apical walls of the progametangia of *Rhizopus sexualis* (Smith) Callen: an electron microscope study. *New Phytologist*, **68**, 133–40.

Hemmes, D. E. & Hohl, H. R. (1973). Mitosis and nuclear degeneration: simultaneous events during secondary sporangia formation in *Phytophthora palmivora*. *Canadian Journal of Botany*, **51**, 1673–5.

Lange, I. (1966). Das Bewegungsverhalten der Kerne in fusionierten Zellen von *Polystictus versicolor* (L.). *Flora (Jena)*, **156**, 487–97.

Lehfeldt, W. (1923). Uber die Entstehung des Paarkernmyzels bei heterothallischen Basidiomyceten. *Hedwigia*, **64**, 30–51.

Moreau, F. & Moreau, A. (1930). Le développement du périthèce chez quelques Ascomycètes. *Revue générale de Botanique*, **42**, 65–98. (Cited Olive, 1953.)

Niederpruem, D. J. (1980*a*). Direct studies of dikaryotisation in *Schizophyllum commune.* I. Live inter-cellular nuclear migration patterns. *Archives of Microbiology*, **128,** 162–71.

Niederpruem, D. J. (1980*b*). Direct studies of dikaryotisation in *Schizophyllum commune.* II. Behaviour and fate of multikaryotic hyphae. *Archives of Microbiology*, **128,** 172–8.

Noble, M. (1937). The morphology and cytology of *Typhula trifolii* (Rostr.). *Annals of Botany (N.S.)*, **1,** 67–98.

Olive, L. (1953). The structure and behaviour of fungus nuclei. *Botanical Review*, **19,** 439–586.

Papazian, H. P. (1958). The genetics of Basidiomycetes. *Advances in Genetics*, **9,** 41–69.

Parmeter, J. R., Sherwood, R. T. & Platt, W. D. (1969). Anastomosis grouping among isolates of *Thanatephorus cucumeris*. *Phytopathology*, **59,** 1270–8.

Raper, J. R. (1966). *Genetics of Sexuality in Higher Fungi.* New York: Ronald Press.

Raudaskoski, M. (1973). Light and electron microscope study of unilateral mating between a secondary mutant and a wild-type strain of *Schizophyllum commune*. *Protoplasma*, **76,** 35–48.

Rayner, A. D. M. & Todd, N. K. (1977). Intraspecific antagonism in natural populations of wood-decaying basidiomycetes. *Journal of General Microbiology*, **103,** 85–90.

Rayner, A. D. M. & Todd, N. K. (1978). Polymorphism in *Coriolus versicolor* and its relation to interfertility and intraspecific antagonism. *Transactions of the British Mycological Society*, **71,** 99–106.

Rayner, A. D. M. & Todd, N. K. (1979). Population and community structure and dynamics of fungi in decaying wood. *Advances in Botanical Research*, **7,** 333–420.

Rijkenberg, F. J. H. & Truter, S. J. (1975). Cell fusion in the aecium of *Puccinia sorghi*. *Protoplasma*, **83,** 233–46.

Robak, H. (1942). Cultural studies in some Norwegian wood destroying fungi. *Meddelser Vestlandets Forstlige Forsoksstation*, **7**(3), 1–248.

Smythe, R. (1973). Hyphal fusions in the basidiomycete *Coprinus lagopus* sensu Buller. 1. Some effects of incompatibility factors. *Heredity*, **31,** 107–11.

Snider, P. J. (1968). Nuclear movements in *Schizophyllum.* In *Aspects of Cell Motility*, 22nd Symposium of the Society for Experimental Biology, ed. P. L. Miller, pp. 261–83. Cambridge, UK: Cambridge University Press.

Snider, P. J. & Raper, J. R. (1958). Nuclear migration in the basidiomycete *Schizophyllum commune*. *American Journal of Botany*, **45**, 538–46.

Todd, N. K. & Rayner, A. D. M. (1978). Genetic structure of a natural population of *Coriolus versicolor*. *Genetical Research (Cambridge)*, **32**, 55–65.

Todd, N. K. & Rayner, A. D. M. (1980). Fungal individualism. *Science Progress (Oxford)*, **66**, 331–54.

Tulasne, A. R. & Tulasne, C. (1863). *Selecta Fungorum Carpologia, Parisiis, Tomus 11* (cited Buller, 1931).

Valk, P. van der & Marchant, R. (1978). Hyphal ultrastructure in fruit-body primordia of the basidiomycetes *Schizophyllum commune* and *Coprinus cinereus*. *Protoplasma*, **95**, 57–72.

Watkinson, S. C. (1978). End-to-side fusions in hyphae of *Pencillium claviforme*. *Transactions of the British Mycological Society*, **70**, 451–3.

Watrud, L. S. & Ellingboe, A. H. (1973). Use of cobalt as a mitochondrial vital stain to study cytoplasmic exchange in matings of the basidiomycete *Schizophyllum commune*. *Journal of Bacteriology*, **115**, 1151–8.

Webster, J. (1980). *Introduction to Fungi*, 2nd edn. Cambridge, UK: Cambridge University Press.

Zalokar, M. (1959). Growth and differentiation in *Neurospora* hyphae. *American Journal of Botany*, **46**, 602–10.

6
The mycelial habit and secondary metabolite production

M. O. MOSS
Department of Microbiology, University of Surrey, Guildford, Surrey GU2 5XH, UK

The terrestrial mycelial fungi amongst the eukaryotes, and the actinomycetes amongst the prokaryotes, share a number of physiological and ecological features. The filamentous morphology of both groups is associated with the efficient utilisation of exogenous, macromolecular, and often insoluble substrates, as well as the ability to produce a remarkably diverse range of secondary metabolites. This chapter examines the possibility that secondary metabolism may be a form of chemical differentiation associated with the particular type of morphological differentiation imposed by a terrestrial ecology.

Filament formation and the aquatic environment

Filament formation occurs in most of the major groups of aquatic algae as well as the Chytridiomycetes and Oomycetes amongst the aquatic fungi. Such filament formation in water is usually associated with a life cycle which includes attachment to a substratum and even the essentially single-celled diatoms have evolved mechanisms for the formation of filaments in both freshwater and marine environments. The same is true to a lesser extent of the fundamentally single-celled group of freshwater green algae, the desmids (Fig. 1). Amongst the prokaryotes the cyanobacteria which include both free-living and attached filamentous forms and sheathed bacteria, such as *Sphaerotilus*, produce such apparently well developed filamentous growth as to be called 'sewage fungus'. However microscopic observation reveals that the individual bacteria retain their separate existence within a secreted sheath.

None of these forms is widely associated with the production of secondary metabolites, but it is necessary to assess the role, if any, of secondary metabolism in their biology when considering any particular association of this process with filamentous organisms.

Filament formation and the terrestrial environment

Although some filamentous green algae have a tendency towards the terrestrial habitat, as far as microorganisms are concerned it is the mycelial fungi amongst the eukaryotes and the actinomycetes

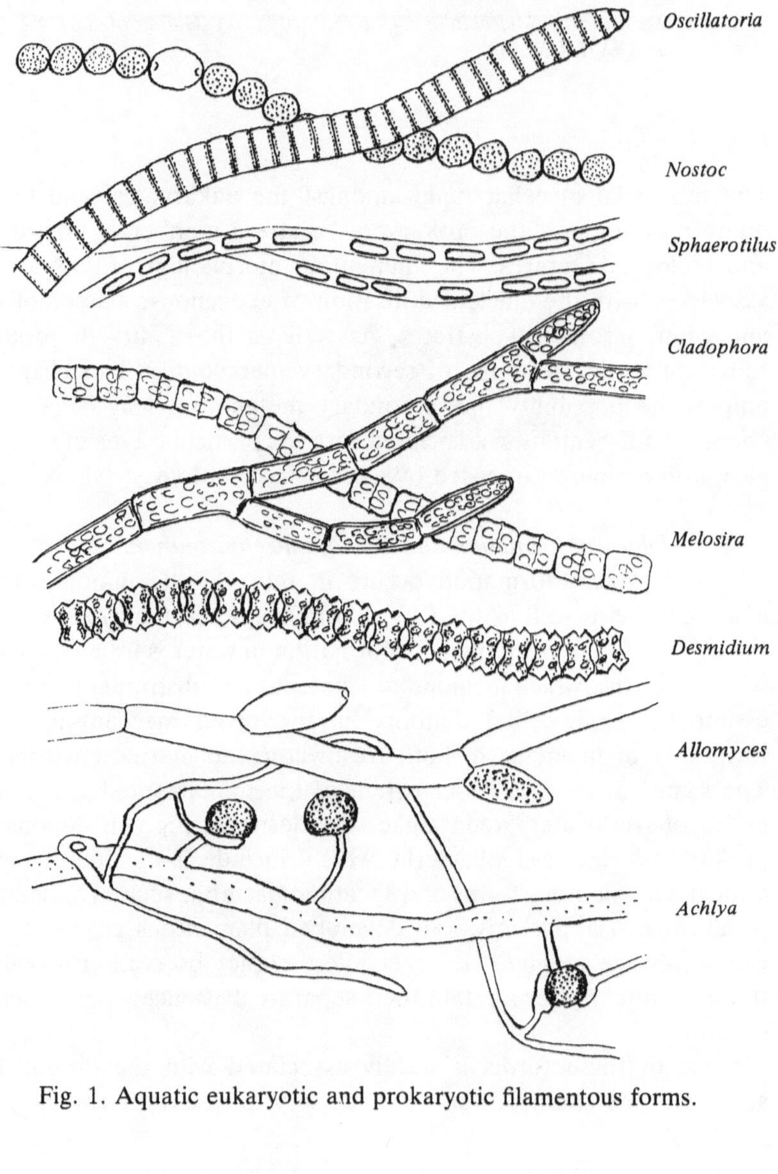

Fig. 1. Aquatic eukaryotic and prokaryotic filamentous forms.

amongst the prokaryotes, which have a particularly well developed filamentous morphology. These two taxonomically distinct groups share a number of physiological and ecological features. As well as the efficient utilisation of exogenous substrates in the solid phase, they show morphological differentiation arising from the polarised growth of filamentous structures (Burnett & Trinci, 1979), sophisticated and very varied means of aerial dispersal, and the ability to produce a remarkably diverse range of secondary metabolites, at least in the laboratory.

Secondary metabolites

Organic compounds of relatively low molecular weight, such as alkaloids and terpenes (Fig. 2), are produced by many, if not all, of the higher plants and have been referred to as secondary metabolites, defined as compounds which have no recognised role in the maintenance of fundamental life processes in the organisms which synthesise them (Bell, 1981). Although it is not possible to make any general statement about the significance of secondary metabolites in higher plants, their remarkable co-evolution with insects has frequently led to complex and specific relationships between them which often involve these secondary metabolites. The pine sawfly larva, for example, stores up the volatile pinenes of its food plant during feeding and regurgitates them as a defence reaction to predators. The frequently specific oviposition response of many insects with a herbivorous larval state is dependent on the production of secondary metabolites by the food plant (Rothschild, 1972).

It is thus clear that, in higher plants, some secondary metabolites have a physiological or ecological role, perhaps as part of the defence

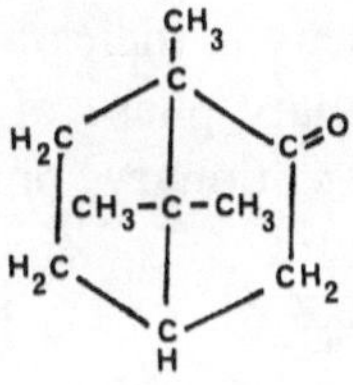

(a)

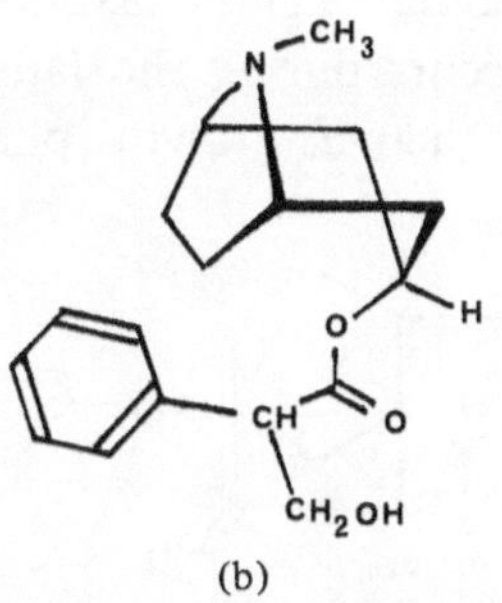

(b)

Fig. 2. Secondary metabolites from higher plants. (a) Camphor, a terpene from *Cinnamomum camphora*. (b) Atropine, an alkaloid from *Atropa belladona*.

mechanisms of a species, perhaps as a recognition signal to a specific group of insects, perhaps as a hormone associated with differentiation (Schreiber, 1979). It is remarkable that two groups of such hormones, the gibberellins and abscisic acid, each with a clear role in the differentiation and growth of plants, are also produced by fungi. However, it should be noted that, although these two compounds, one a diterpene the other a sesquiterpene (Fig. 3), are widely produced amongst plant species, their production by fungi has been reported for only a single species in each case. With gibberellins it is probable that many of the steps in the biosynthetic pathway are the same in plants as in *Gibberella fujikuroi* and its anamorph, *Fusarium moniliforme*. Abscisic acid is produced by the fungus *Cercospora rosicola* (Assante, Merlini & Nasini, 1977), but it should be pointed out that this was first shown for cultures on potato broth but not other widely used microbiological media. Recent studies by Griffin & Walton (1982), showing that phosphate limitation is involved in initiating the synthesis of abscisic acid by *C. rosicola*, confirm that it is a secondary metabolite of this fungus. These studies also show that, in this instance, production of the metabolite may be controlled by, or certainly influenced by, events occurring during exponential growth.

Primary and secondary metabolites

With increasing study the generalisations made about primary and secondary metabolites become somewhat blurred, but it is still useful to think of primary metabolites as a group of compounds, with a relatively low diversity of structure at the monomeric level, universally produced by many forms of life during the active growth phase. By contrast, secondary metabolites have a great diversity of molecular structure and frequently show taxonomic specificity in their production which usually occurs during the later or stationary phase of growth. Such terms as 'active growth phase' and 'stationary phase', or

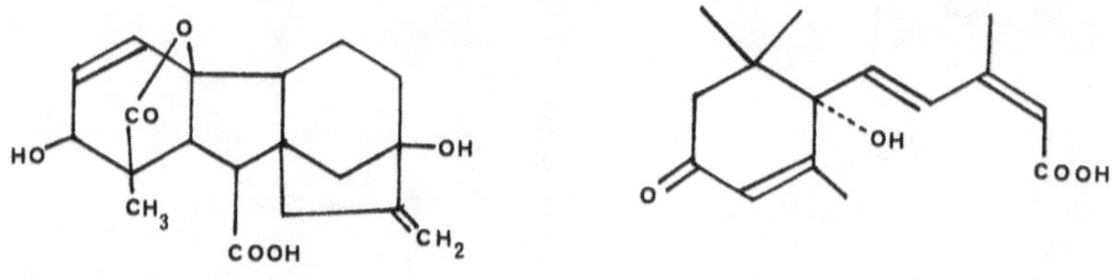

Gibberellic acid Abscisic acid

Fig. 3. Structures of plant hormones also produced as secondary metabolites by moulds.

'trophophase' and 'idiophase' (Bu'Lock, 1980), are associated with the widespread use of batch cultures of micro-organisms in the laboratory, but they emphasise that we are looking at a process of chemical differentiation (see also Jennings; Chapter 7). It has been suggested that in higher plants the expression of secondary metabolism is associated with morphological differentiation (Schreiber, 1979) and it may be useful to re-examine the production of secondary metabolites by micro-organisms in these terms. Interest in the production of secondary metabolites by higher plants has always been based on their application as therapeutic agents, their association with flavour, smell or colour, and in chemotaxonomy. The study of fungal and actinomycete metabolites was initially a challenge to the natural products chemist but was rapidly enhanced by the observation that some have antibiotic properties, some are phytotoxic, and a significant number are potent mycotoxins. In the lichens, they have also been used in chemotaxonomy.

Antibiotics and toxins

A survey of known, but not necessarily useful, antibiotics shows that the majority are produced by actinomycetes. Over 2000 have been catalogued in a number of recent collections of antibiotics from actinomycetes (Umezawa, 1978). Antibiotics produced by other prokaryotes number about 120 and are dominated by metabolites of *Pseudomonas* and *Bacillus* (Korzybski, Kowszyk-Gindifer & Kurylowicz, 1978). The number of antibiotic metabolites associated with fungi is approximately 260 and these seem to be entirely from the terrestrial filamentous fungi of the ascomycetes, basidiomycetes and imperfect fungi, with perhaps one or two from the zygomycetes (cf, Rayner & Webber: Chapter 18).

Despite the undoubted metabolic capability of the actinomycetes, especially members of the genus *Streptomyces*, they have not been implicated in the poisoning of food or animal feeds as have filamentous fungi. Food poisoning is only one aspect of the spoilage of food and even in this wider context the actinomycetes do not seem to be a problem. The number of fully characterised mycotoxins produced by fungi is approximately 250, but it must be noted that antibiotics and mycotoxins are not mutually exclusive groups of secondary metabolites. Compounds such as patulin and citrinin were first studied because of their antibiotic activity, but are justifiably considered as mycotoxins.

Unlike the actinomycetes, the yeasts are well known as agents of spoilage and alteration of foods, and yet there are no records of mycotoxicoses implicating yeasts, unless ethyl alcohol, which is not a secondary metabolite, is considered to be a mycotoxin!

Of course, a number of single-celled prokaryotes do produce toxins, and bacterial food poisoning is a major concern of the food microbiologist. However, such exotoxins as those of *Staphylococcus aureus* and *Clostridium botulinum* and the endotoxins of the enterobacteria are macromolecular compounds and are not secondary metabolites in the sense discussed here. Thus the endotoxin of *Salmonella* is a part of the complex cell envelope of the actively growing cell which is released following death and lysis of the cell in the gut, and the toxin responsible for botulism is a complex protein, the primary structure of which is probably coded for by a plasmid. An organism known as *Pseudomonas cocovenenans* has been shown to produce low-molecular-weight toxic metabolites, such as toxoflavin and bongkrekic acid (Fig. 4), implicated in poisoning following the consumption of contaminated coconut tempeh in Indonesia (Van Damme, Johannes, Cox & Berends, 1960). Indeed members of the genus *Pseudomonas* produce a number of low-molecular-weight pigments and antibiotics which have some of the

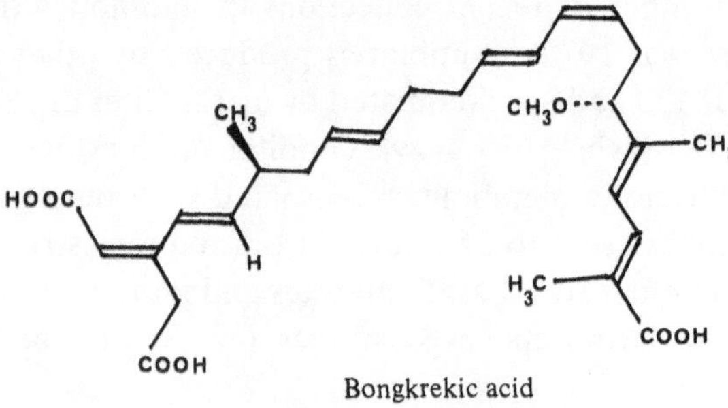

Bongkrekic acid

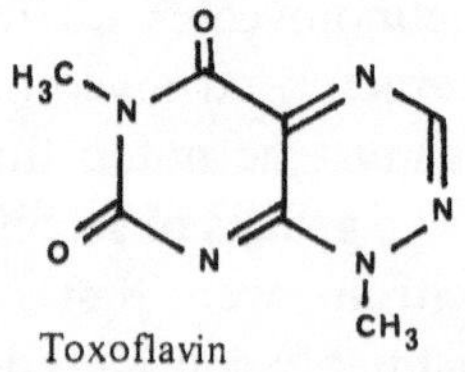

Toxoflavin

Fig. 4. Toxic metabolites of *Pseudomonas cocovenenans.*

characteristics of secondary metabolites and deserve further study in this context. Nevertheless, if the production of low-molecular-weight antibiotics and toxins is an acceptable indication of the diversity of secondary metabolite production, then these compounds would seem to be produced mainly by the terrestrial, filamentous actinomycetes amongst the prokaryotes and the filamentous, terrestrial members of the fungi.

Water moulds and algae

Before exploring further the significance of secondary metabolism and the mycelial habit of fungi and streptomycetes, the assumption that secondary metabolites are not widely produced by other groups of filamentous micro-organisms must be examined more critically. Low-molecular-weight compounds are known to act as hormones triggering

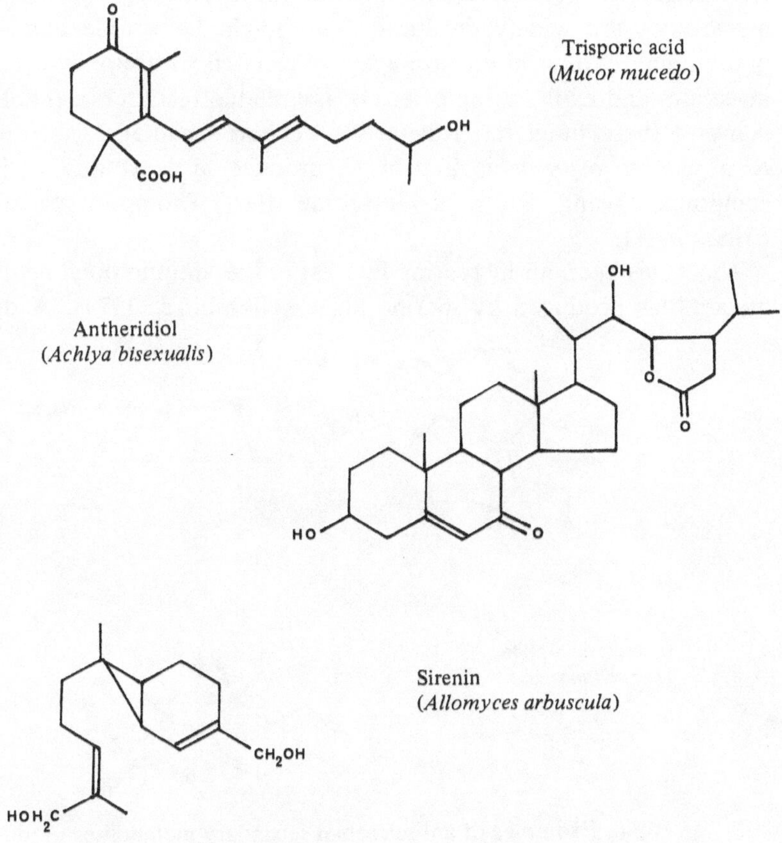

Fig. 5. Structures of fungal hormones involved in sexual reproduction.

stages in the sexual reproduction of chytrids, oomycetes and zygomycetes (Gooday, 1973). Those which have been characterised, such as trisporic acid, antheridiol and sirenin, are all mevalonate-derived metabolites, the first produced by degradation of a carotene, the second a sterol derivative and the third a sesquiterpene (Fig. 5). These compounds are usually produced at very low concentrations and it is likely that their role as hormones requires a tight control of their production which is possibly more closely associated with the alteration of primary, rather than the production of secondary metabolites.

It has to be admitted that attention is usually drawn to the production of secondary metabolites because of a striking physiological or physico-chemical property such as antibiosis, toxicity, colour or fluorescence. There is a paucity of such activity associated with the filamentous water moulds and even the zygomycetes, which have successfully invaded terrestrial environments, do not provide much evidence that secondary metabolites are widely produced. This might be a reflection of the predominantly ruderal life strategies, associated with rapidly extending mycelium and exploitation of easily assimilable resources exhibited by many of these fungi (cf. Rayner & Webber: Chapter 18). However, *Rhizopus oligosporus* is known to produce at least one antibiotic compound (Wang, Ruttle & Hesseltine, 1969; Thompson, Smalley & Eribo, 1981).

There has been an increasing interest in the antimicrobial activity of metabolites produced by marine algae (Glombitza, 1979). A diverse

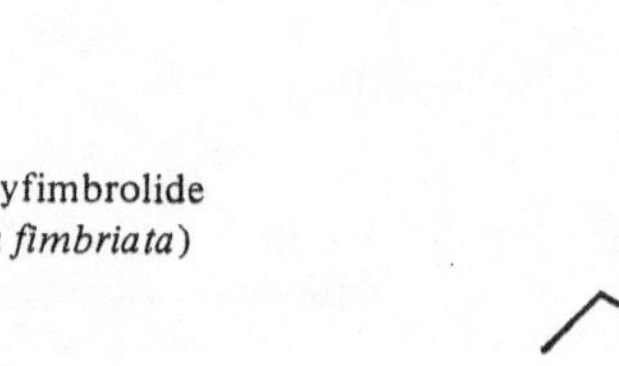

Chondriol
(*Chondria oppositicladia*)

Hydroxyfimbrolide
(*Delisia fimbriata*)

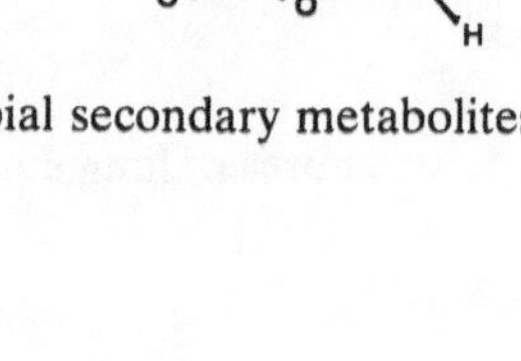

Fig. 6. Examples of antimicrobial secondary metabolites produced by seaweeds.

range of structures, many of them containing bromine and chlorine, are produced by seaweeds with relatively complex thalli, such as the Ulvales amongst the green algae, Bonnemaisoniaceae and Rhodomelaceae amongst the red algae and the Fucales amongst the brown algae. Figure 6 illustrates some typical examples, chondriol from *Chondria oppositicladia* (Fenical, Sims & Radlick, 1973) and hydroxy fimbrolide from *Delisea fimbriata* (Pettus, Wing & Sims, 1977).

A number of algae, such as dinoflagellates, produce toxic metabolites and are responsible for such phenomena as paralytic shellfish poisoning. Otherwise there is little information about secondary metabolites from simpler algae, especially freshwater filamentous algae, although this may simply reflect the fact that they have not been looked for, or that their production is under tight control because they are involved as hormones in reproduction. Nevertheless one must ask why are secondary metabolites so widespread in the filamentous fungi and actinomycetes?

Secondary metabolism and differentiation of filamentous fungi

Martin & Demain (1978) have made the point that secondary metabolite production by filamentous fungi is still not well understood at the enzymological level, despite the importance of some of these metabolites on an industrial scale. They point out in particular that the cellular sites of biosynthesis and where and how the products are secreted are generally unknown.

Secondary metabolism and sporulation are both examples of differentiation, one at the molecular, the other at the morphological level. It is generally accepted that sporulation in fungi is not expressed while balanced, active growth is possible (see also Raudaskoski: Chapter 13; Lysek: Chapter 14). It is also generally accepted that secondary metabolite production only occurs after balanced growth has ceased. It would thus seem that the two phenomena of morphological differentiation and secondary metabolism are at least linked in time. A statement made by Martin & Demain (1978) is worth considering in more detail:

> Growth and secondary metabolism should be understood as processes which compete for key metabolic intermediates rather than as mutually exclusive phenomena.

In animals, and to a lesser extent in higher plants, differentiation and growth are integrated processes and one can consider the great diversity of form and colour amongst these groups as an integration of some aspects of secondary metabolism into the processes of controlled and continuous differentiation. Of course, the fungal hypha itself is a polarised structure (Burnett, 1979) and the transport of materials to support the growing hyphal tip is, in a sense, a process of differentiation (see Jennings: Chapter 7).

There are some examples of mould metabolite formation in which integration with the processes of growth and differentiation can be hinted at. Citrinin formation has been reported to be a biphasic process (Betina, Baunt, Hajnicka & Nadova, 1973) being produced during exponential and stationary phase. Bu'Lock, Shepherd & Winstanley (1969) obtained evidence that 6-methylsalicylic acid synthetase, an enzyme complex involved in the first stages of patulin biosynthesis, is synthesised, although it is not active, during the exponential growth phase of *Penicillium griseofulvum.* It becomes active during the stationary phase although there may be some variation between strains in the exact timing of activation of the enzyme (Light, 1967). Those enzymes involved in the transformation of 6-methylsalicylic acid to patulin are not formed until the stationary phase. In the case of patulin biosynthesis, then, secondary metabolite production would seem to involve the formation of part of the machinery during active growth, but induction of the rest of the machinery and its activation only occurs during the stationary phase.

An interesting situation has been reported in the production of cyclopenin and related metabolites by *Penicillium cyclopium.* Cyclopenase, a key enzyme in their biosynthesis is produced within the conidia and so is spatially linked with morphogenesis (Wilson & Luckner, 1975).

The genome and secondary metabolism

Some understanding of the relationships between secondary metabolism, vegetative growth and morphogenesis will arise from a deeper knowledge of the genetics of these processes. Probably the best known genome of a filamentous fungus is that of *Neurospora crassa* (Perkins, Radford, Newmeyer & Bjorkman, 1982). A single locus referred to as *En (pdx)* is known to modify another gene resulting in the production of a yellow pigment in the medium (Martin, 1967) but, even in this well documented genome, we do not seem to have the definitive identification of any genes producing messenger RNA coding for the enzymes supposedly involved in the biosynthesis of secondary metabolites. Perhaps such genes do not exist! Perhaps the battery of enzymes which we assume must be involved in secondary metabolism are modified forms of the enzymes neatly coded for in the genome for primary metabolism. Sometimes the same enzyme may seem to play a different role in different groups of organisms. Thus phenylalanine ammonia-lyase is considered to be an important enzyme in the pathways leading to the biosynthesis of aromatic secondary metabolites by many basidiomycetes and higher plants – the antibiotic mucidin produced by *Oudemansiella mucida* is a good example (Zouchova, Wurst, Nerud & Musilek, 1982). This same enzyme is produced by the yeast *Rhodotorula glutinis* but, in this instance, its role is as a catabolic enzyme in the utilisation of phenylalanine as a carbon or nitrogen source (Marusick, Jensen & Zamir, 1981).

A mutation in a strain of *Fusarium solani* f. sp. *cucurbitae*, which prevented protoperithecial formation but determined the diffusion of red pigments into the medium, has been obtained by Snyder, Georgopoulos, Webster & Smith (1975). Six nuclear genes of *Fusarium solani*, the anamorph of *Nectria haematococca*, which are involved both with pigment production and in some way with sexual reproduction, have been mapped by Parisot, Maugin & Gerlinger (1981).

Conidiation of *Aspergillus nidulans* is known to involve the induction of two distinct phenol oxidase enzymes not produced during vegetative growth (Clutterbuck, 1977). These enzymes are responsible for the green colour of the spores and the brown colour of the conidiophores and sterigmata characteristic of this species. It is possible that such enzymes may also act on substrates arising from the activity of, for example, the polyketide synthetases thus leading to a diversification of secondary metabolites. These observations are the sort of hint relating secondary metabolite production and morphogenesis.

Polyketides – a group particularly associated with filamentous fungi

The association of polyketide biosynthesis with an enzyme complex produced during active growth, and probably involved in some aspect of primary metabolism, has already been indicated in the earlier discussion of patulin biosynthesis. Ward & Packter (1974) showed that both fatty acid biosynthesis during active growth, and phenolic metabolite (polyketide derived) biosynthesis during stationary phase, are inhibited by the addition of 4-fluorophenylalanine during active growth. It seems increasingly likely that polyketide synthetase is identical, or closely related, to the fatty acid synthetase. Is polyketide production particularly common amongst fungi because of the presence of a multienzyme complex for fatty acid biosynthesis? In contrast, a disaggregated enzyme system occurs in animals and prokaryotes. Despite the considerable diversity of metabolites which may be produced from a polyketide chain of a particular length, there are some rules. Thus, after cyclisation, the C-methyl end is always longer, or as long, and never shorter than the C-carboxyl end. If a part of the chain is reduced then the C-methyl end of the chain is usually the most reduced as though the chain started off as a normal fatty acid chain, but the system ran out of reducing ability as the chain grew. A recently described metabolite of *Achaetomium cristalliferum*, referred to as achaetolide (Bodo, Molho, Davoust & Molho, 1983), illustrates these points (Fig. 7). Is the availability of NADPH, which is required for fatty acid biosynthesis, more difficult to maintain in a polarised filamentous structure once it reaches a particular stage of development?

It has been suggested that a major source of NADPH in filamentous fungi is by the functioning of a mannitol cycle (Hult & Gatenbeck, 1979)

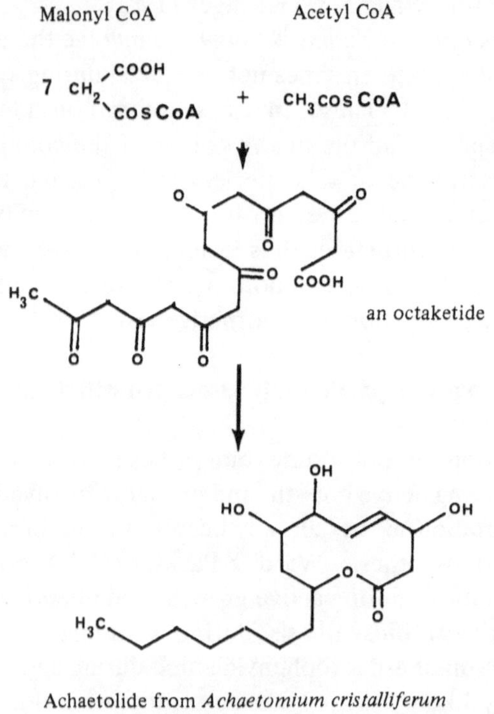

Fig. 7. The structure and biosynthesis of achaetolide, a secondary metabolite of *Achaetomium cristalliferum*.

in which NADPH is generated by the oxidation of mannitol to fructose. The cycle consumes NADH and ATP during the reconversion of fructose to mannitol and so can be considered as a link between catabolic and anabolic metabolism. Both mannitol and the enzymes of the cycle are widespread in filamentous fungi (Hult, Veide & Gatenbeck, 1980). It has been shown that a strain of *Alternaria alternata*, which produces the polyketide alternariol, is less active in the oxidation of mannitol (and by implication less efficient in synthesising fatty acids) than a non-alternariol-producing strain of the same species (Hult & Gatenbeck, 1978). It would be of interest to know what the spatial distribution of mannitol is within the developing hypha. Indeed, a knowledge of the spatial distribution of a range of biochemical activities along the length of the hypha away from the hyphal tip might be the key to understanding the relationship between this particular type of filamentous structure and secondary metabolite production.

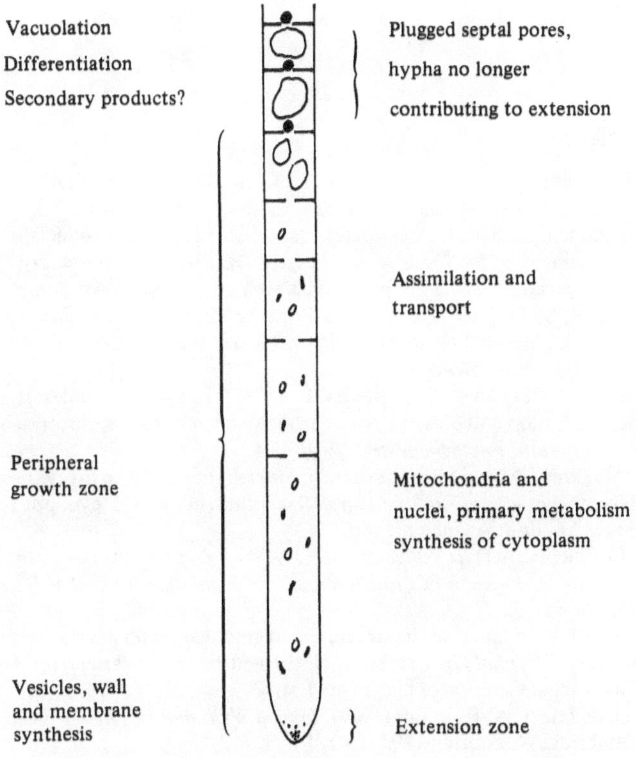

Fig. 8. Differentiation along the polarised structure of the fungal hypha.

Conclusions

The very nature of hyphal extension implies a process of increasing cytological differentiation even during vegetative growth before overt morphogenesis occurs. The hyphal tip is a region of vigorous wall and membrane biosynthesis involving the migration and activity of vesicles. This region is supported by a length of hypha containing cytoplasm packed with mitochondria along which assimilation and transport will be maintained by vigorous anabolic and catabolic processes. Further along the hypha vacuolation and compartmentation by the plugging of septal pores become evident and it is feasible to suppose that multienzyme complexes such as fatty acid synthetases may be too far removed from the sources of anabolic reductants such as NADPH. This is still too naïve a picture but such a system in which active extension growth, cytological differentiation and morphogenesis are separated in space and time (Fig. 8) may have as a natural consequence the kind of chemical differentiation which we recognise as secondary metabolism.

References

Assante, G., Merlini, L. & Nasini, G. (1971). (+)-abscisic acid, a metabolite of the fungus *Cercospora rosicola*. *Experientia*, **33**, 1556–7.

Bell, E. A. (1981). The physiological role(s) of secondary (natural) products. In *The Biochemistry of Plants, a Comprehensive Treatise, vol. 7 Secondary Plant Products*, ed. E. E. Conn, pp. 1–19. New York, London: Academic Press.

Betina, V., Baunt, S., Hajnicka, V. V. & Nadova, A. (1973). Diphasic production of secondary metabolites by *Penicillium notatum* Westling S-52. *Folia Microbiologica*, **18**, 40–8.

Bodo, B., Molho, L., Davoust, D. & Molho, D. (1983). Fungal macrolides: structure determination and biosynthesis of achaetolide, a lactone from *Achaetomium cristalliferum*. *Phytochemistry*, **22**, 447–51.

Bu'Lock, J. D. (1980). Mycotoxins as secondary metabolites. In *The Biosynthesis of Mycotoxins, a Study in Secondary Metabolism*, ed. P. S. Steyn, pp. 1–16. New York, London: Academic Press.

Bu'Lock, J. D., Shepherd, D. & Winstanley, D. J. (1969). Regulation of 6-methylsalicylate and patulin synthesis in *Penicillium urticae*. *Canadian Journal of Microbiology*, **15**, 279–85.

Burnett, J. H. (1979). Aspects of the structure and growth of hyphal walls. In *Fungal Walls and Hyphal Growth*, ed. J. H. Burnett & A. P. J. Trinci, pp. 1–25. Cambridge: Cambridge University Press.

Burnett, J. H. & Trinci, A. P. J., eds (1979). *Fungal Walls and Hyphal Growth*. Cambridge: Cambridge University Press.

Clutterbuck, A. J. (1977). The genetics of conidiation in *Aspergillus nidulans*. In *Genetics and Physiology of Aspergillus*, ed. J. E. Smith & J. A. Pateman, pp. 305–17. London, New York: Academic Press.

Fenical, W., Sims, W. & Radlick, P. (1973). Chondriol, a halogenated acetylene from the marine alga *Chondria oppositicladia. Tetrahedron Letters*, 313–16.

Glombitza, K.-W. (1979). Antibiotics from algae. In *Marine Algae in Pharmaceutical Science*, ed. H. A. Hoppe, T. Levring & Y. Tanaka, pp. 303–42. Berlin: W. de Gruyter.

Gooday, G. W. (1973). Differentiation in the mucorales. In *Microbial Differentiation*, ed. J. M. Ashworth & J. E. Smith, pp. 269–94. Cambridge: Cambridge University Press.

Griffin, D. H. & Walton, D. C. (1982). Regulation of abscisic acid formation in *Mycosphaerella (Cercospora) rosicola* by phosphate. *Mycologia*, **74,** 614–18.

Hult, K. & Gatenbeck, S. (1978). Production of NADPH in the mannitol cycle and its relation to polyketide formation in *Alternaria alternata. European Journal of Biochemistry*, **88,** 607–12.

Hult, K. & Gatenbeck, S. (1979). Enzyme activities of the mannitol cycle and some connected pathways in *Alternaria alternata* with comments on the regulation of the cycle. *Acta Chemica Scandinavica*, **B33,** 239–43.

Hult, K., Veide, A. & Gatenbeck, S. (1980). The distribution of the NADPH regenerating mannitol cycle among fungal species. *Archives of Microbiology*, **128,** 253–5.

Korzybski, T., Kowszyk-Gindifer & Kurylowicz, W. (1978). *Antibiotics – Origin, Nature and Properties.* Washington: American Society for Microbiology.

Light, R. J. (1967). Effects of cycloheximide and amino acid analogues on biosynthesis of 6-methylsalicylic acid in *Penicillium patulum. Archives of Biochemistry and Biophysics*, **122,** 494–500.

Martin, J. F. & Demain, A. L. (1978). Fungal development and metabolite formation. In *The Filamentous Fungi*, vol. 3, *Developmental Mycology*, ed. J. E. Smith & D. R. Berry, pp. 426–50. London: Edward Arnold.

Martin, P. G. (1967). A gene modifying pigment production by *pdx-1* (44602) of *Neurospora crassa. Microbial Genetics Bulletin*, **27,** 9–10.

Marusick, W. C., Jensen, R. A. & Zamir, L. O. (1981). Induction of L-phenylalanine ammonia-lyase during utilization of phenylalanine as a carbon or nitrogen source in *Rhodotorula glutinis. Journal of Bacteriology*, **146,** 1013–19.

Parisot, D., Maugin, M. & Gerlinger, C. (1981). Genetic and epigenetic factors involved in the excretion of naphthoquinone pigments into the culture medium by *Nectria haematococca. Journal of General Microbiology*, **126,** 443–57.

Perkins, D. D., Radford, A., Newmeyer, D. & Bjorkman, M. (1982). Chromosomal loci of *Neurospora crassa. Microbiological Reviews*, **46,** 426–570.

Pettus, J. A., Wing, R. M. & Sims, J. J. (1977). Marine natural products. XII. Isolation of a family of multihalogenated gamma-methylene lactones from the red seaweed *Delisea fimbriata. Tetrahedron Letters*, 41–4.

Rothschild, M. (1972). Some observations on the relationship between plants, toxic insects and birds. In *Phytochemical Ecology*, ed. J. B. Harborne, pp. 1–12. London, New York: Academic Press.

Schreiber, K. (1979). Interrelationships between secondary products and hormones in plants. In *Regulation of Secondary Product and Plant Hormone Metabolism*, FEBS Symposium, vol. 55, ed. M. Luckner & K. Schreiber, pp. 1–11. Oxford: Pergamon Press.

Snyder, W. C., Georgopoulos, S. G., Webster, S. K. & Smith, S. N. (1975). Sexuality and genetic behaviour in the fungus *Hypomyces (Fusarium) solani* f. sp. *cucurbitae. Hilgardia*, **43,** 161–85.

Thompson, D. P., Smalley, A. W. & Eribo, B. E. (1981). Fungal metabolite from members of the genus *Rhizopus. Applied and Environmental Microbiology*, **41,** 1271–3.

Umezawa, H. (1978). *Index of Antibiotics from Actinomycetes,* vol. 2. Tokyo: Japan Scientific Societies Press.

Van Damme, P. A., Johannes, A. G., Cox, H. C. & Berends, W. (1960). On toxoflavin, the yellow poison of *Pseudomonas cocovenenans. Recueil des Travaux Chimiques des Pays-Bas et de la Belgique*, **79,** 255–67.

Wang, H. L., Ruttle, D. I. & Hesseltine, C. W. (1969). Antibacterial compound from a soybean product fermented by *Rhizopus oligosporus. Proceedings of the Society for Experimental Biology and Medicine*, **131,** 579–83.

Ward, A. C. & Packter, N. M. (1974). Relationship between fatty acid and phenol synthesis in *Aspergillus fumigatus. European Journal of Biochemistry*, **46,** 323–33.

Wilson, S. & Luckner, M. (1975). Cyclopenase, ein Lipoproteid der cytoplasma Membran von Konidiosporen des Pilzes *P. cyclopium* Westling. *Zeitschrift für allgemeine Mikrobiologie,* **15,** 45–51.

Zouchova, Z., Wurst, M., Nerud, F. & Musilek, V. (1982). Metabolism of aromatic acids in the antibiotic-producing basidiomycete *Oudemansiella mucida. Folia Microbiologia*, **27,** 446–50.

7
Water flow through mycelia

D. H. JENNINGS
Botany Department, The University, P.O. Box 147, Liverpool L69 3BX, UK

If we have a hollow linear structure in which water flows into (or is drawn out of) one end due to an osmotic gradient across the walls, there will be a flow of solution through the interior of the structure. The distance for which the flow will be maintained and the rate at which it occurs will depend upon leakage of water across the walls away from the point where the osmotic gradient is applied and upon resistance to flow within the structure due to its shape and the viscosity of the solution which is moving. If the structure is vertical, water movement in an upward direction will be impeded by gravity. A tree is the prime biological example of such a structure. Movement of water is brought about by the low water potential in the aerial environment drawing water initially from the cellulose walls of the leaf cells. The flow of water to them is aided by the long relatively continuous conducting system, the low viscosity of the solution conducted and the impermeable side walls (cork) of the structure. In an herbaceous plant, significant movement of solution upwards can be generated by the active concentration of salts in the xylem which results in the so-called root pressure.

Fungi produce linear structures at the level of both the microscopic (hyphae) and the macroscopic, e.g. rhizomorphs, sporophores. How far is there a flow of solution through these structures and can it be of any significance for the biology of the particular fungi? This chapter attempts to provide some answers. Before beginning, it is important to remember that whenever a linear structure traverses a substratum or medium where the osmotic gradient across its walls may differ along its length due to changes either in internal solute concentration or in the water potential of the external medium, there will be water flow within the structure. The study of mycelia in homogeneous media – notably agar plates and submerged cultures – has tended to make us forget this, such that it is implicitly assumed that there is osmotic equilibrium not only through the medium (which may not be true) but also through the mycelium (see Trinci, Chapter 2).

Solution flow along a hypha

Bulk flow

When solution flows along a hypha, there is bulk flow of material certainly as solutes but possibly as discrete and possibly visible material suspended in the solution. De Vries (1885) was the first to observe such visible bulk flow in the sporangiophore of *Phycomyces nitens*, the flow being another example of what was then called protoplasmic streaming. Arthur (1897) examined such streaming experimentally in a number of lower filamentous fungi, and showed that the application of 2 mol l^{-1} potassium nitrate causes vigorous movement to the point of application. Schröter (1905) confirmed from studies on *Mucor stolonifera* (= *Rhizopus nigricans*) and *Phycomyces nitens* that streaming was dependent upon the difference between the water potential of the surrounding solution and that of the aerial environment. When the mycelium was grown submerged in a homogeneous substratum or in air saturated with water vapour no streaming could be observed. An important observation made by both Arthur and Schröter was that, while the central protoplasm, vacuoles and cell-sap in the form of a central cylinder moved in one direction, an outer mass of protoplasm in the form of a cylinder mantle, free of vacuoles, moved in the opposite direction. The likely mechanism bringing about this latter movement and its significance will be discussed later.

Ternetz (1900) showed from studies on the discomycete, *Ascophanus carneus*, that osmotically driven protoplasmic streaming could occur in a higher fungus possessing septa within the hyphae. Buller (1933) confirmed that protoplasmic streaming occurred in a variety of higher fungi. Although not providing any direct evidence that osmotic forces were involved, he did show that streaming towards the growing tips of hyphae was associated with the growth of vacuoles of mature regions and that septa and vacuoles deform in the direction anticipated from the direction of flow of particles. Buller gave values for the velocity of protoplasmic streaming as being in the range 6–12 cm h^{-1} at around 20 °C, with one quoted value of 19.8 cm h^{-1} determined for *Rhizopus nigricans* at 28 °C by Arthur (1897). These velocities tend to be greater than the known velocities of nuclear migration, a process which is thought to be brought about, at least partially, by a contractile mechanism (Jennings, 1976*a*; Jennings, Thornton, Galpin & Coggins, 1974; see Rayner *et al.* Chapter 23). However the velocities for protoplasmic streaming are of the same order of magnitude as, perhaps somewhat slower than, cyclosis in characean algal coenocytes such as *Nitella*, for which there is powerful evidence that the movement is brought about by the shearing of continuously bending filaments of actin (Allen, 1974; Wagner, 1979).

More direct evidence that water can move for considerable distances along undifferentiated hyphae comes from studies on droplet formation at the apical regions of hyphae of *Serpula lacrimans* (Coggins, Jennings & Clarke, 1980; Brownlee & Jennings, 1981*a*). This process can be observed when the mycelium is growing from an agar inoculum over a non-absorbent transparent surface. Hyphal continuity with the inoculum must be maintained for the droplets to increase in volume. The rate of increase, together with hyphal extension, is inhibited by addition to the mycelium on the agar of high concentrations of solutes or of metabolic inhibitors, such as sodium azide or oligomycin. Hyphal extension is also inhibited. On the other hand, inhibitors of microtubule and microfibril function, e.g. colchicine, cytochalasin B, vinblastine sulphate, have no significant effect on either droplet volume increase or hyphal extension. Determination of the osmotic potential of the droplets gave values which were consistent with the ultrafiltration of water through the hyphal membrane by a hydrostatic pressure. This pressure was hypothesised to be generated by solute uptake by the mycelium on the agar. Colotelo (1978) has pointed out the ubiquity of droplet production by fungal hyphae, mycelium and fruiting structures and showed that the droplets may contain one or more hydrolytic enzymes. The droplets, when studied, have been observed to have a membranous covering (Colotelo, Sumner & Voegelin, 1971; Colotelo, 1978).

Spore release throughout the fungal kingdom provides many examples where water movement along hyphae almost certainly occurs at some stage (Ingold, 1971). However, it is only for the growth of sporangiophores of *Phycomyces blakesleeanus* (Grehn, 1932; Cowan, Lewis & Thain, 1972) and release of zoospores in *Phytophthora* and *Pythium* (Gisi & Zentmyer, 1980) that critical experimental evidence is available showing that osmotic forces play a major role in the processes concerned.

Role in primary apical growth. De Vries (1885), Arthur (1897), Ternetz (1900), Schrötter (1905) and Buller (1933) all believed strongly that protoplasmic streaming is an important means of transferring material to points of growth. The probable role in apical extension of micro- and macrovesicles containing wall components, lytic and synthetic enzymes is well established (Burnett & Trinci, 1979). I (Jennings, 1979) have previously suggested that water flow may be involved in the movement to the mycelial tip of these vesicles which are involved in wall synthesis. Although it could be argued that vesicle movement is via a process other than water flow, since protoplasmic streaming cannot be observed in hyphae growing in air saturated with water vapour, lack of observable streaming under the light microscope may not mean that there is no fluid flow – the latter may be occurring at a velocity which is insufficient to move visible particles. The mechanism proposed by me (Jennings, 1979), by which a number of ion pumps or their activity increases basipetaly from the tip, coupled with a higher hydraulic conductivity at the tip itself, would be equally effective in a medium in which the bulk osmotic properties were uniform throughout.

The mechanism just referred to is not the only one possible. Provided that the hydraulic conductivity of the tip is greater than elsewhere along a hypha, water flow can be generated by the metabolism changing basipetaly such that more osmotically active material is produced. Increasing vacuolation within hyphae as they age, and which Buller (1933) felt was very significant in the generation of streaming, may be due to such increased metabolic activity. We need to know, particularly for fungi with coenocytic hyphae such that flow generated some considerable distance away from the tip readily moves towards it, how the metabolism changes with age at a specific point along a hypha. Some idea of the possible changes can be gained by studying over time the metabolism of mycelium as it grows in liquid culture. This has been done for mitochondrial properties in *Neurospora crassa* (Schwitzguebel & Palmer, 1982). However, more precise information about spatial changes is likely to come from histochemistry or, more significantly, from sequential release of protoplasts along a hypha as the wall is lysed from the tip backwards (Isaac, Ryder & Peberdy, 1978; Isaac, Briarty & Peberdy, 1979).

Cyclosis

It must not be assumed from the foregoing that all translocation within hyphae is brought about by bulk flow of solution. As previously indicated, nuclear migration appears to occur via the mediation of microtubules (Jennings *et al.*, 1974). Cyclosis *sensu* that process which can be seen to occur in characean algae, may also occur in hyphae as in the case of the movement of cytoplasm in the reverse direction to osmotically-driven bulk flow seen by Arthur (1897) and others referred to above. In the lower portions of thin or old sporangiophores of *P. blakesleeanus* a flow of particles can readily be observed by phase-contrast or interference microscopy to occur in both directions along seemingly single strands of protoplasm (Oort & Roelofsen, 1932; Bergman *et al.*, 1969) in a manner similar to cyclosis and to the movement of particles along the filopodia which extend out of the exoskeleton of Foraminifera such as *Allogromia* (Allen, 1964; McGee--Russell, 1974). Further, an actin-like protein has been observed and isolated from a morphological mutant of *Neurospora crassa* (Allen, Lowry & Sussman, 1948; Allen & Sussman, 1978). There is good evidence that the movement of phosphorus (as granules of polyphosphate) in hyphae of *Glomus mosseae* appears to occur via cytoplasmic streaming *sensu* cyclosis (Cox *et al.*, 1980; Cooper & Tinker, 1981). Cyclosis may also be important in homeostasis within a hypha particularly at the tip. Bulk flow of solution to the tip, particularly if water is ultrafiltered through its outer membranes, could present problems if the concentration of solutes were to rise to levels which might alter the rate of metabolism (Turian, 1978, 1979, 1981). A stream of cytoplasm in the reverse direction could act as a countercurrent taking solutes in the reverse direction. On the other hand, if the plasma-membrane is not completely semi-permeable some toxic solutes may be removed by the flow of water out of the hyphal tip. This seems to be so in *S. lacrimans* for hydrogen ions (Brownlee & Jennings, 1981*a*).

It is believed now that the maintenance of a current carried by protons or sodium ions through the tip and out of a hypha via a proton or sodium-extruding ATPase transport system is responsible for the maintenance of polarity of growth (Harold, 1977; Jennings, 1979). At first sight water flow would act against such a current. However such a flow will generate a streaming potential which will be countered by an electro-osmotic flow due to the fixed negative charges in the peripheral cytoplasm. Indeed the observed flows in this region of the protoplasm moving against the direction of protoplasmic streaming may be brought about by electro-osmosis and not by the mediation of contractile material such as actin.

Solution flow through mycelium

Transpiration from basidiocarps

There is little doubt that solution flow can take place in mycelia over significant macroscopic distances. Schütte (1956) demonstrated the ability of basidiocarps in transpiration, showing that the velocity of movement of the dyes rose edicol and fluorescein was increased under conditions where evaporation was increased. Plunkett (1956) made the first quantitative study, of which there seem to have been few others subsequently, of transpiration from sporophores. He showed that, at 20 °C with air of 0% relative humidity passing over basidiocarps of *Polyporus brumalis*, the rate of transpiration could be as high as 0.68 g cm^{-2} h^{-1}. At higher and more ecologically-appropriate relative humidities the rate of water loss was much lower (0.12 and 3.5×10^{-3} g cm^{-2} h^{-1} for 75% and 96% relative humidity respectively). The rate of flow of air at the higher humidity was however only one-fortieth of that used for those observations made at the lower humidities. These values compare with 0.5–1.0 g cm^{-2} h^{-1} for the average peak transpiration rate of shade plants measured in their natural environment (Larcher, 1980). Nobel (1975) showed that the tissue resistance for water vapour movement from basidiocarps of, for example, *Agaricus campestris, Boletus brunneus, Coprinus plicatilis, Lycoperdon perlatum*, and *Scleroderma australe* is such that up to a wind velocity of 100 cm s^{-1} (greater than would be generally expected near the ground in the field) the loss of water from them is controlled by the boundary layer resistance. Nevertheless the flux of water through the mycelium to the surrounding air via a basidiocarp can be sufficient to bring about increased growth. Plunkett (1958) showed for *Polyporus brumalis* that, at a rate of water loss of 0.65 (for young primordia) – 0.15 (mature fruiting body) g cm^{-2} h^{-1}, there was a much greater dry matter increase and more potassium in fruit bodies than when the rate of water loss was about 1×10^{-3} g cm^{-2} h^{-1}. All these figures for transpiration rates are minimal estimates for the flux of water into basidiocarps, since no account is taken of water used in growth. It needs to be noted however that Plunkett (1958) found that for *Flammulina velutipes* expansion from primordia was prevented by rates of water loss exceeding quite low values.

Flow through vegetative mycelium

Studies on long-distance translocation through vegetative mycelium have mostly been made with *Serpula lacrimans*. These have demonstrated that both the process itself and vegetative spread are dependent upon a flux of water through the mycelium. Reference has already been made to the role of water movement through the hyphae in the production of droplets at the hyphal apices. It is thought that aggregation of these microscopic droplets gives rise to the large drops characteristically observed on mycelium covering an agar plate or growing on plaster. Clarke, Jennings & Coggins (1980) showed that growth over a nutritionally-inert artificial substratum was dependent on the water potential of the substratum and the environment surrounding the inoculum from which the mycelium spread. The evidence that translocation is brought about by bulk flow of solution comes from studies with radioactive isotopes. Firstly, these move with a similar velocity – independent of chemical composition of the molecule into which they are incorporated – of the order of 25 cm h^{-1}, which is very similar to the velocity of protoplasmic streaming referred to earlier. Secondly, if a metabolic inhibitor such as sodium azide, or a solution of low water potential, is applied to the source of translocation, movement of radioactivity along the mycelium stops at all points (Brownlee & Jennings, 1982*a*). Trehalose is the carbohydrate translocated, being converted to arabitol in what might be called the 'sinks' of translocation (Brownlee & Jennings, 1981*b*). The path is via strands or syrrotia or in the youngest part of the mycelium along certain hyphae. Although the latter are clearly physiologically specialised, they cannot readily be distinguished on morphological grounds from those hyphae not involved in translocation (Brownlee & Jennings, 1982*b*). Schütte (1956) similarly found no morphological differences in basidiocarps between those hyphae translocating dye and those not doing so.

The mechanism bringing about translocation in *S. lacrimans* mycelium appears to be, as described for individual hyphae above, pressure-driven flow of solution along the mycelium. The hydrostatic pressure is generated at the food source by the active uptake of glucose resulting from cellulose breakdown and subsequent conversion of the hexose to trehalose within the hyphae. This latter compound is presumed to be the solute responsible for generating the necessary osmotic gradient across the plasma membrane at the food source.

Besides these studies on *S. lacrimans*, recent studies with mycorrhizal fungi have demonstrated absorption and movement of water along strands over ecologically significant distances (Duddridge, Malibari & Read, 1980; see Read, Chapter 10).

To summarise, it appears that water movement can occur through mycelium as a result either of evaporation or of pressure-driven flow.

Though convincing evidence for the latter is confined to *S. lacrimans*, the frequent presence of droplets on hyphae, fruit bodies and sclerotia (Colotelo, 1978) and on rhizomorphs (I. H. Granlund, unpublished observations) suggests that pressure-driven flow of solution is a widespread phenomenon in fungal mycelium.

Implications

The potential benefits and consequences of water flow through mycelium are as follows. Firstly, the process enables the transport or translocation of nutrients from sources relatively rich in such substances or their precursors (either *in toto* or in certain constituents) to sinks, e.g. hyphal tips or developing sporophores, relatively poor or without an immediately-available supply of these substances. Secondly, there are implications for maintenance of ionic balance and pH within the mycelium. Thirdly, if the flow is generated by hydrostatic pressure, this will aid the production of turgor at the hyphal tips and hence be of considerable value when the fungus is growing through an environment of unfavourable water potential. Finally, a flow of water through the mycelium will draw mobile nutrients to it at those points where water is entering the hyphae.

Translocation; sources, sinks and colony differentiation

It has long been realised that translocation must obviously be a key process in the production of aerial structures such as many types of sporophore. Less frequently appreciated is that if a fungus is to invade nutritionally-poor habitats *en route* to those which are richer, as must frequently be the case in nature as opposed to culture (see Watkinson: Chapter 8), then translocation is also vital. When the process is associated with growth into an environment of low water potential, for instance when a sporophore grows into the air, the hydrostatic pressure which needs to be generated must be such as to: overcome the resistance in the channels of the mycelium; produce a flow of water of sufficient magnitude to meet both that needed for maintenance of metabolic function and that lost by evaporation; and take all the solutes for growth at the required rate and over the necessary distance. In certain instances, as has been shown above for *Polyporus brumalis*, it seems that evaporation-driven flow can augment pressure-driven flow but in others, as with *Flammulina velutipes* basidiocarps and mycelial spread of *Serpula lacrimans* over inert surfaces (Clarke *et al.*, 1980), this may not be so. In these latter cases, there will be a need to generate within the mycelium or absorb from the external medium sufficient solute to produce and maintain the required water potential across the plasmalemma at the source of translocation. The solute has to be available in sufficient quantity to allow it to be replaced as the solution moves away from the food source.

The foregoing has referred to a single solute, since it is unlikely that there will be several solutes involved in the generation of the necessary hydrostatic pressure. This is supported by the fact that trehalose is associated with translocation of carbon in the mycelium of a wide variety of fungi (see below). However the case for the involvement of only a limited number of solutes can be argued theoretically. Inorganic ions seem unlikely candidates not only because of the need for a significant source of supply but, more important, because at high concentrations in the hyphae they will have – unless in a special compartment – untoward effects on metabolism. Equally as the number of organic solutes involved increases, there will be a need for an increasingly complex system of regulation of metabolism. Further the range of compounds which can be synthesised is limited by their effect at high concentrations on the activity of proteins within the hyphae. Alterations in the properties of water by dissolved species will lead to the perturbation of protein conformation (Franks & Eagland, 1975). Thus monovalent ions up to 100–150 mmol l^{-1}, while causing some suppression of the Donnan diffuse double layer of proteins and other macromolecules, can do so without causing conformational instability although higher concentrations do have this effect. Wyn Jones *et al.* (1977) give details of those organic solutes whose concentrations can rise to much higher values than those cited above for monovalent ions with minimal effect on protein conformation. The list includes nitrogen dipoles such as α-amino acids and betains, polyols such as glycerol, arabitol and mannitol and polyol derivatives such as α-galactosyl glycerol. To these can be added sucrose, probably trehalose (but this requires examination), and urea. Thus we can postulate that there is a change in the physiology of parts of a mycelium acting as a source of translocate, from primary metabolism associated with active growth to what is effectively secondary metabolism such that there is a significant flux of carbon directed towards one compound. Bu'Lock *et al.* (1965) termed that phase when primary metabolism predominates as the *trophophase* and that when secondary metabolism predominates as the *idiophase*. The latter is characterised by limited growth. Bu'Lock (1975) has pointed out that secondary metabolism is an aspect of mycelial differentiation, the process of which is implicit in the identification of a mycelium having limited growth (see Moss: Chapter 6).

Holligan & Jennings (1972*a*) pointed out that in cultures of *Dendryphiella salina* the synthesis of arabitol, which under the conditions of their experiments showed very low turnover (Holligan & Jennings, 1972*b*), had many of the characteristics of secondary metabolite production. Panek (1975) and Panek & Matoon (1977) have shown that trehalose synthesis in *Saccharomyces cerevisiae* takes place only in the final stages of growth when glucose is exhausted from the medium. Interestingly glycogen does not contribute to trehalose synthesis; seemingly an amino acid acts as the substrate via gluconeogenesis. From such observations it appears that enhanced synthesis of an appropriate osmotically-active solute can occur even in fungi which are not considered to be capable of translocation or only to translocate to a limited degree.

This concept of a change in metabolism from one which is in an overall primary state to one which has features of secondary metabolism, such that the mycelium becomes the source of translocate, can be considered within the context of a model proposed by Chanter & Thornley (1978). The model they used consisted of three compartments (substrate, mycelium, sporophore) in which the concentration of a single substrate (of which mannitol, trehalose, glucose and sucrose were the putative candidates) controlled the growth of the mycelium and the basidiocarps of *Agaricus bisporus*. Below a certain (threshold) substrate concentration, initiation cannot take place. When the threshold is exceeded the rate of sporophore initiation is assumed to be proportional to the difference between the two concentrations. The parameter values inserted led to a solution for the model which agrees reasonably well with the observed data.

The division of the sequence of events during mycelial growth into the two phases, trophophase and idiophase, implies that primary metabolism is separated temporally from secondary metabolism. However the foregoing implies that in natural environments it is necessary to consider the extent to which spatial separation between the two phases is possible. Recent studies (Fenn, Choi & Kirk, 1981; Grootwassink & Gaucher, 1980) indicate that the two phases can be made to occur simultaneously by manipulation or mutation. Nevertheless it is still conceivable that even under the culture conditions used there was spatial separation of the two phases.

While with the topic of secondary metabolism, it is interesting to note that the onset of ligninolytic activity in the mycelium of the wood-decomposing hymenomycete *Phanerochaete chrysosporium* occurs when there is near depletion of nutrient nitrogen and cessation of net DNA synthesis (Keyser, Kirk & Zeikus, 1978; Reid, 1979). Ligninolytic activity is a nitrogen-based regulated system about which Fenn *et al.* (1981) and Fenn & Kirk (1981) have provided further details. Not only does ligninolytic activity have the properties of secondary metabolism but onset of activity is accompanied by *de novo* synthesis of the secondary metabolite veratryl (3,4-dimethoxybenzyl) alcohol (Keyser *et al.*, 1978; Lundquist & Kirk, 1978). However, the experiments on which this statement is based were carried out with the mycelium growing in liquid culture. It seems likely that if the mycelium were able to spread like *Phanerochaete laevis*, as observed in the experiments carried out by Thompson & Rayner (1982) (see Thompson: Chapter 9) one might anticipate that mycelium associated with the wood inoculum would have low growth rate, high lignolytic activity and high levels of soluble carbohydrate, while that growing away from the wood would have a high growth rate and low lignolytic activity.

A broadly similar situation to that for the production of ligninolytic activity appears to hold for the production of cellulase activity in fungi (Hulme & Stranks, 1971) though the molecular mechanism by which activity of the enzyme is kept low in growing mycelium may differ in different species (Canevascini *et al.*, 1979).

The ideas just presented have a bearing on an earlier statement that I made (Jennings, 1982) on the physiological basis of inoculum potential, that important concept introduced by Garrett (1970) which he defined as 'the energy of growth of a fungus available for colonisation of a substratum to be colonised'. I suggested that the energy within inoculum potential should be partitioned into that which allows growth on the new substratum *and* that which is required to transport sufficient biomass to colonise a new food source competitively. Clearly from what has been said above, the energy required for transport must be expended at the original food source either in solute uptake or solute production within the mycelium.

Bidirectional flow

From what has just been said it is clear that the availability of nutrients in the external medium is likely to determine whether translocation is necessary for growth into that medium. If translocation is taking place, any absorbable nutrients in the medium could influence its pattern of metabolism. Thus the regulatory processes in metabolism will have to cope not only with alterations in nutrient input across the plasma membrane, as would be the case for example with yeast cells growing in liquid batch culture, but also with alterations in the composition of the translocation stream. The input from that stream may be very much less susceptible to membrane regulation at the translocation sink than the entry of nutrients from the external medium (Jennings, 1974). The question therefore arises as to whether or not there is any interaction between source and sink of the translocation stream. Clearly the source, through the pressure-driven flow of solution from it, can influence events at the sink but, if there is to be interaction in the reverse direction, there must be some sort of reverse flow. This might be by cyclosis but there is no reason why there should not be two pressure-driven flows in opposite directions analogous to those which occur in higher plants. The important criteria for the existence of such a situation is that the lateral walls should have low permeability to the solutes being transported and that at the end of each channel there should be unidirectional transport systems in the relevant membranes removing solute from the entering flow of solution.

The presence or absence of bidirectional movement needs investigation. It needs to be carried out with care; any experiment needs to avoid those conditions which might reverse the direction of water flux along a particular channel. Experiments to date which have purported to demonstrate translocation in both directions through fungal mycelium can be criticised on a number of grounds but in particular on failing to avoid the conditions just referred to. Of course, if reverse translocation were present in a mycelium, the back flow of solution away from the mycelial front could remove excess carbon in those situations where a fungus is growing from a food base, such as wood, where there is a very high C:N ratio. Some aspects of this particular problem have already been discussed by Jennings (1982). Finally, reverse translocation could be important in helping to maintain cytoplasmic pH at the mycelial front.

Regulation of pH and ionic balance

There are two accepted ways in which the pH of plant cytoplasm can be controlled: by the pumping of protons or hydroxyl ions across the plasmalemma into the external medium (Raven & Smith, 1973), or by carboxylation to produce organic acids (increase of organic acid anions) and decarboxylation (decrease in anions) in response respectively to a decrease or increase in cytoplasmic pH – the so-called biochemical pH-stat (Davies, 1973). The role of these two processes in cytoplasmic pH control has been reviewed by Smith & Raven (1979). In fungi there is no doubt that the process of pumping out protons occurs through the presence of a proton-extruding ATPase in the plasma-membrane (Jennings, 1976*b*; Slayman & Gradmann, 1975; Borst-Pauwels, 1981). Organic acid metabolism could be important in cytoplasmic pH control but one might anticipate that pH regulation by this means would be confined to those parts of the mycelium which contain vacuoles such that the organic acid anions might be sequestered. If this were not so, relatively high concentrations of such anions might have untoward effects on primary metabolism, *vide* a possible effect of citrate on the rate of glycolysis (Newsholme & Start, 1973).

However for proton or hydroxyl ion extrusion to be possible, there must be a source of cations or anions in the external medium to exchange. This certainly will not be so in aerial sporophores and hence the hydrogen and hydroxyl ions must either be neutralised in some other manner or transported to that part of the mycelium where their extrusion can take place. I should like to suggest that consideration be given to polyol synthesis and breakdown as a means of regulating cytoplasmic pH in fungi. Berry (1981) has pointed out that important sources or sinks for protons are the numerous redox reactions which occur within the cytoplasm. Thus the reaction,

$$\text{Sugar} + \text{NAD(P)H} + \text{H}^+ \rightleftharpoons \text{Polyol} + \text{NAD(P)}^+,$$

can be considered as a means of removing or producing hydrogen ions. The production of the translocatable polyol could mean that this compound is the vehicle by which hydrogen ions are moved through the cytoplasm to sites where the compound can be broken down such that the hydrogen ions released are extruded as protons into the external medium in exchange for (say) potassium ions.

Botton (1978) lists a significant number of publications relating to the promotion of growth of fungi by calcium and the necessity of the ion for the differentiation of reproductive structures in a wide range of lower and higher fungi. These observations together with the demonstration that calmodulin is certainly present in higher fungi (Charbonneau & Cormier, 1979; Anderson *et al.*, 1980; Grand, Nairn & Perry, 1980; Hubbard *et al.*, 1982) indicate that calcium is involved in the modulation of cellular activities in fungi. In view of this, we must anticipate, on the basis of what is known about the regulation by calcium of cellular activities in animals (Duncan, 1976), that there is a need to maintain the concentration of the ion at a relatively low level.

If calcium is swept into and through the mycelium by the flow of water and there is no reverse translocation, there must be some means of removing calcium ions from the cytoplasm. The most likely mechanism is by making calcium insoluble. Oxalic acid is known to be widespread in fungi particularly in the Agaricales (Cochrane, 1958) and the calcium salt is singularly insoluble. Sollins, Cromack, Fogel & Ching Yan Li (1981) have discussed the role of low-molecular-weight compounds, particularly oxalate, in the inorganic nutrition and ecology of fungi and the authors have pointed out that oxalate synthesis may also be important in helping to regulate cytoplasmic pH. Though we know little about the metabolic pathways leading to oxalate in fungi (Casselton, 1976), this view of the role of oxalate is supported by studies on organic acid synthesis in higher plants where it is clear that the process occurs in response to a decrease in intracellular acidity (Smith & Raven, 1979). However, the precipitation of calcium by oxalate will itself lead to release of hydrogen ions. While calcium oxalate is known to be present in basidiocarps, the most striking example of production of crystals of the compound in a fungal aerial structure is in the walls of the sporangiophores of *Mucor mucedo*. Urbanus, van den Ende & Koch (1978) unambiguously demonstrated the presence of the crystals by a variety of techniques and in this particular aerial structure there is no doubt about calcium ions being removed out of the cytoplasm. Furthermore, it may not be coincidental that mycelial cords of several Basidiomycotina are characteristically coated with calcium oxalate (see Thompson: Chapter 9).

Maintenance of turgor at unfavourable water potential

Clarke *et al.* (1980) have provided evidence for the role of pressure-driven water movement through mycelium of *Serpula lacrimans* in allowing it to extend over surfaces of unfavourable water potential. The hydrostatic pressure causing solution flow will assist in the maintenance of turgor but, as well, movement of a solution into growing hyphae will also reduce their water potential through a reduction in the internal osmotic potential. It may be significant that the disaccharide trehalose has been demonstrated to be or postulated to be a major constituent of the carbon translocated into the basidiocarps of *Agaricus bisporus* (Hammond & Nichols, 1976) and *Flammulina velutipes* (Kitamoto & Gruen, 1976), along the strands of *Serpula lacrimans*

(Brownlee & Jennings (1981*b*) and in the mycorrhizal fungus of *Dactylorchis purpurella* (Smith, 1967). In *A. bisporus* the trehalose is converted into mannitol in the basidiocarp and in *S. lacrimans* into arabitol at the mycelial front. When the disaccharide is converted to the hexitol or the pentitol the osmotic pressure of a solution due to the carbohydrates is approximately doubled. Thus at the end of the translocation pathway, trehalose can be metabolised so that one glucose moiety can be utilised in respiration and the provision of carbon skeletons for growth and the other moiety can contribute to maintenance of the osmotic potential.

The extent to which soluble carbohydrates make a major contribution to the osmotic potential for the growing hyphae at the end of a translocation pathway will depend on the extent to which other solutes are either translocated to or can be absorbed by the growing hyphae from the surrounding medium if it contains nutrients. Data on the composition of edible mushrooms (Crisan & Sands, 1978) give some clues as to the strategy whereby basidiocarps generate turgor in hyphae at the end of the translocation stream. Of course the extent to which other solutes brought in by the translocation stream make a contribution to the osmotic potential of the growing hyphae could determine the fate of that soluble carbohydrate which is also translocated. The increase of glycogen in developing caps of basidiocarps of *Coprinus cinereus* (Moore, Elhiti & Butler, 1979) may be due to a need to remove soluble carbohydrates from solution because otherwise there would be an unfavourable osmotic potential.

Effects outside the mycelium

When considering water flux through mycelia, we must realise that the movement of nutrients which is affected by such a flux will not be confined to the mycelium itself. The flux of water will extend out into the surrounding medium which is providing the nutrients for the source of translocation and this flux will draw mobile solutes to the mycelium surface. Bulk water flow through the soil brought about by transpiration is known to be important in drawing ions to the root surface (Nye & Tinker, 1977); and Bray (1954) has indicated how competition between roots for mobile nutrients could be brought about by the competing water flow to adjoining plants. Similar effects between adjacent mycelia seem plausible; whether or not they occur remains to be seen. It is conceivable that the increased growth of grasses associated with basidiocarps in a 'fairy ring' may in part be due to the flow of nutrients brought about by the transpirational flux to the basidiocarps.

Conclusions

The mammalian physiologist is blessed with what can be regarded in many ways as a conceptually simple system for study; that is an organism built of discrete organs with characteristic functions integrated by messages in the form of nerve impulses and hormones. The higher-plant physiologist is less fortunate because the object of study, since it is static, cannot move away from an unfavourable environment. So the higher plant, though built up of the individual organs of root, stem and leaves, responds to the environment in both its growth and everyday physiology in a manner which is much less readily described at a first approximation than is the case for a mammal. Indeed we speak of plasticity as being a very important element in the behaviour of a higher plant (Jennings, 1977). Nevertheless the presence of discrete organs which must act in concert has perforce made higher-plant physiologists consider the processes of integration. While the concepts of hormonal control are now somewhat suspect (Trewavas, 1981, 1982) there is no doubt that water flows via the phloem and xylem are an important element in the coordinated activity of the whole plant.

Many fungi do not produce discrete organs or tissues and a fungus often appears to avoid major environmental perturbations by producing resistant dormant structures or by dispersing itself by means of spores. Whether it is because of these features or lack of much positive evidence, there is no doubt that fungal physiologists even less than their higher-plant counterparts give little thought to the three-dimensional integration of activity within the fungal mycelium. While this chapter has focused inevitably on those fungi which produce macroscopic organs, particularly the basidiocarp, there is sufficient evidence for a wide variety of fungi that the signals which are necessary for integration of activity within a mycelium can be carried by water flow through the various hyphae. I hope that the chapter will encourage others to think about these matters and produce further information.

Finally those who may be still doubtful about the reality of water acting as a signal to alter growth, should note the studies on the avoidance responses of sporangiophores of *Phycomyces*, that is their ability to change direction as they approach another object. The response was first observed by Elfring (1881) and rediscovered by Shropshire (1962). Very recently Gamow & Bottger (1982) have demonstrated by some very elegant experiments that the hypothesis that the sporangiophores respond to some self-emitted gas is correct and that by far the most likely candidate for it is water vapour, which manifests its action through hydration of the cell wall. I doubt whether water vapour would have been put high in the list of putative candidates for the gas involved by those who were concerned with the early stages of formulating the hypothesis for the mechanism of the avoidance action. Indeed Russo, Halloran & Gallori (1977) believed that the gas is ethylene. That water vapour can act in this unexpected manner should make us more receptive to the idea of other roles for water controlling the functioning of fungal mycelia.

One example of a possible other role will suffice here. There have been several references by other authors to hyphae moving chemotropically toward each other. If water is ultrafiltered through a tip or a local area of wall and membrane, as has been observed for *Serpula lacrimans* (as described earlier), then the tropism might be generated by an osmotic gradient in the medium and the directed movement of the hyphal tip brought about by a change in the degree of hydration of a local part of the wall, as is suggested above for the avoidance reaction. This means that the search for the involvement of a specific solute in these tropic movements might be an unproductive endeavour.

I wish to thank Drs Alan Rayner and Wendy Thompson for their valuable comments on the initial drafts of this chapter.

References

Allen, E. D., Lowry, R. J. & Sussman, A. S. (1974). Accumulation of microfilaments in a colonial mutant of *Neurospora crassa*. *Journal of Ultrastructural Research*, **48**, 455–64.

Allen, E. D. & Sussman, A. S. (1978). Presence of an actin-like protein in mycelium of *Neurospora crassa*. *Journal of Bacteriology*, **135**, 713–16.

Allen, N. S. (1974). Endoplasmic filaments generate the motive force for rotational streaming in *Nitella*. *Journal of Cell Biology*, **63**, 270–87.

Allen, R. D. (1964). Cytoplasmic streaming and locomotion in marine foraminifera. In *Primitive Motile Systems in Cell Biology*, ed. R. D. Allen & N. Kamiya, pp. 407–32. New York: Academic Press.

Anderson, J. M., Charbonneau, H., Jones, H. P., McCann, R. O. & Cormier, M. J. (1980). Characterisation of the plant nicotinamide adenine dinucleotide kinase activator protein and its identification as calmodulin. *Biochemistry*, **19**, 3113–20.

Arthur, J. C. (1897). The movement of protoplasm in coenocytic hyphae. *Annals of Botany*, **11**, 491–507.

Bergman, K., Burke, P. V., Cerda-Olmedo, E., David, C. N., Delbruck, M., Foster, K. W., Goodell, E. W., Heisenberg, M., Meissner, G., Zalokar, M., Dennison, D. S. & Shropshire, Jr. W. (1969). *Phycomyces*. *Bacteriological Reviews*, **33**, 99–157.

Berry, M. N. (1981). An electrochemical interpretation of metabolism. *FEBS Letters*, **134**, 133–8.

Borst-Pauwels, G. W. F. H. (1981). Ion transport in yeast. *Biochimica et Biophysica Acta*, **650**, 88–127.

Botton, B. (1978). Influence of calcium on the differentiation and growth of aggregated organs in *Sphaerostilbe repens*. *Canadian Journal of Microbiology*, **24**, 1039–47.

Bray, R. H. (1954). A nutrient mobility concept of soil–plant relationships. *Soil Science*, **78**, 9–22.

Brownlee, C. & Jennings, D. H. (1981*a*). Further observations on tear or drop formation by mycelium of *Serpula lacrimans*. *Transactions of the British Mycological Society*, **77**, 33–40.

Brownlee, C. & Jennings, D. H. (1981*b*). The content of soluble carbohydrates and their translocation in mycelium of *Serpula lacrimans*. *Transactions of the British Mycological Society*, **77**, 615–19.

Brownlee, C. & Jennings, D. H. (1982*a*). Long distance translocation in *Serpula lacrimans*: velocity estimates and the continuous monitoring of induced perturbations. *Transactions of the British Mycological Society*, **79**, 143–8.

Brownlee, C. & Jennings, D. H. (1982*b*). The pathway of translocation in *Serpula lacrimans*. *Transactions of the British Mycological Society*, **79**, 401–7.

Buller, A. H. R. (1933). *Hyphal Fusions and Protoplasmic Streaming in Higher Fungi, together with an account of the production and liberation of Spores in* Sporobolomyces, Tilletia *and* Sphaerobolus. *Researches in Fungi, 5*, London: Longman Green & Co.

Bu'Lock, J. D. (1975). Secondary metabolism in fungi and its relationships to growth and development. In *The Filamentous Fungi*, vol. 1, *Industrial Mycology*, ed. J. E. Smith & D. R. Berry, pp. 33–58. London: Edward Arnold.

Bu'Lock, J. D., Hamilton, D., Hulme, M. A., Powell, A. J., Smalley, H. M., Shepherd, D. D. & Smith, G. N. (1965). Metabolic development and secondary biosynthesis in *Penicillium urticae*. *Candian Journal of Microbiology*, **11**, 765–78.

Burnett, J. H. & Trinci, A. P. J. (1979). *Fungal Walls and Hyphal Growth*. Cambridge: Cambridge University Press.

Canevascini, G., Coudray, M.-R., Rey, J.-P., Southgate, R. J. G. & Rossomando, E. F. (1979). Induction and catabolite repression of cellulase synthesis in the thermophilic fungus, *Sporotrichum thermophile*. *Journal of General Microbiology*, **110**, 291–303.

Casselton, P. J. (1976). Anaplerotic pathways. In *The Filamentous Fungi*, vol. 2, *Biosynthesis and Metabolism*, ed. J. E. Smith & D. R. Berry, pp. 121–36. London: Edward Arnold.

Chanter, D. O. & Thornley, J. H. M. (1978). Mycelial growth and the initiation and growth of sporophores in the mushroom crop: a mathematical model. *Journal of General Microbiology*, **106**, 55–65.

Charbonneau, H. & Cormier, M. J. (1979). Purification of plant calmodulin by fluphenazine-sepharose affinity chromatography. *Biochemical and Biophysical Research Communications*, **90**, 1039–47.

Clarke, R. W., Jennings, D. H. & Coggins, C. R. (1980). Growth of *Serpula lacrimans* in relation to water potential of substrate. *Transactions of the British Mycological Society*, **75**, 271–80.

Coggins, C. R., Jennings, D. H. & Clarke, R. W. (1980). Tear or drop formation by mycelium of *Serpula lacrimans. Transactions of the British Mycological Society*, **75,** 63–7.

Cochrane, V. W. (1958). *Physiology of Fungi*, New York, London: John Wiley & Sons.

Colotello, N. (1978). Fungal exudates. *Canadian Journal of Microbiology*, **24,** 1173–81.

Colotello, N., Sumner, J. L. & Voegelin, W. S. (1971). Presence of sacs enveloping the liquid droplets on developing sclerotia of *Sclerotinia sclerotiorum* (Lib.) De Bary. *Canadian Journal of Microbiology*, **17,** 300–1.

Cooper, K. M. & Tinker, P. B. (1981). Translocation and transfer of nutrients in vesicular-arbuscular mycorrhizas. IV. Effect of environmental variables on movement of phosphorus. *New Phytologist*, **88,** 327–39.

Cowan, M. C., Lewis, B. G. & Thain, J. F. (1972). Uptake of potassium by the developing sporangiophore of *Phycomyces blakesleeanus. Transactions of the British Mycological Society*, **58,** 113–16.

Cox, G., Moran, K. J., Sanders, F., Nockolds, C. & Tinker, P. B. (1980). Translocation and transfer of nutrients in vesicular-arbuscular mycorrhizas. III. Polyphosphate granules and phosphorus translocation. *New Phytologist*, **84,** 649–59.

Crisan, E. V. & Sands, A. (1978). Nutritional value. In *The Biology and Cultivation of Edible Mushrooms*, ed. S. T. Chang & W. A. Hayes, pp. 137–68. London, New York, San Francisco: Academic Press.

Davies, D. D. (1973). Control of and by pH. In *Rate Control of Biological Processes*, 27th Symposium of the Society for Experimental Biology, August 1972, ed. D. D. Davies, pp. 513–29. Cambridge: Cambridge University Press.

De Vries, H. (1885). Über die Bedeutung der Circulation und der Rotation des Protoplasma für Stofftransport in der Pflanze. *Botanische Zeitung*, **43,** 1–6, 16–26.

Duddridge, J. A., Malibari, A. & Read, D. J. (1980). Structure and function of mycorrhizal rhizomorphs with special reference to their role in water transport. *Nature*, **287,** 834–6.

Duncan, C. J. (1976). *Calcium in Biological Systems*, 30th Symposium of the Society for Experimental Biology, September 1975. Cambridge: Cambridge University Press.

Elfring, F. (1881). En obeaktad känslighet hos *Phycomyces. Botanisker Notiser*, **4,** 105–7.

Fenn, P., Choi, S. & Kirk, T. K. (1981). Ligninolytic activity of *Phanerochaete chrysosporium*: physiology of suppression by NH_4^+ and L-glutamate. *Archives of Microbiology*, **130,** 66–77.

Fenn, P. & Kirk, T. K. (1981). Relationship of nitrogen to the onset and suppression of ligninolytic activity and secondary metabolism in *Phanerochaete chrysosporium. Archives of Microbiology*, **130,** 59–65.

Franks, F. & Eagland, D. (1975). The role of solvent interactions in protein conformation. *Critical Reviews of Biochemistry*, **3,** 165–219.

Gamow, R. I. & Bottger, B. (1982). Avoidance and rheotropic responses in *Phycomyces.* Evidence for an "avoidance gas" mechanism. *Journal of General Physiology*, **79,** 835–48.

Garrett, S. D. (1970). *Pathological Root-infecting Fungi.* Cambridge: Cambridge University Press.

Gisi, U. & Zentmyer, G. A. (1980). Mechanism of zoospore release in *Phytophthora* and *Pythium. Experimental Mycology*, **4,** 362–77.

Grand, R. J. A., Nairn, A. C. & Perry, S. V. (1980). The preparation of calmodulins from barley (*Hordeum* sp.) and basidiomycete fungi. *Biochemical Journal*, **185,** 755–60.

Grehn, J. (1932). Untersuchungen über Festalt und Funktion der Sporangien träger bei den Mucorineen. II. Der Wasser und Stofftransport. *Jahrbuch für wissenschaftliche Botanik*, **76**, 167–207.
Grootwassink, J. W. D. & Gaucher, G. M. (1980). *De novo* biosynthesis of secondary metabolism enzymes in homogeneous cultures of *Penicillium urticae*. *Journal of Bacteriology*, **141**, 443–55.
Hammond, J. B. W. & Nichols, R. (1976). Carbohydrate metabolism in *Agaricus bisporus* (Lange) Sing.: changes in soluble carbohydrates during growth of mycelium and sporophores. *Journal of General Microbiology*, **93**, 309–20.
Harold, F. M. (1977). Ion currents and physiological functions in micro-organisms. *Annual Review of Microbiology*, **31**, 181–203.
Holligan, P. M. & Jennings, D. H. (1972*a*). Carbohydrate metabolism in the fungus *Dendryphiella salina*. I. Changes in the levels of soluble carbohydrates during growth. *New Phytologist*, **71**, 569–82.
Holligan, P. M. & Jennings, D. H. (1972*b*). Carbohydrate metabolism in the fungus *Dendryphiella salina*. III. The effect of the nitrogen source on the metabolism of (1–^{14}C)- and (6–^{14}C)-glucose. *New Phytologist*, **71**, 1119–33.
Hubbard, M., Bradley, M., Sullivan, P., Shepherd, M. & Forrester, I. (1982). Evidence for the occurrence of calmodulin in the yeasts *Candida albicans* and *Saccharomyces cerevisiae*. *FEBS Letters*, **137**, 85–8.
Hulme, M. A. & Stranks, D. W. (1971). Regulation of cellulose production by *Myrothecium verrucaria* grown on non-cellulosic substrate. *Journal of General Microbiology*, **69**, 145–55.
Ingold, C. T. (1971). *Fungal Spores: Their Liberation and Dispersal*. Oxford: Clarendon Press.
Isaac, S., Briarty, L. G. & Peberdy, J. F. (1979). The stereology of protoplasts from *Aspergillus nidulans*. In *Advances in Protoplast Research*, Proceedings of 5th International Protoplast Symposium, Szeged, Hungary, pp. 213–19. Budapest: Publishing House of the Hungarian Academy of Sciences.
Isaac, S., Ryder, N. S. & Peberdy, J. F. (1978). Distribution and activation of chitin synthase in protoplast fractions released during the lytic digestion of *Aspergillus nidulans* hyphae. *Journal of General Microbiology*, **105**, 45–50.
Jennings, D. H. (1974). Sugar transport into fungi: an essay. *Transactions of the British Mycological Society*, **62**, 1–24.
Jennings, D. H. (1976*a*). Transport and translocation in filamentous fungi. In *The Filamentous Fungi*, vol. 2, *Biosynthesis and Metabolism*, ed. J. E. Smith & D. R. Berry, pp. 32–64. London: Arnold.
Jennings, D. H. (1976*b*). Transport in fungal cells. In *Encyclopaedia of Plant Physiology*, new series, vol. 2, *Transport in Plants, Part A, Cells*, ed. U. Lüttge & M. G. Pitman, pp. 189–228. Berlin, Heidelberg, New York: Springer-Verlag.
Jennings, D. H. (1977). Introduction. In *Integration of Activity in the Higher Plant*, 31st Symposium of the Society for Experimental Biology, September, 1976, ed. D. H. Jennings, pp. 1–5. Cambridge: Cambridge University Press.
Jennings, D. H. (1979). Membrane transport and hyphal growth. In *Fungal Walls and Hyphal Growth*, ed. J. H. Burnett & A. P. J. Trinci, pp. 279–94. Cambridge: Cambridge University Press.
Jennings, D. H. (1982). The movement of *Serpula lacrimans* from substrate to substrate over nutritionally inert surfaces. In *Decomposer Basidiomycetes*, ed. J. Hedger & J. Frankland, pp. 91–108. Cambridge: Cambridge University Press.
Jennings, D. H., Thornton, J. D., Galpin, M. F. J. & Coggins, C. R. (1974). Translocation in fungi. In *Transport at the Cellular Level*, 28th Symposium of the Society for Experimental Biology, August 1973, ed. M. A. Sleigh & D. H. Jennings, pp. 139–56. Cambridge: Cambridge University Press.

Keyser, P., Kirk, T. K. & Zeikus, J. G. (1978). Ligninolytic enzyme system of *Phanerochaete chrysosporium*: synthesis in the absence of lignin in response to nitrogen starvation. *Journal of Bacteriology*, **135**, 790–7.

Kitamoto, Y. & Gruen, H. E. (1976). Distribution of cellular carbohydrates during development of the mycelium and fruit bodies of *Flammulina velutipes*. *Plant Physiology*, **58**, 485–91.

Larcher, W. (1980). *Physiological Plant Ecology*, 2nd edn. Berlin, Heidelberg, New York: Springer-Verlag.

Lundquist, K. & Kirk, T. K. (1978). *De novo* synthesis and decomposition of veratryl alcohol by a lignin-degrading basidiomycete. *Phytochemistry*, **17**, 1676.

McGee-Russell, S. M. (1974). Dynamic activities and labile microtubules in cytoplasmic transport in the marine foraminiferan *Allogromia*. In *Transport at the Cellular Level*, 28th Symposium of the Society for Experimental Biology, August 1973, ed. M. A. Sleigh & D. H. Jennings, pp. 157–89. Cambridge: Cambridge University Press.

Moore, D., Elhiti, M. M. Y. & Butler, R. D. (1979). Morphogenesis of the carpophore of *Coprinus cinereus*. *New Phytologist*, **83**, 695–722.

Newsholme, E. A. & Start, C. (1973). *Regulation in Metabolism*. London: John Wiley.

Nobel, P. S. (1975). Effective thickness and resistance of the air boundary layer adjacent to spherical plant parts. *Journal of Experimental Botany*, **26**, 120–30.

Nye, P. H. & Tinker, P. B. (1977). *Solute Movement in the Soil-Root System*. Oxford, London, Edinburgh, Melbourne: Blackwell Scientific Publications.

Oort, A. J. P. & Roelofsen, P. A. (1932). Spiralwachstum, Wandbau und Plasmaströmung bei *Phycomyces*. *Proceedings of the Koniklijke Nederlandse Akademie van Wetenschappen*, **35**, 898–908.

Panek, A. D. (1975). Trehalose synthesis during starvation of baker's yeast. *European Journal of Applied Microbiology*, **2**, 39–46.

Panek, A. D. & Matoon, J. R. (1977). Regulation of energy metabolism in *Saccharomyces cerevisiae*. Relationship between catabolite repression, trehalose synthesis and mitochondrial development. *Archives of Biochemistry and Biophysics*, **183**, 306–16.

Plunkett, B. E. (1956). The influence of factors of the aeration complex and light upon fruit-body form in pure cultures of an agric and a polypore. *Annals of Botany*, **20**, 563–86.

Plunkett, B. E. (1958). Translocation and pileus formation in *Polyporus brumalis*. *Annals of Botany*, **22**, 237–49.

Raven, J. A. & Smith, F. A. (1973). The regulation of intracellular pH as a fundamental biological process. In *Ion Transport in Plants*, ed. W. P. Anderson, pp. 271–8. London, New York: Academic Press.

Reid, I. D. (1979). The influence of nutrient balance on lignin degradation by the white-rot fungus *Phanerochaete chrysosporium*. *Canadian Journal of Botany*, **57**, 2050–8.

Russo, V. E. A., Halloran, B. & Gallori, E. (1977). Ethylene is involved in auto-chemotropism of *Phycomyces*. *Planta*, **136**, 61–7.

Schröter, A. (1905). Über Protoplasmastormung bei Mucorineen. *Flora, Jena*, **45**, 1–20.

Schütte, K. H. (1956). Translocation in fungi. *New Phytologist*, **55**, 164–82.

Schwitzguebel, J.-P. & Palmer, J. M. (1982). Properties of mitochondria as a function of the growth stages of *Neurospora crassa*. *Journal of Bacteriology*, **149**, 612–19.

Shropshire, Jr W. (1962). The lens effect and phototropism of *Phycomyces*. *Journal of General Physiology*, **45**, 949–58.

Slayman, C. L. & Gradmann, D. (1975). Electrogenic proton transport in the plasma membrane of *Neurospora*. *Biophysical Journal*, **15**, 968–71.

Smith, F. A. & Raven, J. A. (1979). Intracellular pH and its regulation. *Annual Review of Plant Physiology*, **30**, 289–311.

Smith, S. E. (1967). Carbohydrate translocation in mycorrhizal fungi. *New Phytologist*, **66**, 371–8.

Sollins, P., Cromack, Jr K., Fogel, R. & Ching Yan Li (1981). Role of low-molecular-weight organic acids in the inorganic nutrition of fungi and higher plants. In *The Fungal Community*, ed. D. T. Wicklow & G. C. Carroll, pp. 607–19. New York, Basel: Marcel Dekker.

Ternetz, C. (1900). Protoplasmabewegung und Fruchtkörperbildung bei *Ascophanus carneus. Jahrbuch für wissenschaftliche Botanik*, **35**, 273–312.

Trewavas, A. (1981). How do plant growth substances work? *Plant Cell and Environment*, **4**, 203–28.

Trewavas, A. (1982). Possible control points in plant development. In *The Molecular Biology of Plant Development*, ed. H. Smith & D. Grierson, pp. 7–27. Oxford, London, Edinburgh, New York, Melbourne: Blackwell Scientific Publications.

Turian, G. (1978). The 'spitzenkorper', centre of reducing power in the growing hyphal apices of two septomycetous fungi. *Experientia*, **34**, 1277–9.

Turian, G. (1979). Polarity of elongation growth generated and sustained by anisotropic distribution of the protons ejected from mitochondria into the cytosol of hyphal apices (*Neurospora* model). *Archives des Sciences, Genève*, **32**, 251–4.

Turian, G. (1981). Low pH in fungal bud initials. *Experientia*, **37**, 1278–9.

Urbanus, J. F. L. M., van den Ende, H. & Koch, B. (1978). Calcium oxalate crystals in the wall of *Mucor mucedo. Mycologia*, **70**, 829–42.

Wagner, G. (1979). Actomyosin as a basic mechanism of movement in animals and plants. In *Encyclopedia of Plant Physiology*, new series, vol. 7, *Physiology of Plant Movement*, ed. W. Haupt & M. E. Feinleb, pp. 114–26. Berlin, Heidelberg, New York: Springer-Verlag.

Wyn Jones, R. G., Storey, R., Leigh, R. A., Ahmed, N. & Pollard, A. (1977). A hypothesis on cytoplasmic osmoregulation. In *Regulation of Cell Membrane Activities in Plants*, ed. E. Marrè & O. Ciferri, pp. 121–36. Amsterdam: Elsevier/North Holland Biomedical Press.

8

Morphogenesis of the *Serpula lacrimans* colony in relation to its function in nature

S. C. WATKINSON
Botany Department, University of Oxford, South Parks Road, Oxford OX1 3RA, UK

As is emphasised by several authors in this volume, basidiomycete mycelium often persists for months or years, and during its growth there is differentiation of hyphae and development of structures which perform important functions in the mature colony, such as translocation and development of fruit bodies for spore dispersal. These functions must often be closely integrated with the process of extracting nutritients from a recalcitrant, insoluble resource such as lignocellulose.

The term 'morphogenesis' is here used in the sense proposed by Esser (1977) to mean the sum of individual hyphal differentiation that results in the development of a multihyphal structure. It is characteristic of fungi that they usually possess a programme of morphogenesis (the 'modes' of Gregory: Chapter 1) which may or may not occur, and which is not an indispensable part of the life of the organism. In this 'facultative morphogenesis' they differ from higher plants and animals, in which life cannot be carried on without the development of a multicellular body, and from bacteria in which true multicellular organisation does not occur. This programmed morphogenesis is often set in motion by environmental changes. The ability to respond to a specific environmental condition or set of conditions can itself be seen as part of the programmes of morphogenesis in the fungi. This feature of the fungal kingdom, that a simple signal elicits morphogenesis, gives these organisms a special interest in the wider context of cell biology.

This review is devoted to the morphogenesis of *Serpula lacrimans* and its relationship to the environment and to environmental changes; and to the subsequent functioning of its developed colonies in relation to nutrition. There are several advantages in choosing *S. lacrimans* to study the relationship between the morphogenesis of a fungal colony and its functions in nature. The fungus has been studied experimentally for over a century, as its depredations were early seen as a consequence of its particular biology (e.g. Dickson, 1837). Thus there is a larger literature than usually exists for a single fungal species. Also, it grows under conditions unusually free from biotic competition – which removes a major complication experienced in studying the auto-ecology of most fungi (cf. Frankland: Chapter 11).

S. lacrimans grows from one nutrient resource to another over intervening non-nutrient areas. This ability is common to many Basidiomycetes (cf. Thompson: Chapter 9; Read: Chapter 10) but is unusually well developed in *S. lacrimans* since the distances travelled may be many metres and the regions traversed devoid of nutrients. One might therefore expect that the form of the colony would be regulated, both at first during the laying down of the initial framework, and later, during its development into multihyphal structures, so that reserves of scarce substances are not wasted by overabundant growth in a direction in which there is no further growth substrate. The importance of C:N ratio and the nature of the N source in regulating morphogenesis in Basidiomycetes have been demonstrated by many workers (Watkinson, 1979). I have argued that specific responses to particular environmental conditions can be taken as an indication that these conditions are important in the life of the organism, and more specifically that the responses to N nutrition of *S. lacrimans* must be an important part of its biology. In nature, morphogenetic responses of *S. lacrimans* to changing N conditions might be seen as an essential part of its N economy during the following cycle of events: exploitation of a substratum ('food base') such as wood with production of enzymes and mycelial growth; extension of the colony from this food base over non-nutrient areas; colonisation of a second food base, with enzyme production and mycelial growth; removal, transport and re-utilisation of N from the first food base; and further extension of the colony from the second food base. Evidence bearing on the nature of some of these processes is presented below.

The form of the colony extending from a food base

This has been shown to be affected by the N content of the food base, with an optimum N level, below or above which the area covered by the extending colony is reduced (Fig. 1). All the colonies grew from circular discs over a non-nutrient substratum with 25 mg sucrose as sole C source and from 0 to 1.3 mg N as $NaNO_3$. These nutrients were given in fresh agar discs placed on top of the inoculum discs. Initial growth of the colonies was mainly in the form of aerial mycelium. By five days after addition of the nutrient discs, the effect of the nutrient composition of the food-base had become evident at the growing-colony margin. The much more rapid extension at the optimum N level than that at other levels resulted in a more extensive colony at the end of the experimental period even though extension rates had slowed down towards the end of the experiment. A similar pattern of response has been obtained with glutamine as N source, confirming that the response is to N. This result appears to have been due to variation in the proportion of lateral branches to leaders so that on a N-rich food base most of the mycelium grew over and around the food base and exploited it, while on a food base where N was scarce, but adequate for some growth, there was extension of a few hyphae. Presumably these are scavenging for N with maximum economy. At the lowest N levels no hyphae extended over the substratum. How the amount of N in the food base regulates the relative growth rates of different hyphae is not known. While it must be admitted that circular colonies growing from agar discs are dissimilar to those of irregular shape that develop from natural insoluble cellulosic materials, there is a strong argument that conditions near the mycelial margin are nevertheless close enough to what might be happening in nature. Growth is at a solid/gas interface, at a distance from a food

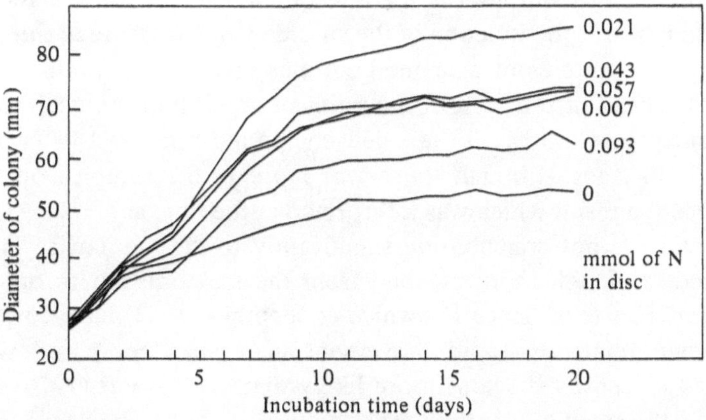

Fig. 1. Linear extension of colonies over a non-nutrient surface from an agar food base containing various $NaNO_3$ levels. Figures are mmol N present as $NaNO_3$ in each disc; volume of discs was 0.5 ml each and they contained, per 1: agar 20 g, KH_2PO_4, 1 g; $MgSO_4{\cdot}7H_2O$, 0.5 g; $FeSO_4{\cdot}7H_2O$, 10 mg; and sucrose 50 g.

base, and the hyphae differentiate into leaders, vessels and tendrils as described by Butler (1958).

Many of the earlier data that exist on the morphogenetic effects of C:N ratio and on the chemical nature of the N source are from experiments where colonies were grown over agar or static liquid media (Watkinson, 1975). However, such experiments illuminate metabolic effects on morphogenesis. The abundance of mycelial strands was proportional to C:N ratio, as was pigment production. Organic N sources such as casein hydrolysate, peptone and sodium aspartate caused a reticulum of thick strands to develop and yellow pigment to be produced while inorganic N sources [$(NH_4)_2SO_4$ and $NaNO_3$] support a thin mycelium in which strands developed early and were mainly radial in orientation. In shaken submerged culture, nitrate-grown mycelium was relatively fragmented and thin-walled compared with that grown on organic N. In static liquid culture, it was observed that relatively large amounts of polysaccharide were secreted into the medium when NO_3^- was the sole N source; also ^{14}C from ^{14}C-labelled glucose fed to the culture reappeared in the medium much faster on NO_3^- media than aspartate indicating release of solutes from hyphae. Incidentally, this solute release may also explain the results of an earlier experiment on phosphorus translocation by stranded and unstranded mycelium (Watkinson, 1975). In a split-plate experiment, agar containing $NaNO_3$ as N source and malt agar were used in the same Petri dish to grow a colony stranded on one side but not on the other. Labelled medium was fed to the inoculum disc at the junction of the stranded and unstranded sides of the colony and samples of agar medium plus mycelium cut out of the cultures at a fixed distance from the point of application on each side. The distance travelled by the labelled compound appeared to be the same on both sides although there was more in the samples on the stranded side, a result which was interpreted at the time as showing that the strands were not contributing significantly to the amount of label transported, although more escaped from the leaky mycelium on the nitrate agar. However, since Brownlee & Jennings (1982) have conclusively shown relatively rapid movement in strands compared with unstranded mycelium it seems more likely that, as Burnett (1976) has suggested, the results of this experiment could be due to faster translocation in stranded mycelium followed by rapid leakage of the label into the nitrate agar used to induce strand formation.

The kind of nutrients available in the natural environment of the fungus are clearly likely to have major effects on its metabolism and also on its morphogenesis, since both are closely linked in fungi. The colony of *S. lacrimans,* like that of many Basidiomycetes, develops as an initial framework resulting from outward growth of hyphae from the food-base inoculum. This has many of the usual features described for 'typical' colonies; the emergence of a hierarchy of branches with regular branching angles and some appearance of apical dominance; formation of clamps and septa; but also in *S. lacrimans* differentiation of hyphae as described by Falck (1912), Butler (1957, 1958), and Coggins, Hornung, Jennings & Veltkamp (1980). Later this framework becomes elaborated; for example faster-growing hyphae appear (Coggins *et al.*, 1980) and strands develop round some leading hyphae while other hyphae autolyse, and the final form of the colony thus becomes quite different from the circular array of separate hyphae seen at first. Both initial growth and later elaborations are subject to environmental control, as the experiments described above show. What, at the cellular, or hyphal, level are the changes that occur during this structural elaboration?

Differentiation of hyphae in mycelial strands

The early development of the colony over a non-nutrient (glass) surface has been described by Butler (1958) and the multihyphal development of strands by Hartig & von Tubeuf (1902), Falck (1912), and Butler (1957, 1958). Mycelial strands at various stages of development, from initials to mature, functioning strands in natural outbreaks of dry rot, have also been the subject of a recent combined light- and electron-microscopic study (Jennings & Watkinson, 1982). In this, the observation of Butler was confirmed that there was hyphal differentiation even before aggregation of hyphae, with differential staining of main and tendril hyphae with Nile blue showing that the relatively wide, thin-walled, empty vessel hyphae contained neutral lipid in the wall (stained pink) and the later associated tendril hyphae contained acidic lipid particles in their cytoplasm (stained dark blue). Transmission electron micrographs of early associations between hyphae showed abundant glycogen rosettes, digestible with β-amylase and between the hyphae an extrahyphal matrix. Two-month-old strands from a culture on a sucrose/nitrate medium were shown by scanning electron microscopy (SEM) to have smooth surfaces from which the individual hyphae could not readily be discerned. Transmission electron microscopy (TEM) showed that the extrahyphal material in such strands was very abundant between and around individual hyphae, the walls of which were considerably interdigitated, although not anastomosed, so that they could be dissected apart by maceration in KOH solution (Corner, 1932). This adhesive material was not analysed; it resembles in appearance and distribution that seen in rhizomorphs of *Sphaerostilbe repens* by Botton & Dexheimer (1977) which was found to be α(1–4) and β(1–4) glucan. A much older strand taken from an outbreak of dry rot in a house was also examined. Strands, probably several months old, were found crossing the floor under the carpet, and the one which was removed for electron microscopy was connected to a developing sporophore and probably functional. In this an extrahyphal matrix was seen to form the main part of the strand, and individual hyphae were only occasionally distinguishable. Scanning electron microscopy of this strand showed an outer layer of thin-walled hyphae, and some degree of zonation of the strand, with wide channels, apparently empty, in the centre; towards the edges these channels were narrower. Transmission electron micrographs showed fibre hyphae with extremely thick walls made of concentric lamellae, and almost-occluded lumina; these hyphae were most frequent towards the edge of the strand. In transverse sections a maximum density of approximately 5 fibres per mm^2 were recorded but there were none in the centre, so that the strand was in effect bounded by a rigid rube made of matrix material stiffened by fibres. Throughout the strand, widely scattered hyphae containing cytoplasm were seen occasionally. The wide central spaces appeared empty of contents in SEM, although in TEM a lattice-like appearance was sometimes seen in them. The spaces appeared to have arisen through the collapse and compression of hyphae.

These observations of the structure of mature strands may help to explain the earlier findings described above in which media with high C:N ratios with aspartate as sole N source were shown to cause development of thick strands and of pigment; while those with NO_3^- caused strands to be thinner, and suppressed pigment production. Perhaps a large excess of carbon over that required for respiration and initial growth is needed for the construction of the massive polysaccharide matrix of mature strands. Jennings (1982) has speculated that synthesis of thick walls 'may be a mechanism which removes excess carbohydrate from the translocation stream', necessary because the fungus inevitably takes up more carbohydrate than it needs since, in wood, cellulose has to be broken down to release N. Both pigment and matrix material seem to be products of secondary metabolism on carbon-rich media in this fungus. The observation that on NO_3^-, strands form earlier and are thinner, and that, as in cultures on media with low C:N ratios, no pigment develops, suggests that assimilation of nitrate decreases production of the extracellular polysaccharide involved in strand thickening. It might be this interference with synthesis that also causes mycelium grown on nitrate as sole N source to be leaky to solutes, thin-walled and easily fragmented in shake culture.

The structure of mature strands described above is consistent with the model described by Jennings (Chapter 7) for the mechanism of water and solute translocation in strands. The development of an inextensible tube encircling the strand has obvious importance for the maintenance of internal hydrostatic pressure (Jennings, Thornton, Galpin & Coggins, 1974) and it is significant that this structure arises in the older part of the strand near the food base where this internal pressure would be greatest.

The nature of the interface between hypha and wood, where solubilised substrate is taken up, has hardly been studied from a physiological point of view although Montgomery (1982) in an excellent review has discussed the problem of enzyme regulation in the heterogeneous environment of wood. The specific nature of the interaction between a particular basidiomycete fungus and the cells of the wood substrate has been amply shown (Liese, 1970; Levy, 1982; Montgomery, 1982), and characteristic types of enzyme action can be deduced from such morphological studies, as well as from purely physiological studies of the decay-causing fungus. To understand the physiology and biochemistry of the attack on wood, it is surely essential to investigate the structure of the interface between wood-cell wall and mycelium since the biochemical processes involved must be spatially heterogeneous, and their co-ordination *in vivo* at least partly dependent on the changing structure of this interface as it is degraded by hyphae moving over it. For example, Ander & Erikssen (1978) have shown synergism in wood breakdown by a white rot fungus between the different enzyme systems that attack cellulose and lignin, materials that are present in layers in wood cell walls and thus are likely to be encountered sequentially by a penetrating hypha. The erosion zones in tracheid walls figured by Liese (1970) from wood attacked by *Trametes pini* could only have been caused by hyphae that attacked first the lignified S_3 layer and then the predominantly cellulose S_2 layer.

It is not yet known how the interface between hyphae of *S. lacrimans* and wood changes during colonisation, utilisation and exhaustion of the substratum, but some observation of the colonisation stage has been made (Helsby, 1976; Jennings & Watkinson, 1982). When *S. lacrimans* colonised blocks of *Pinus sylvestris* in culture the first stage was a fan-like growth of the mycelial margin over the surface of the block from the initial point of contact. Following this, strands developed and thickened in the mycelium connecting the original inoculum to the block. Crystals appeared on the surface of the leading hyphae; this has previously been interpreted as the result of a sudden increase in the nutrient status of the leader hyphae, followed by leakage of substances some of which could be utilised in strand thickening (Watkinson 1971*a*) or released to the medium (Goksøyr, 1958). Sections of the wood showed hyphae attacking medullary ray cells, branching, and forming thin strands inside tracheids. Hyphae were seen apparently passing through those transverse holes which had been dissolved in tracheid walls, and occasional cavities were seen in the S_2 layer. There was very close contact between hyphae and the wood-cell wall, the intervening space being filled with fibrillar material. The surface of the tracheid wall was sometimes seen to be locally disrupted where a hypha was lying over it (Fig. 2). When the internal surfaces of tracheids were viewed by SEM, some hyphae appeared to be embedded in the wall, presenting an almost continuous surface with it. Hyphae in wood cells appeared to be autolysing, as evidenced by the presence of membranous sheets inside them without obvious organelles (Read, personal communication).

The association of part of the colony with its wood substratum is clearly intimate and long-lasting, involving highly developed morphogenetic responses by the hyphae. Formation of the association affects growth and morphogenesis in other parts of the colony. Not only do strands form in the connecting mycelium, but the newly-colonised food base becomes the new centre for radiating mycelial growth. This general phenomenon in the fungi is little understood. Its primary cause is evidently the encountering of a new nutrient source. The intracellular level of specific metabolites has been ascribed a regular role in fungal morphogenesis (Wright, 1973; Watkinson, 1979); for example, high levels of intrahyphal glutamic acid have been shown to induce coremium formation in *Penicillium claviforme*, a process which involves the onset of branching from mycelium behind the colony margin (Watkinson, 1981).

Functions of the mature colony in nitrogen economy

The morphogenetic responsiveness of the colony to N regime seems likely to reflect the need of *S. lacrimans* to regulate N movement and metabolism. Fungi whose sole nutrient source is wood need to conserve N (Levi & Cowling, 1969). The mycelium of *S. lacrimans* is able to extract, and concentrate the N from wood (Table 1). Much of the N in wood is in the form of protein (Baker, Laidlaw & Smith, 1970) and only about 10% is extractable with solvent and protease treatment (Merill & Cowling, 1966), probably because it is masked by layers of

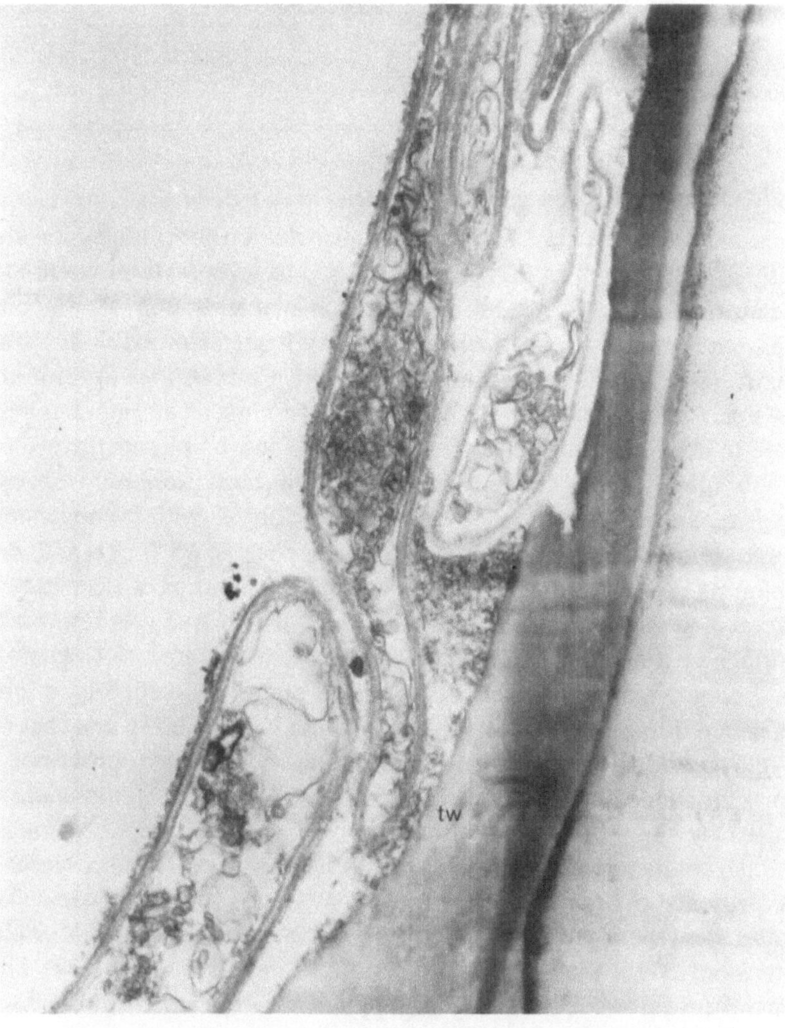

Fig. 2. Interface between hypha and tracheid wall (tw). Disruption of the wall in contact with the hypha. ×40 000. Reproduced by courtesy of L. Jennings from Helsby (now Jennings) (1976.)

Table 1. *Nitrogen content of* Pinus sylvestris *sapwood and of mycelium (mg g^{-1} dry wt of sample)*

Samples of wood of different ages from 13th internode of tree		12–13-year-old wood containing colonising mycelium	Mycelium grown from wood food base
0–3 years old	12–13 years old		
1.05	0.38	0.54	16.4
1.27	0.67	0.55	36.8
			23.6

Each column shows replicate samples.

lignin and cellulose. Hyphae can disrupt these layers (Fig. 2) and this must explain the ability of lignicolous fungi to grow well on wood as a sole nutrient source. Since nitrogen is not available in adequate amounts in all parts of the *Serpula lacrimans* colony it seems probable that this fungus must not only *extract* and *concentrate* it but also *transport* it and *re-use* it. In order to test the necessity of this cycling an attempt has been made to discover whether concentration of N in wood limits the growth of the fungus. The concept of limiting concentration of a specific nutrient, as originally applied to bacterial cultures with homogeneous undifferentiated growth, is not directly applicable to fungal colonies exploiting insoluble substrates, because this exploitation is a complex of several activities, the rate of any one of which may be limited by shortage of a nutrient. Garrett (1982) has demonstrated, for example, that cellulolysis by, and hyphal extension of, several cellulolytic fungi are affected to different degrees by changing N availability and that the effect of adding N on the relative rates of the two processes is characteristic of individual fungal species. Whether *S. lacrimans* is limited by the N levels in wood can only be determined by finding whether the separate processes involved in growth will proceed faster in the presence of more utilisable nitrogen than is found in wood. The matter was investigated by studying dry-weight increase, rates of hyphal extension over a non-nutrient substratum and over a cellulose filter paper substratum with nitrogen supplied uniformly in solution, and cellulolysis measured as percent loss of dry weight of cellulose filter paper. Each process was measured over a range of N concentrations spanning those in wood. The experimental conditions in each case were arranged so that each process studied could be measured in isolation from the rest and so that the amounts of N present could be controlled. It is assumed that these experiments, although remote from natural conditions as Dowding (1982) has pointed out, may nevertheless give insight into what is occurring during natural growth. Dry weight production and cellulolysis could both be increased in the experimental system by adding extra N at a concentration well above those found in wood. The existence of an optimum N concentration of 0.021 mM of N in a 0.5-ml disc for extension from an agar foodbase was described above. When extension was measured over cellulose filter paper as sole carbon source with uniform $NaNO_3$ solution at different concentrations as sole N source, extension was maximal at 0.11 M N, the highest concentration used. The shape of the curve of colony diameter against time also varied with N concentration, showing a statistically significant deviation from linearity (Fig. 3) at intermediate concentrations. This was interpreted as an initial fast extension while enough N was present in the mycelium from the inoculum to sustain rapid growth, then a slowing of extension as this N became insufficient, followed by a renewed burst of extension when the mycelium had concentrated enough N from the solution to enable it to resume cellulolysis and so release enough carbon substrate from the filter paper to continue.

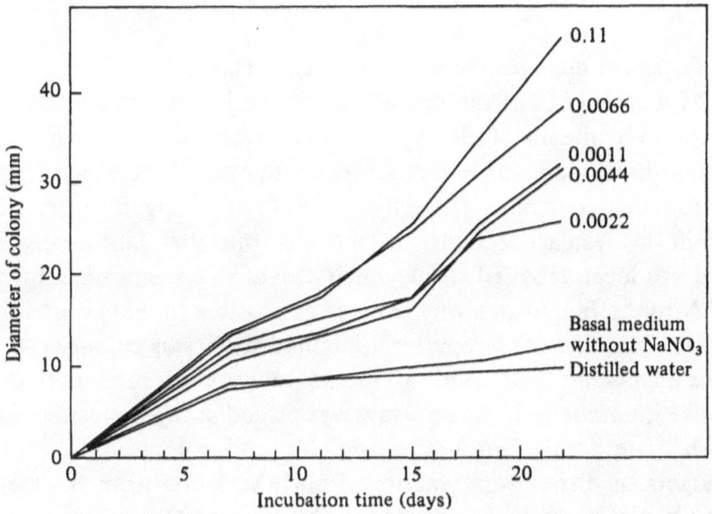

Fig. 3. Linear extension of colonies over cellulose filter paper as sole carbon source. Paper was saturated with 2 ml of a medium containing salts as in Fig. 1, and N as $NaNO_3$ at the molarities indicated. For analysis of variance, see Watkinson *et al.* (1981).

Not only did a N concentration greater than that in wood increase cellulolysis of filter paper, but it did so when supplied at a point in the colony separated from the site of cellulolysis. Altogether, the results suggest that concentration of the N in the mycelium was necessary for the breakdown of wood, and that N could be moved in the mycelium to a site where it was needed for cellulolysis or growth. Based on these results, a hypothesis is now advanced that N is cycled during mycelial progress from food base to food base and that the following processes occur:

(i) Concentration of N in mycelium exploiting a cellulosic substrate.

(ii) Export of N as this substrate becomes exhausted, into growing mycelium.

(iii) Import of N from the connecting mycelium into mycelium colonising a second cellulosic substrate where it is used for enzyme production and growth.

(iv) Re-export of N as it becomes exhausted from this second substrate, for example by resorption of N in the form of enzyme protein and products of autolysis.

According to this scheme, cycling of N is an essential part of the machinery used for exploitation of substrates in which C predominates over N.

Export of nitrogen from a natural substrate

Movement of N from one cellulosic food base into another was demonstrated by means of the experimental arrangement shown in Fig. 4. The food base used was barley straw containing ^{15}N, derived from a barley crop that had been treated with ^{15}N-containing fertiliser. The nitrogen in this was accordingly incorporated into the plant tissues and the latter produced labelled straw which provided a means of supplying labelled N to the fungus in a way as close as possible to that occurring in nature. One gramme of chopped ^{15}N-labelled straw was colonised by *S. lacrimans* mycelium and used to inoculate each experimental dish. Uncolonised, unlabelled, sterile straw was placed at the opposite end of each dish. The fungus grew towards this and colonised it, and at intervals sets of dishes were sacrificed and the straw from both ends dried and ball-milled to a powder for analysis by the Dumas method for total N and for percent enrichment of ^{14}N by ^{15}N. The amount of ^{15}N in the straw used as inoculum fell over the three months' experimental period showing that it was released by the fungus. The proportion of the total N in each dish which was in the second food base increased at each sampling period until at the end of the experiment 81% was in the second food base (Table 2). This result supports parts (ii) and (iii) of the hypothesis. The chemical form in which the N is mobile is as yet unknown.

There is some evidence that a dynamic equilibrium exists between a stored and a mobile form of N. When *S. lacrimans* was inoculated centrally on N-rich nutrient agar plates containing N as 0.2% sodium L-glutamate, and the mycelium harvested at intervals for 8 weeks, the major part of the total N taken up was at first, while the mycelium was growing over the agar, extractable in 80% ethanol and so was probably in the form of free amino acids (Fig. 5). After the mycelium had reached the edges of the plate at two weeks the proportion of its total N extractable in this way fell to one fifth of its previous maximum value and there was an increase in ethanol-insoluble N in the mycelium, suggesting that the soluble amino acids that were accumulated from the

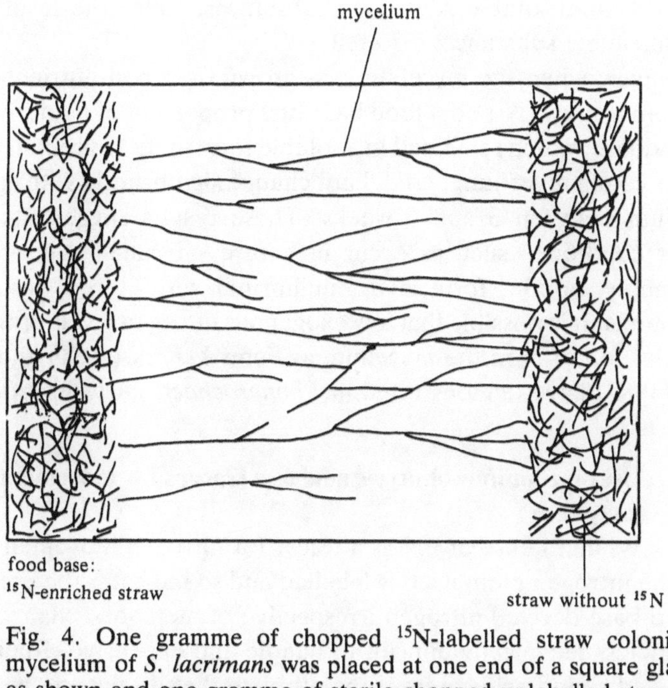

Fig. 4. One gramme of chopped ^{15}N-labelled straw colonised by mycelium of *S. lacrimans* was placed at one end of a square glass dish as shown and one gramme of sterile chopped unlabelled straw at the other. Mycelium grew across the glass surface (approximately 10 cm) and colonised the sterile straw. Ten replicate dishes were sacrificed for analysis at 0, 3.5, 10 and 13 weeks after the sterile straw had been reached.

Table 2. *Movement of N from an already colonised straw food base to a second, fresh one, via connecting mycelium (arranged as in Fig. 4), during 13 weeks following initial colonisation of the second food base.*

Weeks after colonisation of 2nd food base	Total weight of N in sample (mg)		% of total N in 2nd food base
	1st food base	2nd food base	
0	4.20 ± 1.08	3.56 ± 0.41	45.9
3.5	2.04 ± 0.41	4.25 ± 0.81	67.6
10	1.94 ± 0.43	5.16 ± 1.21	72.7
13	1.02 ± 0.25	4.43 ± 0.8	81.3

Numbers are mg N in the entire sample (straw plus mycelium) removed from each end of the dish. Standard deviations are based on separate calculations for each of 10 replicate samples.

fresh nutrient medium had then been mostly converted to protein or other alcohol-insoluble N storage substances. Later the level of N in these insoluble substances also fell.

However, when the mycelium was grown over non-nutrient substratum from a straw (N-poor) food base, the proportion of N, derived from the foodbase, which remained in a soluble form in the mycelium was high even in young mycelium and did not change significantly with age of the mycelium between 3 and 6 weeks. These results suggest that under N-poor conditions such as occur in nature, mycelial N in a soluble, presumably mobile, form is in equilibrium with stored, insoluble N compounds. It is possible that low exogenous nitrogen levels promote the solubilisation of N in the mycelium as Fenn & Kirk (1981) and Kirk & Fenn (1982) have demonstrated in *Phanerochaete chrysosporium*.

Use of α-aminoisobutyric acid as a tracer of N movement in mycelium

While the use of ^{15}N as a tracer for nitrogen movement is ideal since the nitrogen atom itself is labelled and so indicates the presence of all food base-derived nitrogen irrespective of metabolic changes, ^{15}N is nevertheless technically difficult to handle and expensive. Radioactively-labelled amino acids present the problem that if they are, as is usual, labelled with ^{14}C, metabolism of the amino acid may eliminate or reduce the amount of the label independently of utilisation of the N atom. A way of overcoming this is to use a non-metabolised amino acid. The most commonly used substance for this purpose is α-aminoisobutyric acid (AIB) which has been shown to be taken up but not metabolised by *Saccharomyces cerevisiae* (Kotyk & Rihova, 1972) and *Neurospora crassa* (Ogilvie-Villa, De Busk & De Busk, 1981). *S. lacrimans* also takes up this substance; it remains unchanged in the mycelium, as shown by chromatography of ethanolic mycelial extracts; it is not incorporated into protein; it is not utilised as sole nitrogen source in culture (Table 3). Therefore it would appear to remain in the pool of free amino acids inside the hyphae which could move through the mycelium by mass flow. Indeed in experiments where ^{14}C-AIB was fed at one end of a mycelial bridge connecting two agar discs on a glass slide, transfer of radioactivity took place (Table 4).

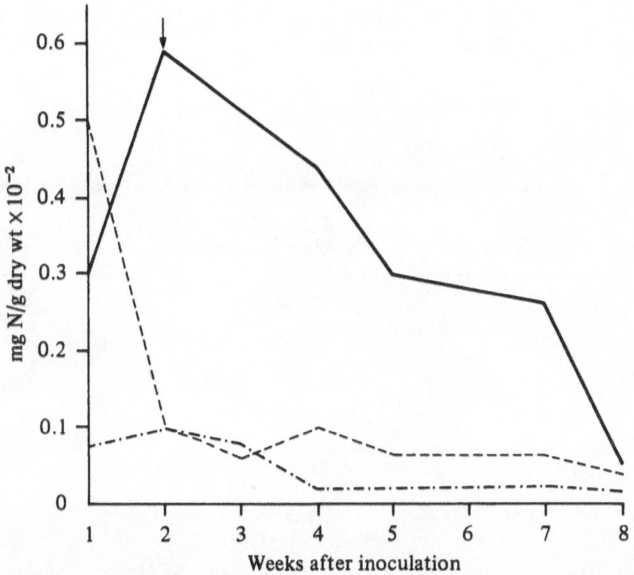

Fig. 5. Distribution of N between ethanol-soluble and -insoluble fractions of mycelium and agar substratum during growth in Petri dishes. Plates were centrally inoculated on medium containing, per l: KH_2PO_4, 1 g; $MgSO_4 \cdot 7H_2O$, 0.5 g; $FeSO_4 \cdot 7H_2O$, 10 mg; sucrose 30 g; and monosodium-L-glutamate, 2 g. Plates were sacrificed for analysis in sets of 3 and all the mycelium used for analysis in each case. The arrow indicates the point at which the plate was covered by mycelium. Total N in samples was estimated by Kjeldahl's method and N in amino groups by that of Yemm & Cocking (1955). The continuous line indicates mycelium after extraction in 80% ethanol; the dashed line, extracted amino acid; and the dash-dot line, ethanol extracts of the agar.

Table 3. *Uptake of ^{14}C from [^{14}C] monosodium-L-glutamate and [^{14}C]-α-aminoisobutyric acid into ethanol-soluble and -insoluble fractions of mycelium*

	Radioactivity present (cpm)	
Labelled amino acid added	insoluble mycelial residue	ethanol-soluble extract
α-aminoisobutyric acid	22	1290
monosodium-L-glutamate	380	2850

Two-week-old cultures on static liquid medium were each given 0.5 μCi ^{14}C in the form of either 2-amino[1-^{14}C]isobutyric acid or monosodium-L-glutamate. After 22 h incubation each mycelial mat was extracted in boiling 80% EtOH and radioactivity in extract and mycelial residue measured. Figures are means of results from each of three replicate culture flasks.

Table 4. *Movement of ^{14}C-labelled α-aminoisobutyric acid (AIB) in mycelium connecting two agar discs*

Hours after adding ^{14}C-AIB	Proportion of total added cpm in inoculum disc (%)
5.5	1.08
	4.17
	2.01
12	5.24
	2.92
	11.66
36	4.9
	1.3
	5.8
50	14.0
	12.2
	16.3

Mycelium was allowed to form a bridge connecting an agar inoculum disc on a glass slide to an agar ring 15 mm away. Labelled AIB was added to the well in the middle of the agar ring on each of 12 slides, which were incubated at 22 °C. At intervals, sets of 3 replicates were sacrificed and radioactivity of disc and ring on each slide measured by scintillation counting of ethanolic extracts. This arrangement measured ^{14}C moved in the opposite direction to that of growth and excludes transport due merely to hyphal extension.

Extension of hyphae over a non-nutrient substratum was inhibited by application of AIB at the central food base, and growth over nutrient agar was also slowed when AIB was incorporated in it. It is likely that AIB competes with glutamate in metabolism (Shive & Skinner, 1963). The inhibitory effects were particularly striking when wood was used as a food base. Extension growth was inhibited in colonies growing from a wood block when AIB solution was applied to the block, showing that its effect operated at the hyphal tips, and strand formation was prevented or much reduced (Fig. 6). Reduction in stranding was as expected on the basis of the hypothesis (Watkinson 1971a, 1975) that strands form as a result of the extra growth that is made possible by nitrogenous nutrients leaking from translocating or senescing leading hyphae. Here, the leaked N compound was presumably non-utilisable so that no stranding growth was possible. Impregnation of wood with AIB also decreased weight losses after subsequent incubation for 3 months; in one experiment, treated wood blocks lost only 2.8% of original dry weight compared with 25% in controls. Since the rate of wood decay by this fungus is probably N-limited under natural conditions (Watkinson, Davison & Bramah, 1981) it is understandable that

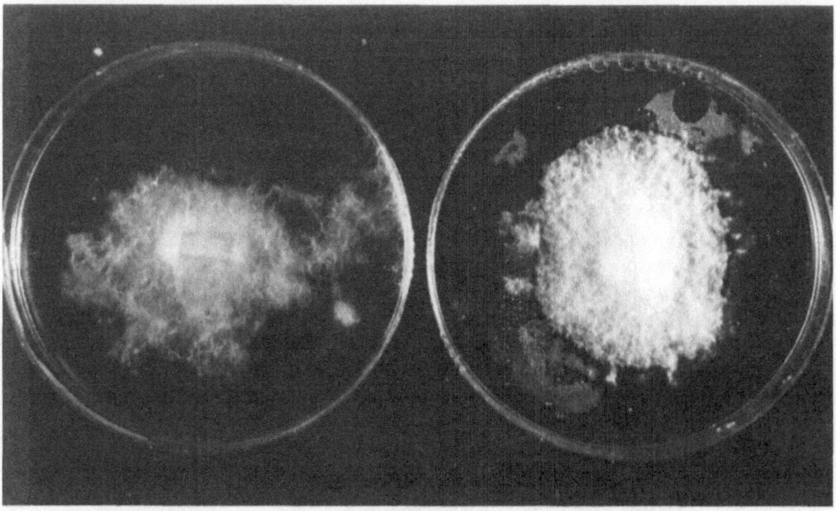

Fig. 6. Appearance of colonies grown from wood block inocula in 20-cm diameter Petri dishes. Aqueous α-aminoisobutyric acid (0.2 ml, 30% w/v) was added to test blocks after mycelium had begun to extend from them. The plates show (1) untreated colonies, and (2) treated colonies. Note spread to edge of dish and strand formation in (1) compared with restricted spread and no stranding in (2).

the presence of an inhibitor of amino acid metabolism interfered with wood decay.

The main conclusion drawn from the results described above is that colony morphogenesis in *Serpula lacrimans* is closely related to its N economy. N is efficiently extracted from lignocellulosic food bases. The N status of the food base affects the rate of colony extension over a non-nutrient substratum, the rate of decay of cellulose, and the amount of mycelium produced and its degree of differentiation. N extracted from the food base is concentrated in the mycelium and moved into further fresh food bases, presumably in soluble form. The chemical form in which it is moved has yet to be determined.

Straw labelled with ^{15}N was kindly supplied by Mr E. R. Mercer of the Agricultural Research Council's Letcombe Laboratory, and samples were analysed by Mr M. G. Johnson in the Mass Spectrometry Section.

References

Ander, P. & Eriksson, K. E. (1976). The importance of phenol oxidase activity in lignin degradation by the white rot fungus, *Sporotrichum pulverulentum. Archives of Microbiology*, **109**, 1–8.

Baker, J. M., Laidlaw, R. A. & Smith, G. A. (1970). Wood breakdown and nitrogen utilisation by *Anobium punctatum* Deg. feeding on Scots Pine sapwood. *Holzforschung*, **24**, 46–53.

Botton, B. & Dexheimer, J. (1977). Ultrastructure des Rhizomorphes du *Sphaerostilbe repens* B et Br. *Zeitschrift für Pflanzenphysiologie*, **85**, 429–43.

Brownlee, C. & Jennings, D. H. (1982). Long distance translocation in *Serpula lacrimans*: velocity estimates and the continuous monitoring of induced perturbations. *Transactions of the British Mycological Society*, **79**, 143–8.

Burnett, J. H. (1976). *Fundamentals of Mycology*, 2nd edn. London: Edward Arnold.

Butler, G. M. (1957). The development and behaviour of mycelial strands in *Merulius lacrymans* (Wulf.) Fr. I. Strand development during growth from a food base through a non-nutrient medium. *Annals of Botany*, **21**, 523–7.

Butler, G. M. (1958). The development and behaviour of mycelial strands in *Merulius lacrymans* (Wulf.) Fr. II. Hyphal behaviour during strand formation. *Annals of Botany*, **22**, 219–36.

Coggins, C. R., Hornung, U., Jennings, D. H. & Veltkamp, C. J. (1980). The phenomenon of "point growth" and its relation to flushing and strand formation in mycelium of *Serpula lacrimans. Transactions of the British Mycological Society*, **75**, 69–76.

Corner, E. J. H. (1932). The fruit body of *Polystictus xanthopus. Annals of Botany*, **46**, 71–111.

Dickson, R. (1837). A lecture on the dry rot, and on the effectual means of preventing it. London: Adlard.

Dowding, P. (1982). Nutrient uptake and allocation during substrate exploitation by fungi. In *The Fungal Community*, ed. D. T. Wicklow & G. C. Carroll, pp. 621–35. New York, Basel: Dekker.

Esser, K., Stahl, U. & Meinhardt, F. (1977). In *Biotechnology and Fungal Differentiation*, ed. J. Meyrath & J. D. Bu'lock, p. 67. London: Academic Press.

Falck, R. (1912). Die Meruliusfäule des Bauholzes. *Hausschwammforschung*, **6**, 1–405.

Fenn, P. & Kirk, T. K. (1981). Relationship of nitrogen to the onset and suppression of ligninolytic activity and secondary metabolism in *Phanerochaete chrysosporium*. *Archives of Microbiology*, **130**, 59–65.

Garrett, S. D. (1982). *Soil Fungi and Soil Fertility*, 2nd edn. Oxford: Pergamon.

Goksøyr, J. (1958). Studies on the metabolism of *Merulius lacrymans* (Jacq.) Fr. I. Carbohydrate consumption, respiration and acid production in surface cultures. *Physiologia plantarum*, **11**, 855–65.

Hartig, R. & von Tubeuf, C. F. (1902). *Der echte Hausschwamm*. Berlin: Springer.

Helsby, L. (1976). Structural Development of Mycelium of *Serpula lacrimans*. D.Phil. Thesis, University of Oxford.

Jennings, D. H. (1982). The movement of *Serpula lacrimans* from substrate to substrate over inert surfaces. In *Decomposer Basidiomycetes*, ed. J. C. Frankland, J. N. Hedger & M. J. Swift, pp. 241–62. Cambridge: Cambridge University Press.

Jennings, D. H., Thornton, J. D., Galpin, M. & Coggins, C. R. (1974). Translocation in fungi. *Symposium of the Society for Experimental Biology*, **28**, 139–56.

Jennings, L. & Watkinson, S. C. (1982). Structure and development of mycelial strands in *Serpula lacrimans*. *Transactions of the British Mycological Society*, **78**, 465–74.

Kirk, T. K. & Fenn, P. (1982). Formation and action of the ligninolytic system in Basidiomycetes. In *Decomposer Basidiomycetes*, ed. J. C. Frankland, J. N. Hedger & M. J. Swift, pp. 67–90. Cambridge: Cambridge University Press.

Kotyk, A. & Rihova, L. (1972). Transport of α-aminoisobutyric acid in *Saccharomyces cerevisiae*: feedback control. *Biochimica et Biophysica Acta*, **288**, 380–9.

Levi, M. P. & Cowling, E. B. (1969). Role of nitrogen in wood deterioration VII. Physiological adaptation of wood-destroying and other fungi to substrates deficient in nitrogen. *Phytopathology*, **59**, 460–8.

Levy, J. F. (1982). The place of Basidiomycetes in the decay of wood in contact with the ground. In *Decomposer Basidiomycetes*, ed. J. C. Frankland, J. N. Hedger & M. J. Swift, pp. 161–79. Cambridge: Cambridge University Press.

Liese, W. (1970). Ultrastructural aspects of woody tissue disintegration. *Annual Reviews of Phytopathology*, **8**, 231–58.

Merrill, W. & Cowling, E. B. (1966). Role of nitrogen in wood deterioration: amounts and distribution of nitrogen in tree stems. *Canadian Journal of Botany*, **44**, 1555–80.

Montgomery, R. A. P. (1982). The role of polysaccharide enzymes in the decay of wood by Basidiomycetes. In *Decomposer Basidiomycetes*, ed. J. C. Frankland, J. N. Hedger & M. J. Swift, pp. 51–65. Cambridge: Cambridge University Press.

Ogilvie-Villa, S., De Busk, M. R. & De Busk, G. A. (1981). Characterisation of 2-aminoisobutyric acid transport in *Neurospora crassa*: a general amino acid permease-specific substrate. *Journal of Bacteriology*, **147**, 944–8.

Shive, W. & Skinner, C. G. (1963). Amino acid analogues. In *Metabolic Inhibitors*, ed. R. M. Hochster & J. H. Quastel, pp. 1–73. New York: Academic Press.

Watkinson, S. C. (1971*a*). Morphogenesis of mycelial strands in *Serpula lacrimans*: a possible effect of nutrient distribution. *New Phytologist*, **70**, 1079–85.

Watkinson, S. C. (1971*b*). Phosphorus translocation in stranded and unstranded mycelium of *Serpula lacrimans*. *Transactions of the British Mycological Society*, **57**, 535–9.

Watkinson, S. C. (1975). The relation between nitrogen nutrition and the formation of mycelial strands in *Serpula lacrimans. Transactions of the British Mycological Society*, **64,** 195–200.

Watkinson, S. C. (1979). Growth of rhizomorphs, mycelial strands, coremia and sclerotia. In *Fungal Walls and Hyphal Growth*, ed. J. H. Burnett & A. P. J. Trinci, pp. 93–113. Cambridge: Cambridge University Press.

Watkinson, S. C. (1981). Accumulation of amino acids during development of coremia in *Penicillium claviforme. Transactions of the British Mycological Society*, **76,** 231–6.

Watkinson, S. C., Davison, E. M. & Bramah, J. (1981). The effect of nitrogen availability on growth and cellulolysis by *Serpula lacrimans. New Phytologist*, **89,** 295–305.

Wright, B. E. (1973). *Critical Variables in Differentiation.* Englewood Cliffs, New Jersey: Prentice-Hall.

Yemm, E. W. & Cocking, E. C. (1955). The determination of amino acids with ninhydrin. *The Analyst*, **80,** 209–13.

9

Distribution, development and functioning of mycelial cord systems of decomposer basidiomycetes of the deciduous woodland floor

WENDY THOMPSON
Botany Department, The University, Liverpool L69 3BX, UK

On the floor of the majority of deciduous woodlands, especially those where logs, branches, stumps and leaf litter are abundant, networks of mycelium in the form of discrete, filamentous aggregations can usually be found (Fig. 1). These may spread for many metres through the litter and upper soil horizons and are often highly branched with many cross links. In the past such aggregations have been termed 'strands' or 'syrrotia' (De Bary, 1887; Falck, 1912; see also Jennings: Chapter 7; Watkinson: Chapter 8; Read: Chapter 10) but, as the former term can be interpreted literally as single hyphae, the term 'mycelial cords' is preferred here.

Production of cords is due to a change from divergent to coherent growth (Gregory: Chapter 1, Mode 7) and as such their constituent hyphae show a degree of co-ordination greater than that which exists between the individual hyphae of a vegetative colony. They are produced by a wide variety of fungi from a range of habitats but are generally best developed in the Basidiomycotina. As will become evident, considerable variation exists regarding the dimensions attained by wood- and litter-inhabiting Basidiomycotina.

Mycelial cords are superficially similar to rhizomorphs such as those produced by *Armillaria* species. Both are aggregations of hyphae which undergo differentiation to varying degrees, but they differ in that rhizomorphs are fully autonomous and grow apically whilst cords are built up more gradually around a mycelial framework. However, intermediate forms between cords and rhizomorphs do exist as in *Phallus impudicus* (Butler, 1966).

As mentioned by Gregory (Chapter 1) it is still not clear what promotes cord initiation and growth under natural conditions, although various stimuli have been shown to elicit initiation in culture (Table 1). Watkinson (1975; see also Chapter 8) has proposed that for *Serpula lacrimans* leakage of nitrogenous nutrients from leading hyphae plays a

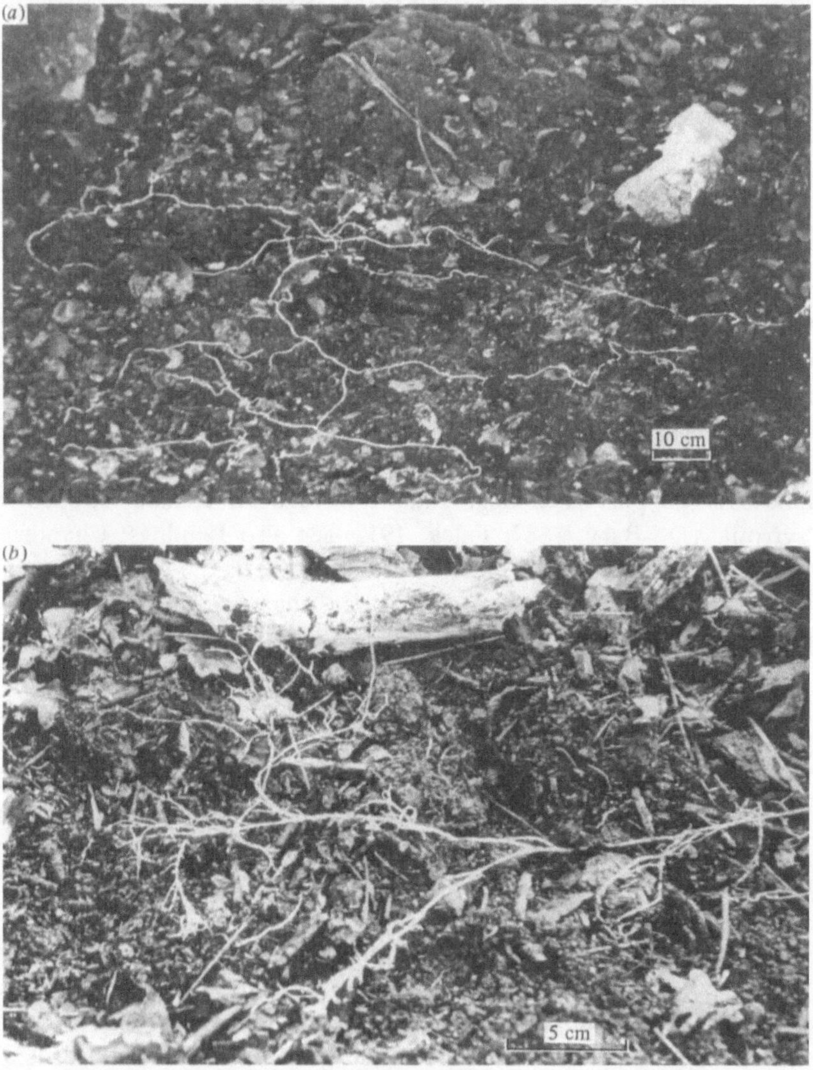

Fig. 1. (a) Excavated cord system of *Phanerochaete laevis* (Thompson & Rayner, 1982*a*). (b) Detail of the periphery of a system of *Phanerochaete velutina*.

Table 1. *Stimuli which elicit the initiation of mycelial cords and rhizomorphs*

Species tested	Stimulus	Source
Agaricus bisporus	Fresh soil placed on compost colonised by this species (in soil tubes)	Garrett, 1954
A. bisporus	Very low nutrient concentrations in the surrounding media	Mathew, 1961
Agaricus campestris	High moisture conditions	Hein, 1930; Styer, 1930
Calvatia sculpta	The presence of a permeable physical barrier (e.g. cellophane) between the mycelium and a medium which would not otherwise support the formation of cords e.g. 2% malt agar or cornmeal agar	Bellotti & Couse, 1980
Coniophora cerebella	Neutral part of the ethyl ether extract of heat-treated pine wood	Glasare, 1970
Helicobasidium purpureum	Translocation in leading hyphae	Valder, 1958
Phymatotrichum omnivoroum	Inocula placed on slides in a sterile, moist chamber	Rogers & Watkins, 1938
Poria xantha	Cu^{2+}	Hirt, 1949
Serpula lacrimans	High C:N ratio; $NaNO_3$	Watkinson, 1975
S. lacrimans	Translocation in leading hyphae	Watkinson, 1971
Xylaria polymorpha	Transfer to cultures of 'producer' fungi, e.g. *Pythium mamillatum, Fusarium oxysporum, Fusarium caeruleum* and *X. polymorpha*	Park, 1963
Armillaria mellea	Certain amino acids	Weinhold & Garraway, 1966
A. mellea	Sufficient nutrition; critical C:N ratio	Garrett, 1953
A. mellea	Ethanol	Weinhold, 1963; Sortkjaer & Allermann, 1972
A. mellea	Indole acetic acid and compounds related to the shikimic acid pathway; amino benzoic acid	Garraway, 1970

Table 1. (*cont.*)

Species tested	Stimulus	Source
A. mellea	Acetate	Sortkjaer & Allermann, 1972
A. mellea	Figwood extract	Weinhold, Hendrix & Raabe, 1962
A. mellea	Moist conditions	Townsend, 1954
A. mellea	Oils and fatty acids	Moody, Garraway & Weinhold, 1968
A. mellea	*Aureobasidium pullulans*	Pentland, 1965
Sphaerostilbe repens	Diffusion of a factor (soluble in water, non-volatile, stable at laboratory temperature and inactive at 100 °C) produced by *Penicillium* spp., *Aspergillus* spp. and *Verticillium* spp. in culture medium	Botton & El-khouri, 1978
S. repens	Continuous presence of calcium in the culture medium	Botton, 1978

part. When the hyphae in a growing colony have taken up nutrients from the culture medium so that most of the nutrient material is in the mycelium, the first-formed hyphae are thought to become more permeable with age, and formation of cords results from further growth around these. However, cord formation resulting from intertwining of undifferentiated hyphae can also occur during outgrowth of *S. lacrimans* into non-sterile sand (A. D. M. Rayner, personal communication).

At least five advantages of mycelial aggregation into cords can be envisaged: protection against deleterious external agencies; channelling of resources (nutrients and water), allowing outgrowth from suitable bases into an environment in which these are deficient; nitrogen conservation; enhancement of inoculum potential; and amplification of the sensitivity of individual hyphae to external stimuli, enabling directed growth responses (see also Jennings: Chapter 7; Watkinson: Chapter 8; Read: Chapter 10; Rayner *et al.*: Chapter 23).

With regard to protection, the buffering effect of aggregation of hyphae into cords may endow a fungus with greater resistance to environmental fluctuations and extremes than could otherwise be contended with by isolated hyphae. Additionally, cords may serve as a means of protection against other organisms, including anti-mycotic-producing bacteria, parasitic or other fungi and grazing microfauna. This protective function may be enhanced by production of calcium oxalate, which, as discussed by Jennings (Chapter 7) may be a characteristic feature of certain actively-translocating systems. Mycelia of terrestrial fungi can contain high calcium concentrations (Stark, 1972; Todd, Cromack & Stormer, 1973; Cromack *et al.*, 1977) and the outermost hyphae of cords of *Phanerochaete velutina, Phanerochaete laevis, Phallus impudicus* and *Tricholomopsis platyphylla* are characteristically heavily encrusted with calcium oxalate (Thompson, 1982). The reduction of pH in the immediate vicinity of cords which is a consequence of calcium oxalate production will tend to suppress bacteria which are generally relatively intolerant of low pH (Gray & Williams, 1971), and might also underly lysis and replacement of other mycelia in combative interactions (cf. Rayner & Webber: Chapter 18).

As has already been indicated by Jennings (Chapter 7) and Watkinson (Chapter 8) aggregation of individual hyphae into cords may allow resources to be channelled so that direct and economical movement of nutrients and water may occur between a colonised 'food base' and the advancing edge of a system and/or vice versa. It seems likely that translocation occurs within mycelial cords as they often extend for many metres between food bases and represent the only connection between the growing mycelial margin and either fruit bodies or food bases. Also they contain elements of wide diameter ('vessel' hyphae, Fig. 2), which from anatomical considerations seem likely to have low resistance to solution flow. These can be very large, up to 60 μm in diameter in *Phanerochaete velutina* for example. There is now very substantial evidence from laboratory studies that cords of *S. lacrimans* translocate water and nutrients (Butler, 1958; Weigl & Zeigler, 1960; Watkinson, 1971; Clarke, Jennings & Coggins, 1980; Hornung & Jennings, 1981; Brownlee & Jennings, 1982; Thompson & Jennings, unpublished). Estimates of the velocities of various radioactive isotopes through mycelia of *S. lacrimans* are all of the order of 25 cm h^{-1} (Jennings: Chapter 7).

The role which cords may have in nitrogen conservation has already been discussed by Watkinson (Chapter 8). Briefly, by growing as aggregated structures, fungi can re-utilise via autolysis the nitrogenous products of old hyphae. Further, pure culture studies indicate that development of cords of *S. lacrimans* depends on high C:N ratios.

Buffering against adverse external factors, channelling and efficient utilisation of resources contribute to another important feature, enhancement of the invasive force, or inoculum potential (see also Rayner *et al.*: Chapter 23). By the same token, in circumstances where suitable substrata are discontinuously distributed, efficiency is gained if wastage of biomass by growth in the wrong direction can be minimised. Whereas individual hyphae may be relatively incapable of long-range responses to suitable external stimuli, their individual responsiveness may be amplified many times by aggregation into cords, promoting directed growth. Evidence that such directed growth may play a role in colonisation processes and mycelial interactions will be given below.

Despite these potential advantages many fungi are, of course, successful without producing cords. Amongst wood decomposers these include *Coriolus versicolor* and *Phlebia merismoides* which colonise via air-borne spores (Carruthers & Rayner, 1979). Although cord formation is not therefore a pre-requisite for success the probability is that it represents an important adaptation whereby, in suitable habitats, colonisation of substrata may be effected from pre-existing vegetative mycelia. Nonetheless, with the exception of the work on *S. lacrimans* and recent studies of sheathing mycorrhizas (Read: Chapter 10), this probability and its consequences have attracted surprisingly little attention. The aim of this chapter is to examine these issues, and their

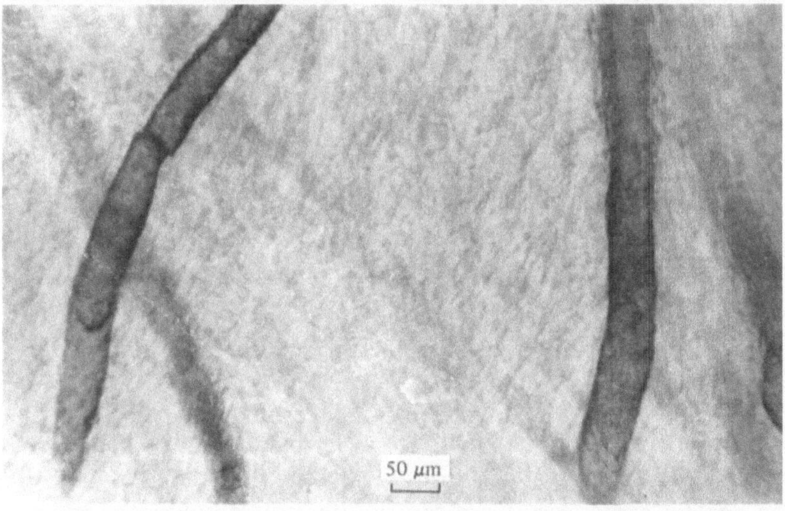

Fig. 2. Micrograph of a mature cord of *P. laevis* showing vessel hyphae (Thompson & Rayner, 1982*a*).

importance for understanding mycelial biology of that somewhat neglected group of fungi mentioned at the outset, which inhabit the deciduous woodland floor as more or less extensive mycelial systems and bring about decomposition of accumulations of woody and other litter.

Distribution

Observations from direct field sampling and excavations

Considerable variation occurs in the form of the mycelium of decomposer basidiomycetes inhabiting the woodland floor, the extent of aggregation ranging from short, woolly mycelial tufts originating from the bases of basidiocarps, as with *Clitocybe flaccida*, to well-defined individual cords up to 5 mm in diameter, which often extend several metres from the associated basidiocarps and also occur freely in the absence of the latter, as with *Phanerochaete velutina.* There seems to be a correlation between mycelial form and ecological role in that it is those fungi which invade wood which form the more clearly-defined cord systems. Although some of these fungi, such as *P. velutina, Phallus impudicus* and *Tricholomopsis platyphylla*, can also decompose leaf litter, this activity is usually marked only in the immediate vicinity of bulky, woody substrata. By contrast, those fungi which specifically rot litter usually form diffuse systems with ill-defined fragile cords or loose aggregations of mycelium. *Mutinus caninus* is intermediate, forming defined cords associated with woody litter components, and more diffuse mycelium in leaf litter or sawdust. This difference in form between the mycelium of litter decomposers and wood decayers may be linked to the nature of the resources they attack. Substrata exploited by litter decomposers are widespread and diffuse and require a correspondingly diffuse, spreading system for exploitation. Wood-decaying fungi on the other hand exploit larger, bulkier resources which are often spatially separated and these fungi produce bulkier and more extensive systems of cords along which translocation must occur. As discussed earlier, these fungi are effectively concentrating their inoculum potential and thereby their invasive force.

Field excavations made by careful removal of soil and litter away from cord systems of wood-decaying species reveal that in all cases such systems are composed of networks of cord, which are generally relatively thick and sparsely branched in older regions and thinner and extensively branched peripherally (Thompson & Rayner, 1983) (Fig. 1). The cords of *P. velutina*, for example, often measure up to 4 mm in diameter in older regions and up to 1.5 mm in younger peripheral parts. The most extensive systems found in terms of total length were those of *Phanerochaete velutina* (up to 31.7 m), *Phallus impudicus* (up to 24.8 m), and *Steccherinum fimbriatum* (22.6 m); systems of other species such as *Tricholomopsis platyphylla* and *Phanerochaete laevis* were usually 15 m or more in total length. Grainger (1962) also found the cord systems of *Phallus impudicus* to be fairly extensive – of the two systems he excavated the longest was over 16 m in overall length. *P. velutina*, *P. impudicus* and *T. platyphylla* are common and widespread in British woodlands and as less-common species and other unidentified fungi also form substantial cord systems it is likely that cord-forming fungi as a group are very important in terms of the amount of living fungal biomass present in the soil and litter of our forests. The dimensions and overall extent of cord systems such as those referred to above would seem to suggest that they have persisted, in some cases for considerable periods of time. It is therefore reasonable to suppose that they may be relatively resistant to various external agencies as discussed earlier.

The ubiquity of cord-forming fungi on the woodland floor is further demonstrated by the extent to which they are present in arrays of suitable substrata. For example, in a study of decomposition of suppressed oak trees in two even-aged plantations at Savernake Forest, Wiltshire, and the Forest of Dean, Gloucestershire (Fig. 3), colonisation of the tree bases was studied by three methods: direct observation of mycelial systems and fruit bodies on the bases; direct incubation of wood slices from the bases in polythene bags; and isolation into pure culture from small wood chips taken from each different decay region (Thompson & Boddy, 1983). Cord-forming fungi and two *Armillaria* species (*A. bulbosa* and *A. ostoyae*) were found to be the most important in terms of frequency and volumes of wood occupied, and were present during all stages of decay. *Phanerochaete velutina* and *Phallus impudicus* were the most important cord-forming species, being found in 26 to 37.5% of sampled trees at these two sites (Table 2). It is interesting to note, however, that at an older oak stand (planted 1929) in the Forest of Dean, observations of fifteen suppressed oak trees indicated that the roots of eight were colonised extensively, and almost exclusively by *Tricholomopsis platyphylla*, one by *P. velutina*, two by *A. bulbosa* and one by both *T. platyphylla* and *A. bulbosa*. Thus at this site *T. platyphylla* was the dominant cord-forming species. However, high frequencies of cords and fruit bodies of this fungus had been observed throughout the younger Forest of Dean site (planted 1943) where only

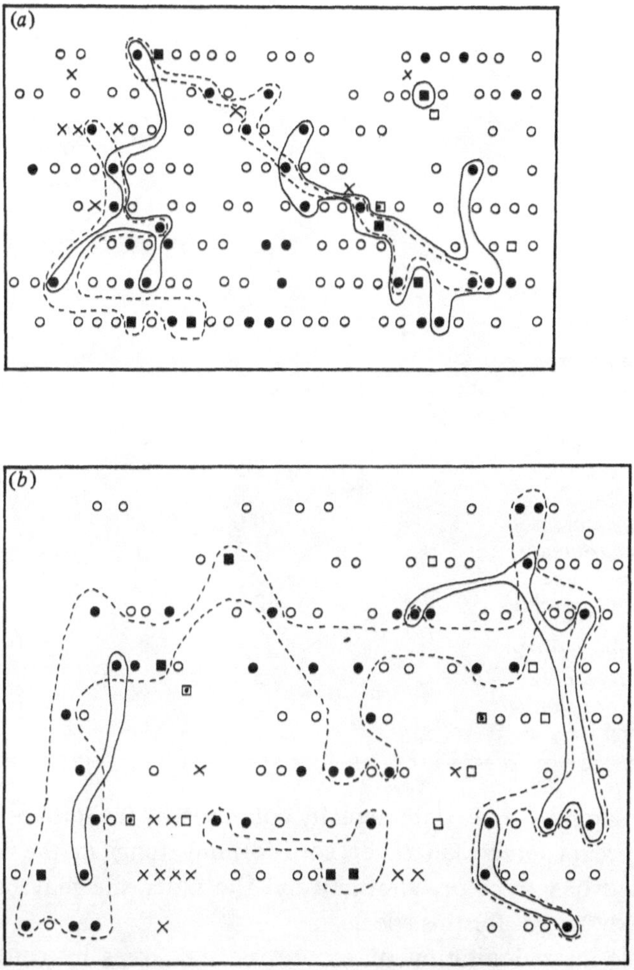

Fig. 3. Distribution of different mycelial types of *P. velutina, A. bulbosa* and *A. ostoyae* in (*a*) a 28 × 9 m site of p. 55 oak at Savernake Forest and (b) a 50 × 11 m site of p. 43 oak at Forest of Dean. Filled circle, sampled dead oaks; hollow circle, living oaks; filled square, sampled stumps of fallen suppressed oaks; hollow square, large oak stumps from trees grown previously on the site; X, trees other than oak; continuous line, delimitation of trees/stumps colonised by the same mycelial type of *P. velutina*; dotted line, delimitation of trees/stumps colonised by the same mycelial type of *A. ostoyae* at Savernake and *A. bulbosa* in the Forest of Dean. (Thompson & Boddy, 1983.)

Table 2. *Frequency of isolation of different fungi from roots of suppressed oak trees*

	Percentage of trees sampled	
Fungus	Savernake	Forest of Dean
Cristella sulphurea	2.5	2.3
Hypholoma fasciculare	7.5	11.9
Mutinus caninus	2.5	9.5
Phallus impudicus	30.0	33.0
Phanerochaete sp. 1	5.0	–
Phanerochaete sp. 2	–	16.6
Phanerochaete velutina	37.5	26.2
Tricholomopsis platyphylla	–	4.7
Unidentified sp.	2.5	14.2
Armillaria bulbosa	–	97.7
Armillaria ostoyae	72.5	–
Bjerkandera adusta	–	7.1
Collybia fusipes	7.5	7.1
Coriolus versicolor	2.5	–
Grifola frondosa	5.0	–
Phlebia merismoides	10.0	14.2
Stereum gausapatum	2.5	2.3
Stereum hirsutum	7.5	4.7
Other basidiomycetes	12.5	30.9
Ascocoryne sarcoides	15.0	2.3
Hypoxylon serpens	5.0	2.3
Xylaria hypoxylon	2.5	2.3
Other fungi	47.5	66.6

4.7% of sampled trees were actually colonised by it. Perhaps then this species occurs later than other cord-forming fungi in the tissues of suppressed oak trees or, alternatively, the older site may have been particularly 'good' for this species.

Considering colonisation of suppressed oak trees by cord-forming fungi and *Armillaria* species in particular (Table 3) it becomes clear just how important these fungi must be in the decay of such 'newly available' substrata. For example, 72.5% of sampled (suppressed) trees at Savernake forest were colonised by cord-forming species whilst the corresponding value for the Forest of Dean was 88%.

A further study which demonstrated the extensive distribution of cord-forming fungi in woodland habitats involved the colonisation, by cord formers, of freshly cut beech logs arranged experimentally on the floor of a mixed deciduous woodland (Thompson, 1982). This study

Table 3. *Percentage of roots of suppressed oak trees colonised by* Armillaria *species and cord-forming fungi*

Fungus	Savernake		Forest of Dean	
	Isolation and direct incubation	Field observations, isolation and direct incubation	Isolation and direct incubation	Field observations, isolation and direct incubation
Armillaria species[a]	42.5	72.5	88.0	97.7
Cord-forming fungi	55.0	72.5	69.0	88.0
Armillaria species and cord-forming fungi	27.5	52.5	69.0	88.0
Neither *Armillaria* species nor cord-forming fungi	30.0	7.5	12.0	2.3

[a] *A. bulbosa* at the Forest of Dean and *A. ostoyae* at Savernake.

indicated that colonisation occurred rapidly throughout the entire site. In fact within 33 months over 75% of all logs had become extensively colonised by cord-forming species, particularly by *P. velutina.*

Population structure

Although direct observations of the type just described provide an idea of the distribution of mycelial cord-forming fungi, for a more complete understanding, knowledge of the spatial distribution of individual genotypes in populations of these fungi is required. Since such information is limited, a detailed analysis of the spatial structure of a population of *T. platyphylla* has been made using the phenomenon of somatic incompatibility between secondary mycelia of this species (Thompson & Rayner, 1982*b*). When genetically and physiologically distinct mycelial types (individuals) are paired in culture they are mutually antagonistic (Rayner & Todd, 1979; Todd & Rayner, 1980), a reaction which provides a basis for the identification of individuals in the field and ultimately a key to the understanding of natural patterns of dispersal, establishment and vegetative spread of such fungi (Rayner & Todd, 1982; see also Frankland: Chapter 11; Rayner *et al.*: Chapter 23). Results from this study revealed that only a limited number of often extensive mycelial types were present throughout a site of 2.08 ha, i.e. of 113 isolates (from wood, cords and fruit bodies) only 22 different mycelial types existed (Fig. 4). In another study the population distributions of *P. velutina* throughout oak stands, containing suppressed individuals, at Savernake Forest and the Forest of Dean, were analysed. Using the technique described above to distinguish individuals, three different mycelial types of *P. velutina* were identified at Savernake and four at the Forest of Dean (Fig. 3).

This situation, where limited numbers of mycelial types of cord-forming species are distributed over very large areas of the forest floor, contrasts markedly with the situation which is found in airborne wood decay species such as *C. versicolor* and *Stereum hirsutum.* Here any one stump or branch may contain large numbers of antagonistic individuals separated by relatively undecayed interaction zones (Rayner & Todd, 1979). Distributions of the former type probably reflect the fact that *T. platyphylla* and *P. velutina* are capable of propagation via vegetative mycelial growth between food bases in the soil. The mycelium which spreads as cords between spatially-separated resources must be adapted to traverse the intervening regions between food bases which may often be hostile to fungal growth in nutritional, competitive and/or physicochemical terms. Such an adaptation endows a single mycelial individual with the potential to occupy a very large territory.

It is important to realise that the pattern of arrival at and initial establishment in resources have important consequences. For example establishment from pre-existing mycelium, as well as being less susceptible than spores to adverse external influences, may be expected to lead to more rapid and extensive initial occupation of available space (i.e.

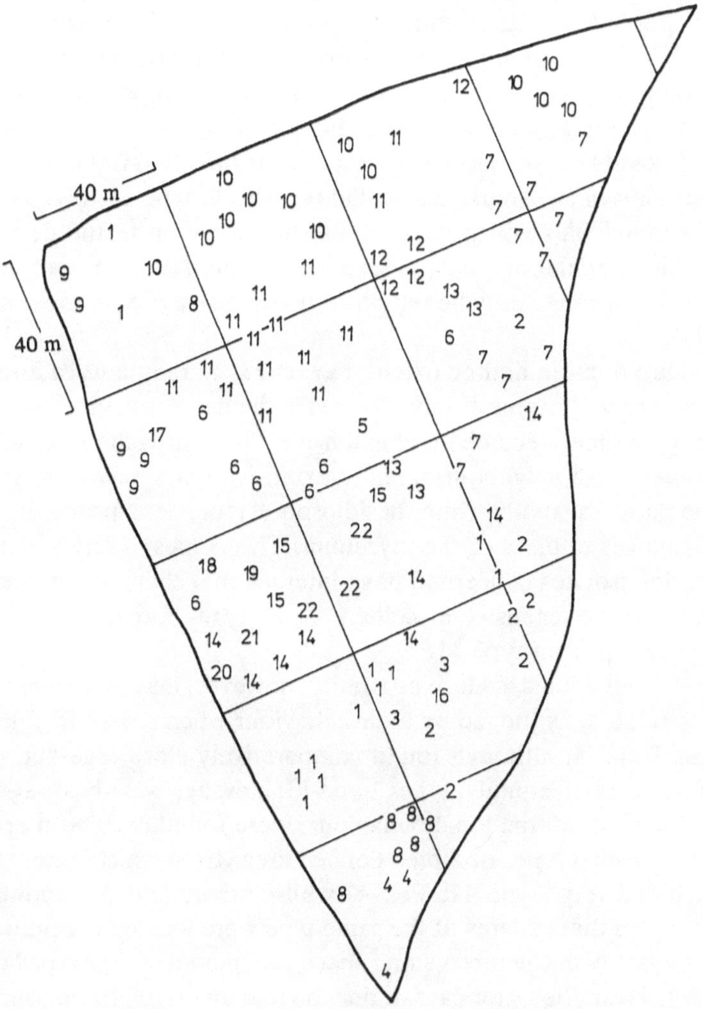

Fig. 4. Distribution of different mycelial types of *Tricholomopsis platyphylla*, indicated by different numbers. (Thompson & Rayner, 1982*b*.)

'resource capture') with consequently fewer individuals becoming established (Carruthers & Rayner, 1979; Rayner & Todd, 1979).

Considering the distribution of mycelial types of *T. platyphylla* (Fig. 4) and given that isolates of the same mycelial type originate from the same individual (as it is unlikely, in a heterogeneous population, that dikaryons sufficiently similar genetically to intermingle in culture should develop from separate colonisations from basidiospores), analysis of their distribution should provide insight into the dynamics of the mycelial population. Unfortunately only limited information was obtained regarding mating-type distribution in the population, due to poor viability of monospore isolates in culture (Thompson & Rayner, 1982*b*). Nevertheless we can, with informed speculation, suggest a variety of possible explanations to account for the observed population structure. Thus very extensive mycelial types (e.g. type 11, Fig. 4) may represent individuals which were established early on in the development of the population and have persisted until the present time; alternatively they may only be temporarily expansive (i.e. at the time of the study).

Individuals of more limited extent may represent remnants of formerly more extensive systems or initial stages in the development of systems which may, in time, become more extensive. As regards their origin it is possible that those neighbouring more extensive types may have arisen from the latter sexually from basidiospores, or, less probably, via somatic changes in parts of the mycelium. This is most likely when the neighbouring isolates concerned have interactional characteristics similar to those of the extensive mycelium. Thus, types 5, 6 and 12 (Fig. 4) may have arisen from type 11.

Different individuals in close proximity, however, may not necessarily always be related, as judged by their behaviour when paired in culture; e.g. types 1 and 3, although found comparatively close together were highly antagonistic. Equally, types 1 and 13, although weakly antagonistic and of similar interactional behaviour, were found over 60 m apart.

Certain mycelial types occupied defined areas from which other types were excluded (e.g. type 11, Fig. 4) whilst others had discontinuous distributions in that isolates of the same type were located considerable distances apart with the intervening space occupied by other types (e.g. type 1, Fig. 4). In the latter case, it may be that an originally continuous mycelium may break up and establish new centres, although the mechanisms underlying such break-up remain obscure.

It is difficult to be certain of the origin and development of mycelial types occurring in peripheral regions of the site (e.g. type 9, Fig. 5). These types may be more extensive outside the limits of the sampling grid and may be in the process of invasion of, or elimination from, the sampling area.

Many of the questions raised above could, of course, be resolved by repeated surveys at the same site, and such consecutive samplings may be important in future population studies. One important factor which probably influenced the observed population structure of *T. platyphylla* is the occurrence of three past thinning operations at the site, the most recent being one year before sampling. These are likely to have had an effect via the sudden generation of new resources for colonisation, but this cannot be gauged with certainty without repeated samplings before and after thinning.

A number of studies, similar to the present, have been made for

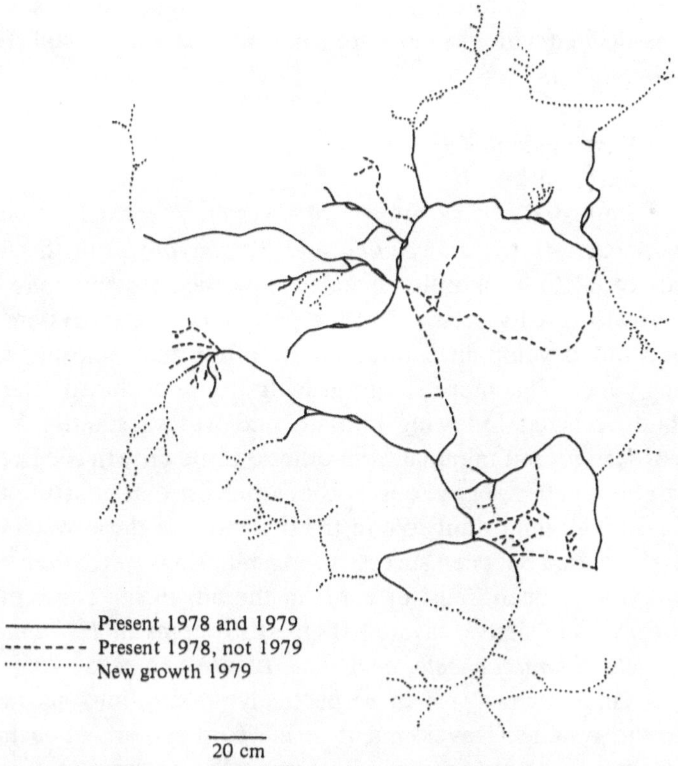

Fig. 5. Diagram illustrating development of a field system of *P. laevis* over a 13-month period. (Thompson & Rayner, 1982*a*.)

several important root-infecting forest pathogens including *Armillaria* spp. (Adams, 1974; Shaw & Roth, 1976; Rishbeth, 1978; Korhonen, 1978). *Phellinus weirii* Murr. (Childs, 1963) and *Phaeolus schweinitzii* (Fr.) Pat. (Barrett & Uscuplic, 1971). These have variously employed cultural characteristics, antagonism (though not always with a sound knowledge of the genetical basis for the reaction), and distribution of mating-type factors (which are not reliable indicators since isolates with the same factors may be derived from separate combinations of sib-related homokaryons) to identify mycelial types (often rather dubiously referred to as 'clones'). These have all served to indicate the different consequences of spread via vegetative mycelium and basidiospores. Thus, the rhizomorphic *Armillaria* species show similar distribution patterns (though normally with fewer types) to *T. platyphylla*, and in some instances isolates of the same type have been located as much as 450 m apart (Anderson, Ullrich, Roth & Filip, 1979). By contrast, different individual trees on the same site colonised by *P. schweinitzii* normally contain different genotypes of the fungus, and this is believed to be associated with basidiospore colonisation from the soil (Barrett & Uscuplic, 1971).

Development

In the field

Repeated excavation of systems of *P. impudicus, Phanerochaete laevis, P. velutina* and *T. platyphylla* in the field over periods of 12 to 14 months revealed how such systems may develop. New growth usually occurred at the periphery of each system but new cords could develop in central areas, often anastomosing with pre-existing cords. This more recent growth often developed after 'sprouting' had occurred following rainfall, and was substantial from some parts of systems but minimal from others. Little growth occurred during May to September. These observations suggest that moisture availability plays an all-important role in the dynamics of these systems. Those parts which died between successive examinations were often central or had previously been 'fans' of cords at the advancing edges of systems when they were first excavated (Fig. 5). Systems of *P. impudicus, T. platyphylla, Phanerochaete laevis* and *P. velutina* made 255, 260, 234 and 535 cm net new growth respectively, taking into account death within the systems. Development of the cord systems of each of these four species was very uneven with some parts growing at greater rates than others.

In the laboratory

In agar culture. Cord-forming fungi show variation in their readiness to produce cords in agar culture. Some species, e.g. *T. platyphylla*, show no signs of corded growth in these conditions; this may be due to the lack of a 'suitable' surface (i.e. one which can support corded growth). For example cord formation can be induced in *Calvatia sculpta* by the presence of a permeable physical barrier between the agar substratum and the growing mycelium (Bellotti & Couse, 1980). Other species, such as *P. velutina*, grow mycelially to the edge of the Petri dish when cords are formed which usually grow back over the mycelium towards the original inoculum. *P. impudicus*, however, produces cords more or less immediately after subculturing on agar media, with a very limited amount of mycelial growth occurring initially. This species resembles *Armillaria* species such as *A. mellea* and *A. bulbosa* which show limited mycelial growth but extensive rhizomorph formation.

Cord-forming species do not always grow uniformly on agar media. Sometimes, for example in colonies of *T. platyphylla*, regions of mycelium develop faster than the mycelium in the majority of the colony. Such changes resemble the phenomenon of point growth, observed in *S. lacrimans*, which occurs on media which produce relatively low growth rates (Coggins, Hornung, Jennings & Veltkamp, 1980). In this fungus point growth is thought to represent a response to environmental stress, although the mechanism of stimulation of growth at a particular part of the mycelium is not known.

In *Hypholoma fasciculare* colonies of two distinct morphological types are commonly found even when isolations are made from the tissue of a single fruit body (as described by Rayner, 1975). The underlying basis of this dimorphism is unclear and the phenomenon needs further investigation, particularly regarding the persistence and cord-forming capacity of each morphological type.

Cord initiation in cultures of *P. velutina, P. laevis* and *T. platyphylla* can be stimulated by beechwood volatiles (see also Table 1). In one instance cords up to 15 mm in length were induced by sterile beechwood shavings suspended 3 cm above growing mycelial colonies of *T. platyphylla* (Thompson, 1982). Such stimulation may be relevant to the phenomenon of directed growth towards potential resources suitable for colonisation.

In soil tubes. Autoclaved beech blocks ($\sim 2\,cm^3$) placed on to the surface of pure cultures of cord-forming fungi readily become permeated with mycelium, and can then be used as inocula in laboratory studies of the outgrowth of cord-forming fungi from food bases into soil.

In one set of experiments six species were allowed to grow, in the dark, through tubes of non-sterile and sterile soil (Thompson & Rayner, 1982*a*, 1983). The patterns of development of cord systems of *P. velutina, P. impudicus, P. laevis* and *T. platyphylla* in tubes of unsterilised soil in the laboratory were similar to that occurring in the field. In all cases extension into the uncolonised soil occurred as much-branched and anastomosed systems of fine cords constituting 'fronts' of more or less constant width which were short-lived. These eventually lysed to become superseded by thick, relatively unbranched cords which were few in number but which maintained a connection between different parts of the expanding mycelium (Fig. 6). When the fungi were grown in sterile (γ-irradiated) soil, mycelial extension of *Steccherinum fimbriatum, Hypholoma fasciculare, P. impudicus, P. velutina* and *P. laevis* was reduced when compared with that occurring in non-sterile soil. *T. platyphylla* grew at similar rates in both sterile and non-sterile soils. Differentiation of the mycelium was affected by soil sterilisation in all the afore-mentioned species but to a lesser extent in *P. impudicus.* Those species which were affected developed luxuriant 'fan' mycelia which showed little differentiation into distinct cords (Fig. 6).

There are two possible explanations of this effect of soil sterilisation – changes in the structure, chemistry and nutrient status of the soil and the elimination of soil organisms. Gross changes in the physicochemical properties of the soil were probably not involved as γ-irradiation and autoclaving of soil gave similar results. However, both methods will result in the release of ammonia nitrogen (Salonius, Robinson & Chase, 1967) and other nutrients from previously-living organisms. As it has been suggested that cords form under conditions of low nutrient availability (Watkinson, 1971), so cord formation would not be expected to occur in sterile soil if this were the case. Alternatively, removal of soil micro-organisms, which may stimulate mycelial differentiation via production of morphogenetically-active substances, direct contact or competition for nutrients, may better explain the effects of soil sterilisation on mycelial differentiation and extension. It is difficult to be sure of the relative importance of each of these factors. However, interaction with other fungi has been shown to stimulate rhizomorph formation in *Sphaerostilbe repens* (Botton & El-Khouri, 1978) and cord formation in *P. velutina* (Rayner & Todd, 1979; see also Rayner & Webber: Chapter 18). Nevertheless *P. velutina* and *P.*

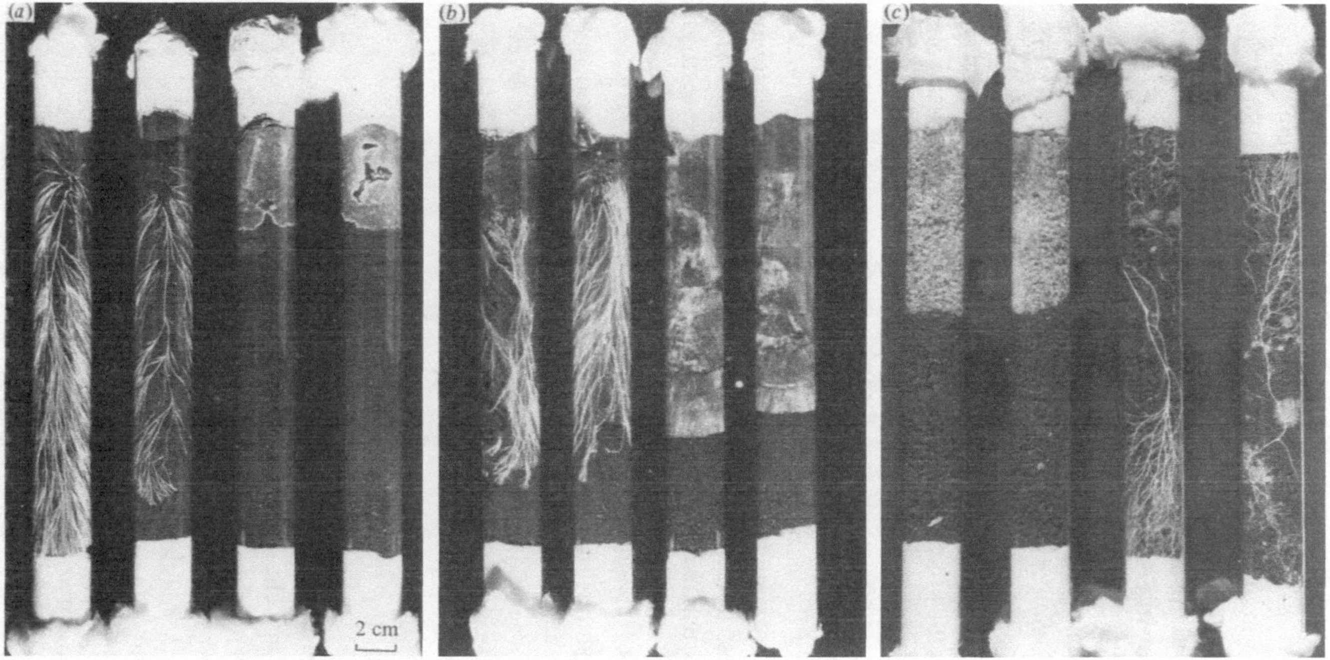

Fig. 6. Growth of cord-forming fungi in soil tubes. (a) *Steccherinum fimbriatum* and (b) *Hypholoma fasciculare* growing through non-sterile soil (left) and irradiated soil (right). (c) *Phanerochaete velutina* growing through irradiated soil (left) and non-sterile soil (right). (Thompson & Rayner, 1983.)

impudicus readily form cords when wood blocks colonised by these species are placed in wells in sterile Perspex cabinets, supplied with sterile distilled water and allowed to grow over Perspex platforms (Fig. 7). Clearly stimulation of cord formation by micro-organisms is most unlikely here and the absence of nutrients on the surface over which the mycelium is growing is the much more likely explanation.

Functioning

Colonisation of substrata

In the field. Detailed studies of the colonisation of various virgin substrata by cord-forming fungi (Thompson, 1982) have demonstrated the rapidity with which the latter can invade and colonise woody substrate. At one site colonisation of beech logs (~ 30 × 10 cm) by *P. velutina* (the most commonly found cord-forming species in this study) occurred within one month of the logs being laid on the woodland floor.

Large areas of the beech logs were initially colonised by *P. velutina* and other cord-forming fungi at those parts in contact with the litter and soil of the forest floor. The mycelia formed small cord systems which were closely associated with the tightly-packed leaf litter beneath the logs. From here cords spread onto and under the bark and over the cut ends of the logs. Genetical analysis of the mycelial types of *P. velutina* isolated throughout the experimental site indicated that a single individual/mycelial type was responsible for all colonisation by this species. This observation emphasises the effectiveness of vegetative propagation as a means of spread through soil and litter.

The speed with which cord-forming fungi invade newly supplied lignicolous substrata suggests that some stimulus may cause directed growth of cords towards the wood.

In the laboratory. The idea of substratum-directed growth to explain the rapid arrival of mycelia of cord-forming fungi at potential food bases has been further explored in the laboratory using an unidentified coprophilous cord-forming species. Autoclaved rabbit pellets, some colonised by the cord former and others left uncolonised, were arranged in glass Petri dishes containing sterile and non-sterile sand so that each pellet was equidistant from each of its neighbours. In sterile sand slow and limited mycelial growth usually occurred for a few millimetres around colonised pellets but no cords were observed. By contrast cord formation in non-sterile sand was extensive (Fig. 8) with the number of cords of diameter greater than 0.25 mm produced in each type of experimental arrangement (i.e. a, b and c, Fig. 8) increasing with the number of pellets colonised originally. In experiments of type b most cords originating from the central colonised pellet grew directly towards those pellets which were originally uncolonised. Indeed in all cases where cord growth occurred this appeared to be directed towards particular pellets rather than at random. As with wood decay cord-forming fungi, growth

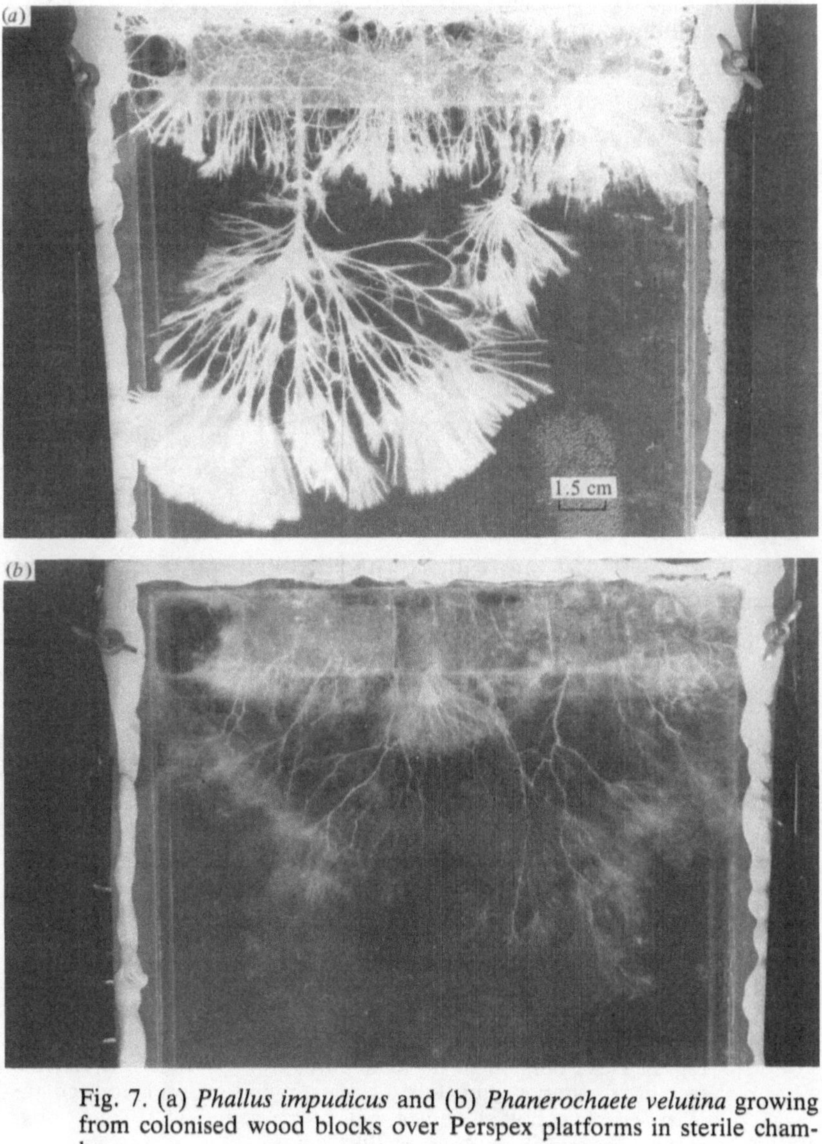

Fig. 7. (a) *Phallus impudicus* and (b) *Phanerochaete velutina* growing from colonised wood blocks over Perspex platforms in sterile chambers.

at the periphery of developing systems occurred as fans of fine cords which grew towards particular pellets and did not spread uniformly through the sand.

This apparently directional growth of cords may represent an efficient way in which a fungus can spread through the soil to potentially colonisable resources without energy being wasted on the multi-directional production of cords. These observations support the idea that certain stimuli, perhaps volatiles, may be important in stimulating and directing cord growth, in which case cord formation may be seen as an effective means of amplifying the sensitivity and responsiveness of individual hyphae and may explain why cord-forming fungi are attracted so rapidly to newly supplied potential food bases.

Resource utilisation and mobilisation

In the soil tube experiments described previously the weight losses of blocks colonised by *P. impudicus, P. velutina, P. laevis, S. fimbriatum* and *H. fasciculare* were significantly greater in irradiated soil than in non-sterile soil. Also when blocks colonised by *P. laevis* were analysed for nitrogen those from sterile soil were found to contain more nitrogen at the end of the experiment than at the beginning or than blocks from unsterilised soil. This extra nitrogen probably originated from the sterile soil and entered the blocks by direct diffusion or translocation by the mycelium. Thus higher nitrogen levels in sterile soil were associated with more decay and with the production of lesser amounts of mycelium, whilst in non-sterile soil lower nitrogen levels were associated with less decay and substantial cord growth. It would appear then that this species, and perhaps other cord-forming species, produced more extensive mycelial cord systems but cause less decay of wood at the food base in non-sterile soil, whilst in sterile soil, where nutrient levels are presumed to be higher, the fungus appears to be

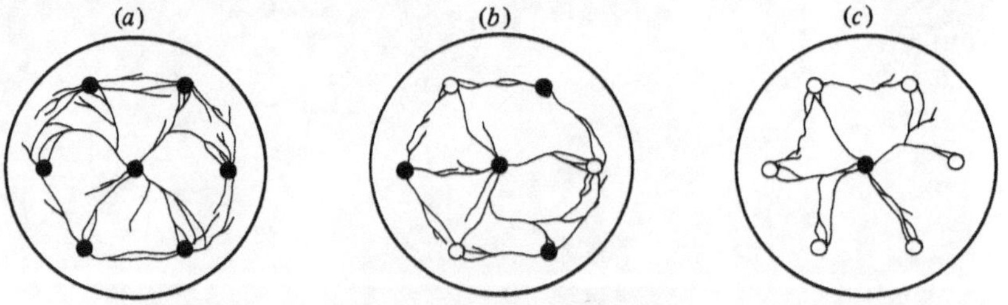

Fig. 8. Development of mycelial cords by a coprophilous basidiomycete in non-sterile sand. Filled circles, rabbit pellets colonised at start of experiment; hollow circles, pellets not colonised at start of experiment. (Thompson & Rayner, 1983.)

more confined to the food base and shows less tendency to explore the surrounding soil.

Interactions

Intraspecific: mycelial recognition phenomena. The existence of somatic incompatibility phenomena amongst cord-forming fungi has already been mentioned within the context of population structure. The occurrence of somatic acceptance and rejection phenomena between like and unlike genotypes respectively, raises the question as to how these might be mediated via aggregations of hyphae as opposed to diffuse mycelia where interactions between individual hyphae initiate the response. Observations of excavated field systems (see above) indicate that anastomoses between cords within the same system occur readily, sometimes over distances of the order of several centimetres. Laboratory studies with *P. velutina* have shown that mycelial cords of different individuals arising from wood-block inocula in non-sterile soil can grow towards one another and come into contact. In all cases when this happens fusions occur but while those between like mycelial types persist, those between different types are followed by discolouration and eventual death of the previously fused regions. This results in an 'avoidance reaction' and a lack of further growth into these areas (Fig. 9).

Interspecific: competition with Armillaria *species.* As has been indicated, colonisation of woody substrata by cord-forming fungi is often effected

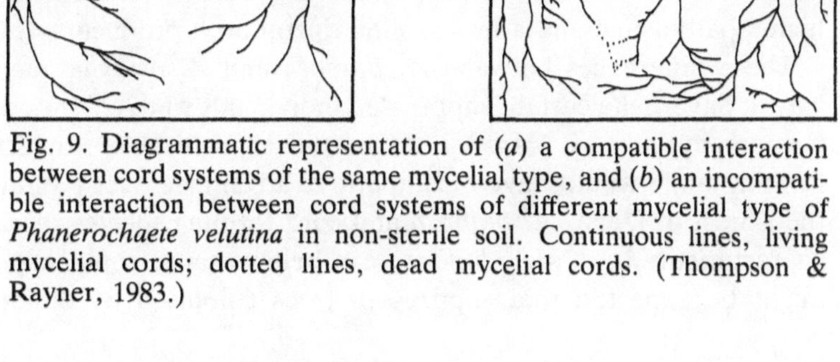

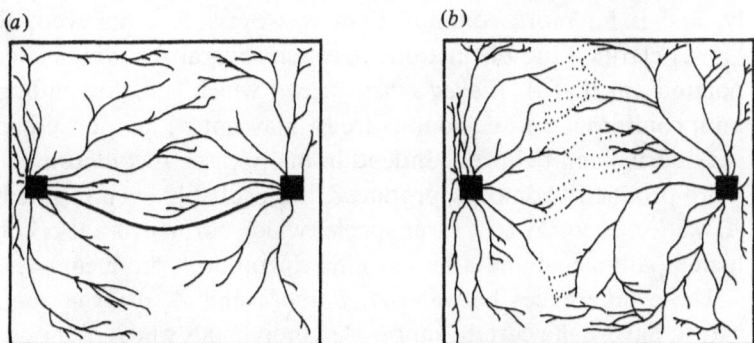

Fig. 9. Diagrammatic representation of (*a*) a compatible interaction between cord systems of the same mycelial type, and (*b*) an incompatible interaction between cord systems of different mycelial type of *Phanerochaete velutina* in non-sterile soil. Continuous lines, living mycelial cords; dotted lines, dead mycelial cords. (Thompson & Rayner, 1983.)

by extensive fans of mycelium, which are typically subcortical if bark is present. Production of the resulting sheath of mycelium allows considerable subsequent primary resource capture in competition with other fungi (see Rayner & Webber: Chapter 18) and hence occupation of large decay columns by one or a few individuals (see Rayner *et al.*: Chapter 23). This may have important practical implications in relation to competition with root pathogens, such as *Armillaria* species. The latter also have an ectotrophic habit and often produce abundant external mycelium and rhizomorphs, but this does not necessarily always imply that large volumes of wood are occupied (Rayner, 1975). In fact when *Armillaria* and cord-forming species occur together subcortically in dead wood the cord formers usually predominate in the actual decay, apparently having a competitive advantage over the former in purely saprotrophic situations (Rayner, 1975).

In the studies of suppressed trees mentioned previously, *A. bulbosa* was found only at one woodland site (the Forest of Dean) and *A. ostoyae* only at the other (Savernake). At Savernake forest 55% of sampled trees were colonised by cord-forming fungi, 42.5% by *A. ostoyae* and 27.5% by both cord-forming species and *A. ostoyae.* However at the Forest of Dean no trees were colonised by cord-forming fungi which were not also colonised by *A. bulbosa* (i.e. 69% were colonised by cord formers, 88% by *A. bulbosa* and 69% by both *A. bulbosa* and cord-forming fungi) (Table 3). These differences probably reflect the different ecological behaviour of the *Armillaria* species. *A. bulbosa* is a competitive saprotroph, although it can behave parasitically, and is far more common than was previously believed. Rishbeth (1982) clarified the distinctions between several *Armillaria* species and pointed out that *A. mellea sensu strictu*, which is highly pathogenic on both coniferous and deciduous trees, may not be as widespread as had previously been believed. Indeed in many cases *A. bulbosa*, which is a more prolific rhizomorph producer, has probably been misidentified as *A. mellea. A. ostoyae*, a rarer species which normally attacks conifers, is highly pathogenic and a less prolific rhizomorph producer.

These differences between *A. bulbosa* and *A. ostoyae* may explain why at Savernake certain suppressed individuals were colonised by cord formers but not *A. ostoyae*, as in the saprotrophic situation cord-forming fungi would have a competitive advantage over pathogens. At the Forest of Dean, *A. bulbosa* and cord-forming species share similar characteristics in that *A. bulbosa* can behave saprotrophically. Thus it might be expected that suppressed trees colonised by either a cord former or an *Armillaria* species might also be colonised by the other, particularly as both cord formers and *A. bulbosa* are capable of producing extensive cord/rhizomorph networks in the soil.

Where cord-forming fungi and *Armillaria* species occurred together in the bases of suppressed trees, both occupied large volumes of wood. Usually, however, *Armillaria* was found in the lowest parts, whilst cord formers occupied greater volumes of wood at collar level. In most cases cord formers occupied peripheral regions, although they did sometimes spread considerable distances towards the central areas of roots (Fig. 10). The observation that cord-forming fungi occupied higher and peripheral regions whilst *Armillaria* species were limited to lower and more central regions may indicate that cord-forming fungi were actively replacing *Armillaria* species. Alternatively the *Armillaria* species may have invaded first (in lower regions) and come to occupy large volumes of wood, being restricted in higher regions by the cord-forming fungi which invaded later peripherally. This hypothesis that cord-forming fungi may limit or replace *Armillaria* species is supported by pairings in culture where all cord-forming species tested (*P. velutina, P. impudicus, T. platyphylla* and *H. fasciculare*) overgrew *A. bulbosa* and *A. ostoyae*. However, although in these pairings the latter were physically over-

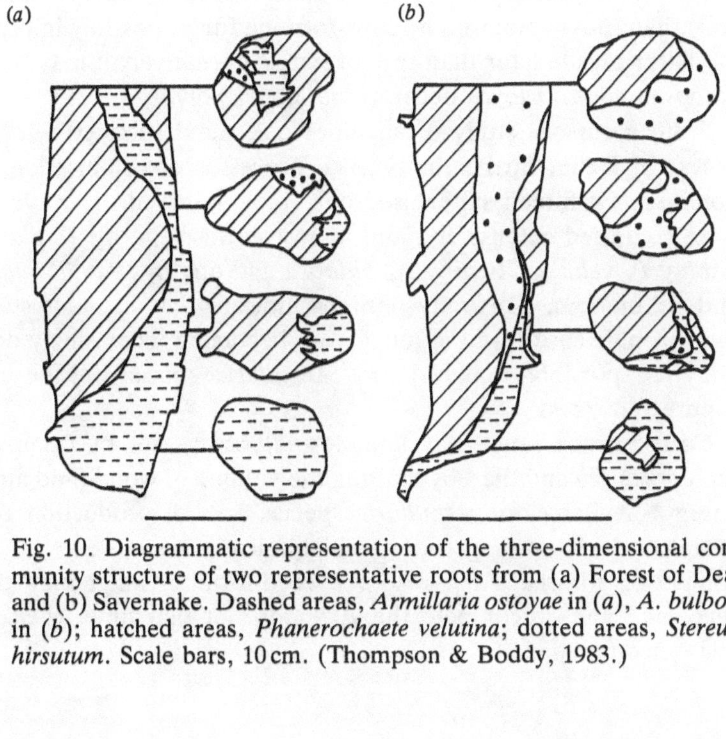

Fig. 10. Diagrammatic representation of the three-dimensional community structure of two representative roots from (a) Forest of Dean and (b) Savernake. Dashed areas, *Armillaria ostoyae* in (*a*), *A. bulbosa* in (*b*); hatched areas, *Phanerochaete velutina*; dotted areas, *Stereum hirsutum*. Scale bars, 10 cm. (Thompson & Boddy, 1983.)

grown they may not necessarily have been replaced. It is possible that the *Armillaria* species can seal themselves off from the overgrowing species, effectively retaining the potential for growth when suitable conditions arise. Also pairings of this type may not represent the situation which occurs in the woody tissues of a dead or dying tree (e.g. Rayner & Todd, 1979).

On the two main sites studied the dead trees had died at different times, some only in the year prior to study whilst others were in advanced stages of decay. The Savernake site consisted of the youngest trees, and in general the majority of dead trees there can be considered to have died fairly recently. This is borne out by the generally higher collar RDs of sampled trees from this site. (RD, relative density, is a means of estimating the state of decay of wood and is calculated as dry weight/volume. Living oak has an RD of 0.55–0.63 g cm^{-3} (Boddy & Thompson, 1983) and more decayed wood has lower RD values.) Fewer roots had been colonised by *A. ostoyae* or cord-forming fungi on this site than by *A. bulbosa* or cord-forming fungi in the Forest of Dean, and considerably fewer were colonised by the two types of fungi together at the former site. Also roots of trees from Savernake were generally at earlier stages of colonisation than at the Forest of Dean site. On both sites roots colonised by *A. bulbosa* or *A. ostoyae* tended to have higher RDs than those colonised by cord-forming fungi, possibly implying that the latter invade later than the former. Alternatively it may be that *A. bulbosa* and *A. ostoyae* decay wood less rapidly.

From a cursory study of an older oak stand (planted 1929) in the Forest of Dean, further hints as to processes of colonisation may be obtained. Of fifteen suppressed oak trees observed the roots of eight were colonised extensively, and almost exclusively, by *T. platyphylla*, one by *P. velutina*, two by *A. bulbosa* and one by both *T. platyphylla* and *A. bulbosa*. This may perhaps indicate that in older stands the saprotrophic cord-forming fungi were becoming increasingly dominant and had possibly replaced any *Armillaria* species in the roots of suppressed trees.

Cord-forming fungi may limit *Armillaria* species by occupying the same resource and thereby limiting the volume of wood (and amount of energy) available for *Armillaria* species for the production of rhizomorphs which may spread to neighbouring trees. Also by actively competing for the same resource cord-forming fungi may be more successful and outcompete the *Armillaria* at that time when no host resistance remains.

Using the technique of somatic incompatibility intra-specific pairings between different mycelial types of the two *Armillaria* species were made in addition to those between *P. velutina* isolates described earlier. The distribution of the two types of *A. ostoyae* found at Savernake and the two of *A. bulbosa* at the Forest of Dean are shown in Fig. 3. Isolates of the same mycelial type of *A. bulbosa* were found in suppressed trees 50 m apart, which again reflects the ability which fungi with a cord-forming/rhizomorphic habit have to spread in the forest floor.

Conclusions

This chapter has been concerned with the importance of mycelial cords in providing inter-communicating systems between spatially discontinuous woody resource units (food bases) on the forest floor, and their role in colonisation and vegetative spread. Understanding the dynamics of these systems provides a new challenge for mycologists, and the principles involved may differ substantially from those applying to diffuse mycelia on or in homogeneous media. Some features, such as branching patterns, formation of anastomoses, and somatic incompatibility, are reminiscent of phenomena exhibited by hyphae in diffuse mycelia but can be viewed on a macro- rather than a micro-scale.

Observations of the distribution of these fungi indicate the large dimensions their individual mycelia may achieve whilst also revealing the great extent to which populations of cord-forming fungi are distributed throughout the woodland floor. Cord-forming basidiomycetes then appear to be a ubiquitous group of saprotrophs in deciduous woodlands and play a major role as agents of decay of fallen wood and litter in forests. They may also be important, in certain situations, as natural and potential competitors against root-infecting fungi such as *Armillaria* species.

I am most grateful to Dr Alan Rayner and Professor David Jennings for their very helpful comments on the manuscript.

References

Adams, D. H. (1974). Identification of clones of *Armillaria* in young-growth ponderosa pine. *Northwest Science*, **48**, 21–8.

Anderson, J. B., Ullrich, R. C., Roth, L. F. & Filip, G. M. (1979). Genetic identification of clones of *Armillaria mellea* in coniferous forests in Washington. *Phytopathology*, **69**, 1109–11.

Barrett, D. K. & Uscuplic, M. (1971). The field distribution of interacting strains of *Polyporus schweinitzii* and their origin. *New Phytologist*, **70**, 581–98.

Bellotti, R. A. & Couse, N. L. (1980). Induction of mycelial strands in *Calvatia sculpta*. *Transactions of the British Mycological Society*, **74**(1), 19–25.

Boddy, L. & Thompson, W. (1983). Decomposition of suppressed oak trees in even-aged plantations. I. Stand characteristics and decay of aerial parts. *New Phytologist*, **93**, 261–76.

Botton, B. (1978). Influence of calcium on the differentiation and growth of aggregated organs in *Sphaerostilbe repens*. *Canadian Journal of Microbiology*, **24**, 1039–47.

Botton, B. & El-Khouri, M. (1978). Synnema and rhizomorph production in *Sphaeorstilbe repens* under the influence of other fungi. *Transactions of the British Mycological Society*, **70**, 131–6.

Brownlee, C. & Jennings, D. H. (1982). Pathway of translocation in *Serpula lacrimans*. *Transactions of the British Mycological Society*, **79**(3), 401–7.

Butler, G. M. (1958). The development and behaviour of mycelial strands in *Merulius lacrymans* (Wulf.) Fr. II. Hyphal behaviour during strand formation. *Annals of Botany*, **22**, 219–36.

Butler, G. M. (1966). Vegetative structure. In *The Fungi: An Advanced Treatise*, vol. 2, ed. G. C. Ainsworth & A. Sussman, pp. 83–112. London: Academic Press.

Carruthers, S. M. & Rayner, A. D. M. (1979). Fungal communities in decaying hardwood branches. *Transactions of the British Mycological Society*, **72**, 283–9.

Childs, T. W. (1963). *Poria weirii* root rot. Symposium: Root Diseases. *Phytopathology*, **53**, 1124–7.

Clarke, R. W., Jennings, D. H. & Coggins, C. R. (1980). Growth of *Serpula lacrimans* in relation to water potential of substrate. *Transactions of the British Mycological Society*, **75**, 271–80.

Coggins, C. R., Hornung, U., Jennings, D. H. & Veltkamp, C. J. (1980). The phenomenon of 'point growth' and its relation to flushing and strand formation in mycelium of *Serpula lacrimans*. *Transactions of the British Mycological Society*, **75**(1), 69–76.

Cromack, K. Jr., Sollins, P., Todd, R. L., Fogel, R., Todd, A. W., Fender, W. M., Crossley, M. E. & Crossley, D. A. Jr (1977). The role of oxalic acid and bicarbonate in calcium cycling by fungi and bacteria: some possible implications for soil animals. *Soil Organisms as Components of Ecosystems. Ecological Bulletin* (Stockholm), **25**, 246–52.

DeBary, A. (1887). *Comparative Morphology and Biology of the Fungi, Mycetozoa and Bacteria*, pp. 22–9. Oxford: Clarendon Press.

Falck, R. (1912). Die *Merulius*-Faule des Bauholzes. *Hausschwammforschungen*, **6**, 1–405.

Garraway, M. O. (1970). Rhizomorph initiation and growth in *Armillaria mellea* promoted by *o*-aminobenzoic and *p*-aminobenzoic acids. *Phytopathology*, **60**, 861–5.

Garrett, S. D. (1953). Rhizomorph behaviour in *Armillaria mellea* (Vahl) Quél. I. Factors controlling rhizomorph initiation by *Armillaria mellea* in pure culture. *Annals of Botany*, **65**, 63–79.

Garrett, S. D. (1954). Function of the mycelial strands in substrate colonization by the cultivated mushroom *Psalliota hortensis*. *Transactions of the British Mycological Society*, **37**, 51–7.

Glasare, P. (1970). Volatile compounds from *Pinus silvestris* stimulating the growth of wood-rotting fungi. *Archiv für Mikrobiologie*, **72**, 333–43.

Grainger, J. (1962). Vegetative and fructifying growth in *Phallus impudicus*. *Transactions of the British Mycological Society*, **45**, 145–55.

Gray, T. R. G. & Williams, S. T. (1971). *Soil Micro-organisms*, p. 43. London and New York: Longman.
Hein, I. (1930). Studies on the mycelium of *Psalliota campestris*. *American Journal of Botany*, **17**, 197–211.
Hirt, R. R. (1949). An isolate of *Poria xantha* on media containing copper. *Phytopathology*, **39**(1), 31–6.
Hornung, U. & Jennings, D. H. (1981). Light and electron microscopical observations of surface mycelium of *Serpula lacrimans*: stages of growth and hyphal nomenclature. *Nova Hedwigia*, **34**, 101–26.
Korhonen, K. (1978). Interfertility and clonal size in the *Armillariella mellea* complex. *Karstenia*, **18**, 31–42.
Mathew, K. T. (1961). Morphogenesis of mycelial strands in the cultivated mushroom, *Agaricus bisporus*. *Transactions of the British Mycological Society*, **44**, 285–90.
Moody, A. R., Garraway, M. O. & Weinhold, A. R. (1968). Stimulation of rhizomorph production in *Armillaria mellea* with oils and fatty acids. *Phytopathology*, **58**, 1060–1.
Park, D. (1963). Evidence for a common fungal growth regulator. *Transactions of the British Mycological Society*, **46**(4), 541–8.
Pentland, G. D. (1965). Stimulation of rhizomorph development of *Armillaria mellea* by *Aureobasidium pullulans* in artificial culture. *Canadian Journal of Microbiology*, **11**, 345–50.
Rayner, A. D. M. (1975). Fungal colonization of hardwood tree stumps. Ph.D. Thesis, University of Cambridge, UK.
Rayner, A. D. M. & Todd, N. K. (1979). Population and community structure and dynamics of fungi in decaying wood. *Advances in Botanical Research*, **6**, 333–420.
Rayner, A. D. M. & Todd, N. K. (1982). Ecological genetics of basidiomycete populations in decaying wood. In *Decomposer Basidiomycetes – Their Biology and Ecology*, ed. J. C. Frankland, J. N. Hedger & M. J. Swift, pp. 129–42. Cambridge, UK: Cambridge University Press.
Rishbeth, J. (1978). Infection foci of *Armillaria mellea* in first rotation hardwoods. *Annals of Botany*, **42**, 1131–9.
Rishbeth, J. (1982). Species of *Armillaria* in southern England. *Plant Pathology*, **31**, 9–17.
Rogers, C. H. & Watkins, G. M. (1938). Strand formation in *Phymatotrichum omnivorum*. *American Journal of Botany*, **25**, 244–6.
Salonius, P. O., Robinson, J. B. & Chase, F. E. (1967). A comparison of autoclaved and gamma-irradiated soils as media for microbial colonisation experiments. *Plant and Soil*, **27**, 239–48.
Shaw, C. G. III & Roth, L. F. (1976). Persistence and distribution of a clone of *Armillaria mellea* in a ponderosa pine forest. *Phytopathology*, **66**, 1210–13.
Sortkjaer, O. & Allermann, K. (1972). Rhizomorph formation in fungi. I. Stimulation by ethanol and acetate and inhibition by disulfiram of growth and rhizomorph formation in *Armillaria mellea*. *Physiologia plantarum*, **26**, 376–80.
Stark, N. (1972). Nutrient cycling pathways and litter fungi. *Bioscience*, **22**, 355–60.
Styer, J. F. (1930). Nutrition of the cultivated mushroom. *American Journal of Botany*, **17**, 983–94.
Thompson, W. (1982). Biology and ecology of mycelial cord-forming basidiomycetes in deciduous woodlands. Ph.D. Thesis, University of Bath, UK.
Thompson, W. & Boddy, L. (1983). Decomposition of suppressed oak trees in even-aged plantations. II. Colonization of tree roots by cord- and rhizomorph-producing basidiomycetes. *New Phytologist*, **93**, 277–91.
Thompson, W. & Rayner, A. D. M. (1982*a*). Structure and development of mycelial cord

systems of *Phanerochaete laevis* in soil. *Transactions of the British Mycological Society*, **78**(2), 193–200.

Thompson, W. & Rayner, A. D. M. (1982*b*). Spatial structure of a population of *Tricholomopsis platyphylla* in a woodland site. *New Phytologist*, **92,** 103–14.

Thompson, W. & Rayner, A. D. M. (1983). Extent, development and functioning of mycelia cord systems in soil. *Transactions of the British Mycological Society*, **81,** 333–45.

Todd, N. K. & Rayner, A. D. M. (1980). Fungal individualism. *Science Progress, Oxford,* **66,** 331–54.

Todd, R. L., Cromack, K. Jr & Stormer, J. C. Jr (1973). Chemical exploration of the micro-habitat by electron probe micro-analysis of decomposer organisms. *Nature*, **243,** 544–6.

Townsend, B. B. (1954). Morphology and development of fungal rhizomorphs. *Transactions of the British Mycological Society*, **37,** 222–33.

Valder, P. G. (1958). The biology of *Helicobasidium purpureum* Pat. *Transactions of the British Mycological Society*, **41,** 283–308.

Watkinson, S. C. (1971). The mechanism of mycelial strand induction in *Serpula lacrimans*: a possible effect of nutrient distribution. *New Phytologist*, **70,** 1079–88.

Watkinson, S. C. (1975). The relation between nitrogen nutrition and formation of mycelial strands in *Serpula lacrimans. Transactions of the British Mycological Society*, **64,** 195–200.

Weigl, N. & Ziegler, H. (1960). Wasserhaushalt und Stoffleitung bei *Merulius lacrymans* (Wulf) Fr. *Archiv für Mikrobiologie*, **37,** 124–33.

Weinhold, A. R. (1963). Rhizomorph production by *Armillaria mellea* induced by ethanol and related compounds. *Science*, **142,** 1065–6.

Weinhold, A. R. & Garraway, M. O. (1966). Nitrogen and carbon nutrition of *Armillaria mellea* in relation to growth promoting effects of ethanol. *Phytopathology*, **56,** 108–12.

Weinhold, A. R., Hendrix, F. F. & Raabe, R. D. (1962). Stimulation of rhizomorph growth of *Armillaria mellea* by indole-3-acetic acid and figwood extract. *Phytopathology*, **52,** 757.

10
The structure and function of the vegetative mycelium of mycorrhizal roots

D. J. READ
Department of Botany, The University of Sheffield, Sheffield, S10 2TN, UK

Examination of the fossil roots of some of the earliest land plants has revealed the presence of vegetative fungal structures which are strikingly reminiscent of those found in present day vesicular-arbuscular (VA) mycorrhizas (Nicolson, 1975). Today, very few plants growing in natural plant communities are free from mycorrhizal infection. The presence of a supply of simple available carbon compounds either inside or close to the roots of autotrophic higher plants would clearly always have been a feature of great nutritional significance to heterotrophic micro-organisms the growth of which was otherwise limited in the soil environment, consisting as it does mostly of less-available carbon sources. If infection in turn provided the autotroph with improved access to minerals which were previously present in growth-limiting quantity, the vigour of both partners in a competitive situation would be improved. Such mutualistic associations were clearly favoured by selection pressures early in the colonisation of the terrestrial environment and they are now recognisable in the form of a number of distinctive root–fungus associations each characterised by a particular pattern of development and distribution of its vegetative mycelium and each most commonly, though not exclusively, associated with a particular complex of soil conditions and climate. Thus, those members of the Ericaceae which normally become dominant plants in acid mor-humus soils of cold sub-arctic and sub-alpine regions have a characteristic internal or endo-infection formed by ascomycetous fungi which is termed an 'ericoid' infection (Read, 1983). In contrast, on the mineral soils of higher pH, which occur in warmer and drier climates with high rates of organic matter turnover, grasses and herbs predominate and these mostly retain what may be regarded as the primitive VA type of infection caused by zygomycetous fungi of genera such as *Glomus*, *Gigaspora* and *Acaulospora*. In this mycorrhizal type, in addition to a prominent internal phase consisting of vesicles for storage and arbuscules through which nutrient exchange with the host takes place, there is an important external mycelial phase.

Between the peaty heathland soils and the drier mineral soils of the grasslands lie the intermediate mor, moder and brown soils which frequently support forests. The dominant trees of these boreal and temperate forest zones – which are often members of the Pinaceae and Fagaceae in the northern hemisphere and of the Fagaceae and Myrtaceae in the southern hemisphere – have distinctive ecto- or sheathing mycorrhizas formed largely by Basidiomycetes, the vegetative mycelium of which is predominantly external, forming a compact sheath around the lateral roots and a more or less extensive mycelial phase in the surrounding soil. These broad relationships between environmental circumstances and mycorrhizal type are presented diagramatically in Fig. 1.

Since the mycorrhizal types can be seen to occupy specific positions along an environmental gradient of latitude and altitude it is logical to suppose that in each case their structure and function have evolved in response to the particular conditions of soil and climate prevailing at

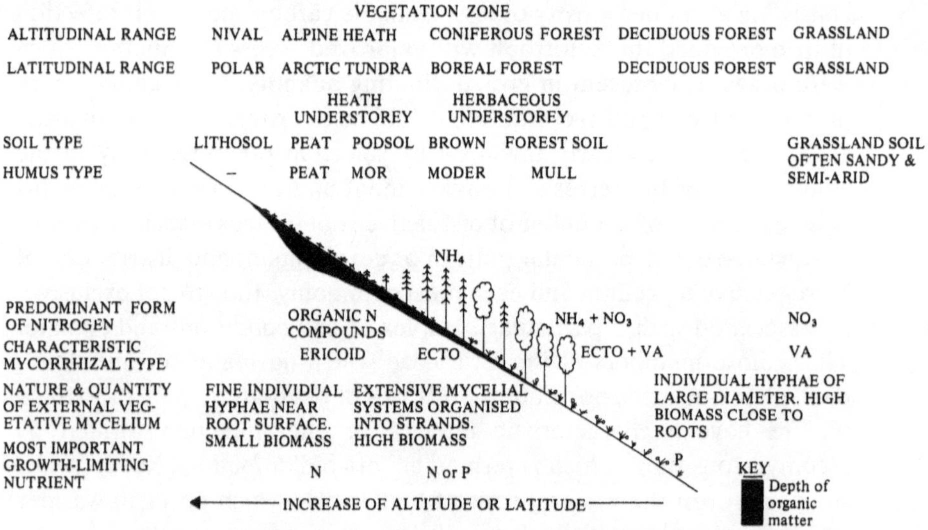

Fig. 1. Diagrammatic presentation of the postulated relationship between latitude or altitude, climate, soil and mycorrhizal type, and development of the vegetative mycelium associated with mycorrhizas.

that point on the gradient. As shown in Fig. 1 demarcation between zones is not always sharp, and intermediate areas occur in which plants bearing two types of mycorrhiza can grow together. Nonetheless the model provides a useful basis from which to analyse the relationship between structure and function of each mycelial type and the vegetation and soil type with which it is associated.

Distribution, structure and function of vegetative mycelium of ericoid mycorrhizas

Ericaceous species become dominant, and indeed often occur as the only plants, in heathland communities which dominate the cold and wet end of the environmental gradient (Fig. 1). Thus, they are best represented in the taiga zone around the northern polar ice cap and immediately below the nival zone in alpine areas. As a result of low soil temperatures, high rain or snowfall and low evaporation rates, soil moisture levels are high and decomposition and mineralisation processes are inhibited. Under these circumstances, organic matter accumulates to give deposits of peat or mor-humus to which the ericoid roots are largely restricted. The younger parts of these roots are devoid of root hairs but are themselves of very narrow dimensions and a dense fibrous root system is produced (Read, 1983).

The young cortical cells of these roots are filled with the much-branched vegetative mycelium of the ascomycetous fungus *Pezizella ericae* (Fig. 2). Examination of micrographs reveals that a very high proportion of the root biomass is made up of fungal material and that most of the vegetative mycelium is in the internal position (Fig. 2a,c). Careful analysis of the relative proportions of fungus to host material in each individual cell of such micrographs indicates that over 70% of the volume of the host cell is occupied by the fungus at the time when the infection unit is mature. The active life of each individual intracellular infection is of the order of four to five weeks and it is terminated by the breakdown of host cytoplasm which envelopes the internal mycelium in the healthy association (Duddridge & Read, 1982).

The external mycelium is relatively weakly developed (Fig. 2b). A very loose weft of hyphae is seen to cover the surface of the growing root and from this hyphae penetrate into each host cortical cell to form the individual infection units (Fig. 2c). Estimated numbers of entry points range from 250 to 2000 per cm of root in seedling roots of *Calluna vulgaris* (Read & Stribley, 1975). Surface views demonstrate clearly that most of the hyphae run parallel to the surface of the root and that some

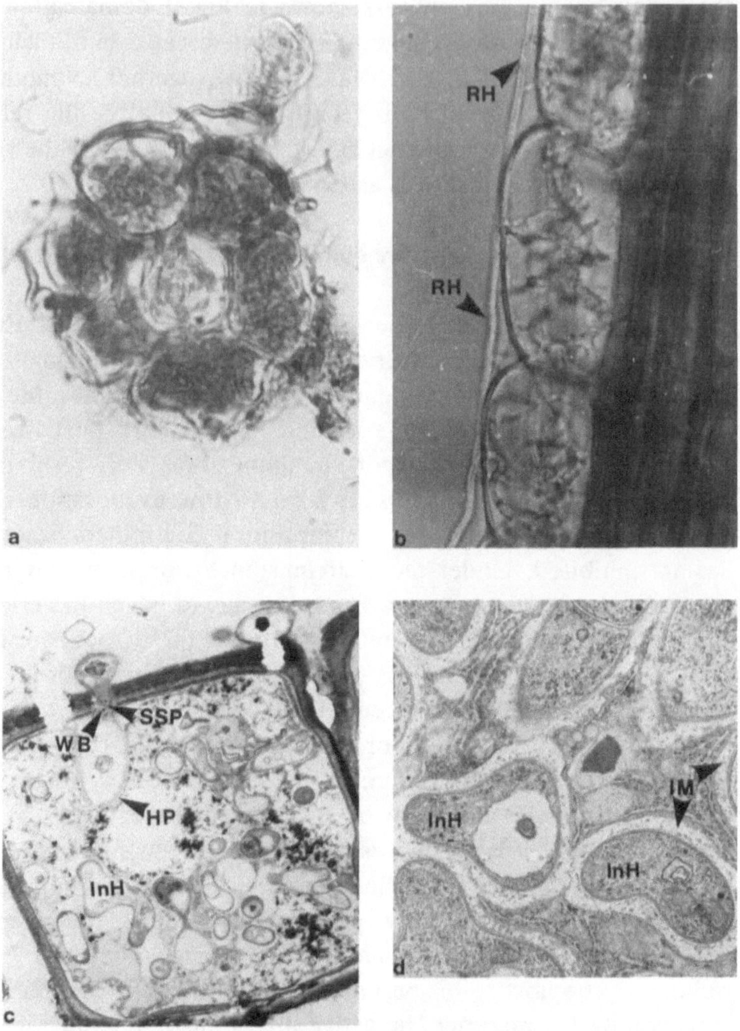

Fig. 2. (a) Transverse section of 'ericoid' root of *Calluna vulgaris* showing cortical cells filled with vegetative mycelium and the loose external weft of hyphae. × 800. (b) Early stage of infection of root of *Calluna vulgaris*. A 'runner' hypha (RH) growing parallel to the root surface produces lateral branches which penetrate the cortical cells and branch freely to produce the hyphal coils typical of ericoid infection. × 1000. (c) Transmission electron micrograph (TEM) of a root cortical cell of *Rhododendron ponticum* showing penetration of the cell wall by a hypha with simple septal pore (SSP) and Woronin body (WB) indicating ascomycetous affinities. Numerous internal hyphae (InH) can be seen surrounded by host plasmalemma (HP). × 2000. (d) High-magnification TEM of intracellular infection of the ericoid type.

of these which have rather larger diameters act as pioneers or runner hyphae from which lateral branches penetrate newly differentiated host cortical cells (Fig. 2b). Extension of mycelium outwards from the root surface is extremely restricted. It is difficult to obtain precise measurements of radial extension of hyphae in peat soils but recovery from sandy heathland soils can be relatively complete. In these circumstances hyphae normally penetrate little more than 1 cm from the root and the majority are restricted to the zone within 0.5 cm of its circumference. The external hyphae do not normally become organised into aggregates either on the surface of the root or in the surrounding soil. The effect of this distribution pattern is to produce a loose weft of mycelium around the root, the density of which is proportional to the intensity of internal infection. This weft is, however, never of a density comparable with that seen in ecto-mycorrhizas and, indeed, is not sufficient to form a complete cover over the surface of the root. The ratio of the biomass of external mycelium to that of internal mycelium is of the order of 0.1.

The predominantly internal development of the mycelium of ericoid mycorrhizas is probably of considerable significance from the functional point of view. In these systems the function of soil exploitation is to a large extent fulfilled by the diffuse and extensive 'hair-root' system. The colloidal organic matter into which the roots grow has a water-holding capacity so great that the roots are normally bathed in a dilute solution of carboxylic acids containing free and adsorbed nutrients. Under these circumstances investment of carbon in a large external network of hyphae would yield little advantage to the host.

Experiments on the endophyte in pure culture indicate that it may be relatively poorly adapted to outward extension into very acidic soil environments, since the optimum pH for growth is between 6 and 7 (Pearson & Read, 1975) and virtually no growth occurs between pH 3 and 4. The pH for optimal growth is thus closer to that of the vacuole of ericoid root cells than to that of heathland soils which are commonly in the range 3 to 4. This must be one of the reasons for the predominantly internal development of the fungus.

The mycorrhizal fungus is capable of producing acid-phosphatases, the level of activity of the enzyme being greatly stimulated at low

---

Caption for Fig. 2 (*cont.*)
A distinct electron-lucent zone, the inter-facial matrix (IM), surrounds all intracellular hyphae (InH). This zone forms a barrier across which all nutrient exchanges must take place and in which complexing of metallic elements is likely to occur. × 15 000.

external levels of inorganic phosphate (Pearson & Read, 1975). Organic phosphates, supplied in the form of ferric or aluminium inositol phytate are readily used as substrates for growth (Mitchell & Read, 1981). Organic nitrogen sources, in the form of amino acids, have been shown to be used as readily as ammonium by the endophyte in pure cultures (Pearson & Read, 1975) and later work (Stribley & Read, 1980) revealed that a range of amino acids could be used as nitrogen sources by mycorrhizal ericaceous plants whereas they were not readily used by the plants in the absence of the fungus. The mycorrhizal fungus thus provides the plant with a capacity to assimilate organic compounds of the kind which would be expected to be present in the colloidal matrix surrounding its roots.

A further attribute of the endophyte which may be of equal or greater importance to the host plant is its heavy-metal resistance. The solubility of potentially toxic metal elements such as aluminium, copper and zinc is greatly increased at the low pH levels prevailing in heathland soils and these elements pose a threat to the survival of plants under these circumstances (Rorison, 1973). Studies have revealed that the ericoid mycorrhizal fungus will grow in metal concentrations an order of magnitude greater than those survived by the host plant grown in the absence of the endophyte (Bradley *et al.*, 1982). It is likely that as shown in other fungi (Ashida, Higachi & Kikuchi, 1963; Paton & Budd, 1972; Duddridge & Wainwright, 1980) the resistance is based on the capacity of the fungus to complex metallic elements in its cell walls. The functional groups responsible for metal binding in the fungal cell wall are thought to be carboxylic-acid moieties of pectin-like substances (Tuominen, 1967). It is interesting that such substances are also present in quantity in the inter-facial matrix (Fig. 2d) between fungus and host cytoplasm in ericoid mycorrhizas (Duddridge & Read, 1982). The hyphal complexes within the host-cortical cell thus provide very large surface areas upon which metals can be complexed. These adsorption mechanisms lead to exclusion of metal ions from the shoot of the host plant which avoids the toxicity otherwise arising from their accumulation. In the absence of a mycorrhizal endophyte ericaceous plants have very little resistance to heavy metals which freely traverse the roots and accumulate in the shoots (Bradley *et al.*, 1981).

Distribution, structure and function of the vegetative mycelium of ectomycorrhizas

Environmental amelioration along the gradient in a southerly or downward direction arises through increasing mean summer temperatures and evapotranspiration rates. With an increased seasonality of climate and some surface drying of soil, coniferous trees replace ericaceous shrubs as dominants in the plant community though in the early part of the transition ericaceous shrubs, notably shade-tolerant members of the genus *Vaccinium*, continue to occur as dominants in the field layer. In contrast to the situation found in ericaceous plants, ectomycorrhizal conifers produce very large quantities of external mycelium, some of which is in the form of the sheath, the remainder being in the form of extensive fans of mycelium growing outwards from the sheath to exploit the soil over distances often in excess of 20 cm from the root (Skinner & Bowen, 1974*a*; Duddridge, Malibari & Read, 1980). Because of their delicacy these mycelial systems are normally broken when soil is excavated but analysis of their development is possible using root chambers with transparent walls (Brownlee, Duddridge, Malibari & Read, 1983) (Fig. 3a). In the case of *Suillus bovinus* mycelium associated with roots of *Pinus sylvestris* or *P. contorta*, the fans move through unsterile peat at rates of up to 3 mm day^{-1} in growth-chamber environments with temperatures of 20 °C day and 15 °C night. This may be compared with a growth rate of 7–8 mm day^{-1} recorded in the strand-forming wood decomposer *Serpula lacrimans* by Jennings (1982). In contrast to the mycelia of ericoid mycorrhizal systems, their growth is totally inhibited by waterlogging, indicating a sensitivity to anoxic conditions, and is generally in the direction of drier surface layers of the soil. This requirement for oxygen may be at least in part responsible for the exclusion of the ectomycorrhizal habit from the colder and wetter end of the gradient. Oxygen supplied as a result of internal transport from the air via plant intercellular spaces may be sufficient to initiate and sustain sheath formation (Read & Armstrong, 1972) but is unlikely to be sufficient to support outward growth of mycelium into the soil.

Hyphae within the mycelial fans often associate together to form more or less differentiated structures (Duddridge *et al.*, 1980) which have been variously described as 'strands', 'ropes' or 'cords'. These structures are not true rhizomorphs *sensu* Garrett (1956) since they lack a differentiated meristem and apical growth. Their structure and mode of growth is analogous to that seen in *Serpula lacrimans* (Falck, 1912; Butler, 1957).

Vessel hyphae of large internal diameter and generally lacking cytoplasm in the mature condition occupy the core of the strands (Duddridge *et al.*, 1980; Foster, 1981) and these are surrounded by densely cytoplasmic hyphae of smaller diameter. (Fig. 4 inset.)

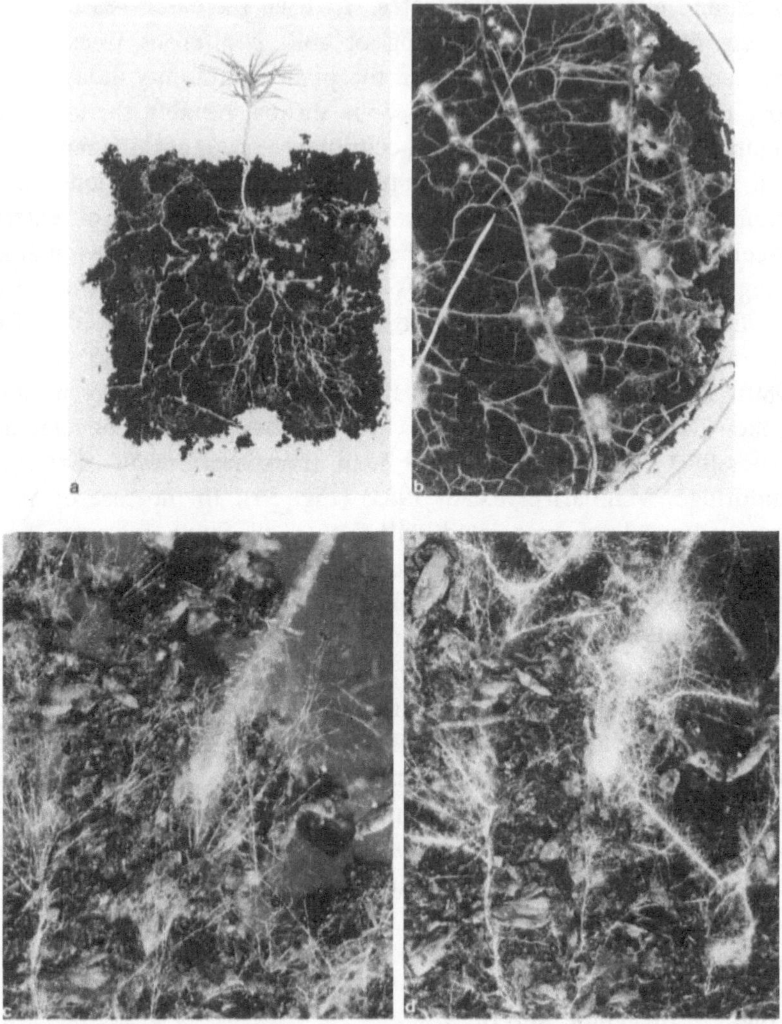

Fig. 3. (a) Seedling of *Pinus sylvestris* infected with the mycorrhizal fungus *Suillus bovinus* then grown for six weeks in unsterile peat in a transparent Perspex root chamber. Mycelial fans have developed from infected mycorrhizal laterals and exploited the peat throughout the chamber. × 0.2. (b) More detailed view of a similar mycorrhizal system showing the intensive exploitation of peat by mycorrhizal strands which arise from, and subsequently interconnect, mycorrhizal lateral roots. The actively-growing white primary root (left centre) alone remains infected. × 0.8. (c), (d) Sequential view of a growing lateral root of *Pinus contorta* as it comes into contact with advancing fans of *Suillus bovinus* mycelium in a root chamber. Figure 3c shows the root three days after contact with the hyphae. Some enhancement

Members of fungal genera such as *Suillus*, *Thelephora*, *Pisolithus* and *Lactarius*, which are among the commonest mycorrhizal associates of the Pinaceae, form particularly prolific differentiated fan systems of this kind. In the upper layers of soil under coniferous trees, the vegetative mycelium of these fungi probably represent by far the greatest component of microbial biomass clearly receiving an abundant supply of free carbon from the host to fuel its extension into the soil. It has even been suggested (Gadgil & Gadgil, 1971) that this growth is so vigorous that it leads to the competitive inhibition of decomposer organisms and reduction of decomposition rates, but this observation needs corroboration.

With further advance along the gradient of climate and soil, deciduous tree species become more important, members of the Fagaceae, notably *Quercus* and *Fagus*, dominating the ecosystem as conifers are excluded. Because of the relatively rapid rate of decomposition of the litter of these trees and of a higher base status of the leaves entering the litter, soil conditions under this type of vegetation are much ameliorated. Organic content and acidity are reduced and base status is increased. First moder and eventually brown forest soils with mull characteristics are encountered. The characteristic trees of such ecosystems are still ecto-mycorrhizal and some of their fungal associates, notably members of such genera as *Amanita* and *Lactarius* produce considerable quantities of external mycelium in addition to a sheath, but there is a trend towards decreased production of external mycelium along the gradient. In extreme cases, as in some beech forests, connections from sheath to the surrounding soil appear to be absent altogether (Harley, 1978). Clearly in such a circumstance the amount of vegetative mycelium in the soil will be much less than in the coniferous system and its function, assuming that one exists, might be quite different.

Since the pioneering work of Melin using pure culture syntheses of mycorrhizas we have known that one of the major functions of the external mycelium of coniferous roots is the absorption and translocation

---

Caption for Fig. 3 (*cont.*)
of fungal growth is evident at the apex but the main root axis is uninfected and has root hairs. Little differentiation of the mycelium within the fans has occurred. Figure 3d shows the same root two weeks later. Infection has established at the root apex and arrested its extension, and two lateral roots behind the apex have also become infected. Following infection vigorous strand development occurs in some fans which are closely associated with the infected root. × 2.

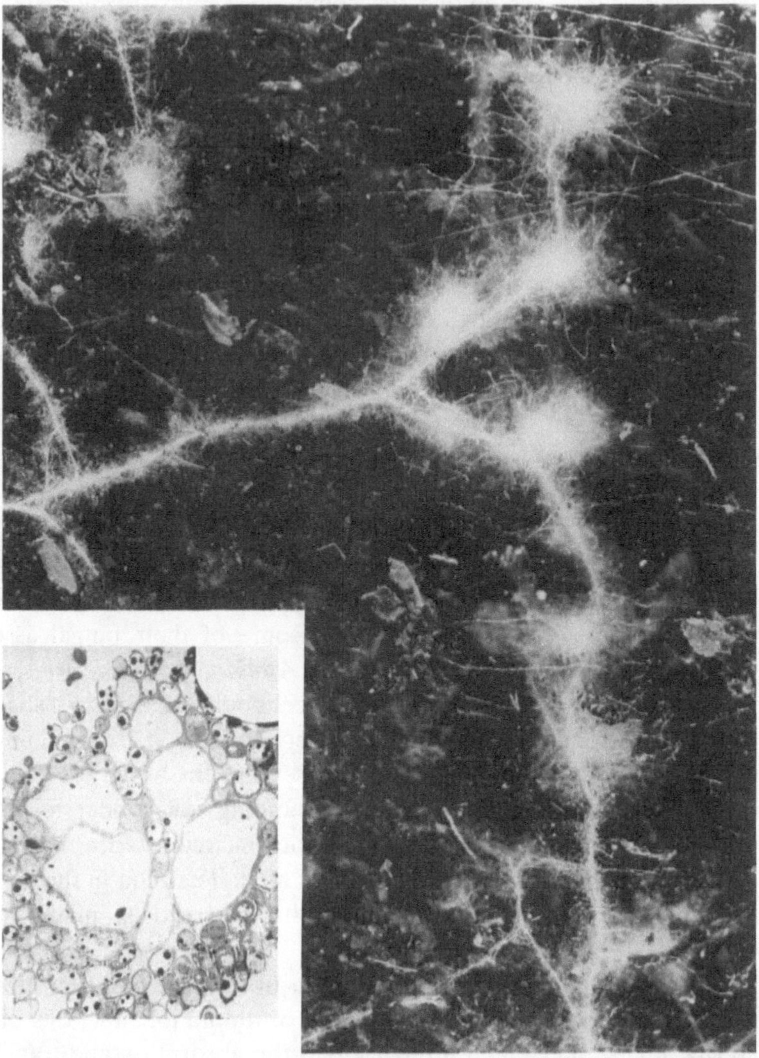

Fig. 4. Mature mycorrhizal roots of *Pinus contorta* infected with *Suillus bovinus*. The mycorrhizal laterals are interconnected by a single prominent strand while most of the diffuse hyphal network which makes up the fan of colonising systems has broken down. The outer surface of the strand is seen to be made up of loose hyphae which have intimate contact with the peat. × 50. (inset) TEM of transverse section through a mature strand of *Suillus bovinus* taken from a system similar to that shown in Fig. 4. Note differentiation into central, cytoplasm-free, vessel hyphae and loose surrounding densely-cytoplasmic sheath hyphae. × 1800.

of the major plant nutrients phosphorus (Melin & Nilsson, 1950) and nitrogen (Melin & Nilsson, 1952). Subsequent studies of detached roots of beech by Harley and his co-workers (Harley, 1969) have demonstrated that the ectomycorrhizal sheath is the site of exchange of phosphate absorbed from soil and of carbohydrates produced by the host. More recently attempts have been made in our laboratory to investigate the function of ecto-mycorrhizal mycelial systems again using observation chambers. Such chambers have the advantage that the normal source-sink relationship of the atmosphere-plant-mycorrhiza-soil pathway need not be broken even while direct visual, microscopic or experimental observations are being made. Examination of coniferous root systems in such chambers reveals that the roots occupy a comparatively small proportion of this soil in comparison with fungal mycelium which develops from its roots (Fig. 3b). In this case, therefore, in contrast to that seen in the ericoid root, the function of the roots appears to have been taken over by the fungus. One of the major functions of the mycelial fans, now demonstrated both in observation chambers (Brownlee *et al.*, 1983) and in the field (Fleming, 1983), is to act as inoculum which will infect roots either of the same plant, of another plant of the same species, or even, depending on the specificity of the fungus concerned, roots of other host species. Stages in the colonisation of an uninfected root after it has contacted a mycelial fan are shown in Fig. 3c,d. It can be observed that major strand development occurs only after contact with the root and the formation of mycorrhizas. Much of the rest of the fan system may subsequently break down leaving large strands connecting mycorrhizal roots (Fig. 4) and forming a potential supply line for compounds moving to or from, and between, roots (cf. Thompson, Chapter 9).

It has been shown experimentally that strands will absorb phosphorus and water from the soil and provide pathways for their transport to seedlings over considerable distances (Skinner & Bowen, 1974*a,b*; Duddridge *et al.*, 1980). Since this mycelial system anastomoses and connects a number of plants it follows that transport will be multi-directional probably along gradients of water potential, and that numerous plants interconnected by the same mycelial system can receive mineral nutrients and water absorbed by the fungus. If, as seems likely, most of the water and dissolved phosphate travel together, it will be through the vessel hyphae in the centre of the mycelial strands because their hydraulic conductivity will be much greater than that of the densely cytoplasmic hyphae which surround the vessels and make up the rest of the mycelial fan. The ecological and physiological significance of these functional characteristics can be profound. Not only does the mycelial system provide a widely dispersed inoculum by which seedlings become infected early in their development, but the strands form

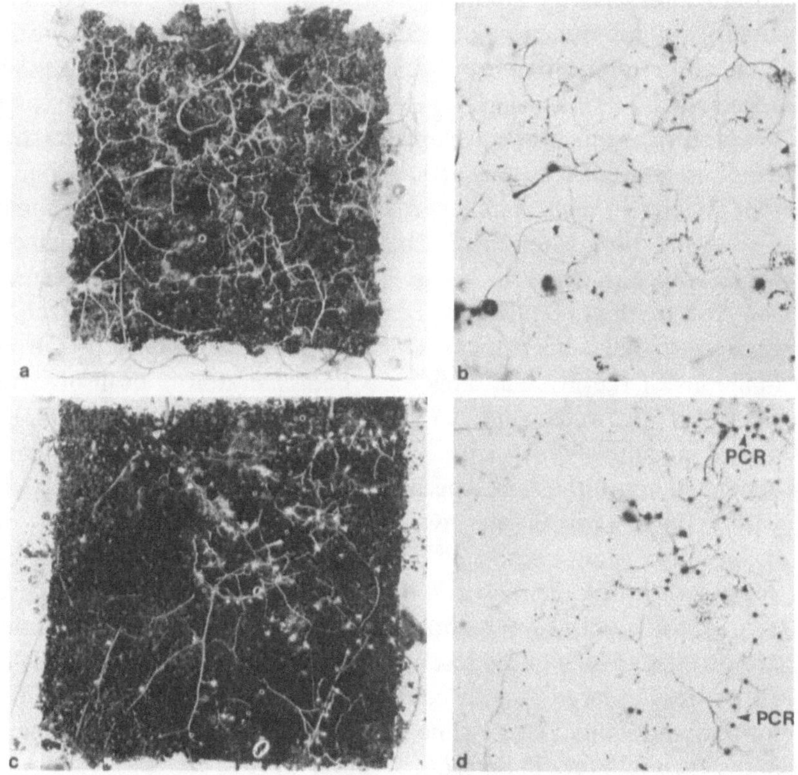

Fig. 5. (a) Roots and associated mycorrhizal mycelial system of a single seedling of *Pinus contorta* infected with *Suillus bovinus* in a root chamber. Most of the soil is exploited by mycelial strands. × 0.3. (b) Autoradiograph of chamber shown in Fig. 5a taken after exposure of the seedling shoot to $^{14}CO_2$ for 24 h and application of the film for a subsequent 24-h period. Note accumulation of label in clusters of mycorrhizal lateral roots and in the network of strands exploiting the peat. × 0.3. (c) Roots and associated mycorrhizal mycelial system of an interspecific association of host seedlings of *Pinus sylvestris* (left) and *P. controta* (right) in which an infection initiated on *P. sylvestris* by *Suillus bovinus* has spread to a younger *P. contorta* seedling. × 0.3. (d) Autoradiograph of the interspecific association shown in Fig. 5c taken after feeding the *P. sylvestris* shoot with 14_{CO_2} for 24 h. Extensive transfer of labelled assimilate has occurred to mycorrhizal laterals on the roots of *P. contorta* (PCR).

functional connections through which seedlings can be supplied with nutrients and water during the critical establishment phase.

The importance of carbon transfer to developing seedlings must also not be underestimated. Autoradiographic analyses of mycelial systems after the application of $^{14}CO_2$ to a single seedling shoot, reveal that current assimilate accumulates in the mycorrhizal roots and from these moves rapidly outwards into the external network of hyphae (Fig. 5a,b). Assimilate moves to the edge of the growing fan in developing systems but in cases where mycelial interconnections between plants have developed (Fig. 5c) label accumulates in the mycorrhizal roots of 'receiver' plants (Fig. 5d). Such transfer occurs in both intra- and inter-specific combinations of host plants. In order to sustain an efficient mycorrhizal system an isolated seedling would have to invest significant quantities of carbon in the fungus. If the carbon can, as appears likely, be supplied from alternative sources such as more mature seedlings through fungal mycelia this burden is reduced or eliminated. Thus the seedling, which may often in a forest environment be growing in deep shade, might become relatively independent of photosynthate from its own shoots and is supported by transfer of assimilates produced in the relatively well-illuminated crowns of over-story trees. Seen from this point of view the vegetative mycelium of ecto-mycorrhizas is of critical importance not only for the survival of seedlings but for the whole process of nutrient cycling in forest ecosystems.

It is important to recall, however, that some ectomycorrhizal systems – like those in many beech forests, while having thick sheaths do not produce the extensive mycelial phase in the soil (Harley, 1978). The ecological significance of mycorrhizal roots which are not connected to the surrounding soil by hyphae must be re-assessed. A clue to the differences in structure and function between the two types may come from study of the soil. Litter added to the mor-humus soil of coniferous forest breaks down slowly through the growing season and in these circumstances it may be necessary for absorptive fungal surfaces to be in close proximity to the sites of release if nutrients are not to be lost to competing saprotrophic fungi. Nutrient release from deciduous leaves like beech is, in contrast, a relatively rapid process. Whereas only 1–2% of the dry weight of newly fallen spruce needles can be leached by water, the value for litter of deciduous trees is nearer 20% (Nykvist, 1963). A capacity for efficient absorption at the time of the 'flush' of nutrient release coupled with a storage function would here be an advantage while the maintenance of an extensive mycelial system in the soil throughout the growing season would be wasteful of resources. The racemosely branched mycorrhizal roots of beech form a layer immediately below the recently fallen litter where their thick sheaths are in an ideal position to absorb nutrients leached from above.

The fungi of ectomycorrhizal beech roots produce large quantities of surface-bound phosphatases and the activity of these enzymes is many times greater in infected than in uninfected roots (Williams & Alexander, 1975). Bartlett & Lewis (1973) showed that the phosphatase hydrolysed a number of organic and inorganic phosphates, the released orthophosphate being absorbed by the sheath. The phosphate-storage capacity of the sheath is large and estimates suggest that in beech, absorption rates of phosphate are five times greater in mycorrhizal than in non-mycorrhizal roots, on an area basis, and over twice as great on a mass basis (Harley & McCready, 1950). Similar results have been obtained with pine (Bowen & Theodoru, 1967). Phosphate is stored in the fungal sheath in the form of polyphosphate granules. Using histochemical staining techniques Chilvers & Harley (1980) showed that P accumulation was accompanied by deposition of these granules in the sheath and that the number of granules produced approximately paralleled the rate of phosphate uptake. More recently Harley & McCready (1981) have demonstrated that with low levels of external phosphate, similar to those found in the field, about 40% of the total absorbed P is converted into polyphosphate. At higher concentrations such as those which might prevail for short periods during seasonal flushes of nutrient release, up to 97% of P storage may be in the form of polyphosphate. Over subsequent periods of time, granule numbers in the sheath decline; this suggests that the phosphate is being mobilised and passed into root tissue. While most of the research emphasis to date has been devoted to beech it seems likely that the same pattern of phosphate storage and release is found in all ectomycorrhizal species. The major distinctions within these mycorrhizal systems thus involve differences in mode of nutrient capture and these in turn are determined by the physicochemical characteristics of the resource material. Scavenging by an extensive mycelial network is a feature of slow-release situations such as those found in mor-humus soils dominated by pine while efficient absorption and retention without extensive hyphal ramification beyond the sheath is a feature of environments with pronounced nutrient flushes such as are found in moder and brown forest soils under deciduous species.

The direction of flow of water and of its dissolved nutrients is probably governed by a simple source-sink relationship in which the major determining factor will be the transpiration flux from the host shoot. Under most environmental circumstances the fall in water potential at the leaf-air interface will be greater than at any other part of the pathway so water will move to the tree, which is also the food base. The densely-cytoplasmic mycorrhizal sheath might be expected to impose a significant resistance to flow in the pathway but calculations have shown that its hydraulic conductivity is large and that it imposes no significant impedance to flow (Sands *et al.*, 1982). Both the direction of water flow and the driving mechanism appear to be distinct from that observed in the saprotrophic strand-forming fungus *Serpula lacrimans* in which water moves away from the food base towards the growing hyphal front along a gradient of hydrostatic pressure generated by high internal solute concentrations (Jennings, Chapter 7).

The direction and mechanism of assimilate transfer may, on the other hand, be very similar in the two fungal systems. While water and mineral nutrients travel towards the autotroph in the mycorrhizal system, carbon flows in the opposite direction. In the case of *Serpula* also, assimilates move from food base to the growing hyphal front (Jennings, 1982) and we may assume that in both systems some of these are used for construction of new hyphal walls, thus maintaining a concentration gradient. It is interesting that the major compound involved in the assimilate flux in strands of ectomycorrhizal mycelium is trehalose (D. J. Read, unpublished), which suggests that in qualitative terms also, the assimilate transport system is similar to that occurring in *Serpula* (Jennings, Chapter 7).

Since bidirectional flow must occur in the strands of ectomycorrhizal mycelium it has been proposed (Brownlee *et al.*, 1983) that the transport system is both structurally and functionally analogous to that found in the plant. Water and dissolved salts move towards the transpiring surface through vessel hyphae of large diameter which lack cytoplasm and are thus comparable with xylem vessels, while assimilates move towards the hyphal front through the cytoplasmic hyphae surrounding the vessels which are thus analogous to sieve tubes. Evidence to support this hypothesis is currently being sought using micro-autoradiographic techniques.

Distribution, structure and function of the vegetative mycelium of VA systems

VA mycorrhizas become dominant in plants at the warmer and drier end of the environmental gradient. As in the case of the heath-forest boundary, however, there is again an overlap in that the forest floor vegetation of the more base-rich coniferous forests and of most deciduous forests consists of herbs with VA mycorrhizal infection. Climatic conditions in areas supporting VA mycorrhizal plants as dominants are such that turnover of organic matter is rapid and evaporation rates are high. This leads to a net upward movement of salts, replacing the leaching tendency experienced in upper and most intermediate parts of the gradient and leading to a higher pH. Grasslands often form the characteristic vegetation in such areas, the dominant graminaceous plants occurring in mixture with herbs and occasional shrubs. Practically all plants in such associations have VA mycorrhizas though members of some plant families like the Cruciferae, Polygonaceae, Juncaceae and Cyperaceae are normally free of this type of infection. The quantity of external mycelium varies as in the case of ectomycorrhizas with soil type and host, and though it does not reach the levels either in terms of biomass or differentiation seen in the ectomycorrhizas of coniferous mor-humus soil it can, nonetheless, form an extensive network. Nicolson (1959) observed particularly extensive wefts of external mycelium around heavily infected roots of grasses in stabilised parts of sand dunes. Estimates of hyphal length vary between 0.8 and 1.3 m per cm infected root (Sanders, Tinker, Black & Palmerley, 1977) and lengths of up to 55 m have been recorded per gram of soil associated with roots in grassland (Tisdall & Oades, 1979). Such a network must clearly represent a large increase in the potential absorptive surface of the root. The hyphae making up the external mycelium have distinctive forms and dimensions. There is a striking variation in diameter from 2 to 30 μm accompanied by great variation in wall thickness. In mature mycelial systems, thick-walled hyphae normally of diameters between 20 and 30 μm give permanence to the external mycelial complex. These have characteristic elbow-like bends which were first described by Peyronel (1924), and termed angular-projections by Butler (1939). Finer thick-walled hyphae branch from these and the ultimate branches are relatively ephemeral thin-walled structures with a diameter of 2–7 μm. Mosse (1959) reported that 75% of the external hyphae associated with apple roots were of the coarse type, in this case with a diameter up to 20 μm. From the external mycelium lateral branches, often formed at right angles to the main hyphae and running in parallel to the roots as in ericaceous mycorrhizas, penetrate the outer cortical cells of the root. The number of entry points varies between 2.6 and 21.1 mm^{-1} of root length in strawberry, and between 4.6 and 10.7 in apple (Mosse, 1959). Internally the mycelium differentiates into vesicles

(Fig. 6a) which are largely filled with lipid reserves (Cooper & Lösel, 1978) and finely divided intracellular arbuscules (Fig. 6b) each branch of which is closely enveloped in host-plasmalemma.

Aspects of the physiology of VA mycorrhizas have been reviewed recently (Gianinazzi-Pearson & Gianinazzi, 1983), and only the function of the vegetative mycelium will be examined here. The evidence indicates that fungal enhancement of phosphorus uptake is the main reason for improvement of growth and yield in VA mycorrhizal plants. As in ecto-mycorrhizal systems, infected roots can absorb significantly more phosphate from culture solutions (Gray & Gerdemann, 1969) and from soils (Mosse *et al.*, 1973; Sanders & Tinker 1973; Gianinazzi-Pearson, Fardceau, Asimi & Gianinazzi, 1981) than uninfected roots. This difference is particularly well marked when levels of available P are low enough to limit growth. The VA mycorrhizal root system has a higher affinity for phosphate ions than the uninfected root (Cress, Throneberry & Lindsey, 1979) which in turn leads to significantly greater P inflow rates (Sanders & Tinker, 1973; Smith, Nicholas & Smith, 1979; Smith, 1982) and the vegetative mycelium extending outwards into the soil is known to be the major absorbing surface (Hattingh, Gray & Gerdemann, 1973; Pearson & Tinker, 1975; Cooper & Tinker 1978).

Once absorbed, phosphate is translocated through VA hyphae at rates calculated to be in the range $0.1–3.8 \times 10^{-9}$ mol. cm s^{-1} (Sanders & Tinker 1973; Pearson & Tinker, 1975). Tinker (1975) has proposed that the transport mechanism involves a combination of bulk flow and cytoplasmic streaming, the concentrations of P in the hyphal cytoplasm being regulated by loading or unloading of polyphosphate (poly P) in vacuoles. As in ectomycorrhizas the presence of poly P granules in the fungal vacuoles has been confirmed by electron microscopy (Cox, Sanders, Tinker & Wild, 1975; White & Brown, 1979; Strullu, Gourret, Garrec & Fourcy, 1981*a*; Strullu, Gourret & Garrec, 1981*b*). Again as in ectomycorrhizas, it has been shown that poly P rapidly accumulates in the external hyphae of P-starved mycorrhizal plants which are supplied with orthophosphate and that as much as 40% of the fungal P may be stored in the form of poly P in the hyphal vacuoles (Callow, Capaccio, Parish & Tinker, 1978). The arrival of this polyphosphate in the internal mycelium is associated with an increase of polyphosphate kinase activity and this activity appears to be restricted to the internal phase of fungal growth (Capaccio & Callow, 1982). Polyphosphate granules disappear from the vacuoles in the fine arbuscule branches (Cox *et al.*, 1975; Strullu *et al.*, 1981*b*) and it seems likely that the

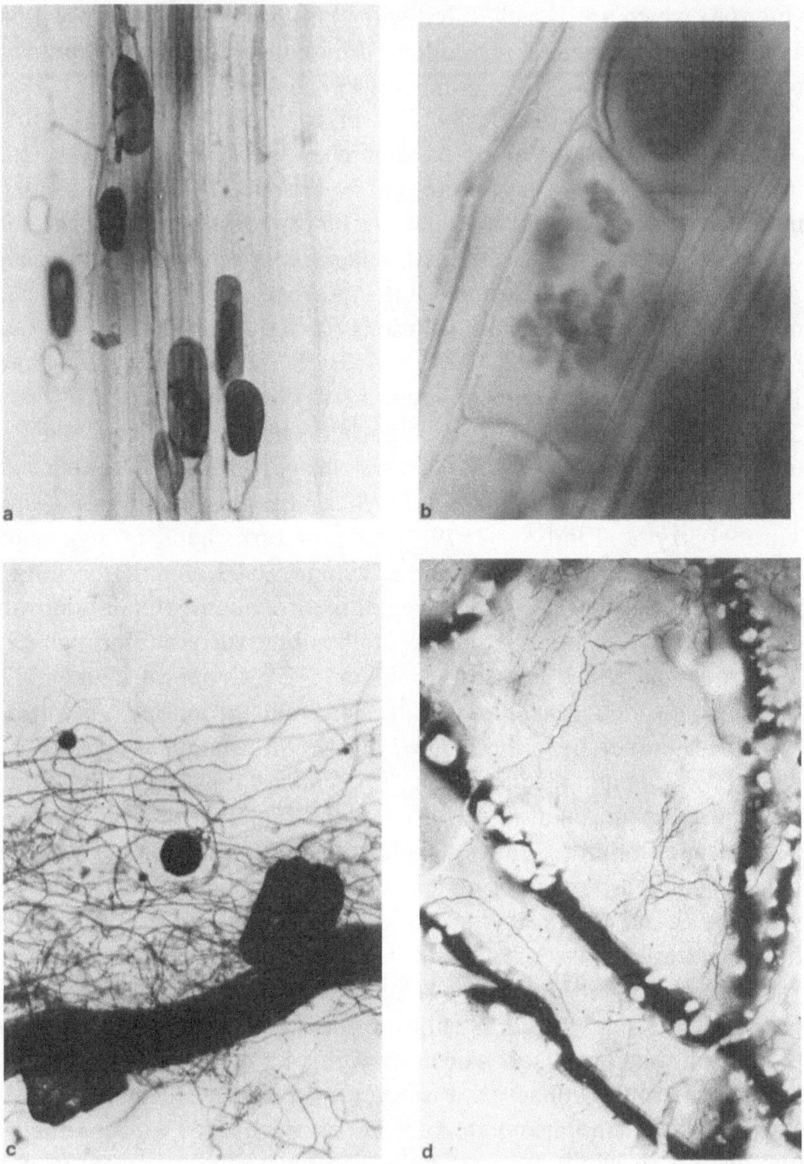

Fig. 6. (a) Vesicles of the VA mycorrhizal fungus *Glomus tenue* in cortical cells of the roots of *Festuca ovina*. × 100. (b) Arbuscule of a VA mycorrhizal fungus in root cortical cell of *Festuca ovina*. × 300. (c) Vegetative mycelium and chlamydospores of the VA mycorrhizal fungus *Glomus clarum* infecting lateral roots of *Festuca ovina* growing in dune-sand. Plants were inoculated with spores of the fungus and grown for six weeks before the soil was gently removed under a slow

released orthophosphate is actively transported from fungus to host across the interface between the arbuscule and the host plasmalemma (Cox & Tinker, 1976). Marx, Dexheimer, Gianinazzi-Pearson & Gianinazzi (1982) have shown that ATPase activity associated with the host plasmalemma is concentrated around the fine branches of the arbuscular hyphae.

In addition to the well established role of the mycorrhizal mycelium in absorption and transfer of phosphorus, interest is increasingly being devoted to its rôle in uptake of water (Safir, Boyer & Gerdemann, 1972; Allen, Smith, Moore & Christensen, 1981; Allen 1982; Hardie & Leyton, 1981) and trace elements (Lambert, Baker & Cole, 1979). On theoretical grounds, if mycorrhizal enhancement of nutrient uptake occurs as a result of the capacity of the external mycelium to exploit larger volumes of soil then plant growth response should increase in proportion to the quantity of external mycelium produced. Evidence that this is the case has now been obtained in studies employing a range of endophytes and host species (Saunders *et al.*, 1977; Graham, Lindermann & Menge, 1982).

It is now realised that as well as playing a fundamental role in plant nutrition, the external mycelium of VA mycorrhizas in natural vegetation systems is more important than spores as a source of inoculum (Read, Koucheki & Hodgson, 1976). The implication of this observation is the same as that made in the ectomycorrhizas, namely that for a period after infection of a root has taken place, the vegetative mycelium forms a direct interconnection between the infected plants. The external mycelium of VA mycorrhizas is less structurally differentiated (Fig. 6c) than that of ectomycorrhizas which makes it more difficult to examine the function of these interconnections experimentally. However, a number of studies have shown that phosphorus may be transferred either directly or indirectly from plant to plant through VA hyphae (Heap & Newman, 1980; Chiariello, Hickman & Mooney, 1982; Whittingham & Read, 1982). It has been shown also that the flux of nutrients from source to sink plants can be sufficient to promote growth

---

Caption for Fig. 6 (*cont.*)
jet of water. Note the extensive, but largely undifferentiated, external mycelium. × 50. (d) Autoradiograph of a similar root system after feeding the host shoot with $^{14}CO_2$. Note extensive transfer of label from roots to external vegetative mycelium. White patches on roots are caused by adhering sand grains which interrupt isotope emission. × 50.

responses of the sinks (Whittingham & Read, 1982). As in the ectomycorrhizas, transfer of carbon from source plants to sink seedlings at both the intra and interspecific levels have been demonstrated (Francis & Read, 1984). This transfer is greatly enhanced if sink seedlings are shaded, indicating again that in the natural environment inter-plant mycelial connections may provide direct channels of support for seedlings in the critical establishment phases. Preliminary evidence that the transfer of carbon is by the direct mycelial pathway rather than by leakage of assimilates from the root of source plants and later absorption by scavenging hyphae is now available in the form of autoradiographs (Fig. 6d) showing labelled hyphal 'bridges' between roots (Francis & Read, 1984).

The occurrence of the VA mycorrhizal type at the warm, dry end of the environmental gradient is probably of importance in a functional sense. Under the circumstances of high turnover of nutrients which prevail in such a climate, plant growth is potentially rapid at least after periods of rain, but may be limited by the levels of availability of the least-mobile nutrient elements. Because of its strong affinity with adsorption sites in soil minerals, phosphorus is frequently the element which most limits growth in mineral soils (Tinker, 1975, 1978; Gianinazzi-Pearson & Diem, 1982), and zones of P depletion are known to occur around growing roots (Nye & Tinker, 1977). Clearly any structure which has the capacity to grow through and beyond such depletion zones has the potential to remove the growth limitations and would provide the plant with considerable competitive advantage. Baylis (1975) has proposed that selection pressures in nutrient-deficient habitats have led to the evolution of two major alternative strategies for nutrient capture in nature. Both provide the plant with the capacity to 'forage' for nutrients like phosphorus beyond the depletion zone. In the first, plants invest carbon in the production of a diffuse root system with extensive root-hair development; in the second, carbon is invested in a mycorrhizal fungus the external mycelium of which has the function of soil exploitation. In the latter type root hairs may be totally lacking and plants are obligately mycorrhizal. The two distinct strategies have been observed in plants of tropical rain forests (St John, 1980). Many plants, for example some members of the Gramineae, appear to occupy an intermediate position having both a diffuse root system with root hairs, and extensive VA infection. It is difficult to envisage what nutritional advantage the mature plant may gain from mycorrhizal infection in such conditions though infection in the seedling stage, as indicated earlier, may be of great significance, and retention of infection would maintain the integration of the population of mycorrhizal plants so that nutrients can move from areas of relatively good to areas of relatively weak supply thus maintaining vigour at the community rather than at the individual level.

General discussion

It is almost one hundred years since Frank (1885) provided the first description of a root-fungus association which he termed a mycorrhiza in the Cupuliferae. Since this time we have made considerable advances in our understanding of this symbiosis. The early studies typified by those of Melin (1923) and Hatch & Doak (1933) on ectomycorrhizas, and Gallaud (1905) and Peyronel (1924) on VA mycorrhizas, placed major emphasis on description of the structural features of the associations. Later, vital information on their function was obtained using aseptic synthesis techniques (Melin & Nilsson, 1950, 1952, 1953), field and pot culture studies (Hatch, 1937), and laboratory analytical procedures with detached roots (Harley, 1969; Harley & Smith, 1983). These and other similar studies have laid the basis for an understanding of mycorrhizal structure and function at the level of the individual plant or root. As a result of these pioneering studies we are now in a position to ask questions about the role of mycorrhizas in plant communities and ecosystems. Such enquiries promise to bring exciting developments. It is increasingly apparent that many plants in both ectomycorrhizal and VA mycorrhizal associations are interlinked by the vegetative mycelium of their fungal associates. The general lack of host specificity shown by mycorrhizal fungi means that these interconnections can occur at both the intra- and interspecific level throughout the plant community. The implication of the existence of such functional pathways for subjects such as nutrient cycling, seedling establishment and competition between plants both at the individual plant and community level are profound and further emphasis on this area of research is greatly needed. The biology of the vegetative mycelium of mycorrhizas must become a subject of research not only for mycologists but for all those who wish to obtain a full understanding of the dynamics of terrestrial ecosystems.

I thank the NERC, the SERC and the Forestry Commission for financial support. Thanks are also due to Mr G. Woods and Mr D. Hollingworth for photographic assistance, to Dr C. Brownlee, Dr J.

Duddridge and Mr R. Francis for providing some previously unpublished photographs, and to Miss J. Bird for patiently typing the manuscript.

References

Allen, M. F. (1982). Influence of vesicular-arbuscular mycorrhizae on water movement through *Bouteloua gracilis*. *New Phytologist*, **91**, 191–6.

Allen, M. F., Smith, W. K., Moore, T. S. & Christensen, M. (1981). Comparative water relations and photosynthesis of mycorrhizal and non-mycorrhizal *Bouteloua gracilis*. *New Phytologist*, **88**, 683–93.

Ashida, J., Higachi, N. & Kikuchi, T. (1963). An electron microscopic study on copper precipitation by copper resistant yeast cells. *Protoplasma*, **57**, 27–32.

Bartlett, E. & Lewis, D. H. (1973). Surface phosphatase activity of mycorrhizal roots of beech. *Soil Biology and Biochemistry*, **5**, 249–57.

Baylis, G. T. S. (1975). The magnolioid mycorrhiza and mycotrophy in root systems derived from it. In *Endomycorrhizas*, ed. F. E. Sanders, B. Mosse & P. B. Tinker, pp. 373–89. Leeds, London: Academic Press.

Bowen, G. D. & Theodoru, C. (1967). Studies on phosphorus uptake by mycorrhizas. *14th IUFRO Congress*, **5**, 116–38.

Bradley, R., Burt, A. J. & Read, D. J. (1981). Mycorrhizal infection and resistance to heavy metal toxicity in *Calluna vulgaris*. *Nature*, **292**, 335–7.

Bradley, R., Burt, A. J. & Read, D. J. (1982). The biology of mycorrhiza in the Ericaceae. VIII. The role of mycorrhizal infection in heavy metal resistance. *New Phytologist*, **91**, 197–209.

Brownlee, C. D., Duddridge, J. A., Malibari, A. & Read, D. J. (1983). The structure and function of mycelial systems of ectomycorrhizal roots with special reference to their role in forming inter-plant connections and providing pathways for assimilate and water transport. In *Proceedings IUFRO Conference on Tree Roots and Their Mycorrhizas*, ed. D. Atkinson *et al.*, pp. 433–43.

Butler, E. J. (1939). The occurrence and systematic position of the vesicular-arbuscular type of mycorrhizal fungi. *Transactions of the British mycological Society*, **22**, 274–307.

Butler, G. M. (1957). The development and behaviour of mycelial strands of *Merulius lacrymans*. *Annals of Botany, New Series*, **21**, 523–37.

Callow, J. A., Capaccio, L. C. M., Parish, G. & Tinker P. B. (1978). Detection and estimation of polyphosphate in vesicular-arbuscular mycorrhizas. *New Phytologist*, **80**, 125–34.

Capaccio, L. C. M. & Callow, J. A. (1982). The enzymes of polyphosphate metabolism in vesicular-arbuscular mycorrhizas. *New Phytologist*, **91**, 81–91.

Chiariello, N., Hickman, J. C. & Mooney, H. A. (1982). Endomycorrhizal role for interspecific transfer of phosphorus in a community of annual plants. *Science*, **217**, 941–3.

Chilvers, G. A. & Harley, J. L. (1980). Visualization of phosphate accumulation in beech mycorrhizas. *New Phytologist*, **84**, 319–26.

Cooper, K. M. & Lösel, D. M. (1978). Lipid physiology of vesicular-arbuscular mycorrhiza, I. Composition of lipids in roots of onion, clover and ryegrass infected with *Glomus mosseae*. *New Phytologist*, **80**, 143–51.

Cooper, K. M. & Tinker, P. B. (1978). Translocation and transfer of nutrients in vesicular-arbuscular mycorrhizas. II. Uptake and translocation of phosphorus, zinc and sulphur. *New Phytologist*, **81**, 43–52.

Cox, G., Sanders, F. E., Tinker, P. B. & Wild, J. A. (1975). Ultra-structural evidence relating to host-endophyte transfer in a vesicular-arbuscular mycorrhiza. *Endomycorrhizas,* ed. F. E. Sanders, B. Mosse & P. B. Tinker, pp. 297–312. London, New York: Academic Press.

Cox, G. & Tinker, P. B. (1976). Translocation and transfer of nutrients in vesicular arbuscular mycorrhizas. I. The arbuscule and phosphorus transfer: a quantitative study. *New Phytologist*, **77**, 371–8.

Cress, W. A., Throneberry, G. O. & Lindsey, D. L. (1979). Kinetics of phosphorus absorption by mycorrhizal and non-mycorrhizal tomato roots. *Plant Physiology*, **64**, 484–7.

Duddridge, J. & Wainwright, M. (1980). Heavy metal accumulation by aquatic fungi and reduction in viability of *Gammarus pulex* fed Cd^{21} contaminated mycelium. *Water Research*, **14**, 1605–11.

Duddridge, J. A., Malibari, A. & Read, D. J. (1980). Structure and function of mycorrhizal rhizomorphs with special reference to their role in water transport. *Nature*, **287**, 834–6.

Duddridge, J. A. & Read, D. J. (1982). An ultrastructural analysis of the development of mycorrhizas in *Rhododendron ponticum. Canadian Journal of Botany*, **60**, 2345–6.

Falck, R. (1912). Die *Merulius* – Faule des Bauholzes. *Hausschwammforschungen*, **6**, 1–405.

Fleming, V. (1983). Succession of mycorrhizal fungi on birch: infection of seedlings around mature trees. *Plant and Soil,* **71**, 262–7.

Foster, R. C. (1981). Mycelial strands of *Pinus radiata:* ultrastructure and histochemistry. *New Phytologist*, **88**, 705–12.

Frank, A. B. (1885). Uber die auf Wurzelsymbiose beruhende. Ernahoung gewisser Baume durch unterirdische Pilze. *Berichte der Deutschen botanischen Geselschaft*, **3**, 128–45.

Francis, R. & Read, D. J. (1984). Direct transfer of carbon between plants connected by vesicular–arbuscular mycorrhizal mycelium. *Nature*, **307**, 53–6.

Gadgil, R. & Gadgil, P. (1971). Mycorrhiza and litter decomposition. *Nature*, **233**, 133.

Gallaud, I. (1905). Etudes sur les mycorrhizes endotrophes. *Revue Général de Botanique*, **17**, 5–48.

Garrett, S. D. (1956). *Biology of Root-infecting fungi.* Cambridge: Cambridge University Press. 293 pp.

Gianinazzi-Pearson, V., Fardceau, J. C., Asimi, S. & Gianinazzi, S. (1981). Source of additional phosphorus absorbed from soil by vesicular-arbuscular mycorrhizal soybeans. *Physiologie Végétale*, **19**, 33–43.

Gianinazzi-Pearson, V. & Diem, H. G. (1982). Endomycorrhizae in the tropics. In *Microbiology of Tropical Soils. Implications in Soil Management*, ed. Y. R. Dommergues & H. G. Diem, pp. 209–57. The Hague: Martinus Nijhoff.

Gianinazzi-Pearson, V. & Gianinazzi, S. (1983). The physiology of vesicular-arbuscular mycorrhizal roots. In Proceedings of IUFRO Conference on *Tree Roots and Their Mycorrhizas*, ed. D. Atkinson *et al.*, pp. 197–209.

Graham, J. H., Lindermann, R. G. & Menge, J. A. (1982). Development of external hyphae by different isolates of mycorrhizal *Glomus* spp. in relation to root colonisation and growth of troyer citrange. *New Phytologist*, **91**, 183–90.

Hardie, K. & Leyton, L. (1981). The influence of VA mycorrhizas on growth and water relations of red clover. I. In phosphate deficient soil. *New Phytologist*, **89**, 599–608.

Harley, J. L. (1969). *The Biology of Mycorrhiza.* London: Leonard Hill.

Harley, J. L. (1978). Ectomycorrhizas as nutrient absorbing organs. *Proceedings of the Royal Society of London,* **B203,** 1–21.

Harley, J. L. & McCready, C. C. (1950). Uptake of phosphate by excised mycorrhizas of beech. *New Phytologist,* **49,** 388–97.

Harley, J. L. & McCready, C. C. (1981). Phosphate accumulation in *Fagus* mycorrhizas. *New Phytologist,* **89,** 75–80.

Harley, J. L. & Smith, S. E. (1983). *Mycorrhizal Symbiosis.* London, New York: Academic Press.

Hatch, A. B. (1937). The physical basis of mycotrophy in the genus *Pinus. Black Rock Forest Bulletin,* **6,** 168 pp.

Hatch, A. B. & Doak, K. D. (1933). Mycorrhizal and other features of the root system of *Pinus. Journal of the Arnold Arboretum,* **14,** 85–99.

Hattingh, M. J., Gray, L. E. & Gerdemann, J. W. (1973). Uptake and translocation of ^{32}P-labelled phosphate to onion roots be endomycorrhizal fungi. *Soil Science,* **116,** 383–7.

Heap, A. J. & Newman, E. I. (1980). The influence of vesicular arbuscular mycorrhiza on phosphorus transfer between plants. *New Phytologist,* **85,** 173–80.

Jennings, D. H. (1982). The movement of *Serpula lacrimans* from substrate to substrate over nutritionally inert surfaces. *Decomposer Basidiomycetes,* 4th Symposium of the British Mycological Society, ed. J. Frankland, B. Hedger & M. Swift), pp. 91–108. Cambridge: Cambridge University Press.

Lambert, D. H., Baker, D. E. & Cole, H. (1979). The role of mycorrhizae in the interactions of phosphorus with zinc, copper and other elements. *Journal of the Soil Sciences Society of America,* **43,** 976–80.

Marx, C., Dexheimer, J., Gianinazzi-Pearson, V. & Gianinazzi, S. (1982). Enzymatic studies on the metabolism of vesicular-arbuscular processes in the host-arbuscule interface. *New Phytologist,* **90,** 37–43.

Melin, E. (1923). Experimentelle Untersuchungen über die Konstitution und Okologie der Mykorrhizen von *Pinus sylvestris* and *Picea abies. Mykologische Untersuchungen,* **2,** 73–331.

Melin, E. & Nilsson, H. (1950). Transfer of radioactive phosphorus to pine seedlings by means of mycorrhizal hyphae. *Physiologia Plantarum,* **3,** 88–92.

Melin, E. & Nilsson, H. (1952). Transfer of labelled nitrogen from an ammonium source to pine seedlings through mycorrhizal mycelium. *Svensk Botanisk Tidskrift,* **46,** 281–5.

Melin, E. & Nilsson, B. H. (1953). Transfer of labelled nitrogen from glutamic acid to pine seedlings through the mycelium of *Boletus variegatus* (SW). *Nature,* **171,** 434.

Mitchell, D. T. & Read, D. J. (1981). Utilization of inorganic and organic phosphates by the mycorrhizal endophytes of *Vaccinium macrocarpon* and *Rhododendron ponticum. Transactions of the British Mycological Society,* **76,** 255–60.

Mosse, B. (1959). Observations on the extra-matrical mycelium of a vesicular-arbuscular endophyte. *Transactions of the British Mycological Society,* **42,** 439–48.

Mosse, B., Hayman, D. S. & Arnold, D. J. (1973). Plant growth responses to VA mycorrhizas. V. Phosphate uptake by three plant species from P-deficient soils *labelled with* ^{32}P. *New Phytologist,* **72,** 809–15.

Nicolson, T. H. (1959). Mycorrhiza in the Gramineae I. Vesicular arbuscular endophytes with special reference to the external phase. *Transactions of the British Mycological Society,* **42,** 421–38.

Nicolson, T. H. (1975). Evolution of vesicular arbuscular mycorrhizas. In *Endomycorrhizas*, ed. F. E. Sanders, B. Mosse & P. B. Tinker, pp. 25–34. London, New York: Academic Press.

Nye, P. & Tinker, P. B. (1977). Solute movement in the soil-root system. *Studies in Ecology 4*. Oxford: Blackwell Scientific Publications.

Nykvist, N. (1963). Leaching and decomposition of water soluble organic substances from different types of leaf and needle litter. *Studia Forestalia Suecica*, **3**, 1–31.

Paton, C. H. & Budd, K. (1972). Zinc uptake in *Neocosmospora vasinfecta*. *Journal of General Microbiology*, **72**, 173–84.

Pearson, V. & Read, D. J. (1975). The physiology of the mycorrhizal endophyte of *Calluna vulgaris*. *Transactions of the British Mycological Society*, **64**, 1–7.

Pearson, V. & Tinker, P. B. (1975). Measurement of phosphorus fluxes in the external hyphae of endomycorrhizas. In *Endomycorrhizas*, ed. F. E. Sanders, B. Mosse & P. B. Tinker, pp. 277–87. London, New York; Academic Press.

Peyronel, B. (1924). Prime recherche sulle micorize endotrofiche and sulla microflora radicola normale delle fanergames. *Rivista di Biologia*, **5**, 463–85.

Read, D. J. (1983). The biology of mycorrhizas in the Ericales. *Canadian Journal of Botany*, **61**, 985–1004.

Read, D. J. & Armstrong, W. (1972). A relationship between oxygen transport and the formation of the ectotrophic mycorrhizal sheath in conifer seedlings. *New Phytologist*, **71**, 49–53.

Read, D. J., Koucheki, H. K. & Hodgson, J. G. (1976). Vesicular arbuscular mycorrhizas in natural vegetation systems. I. The occurrence of infection. *New Phytologist*, **77**, 641–53.

Read, D. J. & Stribley, D. P. (1975). Some mycological aspects of the biology of mycorrhizas in the Ericaceae. In *Endomycorrhizas*, ed. F. E. Sanders, B. Mosse & P. B. Tinker, pp. 105–17. London, New York: Academic Press.

Rorison, I. H. (1973). The effect of extreme soil acidity on the nutrient uptake and physiology of plants. In *Acid Sulphate Soils*, ed. H. Dost, pp. 223–53. Wageningen: International Institute of Land Reclamation and Improvement, Publication 18, vol. 1.

Safir, G. R., Boyer, J. S. & Gerdemann, J. W. (1972). Nutrient status and mycorrhizal enhancement of water transport in soybean. *Plant Physiology*, **49**, 700–3.

Sanders, F. E. & Tinker, P. B. (1973). Phosphate flow into mycorrhizal roots. *Pesticide Science*, **4**, 385–95.

Sanders, F. E., Tinker, P. B., Black, R. L. B. & Palmerley, S. M. (1977). The development of endomycorrhizal root systems. I. Spread of infection and growth promoting effects with four species of vesicular-arbuscular endophytes. *New Phytologist*, **78**, 257–68.

Sands, R., Fiscus, E. L. & Reid, C. P. P. (1982). Hydraulic properties of pine and bean roots with varying degrees of suberization, vascular differentiation and mycorrhizal infection. *Australian Journal of Plant Physiology*, **9**, 559–69.

Skinner, M. F. & Bowen, G. D. (1974*a*). The penetration of soil by mycelial strands of pine mycorrhizas. *Soil Biology and Biochemistry*, **6**, 57–61.

Skinner, M. F. & Bowen, G. D. (1974*b*). The uptake and translocation of phosphate by mycelial strands of pine mycorrhizas. *Soil Biology and Biochemistry*, **6**, 53–6.

Smith, S. E. (1982). Inflow of phosphate into mycorrhizal and nonmycorrhizal plants of *Trifolium subterraneum* at different levels of soil phosphate. *New Phytologist*, **90**, 293–303.

Smith, S. E., Nicholas, D. J. D. & Smith, F. A. (1979). The effect of early mycorrhiza infection on nodulation and nitrogen fixation in *Trifolium subterraneum*. *Australian Journal of Plant Physiology*, **6**, 305–11.

St John, T. V. (1980). Root size, root hairs and mycorrhizal infection: a re-examination of Baylis's hypothesis with tropical trees. *New Phytologist*, **84**, 483–7.

Stribley, D. P. & Read, D. J. (1980). The biology of mycorrhiza in the Ericaceae. VII. The relationship between mycorrhizal infection and the capacity to utilize simple and complex organic nitrogen sources. *New Phytologist*, **86**, 365–71.

Strullu, D. G., Gourret, J. P., Garrec, J. P. & Fourcy, A. (1981*a*). Ultrastructure and electron-probe microanalysis of the metachromic vacuolar granules occurring in *Taxus* mycorrhizas. *New Phytologist*, **87**, 537–45.

Strullu, D. G., Gourret, J. P. & Garrec, J. P. (1981*b*). Microanalyse des granules vacuolaires des ectomycorhizes, endomycorhizes et endomycothalles. *Physiologia Végétale*, **19**, 367–78.

Tinker, P. B. (1975). Effects of vesicular arbuscular mycorrhizas on higher plants. *Symposium of the Society for Experimental Biology*, **29**, 325–49.

Tinker, P. B. (1978). Effects of vesicular arbuscular mycorrhizas on plant nutrition and plant growth. *Physiologia Végétale*, **16**, 743–51.

Tisdall, J. M. & Oades, J. M. (1979). Stabilization of soil aggregates by the root systems of ryegrass. *Australian Journal of Soil Research*, **17**, 429–41.

Tuominen, Y. (1967). Studies on the strontium uptake of the *Cladonia alpestris* thallus. *Annales Botanica Fennicae*, **4**, 1–28.

White, J. A. & Brown, M. F. (1979). Ultrastructural and X-ray analysis of phosphorus granules in a vesicular-arbuscular mycorrhizal fungus. *Canadian Journal of Botany*, **57**, 2812–18.

Whittingham, J. & Read, D. J. (1982). Vesicular-arbuscular mycorrhizas in natural vegetation systems. III. Nutrient transfer between plants with mycorrhizal interconnections. *New Phytologist*, **90**, 277–84.

Williams, B. & Alexander, I. J. (1975). Acid phosphatase localised in the sheath of beech mycorrhizas. *Soil Biology and Biochemistry*, **7**, 195–8.

11
Autecology and the mycelium of a woodland litter decomposer

JULIET C. FRANKLAND
Institute of Terrestrial Ecology, Merlewood Research Station, Grange-over-Sands, Cumbria, LA11 6JU, UK

Autecology has not been the forte of mycologists, apart from studies of certain commercially important species, notably pathogens. Central to an autecological approach is the need to know how the mycelium varies and is distributed under the influence of both genetic and environmental factors. Neglect of the mycelium has therefore gone hand in hand with the neglect of autecology. In this chapter, information on the mycelium of a single species, *Mycena galopus*, has been brought together to illustrate some of the themes developed elsewhere in the volume. Given the problems of identifying and isolating a mostly white and non-descript mycelium, a variety of approaches has been used by investigators, and the basidiocarp itself has often provided valuable, even if indirect, evidence.

M. galopus is one of many small agarics which grow on woodland leaf litter and produce troops of basidiocarps. It is a typical example of a saprotrophic fungus which colonises a spatially continuous resource with a diffuse spreading mycelium, in contrast to a wood decomposer such as *Tricholomopsis platyphylla* with far-ranging discrete cords (Thompson: Chapter 9). The morphology and habit of the basidiocarps are well known (Kühner, 1938; Smith, 1947; Charbonnel, 1977), but taxonomic monographs however excellent usually leave the vegetative mycelium to the imagination. The observations described here should begin to give an insight into the nature and ecology of the *whole* fungus.

Maas Geesteranus (1980) retained *M. galopus* (Tricholomatales) in the Lactipedes section of *Mycena* with other species possessing coloured or white latex in the stipe – a very useful diagnostic feature. Three varieties of this species all with white latex: *galopus, candida* and *leucogala* had been recognised by Pearson (1955). The last variety was subsequently classified as a separate species (Dennis, Orton & Hora, 1960). In this chapter, '*M. galopus*' is assumed to be variety *galopus* unless stated otherwise, but *M. leucogala* is doubtfully a distinct species rather than part of one variable taxon. Forms intermediate between the typical black basidiocarps of *M. leucogala* on burnt ground and those of the grey brown or pure white varieties of *M. galopus* are common, and mycelia of the three taxa appear to be serologically identical (Chard, Gray & Frankland, 1983), but interbreeding has not been confirmed.

General characters and techniques

In culture, the mycelium is white or cream in colour, and appressed or shortly floccose with a silky or woolly texture (terminology according to Stalpers, 1978). Dark pigments are sometimes produced in unfavourable conditions, and mycelial threads of aggregated hyphae about 1 mm in diameter are formed on both nutrient agar and litter, but the mycelium and individual hyphae lack truly distinctive morphological features even when examined by scanning electron microscopy (Newell, 1980). It can be cultured satisfactorily on potato and malt extract media, but it is partially or wholly heterotrophic for thiamine (Lindeberg, 1946), and growth is much improved by additions of yeast extract and hydrolysed casein (Fries, 1949). Particularly good healthy production of mycelium has been obtained in Oelbe's (1982) synthetic medium, containing ammonium tartrate, malt and yeast extract (S. Morton & J. C. Frankland, unpublished); nitrates do not appear to be utilised when present as a sole source of nitrogen. From analysis of litter and mycelium stripped from it, phosphorus is probably the mineral element most likely to limit growth (Frankland, Lindley & Swift, 1978). These requirements could have far-reaching ecological implications for this fungus which may need to decompose considerable quantities of litter to obtain certain amino acids and other organic nutrients in limited supply.

Secondary mycelia have been produced in certain pairings of sib and non-sib monospore isolates, indicating a heterothallic system in which outbreeding was favoured, but some infertility between populations in different woodlands occurred. The 'bow-tie' phenomenon as described in *Stereum hirsutum* (Coates, Rayner & Todd, 1981) appeared frequently in sib matings, with uni- or bilateral dikaryons developing from the interaction zone (J. C. Frankland, unpublished).

Slight morphological differences between the homo- and dikaryons could be seen in culture, and the former were often more pigmented and slower growing. Although it is difficult to see any ecological significance, there is also an intriguing account of on/off luminosity in primary and secondary mycelia of *M. galopus* by Bothe (1935); some homokaryons and synthesised dikaryons were luminous, and others not. However, Chard (1981) did not find any difference in the antigenicity of a homokaryon and dikaryotic isolates of this species in immunodiffusion tests.

Further information on the mycelium has come from experiments with axenic litter cultures, which bridge the gap between entirely artificial media and the field situation. It suggests that *M. galopus* in the absence of competitors is a relatively efficient fungus, capable of maintaining steady decomposition rates over long periods. Using hexosamine assay to determine mycelial biomass (impossible in the field), the efficiency (amount of mycelium produced per unit weight of substrate decomposed) on *Betula* and *Fraxinus* litter at field temperatures was 28–34% (Frankland *et al.*, 1978), compared with 37% on a glucose broth (Mikola, 1956). The long-term decomposition rate in similar model systems of *Quercus* litter at 11–15 °C was calculated to be 0.1% day^{-1} (Hering, 1982), and 0.3% day^{-1} on *Fagus* at 25 °C (Lindeberg, 1947).

More ingenuity is needed to differentiate the mycelium when competitors are introduced. Newell (1984*a*), after practice, achieved 95% accuracy in distinguishing young mycelia of *M. galopus* and *Marasmius androsaceus* on individual *Picea* needles in laboratory cultures by their growth form, including such features as closeness of the mycelial turf to the needle surface. In the field, she traced basidiocarps to their origin so that active mycelium could be located in the litter.

An antiserum which appeared in immunodiffusion tests to be species-specific has been raised against *M. galopus* mycelium (Chard, 1981). This could be useful in small mixed populations if it can be combined with fluorescent-antibody staining without cross-reactions. Membrane filtration was found to be a feasible means of estimating the presence and biomass of fluorescent hyphae of *M. galopus* when tests were made on *Quercus* leaves bearing basidiocarps (Frankland, Bailey, Gray & Holland, 1981).

Simple techniques which I have used for field identification include: plating of litter fragments on nutrient media (washing not always necessary) and comparison of the isolates with known cultures; tagging of mycelium grown on media containing a fluorescent brightener, Calcofluor White, for later retrieval; and incubation of litter to induce fruiting. The latter was a particularly successful means of identifying the mycelium on *Quercus* leaves. Abortive but recognisable basidiocarps containing white latex were produced after at least 8–12 weeks in a damp atmosphere. In this way, the presence of the basidiomycete in a developing fungal community was based on recordings of the mycelium, not of the basidiocarps.

Habitats and resources

In distribution, *M. galopus* is predominantly a north temperate species, usually found in woodlands but also occurring in open habitats such as hedgerows and lawns. It was one of the species recorded most frequently on British Mycological Society forays in the UK over a 12-year period (Rayner, 1979). In North America, it is particularly abundant in the moist areas of the Pacific coast from Washington to California but rare in the mid-west and eastern states. More southern locations include Algeria, Madagascar (single record) and Mexico (R. Watling, personal communication), and a new variety, *mellea*, has been described in Australia (Grgurinovic & Holland, 1982).

The basidiocarps appear to be particularly vulnerable to exposure. Hering (1975) found that they were scarce in two of the highest natural *Quercus* woodlands (305–460 m) in the British Isles (with a relatively low tree line) compared with numbers in similar plant communities at much lower altitudes. I have also observed a marked difference in the frequency of fruiting in high- and low-level plantations of *Picea sitchensis*. It is not known to what extent the vegetative mycelium follows suit; it may well have a greater altitudinal range. Under Sitka spruce at 340 m (National Grid reference: SD 79/729972) the mycelium was always abundant, but fruiting was ubiquitous only in the occasional climatically 'good' year. In other years the basidiocarps could be found only in the lee of a tree stump or similar shelter from the prevailing wind. These observations could be explained by the presence of perennial mycelium more tolerant of, or more protected from, harsh environmental conditions. Hintikka (1964) suggested that *M. galopus* was one of the psychrotolerant species producing mycelium abundantly in Finnish forests in late autumn, and active in winter even under snow although capable of growth at 20–25 °C. This has been supported by the results of some *in vitro* experiments (Frankland, 1982; Newell, 1984*a*). Although the optimum temperature for growth of this species on solid agar media was 21–22 °C, it decomposed *Quercus* litter to a significant extent at 4 °C, grew at temperatures as low as −2 °C, and survived at −12 °C.

M. galopus is, as Parker-Rhodes (1954) described it, a 'quisquilicolous' woodland species. It is not associated with any particular soil type, occurring as it does in a wide variety of broad-leaved and coniferous forests. In a study of *Pteridium* decomposition, it was found to be prominent on six different but adjacent soil types, including moder-type humus, mull and peat (Frankland, 1976). It seems likely that more mycelium will occur in the deeper organic layers of a mor site than in a mull, but Hering (1982) recorded similar quantities (fresh weights) of the basidiocarps on these two humus types when he surveyed comparable sites. It is most common on broadleaf- and needle-tree litter, but it colonises many other plant materials in its path including small woody components such as twigs and beech husks. Particularly in dry forests, the basidiocarps often arise from moist mats of moss, but the origin of the stipe can often be traced back to fragments of tree litter.

The lack of resource-specificity is confirmed by the large number of litter types on which *M. galopus* has been grown, including litter of several tree species, herbs, grasses, ferns and mosses (Hintikka, 1961; Hering, 1967; Frankland, 1969, 1975). However, it did not utilise *Mercurialis* (pH 6.4), and growth on *Calamagrostis* (pH 7.2) and *Cirriphyllum* (pH 8.2) was delayed. Hintikka attributed this to an initially unfavourable pH followed by amelioration of the condition through fungal activity. In the field, he recorded *M. galopus* growing on litter at pH 4–5.2, and in pure culture the optimum was pH 4–5, some growth occurring over a much wider range. A decrease in pH during decomposition by this fungus has been recorded in most of the litters tested; in bracken it fell by as much as one unit (pH 5.3 to 4.3) in 12 months.

The bulkier woody components of the forest floor are rarely colonised, probably because the mycelium is not adapted to conditions in the interior where aeration is restricted and various inhibitory substances tend to accumulate. In contrast to several typical lignicolous basidiomycetes, growth was almost totally inhibited by a concentration of 30% CO_2 in the atmosphere, and it was intolerant of relatively low concentrations of acetates in laboratory experiments (Hintikka, 1969, 1982; Hintikka & Korhonen, 1970).

A secondary colonist

M. galopus appears to have the potential to attack all the major constituents of plant litter. Lignin, α-cellulose, hemicelluloses, protein and soluble carbohydrates in plant litter, and purified xylan and pectin, have all been degraded by it in the laboratory (Hintikka, 1961, 1970;

Hering, 1967; Frankland, 1969). Further evidence of its capabilities as a decomposer comes from the detection of enzyme production, including that of polyphenol oxidases, cellulases and catalase (Lindeberg, 1948; Lamaison, 1976), but the extent to which these enzymes are exocellular needs further investigation.

The lack not only of resource specificity but also of substrate specificity (as shown in the absence of competitors) may suggest an even wider rôle for this saprotroph than is realised in nature. It holds a major position with other basidiomycetes in litter, but it is above all a *secondary* colonist, predominating at later states of decomposition and degrading cellulose and lignin to form a typical white rot. This pattern of behaviour follows a 'combative ecology strategy' (see Rayner & Webber: Chapter 18). The exact time at which a fungus arrives on a resource is usually difficult to determine. I have recorded the mycelium in the phylloplane of *Fraxinus excelsior* but only rarely; it probably originated from aerial spores before other micro-organisms closed in, and it may well have failed to survive. First recordings were usually obtained at a much later stage in the decomposition. On bracken petioles, *M. galopus* appeared in the first year of decomposition, and it was not fully established until the second year, after other species had destroyed much of the epidermis, phloem and non-lignified cortex and had opened up the interior (Frankland, 1966). These other organisms may have actually paved the way for the basidiomycete. Such a mechanism of species replacement corresponds with the *obligatory succession* and *facilitation model* described for higher plants (Frankland, 1981), and with '*stress alleviation*' during the development of a fungal community as discussed by Cooke & Rayner (1984) – see Rayner & Webber: Chapter 18. Similarly, it has been argued that *M. galopus* primes the resource for the next species when it produces soluble carbohydrates by hydrolysis of cellulose. However, such preconditioning, if it occurs, is very difficult to prove (Rayner & Webber: Chapter 18).

A rapid decline in the lignin and cellulose contents followed the colonisation of bracken petioles, and 'bore-holes' resembling those formed by wood decomposers were formed in the fibre walls around the hyphae. Confirmation that *M. galopus* was the agent responsible for this aggressive attack was obtained from axenic cultures on gamma-irradiated bracken; cavities of the same type were formed in the fibres, and breakdown of 25% lignin, 32% α-cellulose and 54% hemicellulose had occurred in one year.

Table 1. *Decomposition by* Mycena galopus *of α-cellulose in leaf litter of different tannin contents*

Litter	Soluble tannins (mean % ± S.E., $n = 5$)	Loss[a] of α-cellulose (%)
Recently fallen		
Quercus petraea	9.5 ± 0.1	5.0 NS
Corylus avellana	5.8 ± 0.1	23.6*
Fraxinus excelsior	3.8 ± 0.2	27.3*
Betula pendula	2.8 ± 0.1	21.7***
After 6 months in the litter layer		
Q. petraea	2.9 ± 0.0	43.4*
C. avellana	1.4 ± 0.1	40.0***

[a] Loss from gamma-irradiated samples inoculated with *M. galopus* compared with control samples after 6 months at 11 °C; absolute values.
NS: Not significantly different from controls.
Significance of difference from controls: *, $P \leq 0.05$; **, $P \leq 0.01$; ***, $P \leq 0.001$.

The secondary colonisation by *M. galopus* of four broadleaf litter types (*Betula, Corylus, Fraxinus* and *Quercus*) has also been followed by taking samples directly from the litter layer in a UK woodland with mull humus (Frankland, Bailey & Costeloe, 1979). The basidiomycete became prominent earlier than on bracken, the exact timing depending on the species of litter. On *Fraxinus*, a quickly-decomposing litter already well attacked before leaf-fall, the period extended from 0 to 6 months; on *Betula* and *Corylus* from 6 to 12 months, and on *Quercus*, the most slowly decomposing litter, from 6 to 18 months or more after leaf fall. Removal of tannins may be a prerequisite. Growth of the mycelium has been shown to be inhibited by tannins in *Quercus* litter, which has a relatively high content of these substances (Harrison, 1971), and the ability of the fungus to degrade α-cellulose in sterilised litter was greater in 'weathered' than recently fallen leaves with higher tannin contents (Table 1).

In the compacted litter of a coniferous forest decomposing under mor conditions and bound together by fungal mycelium, the relationship between individual resource units and a particular species is less obvious, but the location of the mycelium was deduced by tracing the origin of basidiocarps. In this way, *M. galopus* was found to fruit consistently from the F_1 horizon of a *Picea sitchensis* plantation (Newell, 1948*a*), and it could be assumed that active mycelium predominated on needles which were still whole but lacking mesophyll and extensively colonised by other fungi. Again, a preparatory phase of decomposition appears to have been necessary.

Competitive interactions

Success for *M. galopus* in the capture of secondary resources must depend on a complex network of interactions. It grows in close proximity not only to many microfungi but also to other saprotrophic agarics, including closely related species with overlapping niches. Tightly mixed clumps of the basidiocarps with those of *Collybia, Cystoderma, Marasmius* and other *Mycena* species have been described (Newell, 1980; Swift, 1982), and the mycelia of such neighbours must be in positions where they will compete for limited nutrients and space. How then is a competitive balance maintained?

As a secondary colonist, *M. galopus* is likely to be more adapted to combative competition than a pioneer (see Frankland, 1981; Cooke & Rayner, 1984; Rayner & Webber: Chapter 18). The basidiospores germinate readily on laboratory media, but subsequent growth is relatively slow, especially on litter, unlike that typical of some earlier, non-combative, colonisers with diffuse scavenging mycelia. Invasion by migrating mycelia rather than spores, and a relatively high inoculum potential, provided in *M. galopus* by densely branched and often aggregated hyphae, should favour establishment on a resource which is no longer virgin territory. Once established, it needs to withstand its neighbours for long periods while exploiting refractory substances. Mechanisms underlying this combative type of competition can be grouped in terms of antibiosis and contact reactions between hyphae and between mycelia (Rayner & Webber: Chapter 18). Features indicative of such mechanisms were seen in cultures of *M. galopus*, but few facts are known regarding their expression in the field. Antibiotics have not been detected, but there is an unconfirmed record of antibiotic activity against *Escherichia coli* (Mathieson, 1946), and an unusual report of viral inhibition by a liquid medium in which the fungus had been grown (Villard, Oddoux & Porte, 1979; Villard, Porte & Oddoux, 1982). Contact antagonism is common, and interaction zones occurred regularly on both nutrient media and litter in pairings against several basidiomycetes frequently associated with *M. galopus* (J. C. Frankland, unpublished). In these zones, the hyphae grew abnormally, often producing pigments and mycelial barrages were formed. *Mycena epipterygia* eventually grew over *M. galopus* after an initial check, but deadlock between a pair was more usual. This strongly interactive character could be an important means of holding a territory and preventing replacement by a competitor (see Rayner & Todd, 1979; Frankland, 1981), although 'zone lines' in field litter occupied by this species were rarely as distinct as those produced by wood fungi.

Acidification of the resource by the fungus (p. 245) could also be an advantage by restricting bacteria, but some laboratory evidence suggests that streptomycetes in *Picea* litter can antagonise *M. galopus* (Dickinson, Dawson & Goodfellow, 1981). On *Quercus* litter, however, living and dead hyphae were remarkably resistant to attack by the natural microflora, 25% of hyphae tagged with brightener surviving for more than a year in the field (Frankland, 1975; Frankland *et al.*, 1979).

The final outcome of competition will depend on the balance between competitive ability and the inoculum potential – the energy of growth available for colonisation. This is illustrated by the hypothesis proposed to explain the relative distributions of *M. galopus* and *Marasmius androsaceus* (possessing rhizomorphs) in a plantation of *Picea sitchensis* in Grizedale Forest, in the English Lake District (Newell, 1980, 1984*a,b*). The *Mycena* might appear to be a weak contender as regards inoculum potential compared with species which form rhizomorphs or cords, unless the involvement of factors such as grazing pressure which control hyphal density and age is considered, as by Newell.

In the laboratory, *M. androsaceus* appeared to have superior competitive abilities. Its mycelial production and growth rate on a nutrient medium was about 30–50% more than that of the *Mycena*, and its colonisation and decomposition of sterile spruce litter were also significantly greater. When litter inoculated with one or other of the two species was mixed in different ratios and 'seeded' with sterilised spruce needles (equivalent to a primary resource), colonisation of the latter by *Marasmius* was even greater than expected (Fig. 1).

Nevertheless, in the plantation where the two species were close enough to interact, the basidiocarps were equally abundant. In mixed-species clumps, however, the mean depth at which *M. galopus* fruited increased from 6 mm in single-species clumps to 10 mm ($P \leqslant 0.05$), whereas that of *M. androsaceus* remained at 1 mm, suggesting that the mycelia were zoned vertically (Fig. 2). This discovery led to an examination of the effect of grazing by a small yellow collembolan, *Onychiurus latus*. It was the most abundant mycophagous arthropod on

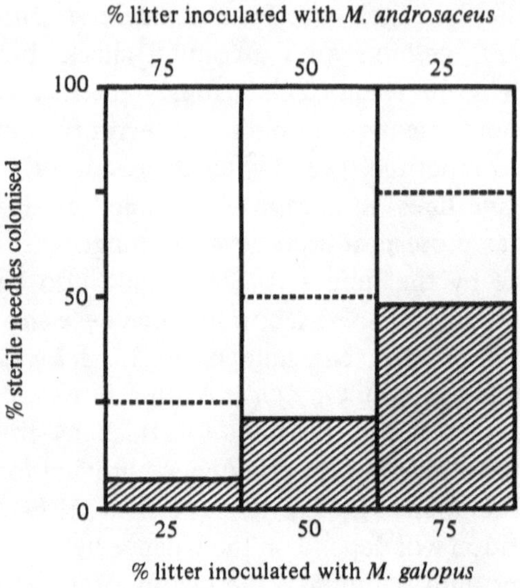

Fig. 1. Colonisation of sterile spruce litter by *Mycena galopus* and *Marasmius androsaceus* after 24 days at 11 °C in mixed cultures, initially 'seeded' with the two fungi on litter in three different ratios (from Newell, 1980). Hatched area, *M. galopus*; open area, *M. androsaceus*; --- expected result if colonising abilities had been equal. (From Newell, 1980.)

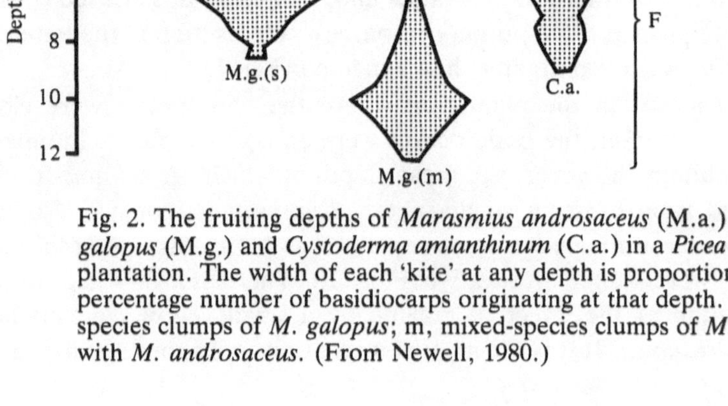

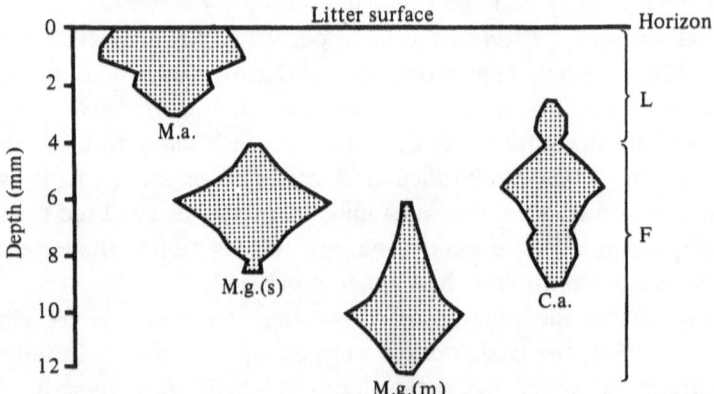

Fig. 2. The fruiting depths of *Marasmius androsaceus* (M.a.), *Mycena galopus* (M.g.) and *Cystoderma amianthinum* (C.a.) in a *Picea sitchensis* plantation. The width of each 'kite' at any depth is proportional to the percentage number of basidiocarps originating at that depth. s, single-species clumps of *M. galopus*; m, mixed-species clumps of *M. galopus* with *M. androsaceus*. (From Newell, 1980.)

the site and its gut consistently contained basidiomycete mycelium. Migration of the animal up and down the profile was related to its life cycle and the moisture content of the litter.

O. latus showed a marked preference for the mycelium of *M. androsaceus* both in the laboratory (Fig. 3) and the field, and a population of 10 specimens per gramme of air-dried litter in mixed cultures of the two fungi on litter could alter the outcome of competitive colonisation in favour of *M. galopus* (Fig. 4). The mycelial characters affecting the palatability and selective grazing of these species are not known.

In field tests, an increase in the density of the collembolan resulted in an increase of *Mycena* basidiocarps and a decrease of those of *Marasmius*, whereas a decrease in density had the reverse effect.

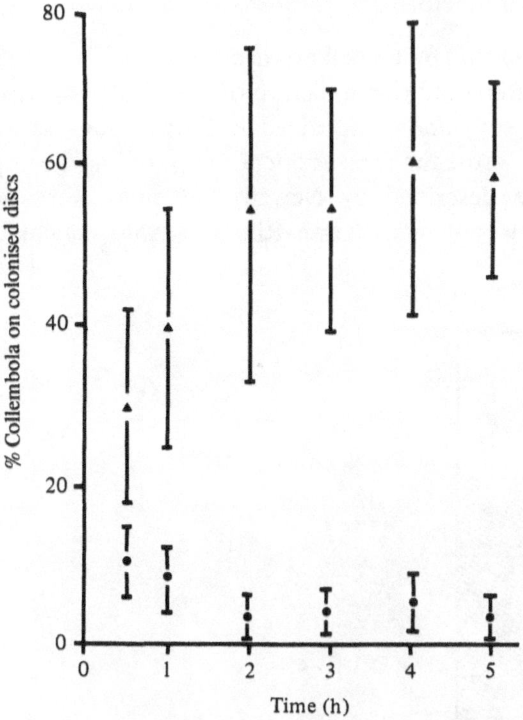

Fig. 3. The percentage number (± S.E.) of *Onychiurus latus* which had migrated onto agar discs covered with mycelium of *Mycena galopus* (●) or *Marasmius androsaceus* (▲) after increasing time intervals. Total number of animals released = 250; each of 10 replicate dishes contained 25 animals and both food sources. After 5 h, 5% of the aerial mycelium of *M. galopus* had been grazed and 72% of that of *M. androsaceus*. (From Newell, 1984*a*.)

Newell's experiments and observations led to the following hypothesis to explain the zonation of the mycelia: selective grazing of *M. androsaceus* mycelium altered the outcome of competition in the F_1 horizon sufficiently to allow *M. galopus* to predominate in that horizon, but in the L horizon the density of the Collembola was not great enough for a sufficiently long period of time to alter the competitive balance, with the result that the *Marasmius* mycelium was more abundant. The hypothesis needs to be tested further bearing in mind the differences between a sterile and an inhabited secondary resource, but it is one example of how a competitive balance could be maintained, and it indicates the major importance of animal/fungal interactions in the ecology of a mycelium. More recent research (Dix, 1984) showing that *M. galopus* is less tolerant than *M. androsaceus* of low water potentials introduces another relevant factor.

Spatial distribution and individual mycelia

The interaction study built up a picture of the *vertical* distribution of the mycelium of *M. galopus* in Grizedale Forest, but less was known about its logistics across the forest floor. Clumps of basidiocarps 1–3 m in diameter were described by Newell, but rings were not observed. In deciduous woodlands, Parker-Rhodes (1954) concluded

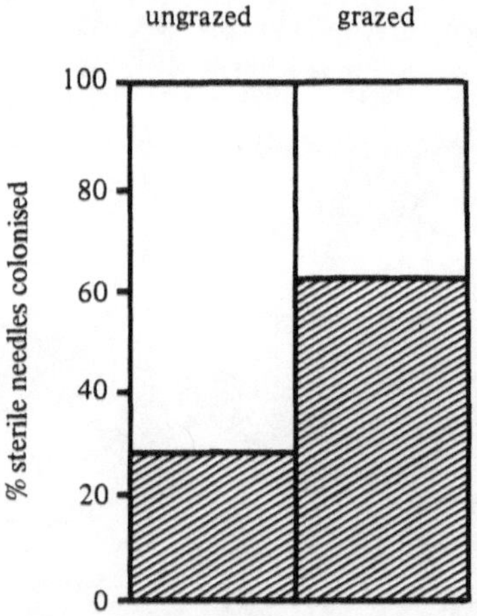

Fig. 4. The colonising abilities of *Mycena galopus* (hatched areas) and *Marasmius androsaceus* (open areas) with and without grazing by *Onychiurus latus* (20 per replicate), after 24 days at 11 °C in mixed cultures. These initially contained equal quantities of litter inoculated with each of the two fungi and 10 sterile spruce needles per replicate. (From Newell, 1984*b*.)

Table 2. *Percentage number of basidiocarps occurring in each one-third band of a permanent plot in three successive years*

Year Band	1	2	3
1	30.1	38.4	36.7
2	45.9	47.3	49.5
3	24.2	14.4	13.8

that populations were polymorphic and Swift (1982) found possible evidence of slight annual 'movement' in the spatial pattern of fruiting. More intensive mapping of the basidiocarps combined with preliminary genetical analysis of a population in the spruce plantation has, however, given some indication of the size and distribution of individual mycelia (J. C. Frankland, unpublished).

The positions of the basidiocarps in a permanent plot (10 × 4.5 m) were mapped using a grid system of 5 mm coordinates; the caps were painted to prevent stimulation of fruiting by their removal or repeated recordings. Data accumulated over five fruiting seasons were then analysed by computer programs. Complete absence of a ground flora greatly facilitated the recording, but the irregular and close spacing of the spruce trees was likely to have complicated the distribution patterns more than did the regular tree plantings examined by mycorrhizalists in similar mapping projects (Mason, Last, Pelham & Ingleby, 1982).

In 'good' seasons at least 1000–2000 basidiocarps were produced on the plot, indicating the presence of abundant mycelium in the litter, especially if it is assumed that the ratio of the production of basidiocarps : production of vegetative mycelium was 1 : 10, as calculated for *Quercus* litter (Frankland, 1975).

The basidiocarps were not distributed randomly over the plot. This was tested by dividing the plot into thirds and calculating the number of basidiocarps in each one-third band as a percentage of the total in the plot in three successive years. The percentages in a particular band were approximately constant but significantly different between bands, indicating non-randomness (Table 2).

M. galopus basidiocarps are ephemeral, usually lasting only 2–3 days. Distinctive spatial patterns were not therefore picked out by on-the-spot observations, but accumulated annual data when plotted and superimposed revealed some arcs or partial annuli of basidiocarps, each apparently related to the position of a tree. Two of these arcs are illustrated in Fig. 5. They might have been expected if the species had been mycorrhizal instead of saprotrophic, but various factors such as inhibitors in stem flow, nutrients in drip from the canopy, and litter depth could have been responsible and need investigation.

Slight annual changes in the overall position of fruiting zones of the mycelium even if not of whole colonies were sometimes suggested if 0.5 m^2 subplots were compared in different years as in Fig. 6. However, comparison of coincident basidiocarp positions over the whole plot in successive and non-successive years indicated stability (Fig. 7). The number of basidiocarps which coincided within ranges of 0–20 mm with positions mapped in a previous year is represented by histograms. In the example of successive years, only seven basidiocarps (total number: 993) coincided exactly with a basidiocarp of the previous year. Coincidences increased as the range increased, but, if due weightings were

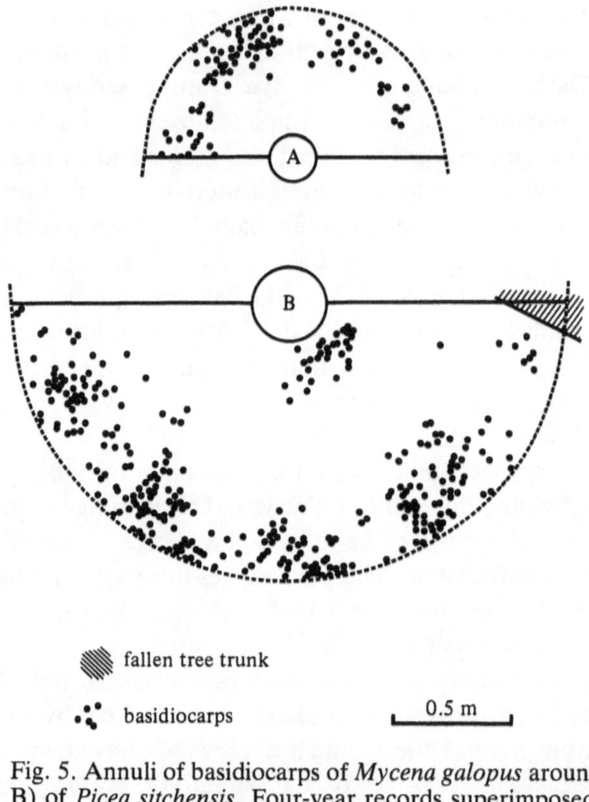

Fig. 5. Annuli of basidiocarps of *Mycena galopus* around two trees (A, B) of *Picea sitchensis*. Four-year records superimposed. Dashed line, arbitrary circular boundary with the tree as centre, cutting off basidiocarps close to neighbouring trees.

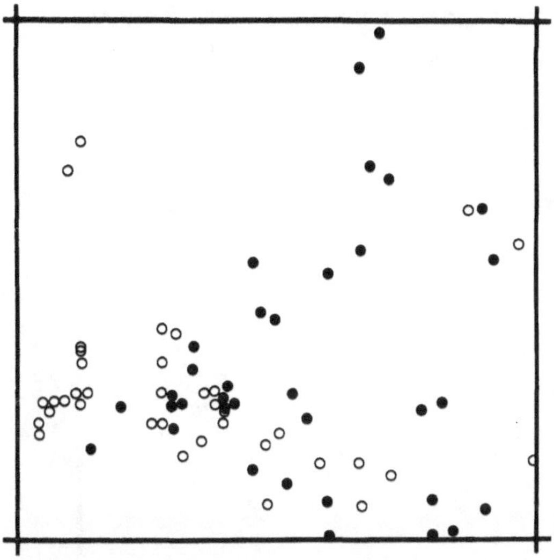

Fig. 6. Mapped positions of *Mycena galopus* basidiocarps on a 0.5 m^2 spruce plot, suggesting a slight shift in time; 1978 (filled circles) and 1981 (hollow circles) records. The distribution pattern of records from the intervening period was 'intermediate' between those of 1978 and 1981.

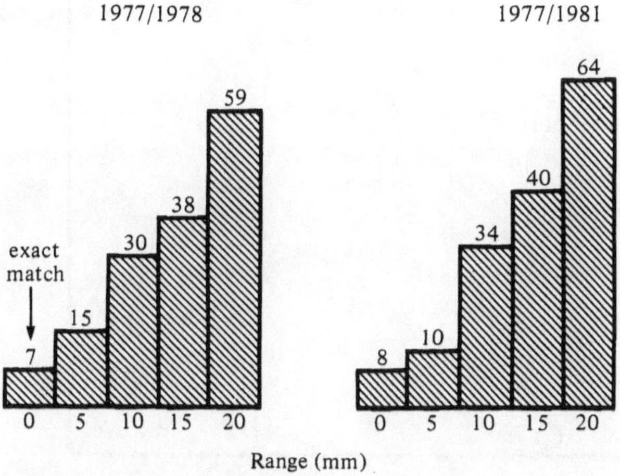

Fig. 7. The numbers of basidiocarps of *Mycena galopus* in a spruce plot which coincided in successive and non-successive years within ranges of 0–20 mm of 1977 positions.

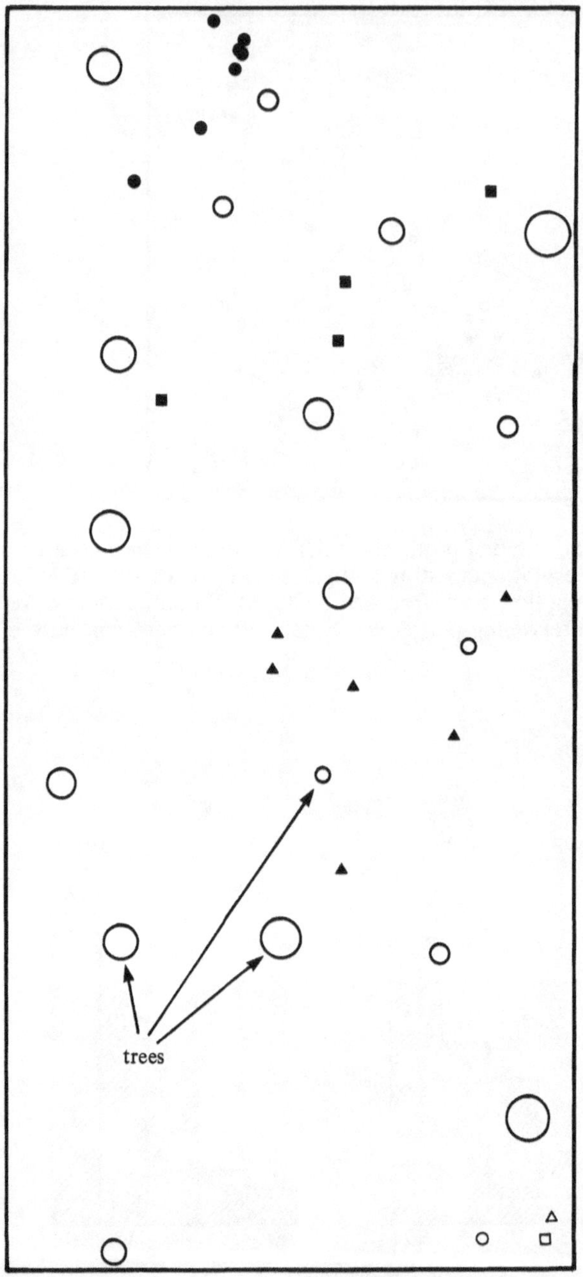

Fig. 8. Individual mycelia of *Mycena galopus* in a 10 × 4.5 m plot of *Picea sitchensis*. Dikaryotic isolates which fused imperceptibly are represented by the same symbol (filled circles, squares and triangles).

made for seasonal differences, they did not decrease in comparisons of non-successive years. All the evidence pointed to perennial or at least renewable mycelium occupying the same locations from year to year.

It is now clear that distinct individual mycelia exist in populations of heterothallic wood-decomposing basidiomycetes (Rayner & Todd, 1979; Thompson & Rayner, 1982). On this basis, the population structure of this litter decomposer on the spruce plot was investigated by examining the interactions between mycelial isolates using similar procedures. Only preliminary findings are reported here, and mating-type factors need further investigation. Forty dikaryotic mycelia were obtained from stipe tissue by dissecting hyphae from the interior of the stipe, or by plating stipes on 2% malt agar, after surface sterilisation in 10% (v/v) aqueous sodium hypochlorite.

As in similar studies, the isolates could be grouped according to whether or not they intermingled or exhibited various degrees of antagonism when paired. Three relatively large individual mycelia or mycelial types appeared to be recognisable by such groupings (Fig. 8), surrounded on the plot by numerous smaller individuals, which it could be surmised had arisen from spores or formed the edge of individuals outside the plot. Sampling was too limited to cover completely the areas occupied by arcs of basidiocarps or to relate them with any certainty to particular mycelial types. The greatest distance on the plot between two basidiocarps producing intermingling isolates was approximately 2.5 m.

The number of individual mycelia on the small spruce plot appears to be proportionally far greater than that of *Tricholomopsis platyphylla* found on an area of 2.1 ha by Thompson & Rayner (1982) (see also Chapter 9; Fig. 4). This possibly reflects a different colonisation strategy on a coniferous litter resource which would be constantly changing as new supplies fall from the canopy. The large perennial individuals on the plot are consistent with the theory of a *K*-selection strategy for basidiomycetes (Swift, 1982; Rayner & Webber: Chapter 18).

It is easy to see that *M. galopus* is common in many temperate woodlands. The overall attainments of the whole fungus as a competitive colonist are less obvious, but often striking. In one UK woodland,

---

Caption for Fig. 8 (*cont.*)
One of several small incompatible groupings is shown in the bottom right-hand corner of the plot (small hollow circle, square and triangle).

more than 80% of the leaves of the dominant *Quercus*, two years after leaf-fall, were found to be colonised by the mycelium when it was induced to fruit (Frankland, Bailey & Costeloe, 1979), and it has been estimated that this fungus decomposes a considerable proportion of the annual litter fall in British woodlands (Hering, 1972). Smith (1947) in his monograph described it as looking 'like a very ordinary slender gray or blackish *Mycena*'. Maybe, but it is a remarkably successful species.

I am very grateful to David Lindley for statistical advice, and to Andrew Bailey, Christine Benson, Pamela Eccles, Stephen Morton and Jan Poskitt for their assistance. I also thank the Forestry Commission for permission to sample in Grizedale Forest.

References

Bothe, F. (1935). Genetische Untersuchungen über die Lichtentwicklung der Hutpilze. *Archiv für Protistenkunde*, **85**, 369–83.

Charbonnel, J. (1977). Etudes microscopiques IV. Le genre *Mycena. Documents Mycologiques (Lille)*, **7**, 1–70.

Chard, J. M. (1981). Development of an immunofluorescence technique for studying the ecology of *Mycena galopus* (Pers. ex Fr.) Kummer. Ph.D. Thesis, University of Essex, UK.

Chard, J. M., Gray, T. R. G. & Frankland, J. C. (1983). Antigenicity of *Mycena galopus*. *Transactions of the British Mycological Society*, **81**, 503–11.

Coates, D., Rayner, A. D. M. & Todd, N. K. (1981). Mating behaviour, mycelial antagonism and the establishment of individuals in *Stereum hirsutum*. *Transactions of the British Mycological Society*, **76**, 41–51.

Cooke, R. C. & Rayner, A. D. M. (1984). *Ecology of Saprotrophic Fungi.* London, New York: Longman.

Dennis, R. W. G., Orton, P. D. & Hora, F. B. (1960). New check list of British agarics and boleti. *Supplement to Transactions of the British Mycological Society*, **43**, 1–225.

Dickinson, C. H., Dawson, D. & Goodfellow, M. (1981). Interactions between bacteria, streptomycetes and fungi from *Picea sitchensis* litter. *Soil Biology and Biochemistry*, **13**, 65–71.

Dix, N. J. (1984). Minimum water potentials for the growth of some litter decomposing agarics and other basidiomycetes. *Transactions of the British Mycological Society* (in press).

Frankland, J. C. (1966). Succession of fungi on decaying petioles of *Pteridium aquilinum*. *Journal of Ecology*, **54**, 41–63.

Frankland, J. C. (1969). Fungal decomposition of bracken litter. *Journal of Ecology*, **57**, 25–36.

Frankland, J. C. (1975). Fungal decomposition of leaf litter in a deciduous woodland. In *Biodegradation et Humification*, ed. G. Kilbertus, O. Reisinger, A. Mourey & J. A. Cancela da Fonseca, pp. 33–40. Rapport 1er Colloque International, 1974, Nancy. Sarreguemines: Pierron.

Frankland, J. C. (1976). Decomposition of bracken litter. *Botanical Journal of the Linnean Society*, **73**, 133–43.

Frankland, J. C. (1981). Mechanisms in fungal successions. In *The Fungal Community –*

Its Organization and Role in the Ecosystem, ed. D. T. Wicklow & G. C. Carroll, pp. 403–26. New York: Marcel Dekker.

Frankland, J. C. (1982). Biomass and nutrient cycling by decomposer basidiomycetes. In *Decomposer Basidiomycetes: Their Biology and Ecology*, ed. J. C. Frankland, J. N. Hedger & M. J. Swift, pp. 241–61. Cambridge: Cambridge University Press.

Frankland, J. C., Bailey, A. D. & Costeloe, P. L. (1979). Ecology of a woodland toadstool, *Mycena galopus. Annual Report, 1978, Institute of Terrestrial Ecology*, pp. 26–8. Cambridge: Institute of Terrestrial Ecology.

Frankland, J. C., Bailey, A. D., Gray, T. R. G. & Holland, A. A. (1981). Development of an immunological technique for estimating mycelial biomass of *Mycena galopus* in leaf litter. *Soil Biology and Biochemistry*, **13**, 87–92.

Frankland, J. C., Lindley, D. K. & Swift, M. J. (1978). A comparison of two methods for the estimation of mycelial biomass in leaf litter. *Soil Biology and Biochemistry*, **10**, 323–33.

Fries, N. (1949). Culture studies in the genus *Mycena. Svensk Botanisk Tidskrift*, **43**, 316–42.

Grgurinovic, C. A. & Holland, A. A. (1982). Descriptions and notes for *Mycena galopus* var. *mellea* and *Mycena sanguinolenta. Brunonia*, **5**, 103–7.

Harrison, A. F. (1971). The inhibitory effect of oak leaf litter tannins on the growth of fungi in relation to litter decomposition. *Soil Biology and Biochemistry*, **3**, 167–72.

Hering, T. F. (1967). Fungal decomposition of oak leaf litter. *Transactions of the British Mycological Society*, **50**, 267–73.

Hering, T. F. (1972). Fungal associations in broad-leaved woodlands in north-west England. *Mycopathologia et Mycologia applicata*, **48**, 15–21.

Hering, T. F. (1975). The macrofungal flora of two natural oakwoods in Cumberland. *Naturalist*, No. 932, 21–3.

Hering, T. F. (1982). Decomposing activity of basidiomycetes in forest litter. In *Decomposer Basidiomycetes: Their Biology and Ecology*, ed. J. C. Frankland, J. N. Hedger & M. J. Swift, pp. 213–25. Cambridge: Cambridge University Press.

Hintikka, V. (1961). Das Verhalten einiger *Mycena* – Arten zum pH sowie deren Einfluss auf die Azidität der Humusschicht der Wälder. *Karstenia*, **5**, 107–21.

Hintikka, V. (1964). Psychrophilic basidiomycetes decomposing forest litter under winter conditions. *Communicationes Instituti Forestalis Fenniae*, **59.2**, 1–20.

Hintikka, V. (1969). Acetic acid tolerance in wood- and litter-decomposing Hymenomycetes. *Karstenia*, **10**, 177–83.

Hintikka, V. (1970). Studies on white-rot humus formed by higher fungi in forest soils. *Communicationes Instituti Forestalis Fenniae*, **69.2**, 1–68.

Hintikka, V. (1982). The colonisation of litter and wood by basidiomycetes in Finnish forests. In *Decomposer Basidiomycetes: Their Biology and Ecology*, ed. J. C. Frankland, J. N. Hedger & M. J. Swift, pp. 227–39. Cambridge: Cambridge University Press.

Hintikka, V. & Korhonen, K. (1970). Effects of carbon dioxide on the growth of lignicolous and soil-inhabiting hymenomycetes. *Communicationes Instituti Forestalis Fenniae*, **69.5**, 1–29.

Kühner, R. (1938). Le genre *Mycena. Encyclopédie Mycologique*, **10**, 1–710.

Lamaison, J.-L. (1976). Intérêt chimiotaxinomique de l'équipement enzymatique des macromycètes. *Bulletin Société Botanique de France*, **123**, 119–36.

Lindeberg, G. (1946). Thiamin and growth of litter-decomposing Hymenomycetes. *Botaniska Notiser*, **1**, 89–93.

Lindeberg, G. (1947). On the decomposition of lignin and cellulose in litter caused by soil-inhabiting Hymenomycetes. *Arkiv för Botanik*, **33A**, 1–16.
Lindeberg, G. (1948). On the occurrence of polyphenol oxidases in soil-inhabiting basidiomycetes. *Physiologia Plantarum*, **1**, 196–205.
Maas Geesteranus, R. A. (1980). Studies in Mycenas-15. *Personoia*, **11**, 93–120.
Mason, P. A., Last, F. T., Pelham, J. & Ingleby, K. (1982). Ecology of some fungi associated with an ageing stand of birches (*Betula pendula* and *B. pubescens*). *Forest Ecology and Management*, **4**, 19–39.
Mathieson, J. (1946). Antibiotics from Victorian basidiomycetes. *Australian Journal of Experimental Biology and Medical Science*, **24**, 57–62.
Mikola, P. (1956). Studies on the decomposition of forest litter by basidiomycetes. *Communicationes Instituti Forestalis Fenniae,* **48.2**, 1–22.
Newell, K. (1980). The effect of grazing by litter arthropods on the fungal colonization of leaf litter. Ph.D. Thesis, University of Lancaster, UK.
Newell, K. (1984*a*). Interaction between two decomposer basidiomycetes and a collembolan under Sitka spruce: distribution, abundance and selective grazing. *Soil Biology and Biochemistry* (in press).
Newell, K. (1984*b*). Interaction between two decomposer basidiomycetes and a collembolan under Sitka spruce: grazing and its potential effects on fungal distribution and litter decomposition. *Soil Biology and Biochemistry* (in press).
Oelbe, M. (1982). Untersuchungen über einige Kohlenhydratabbauende Enzyme des Mykorrhizapilzes *Tricholoma aurantium.* M.Sc. Thesis, University of Göttingen.
Parker-Rhodes, A. F. (1954). Deme structure in higher fungi: *Mycena galopus. Transactions of the British Mycological Society*, **37**, 314–20.
Pearson, A. A. (April–June 1955). *Mycena. Naturalist*, 41–63.
Rayner, A. D. M. & Todd, N. K. (1979). Population and community structure and dynamics of fungi in decaying wood. *Advances in Botanical Research*, **7**, 334–420.
Rayner, R. W. (1979). The frequencies with which basidiomycete species other than rusts and smuts have been recorded on B.M.S. forays. *Bulletin of the British Mycological Society*, **13**, 110–25.
Smith, A. H. (1947). *North American Species of* Mycena. Ann Arbor: University of Michigan Press. Reprint (1971), Vaduz: Cramer.
Stalpers, J. A. (1978). Identification of wood-inhabiting fungi in pure culture. *Studies in Mycology, Baarn,* No. 16, 1–248.
Swift, M. J. (1982). Basidiomycetes as components of forest ecosystems. In *Decomposer Basidiomycetes: Their Biology and Ecology*, ed. J. C. Frankland, J. N. Hedger & M. J. Swift, pp. 307–37. Cambridge: Cambridge University Press.
Thompson, W. & Rayner, A. D. M. (1982). Spatial structure of a population of *Tricholomopsis platyphylla* in a woodland site. *New Phytologist*, **92**, 103–14.
Villard, J., Oddoux, L. & Porte, M. (1979). Recherche d'une activité anti-Coxsackievirus *in vitro* dans le filtrat de culture de quelques Agaricales. *Annales pharmaceutiques françaises*, **37**, 143–52.
Villard, J., Porte, M. & Oddoux, L. (1982). Activité antivirale du filtrat de culture de quelques Agaricales sur l'infection expérimentale au coxsackievirus B. *Annales pharmaceutiques françaises*, **40**, 69–73.

12

The micro-environment of basidiomycete mycelia in temperate deciduous woodlands

LYNNE BODDY

Department of Microbiology, University College, Newport Road, Cardiff, CF2 1TA, UK

Introduction: resource relationships

Throughout this volume emphasis is placed on the mycelium as the basic unit of the fungus by which it obtains its nutrition and achieves vegetative spread. As such it is at this level that the most significant interactions between fungi and the biotic and abiotic environment occur. To consider all such interactions in the very many environments in which fungal mycelia are found would be an enormous undertaking. Thus, the approach adopted here will be to consider a particular case, that of basidiomycete mycelia in temperate deciduous woodlands.

The fruit bodies of Basidiomycotina in woodlands readily attract attention and it has long been known that these organisms are often the major agents of wood decomposition in both natural and man-made environments (Cartwright & Findlay, 1958; Swift, 1977; Rayner & Todd, 1979), although their importance has sometimes been obscured (Boddy & Rayner, 1983*c*; Cooke & Rayner, 1984). More recently their major role in the decomposition of non-woody litter has been emphasised (Frankland, 1982; Hering, 1982; Hintikka, 1982; Swift, 1982; Cooke & Rayner, 1984). Further, they interact with higher plants as parasites and as mutualistic symbionts forming mycorrhizal partnerships. However, despite their obvious importance to the functioning of the woodland system, their actual ecological roles and mycelial biology have received relatively little attention.

In woodlands a large proportion of organic matter, available to decomposers, derives from trees with 20–30% of the above ground input being woody, e.g. branches, trunks, twigs and the remainder leaves, flowers and fruits (Bray & Gorham, 1964; Boddy & Swift, 1983). The below-ground input is less easy to quantify but consists of non-woody tissues, e.g. root hairs, sloughed cells, minor roots, along with large woody roots. Higher plants and bryophytes in ground vegetation also contribute to organic matter input as do animal remains and faeces. Although the input of woody material is much less than that of leaves, its more durable nature usually leads to considerable accumulation on the forest floor (Ovington, 1962; Swift, Boddy & Healey, 1984).

Thus the resources for growth of heterotrophs in this environment are considerable, ranging from bulky to finely divided, living to non-living, and are exploited by all of the three main nutritional modes of fungi, i.e. saprotrophy, necrotrophy and biotrophy. Relationships between basidiomycete mycelia and these resources can be considered to be of two types: those which are restricted to individual component units of the litter, e.g. individual leaves, twigs, flowers, branches, and those which are not restricted by physical bounds and can ramify throughout entire litter systems (Cooke & Rayner, 1984). Examples of the former include species of *Marasmius* and *Mycena* which are restricted respectively to the petiole and lamina of fallen leaves (Fig. 1*c*,*d*), species of *Mycena* on beech cupules (Fig. 1*b*); wood-rotting fungi such as *Coriolus versicolor* and *Stereum hirsutum* which are restricted to the wood unit which they occupy. Cord-forming fungi (Thompson: Chapter 9) and litter-decomposing organisms such as *Clitocybe nebularis* and *Marasmius wynnei* (Fig. 1*a*) are non-component restricted and ramify through the litter for considerable distances.

Many species and even genera appear to be specifically associated with a particular type of resource/substratum. For instance mycorrhizal fungi are predominantly members of the genera *Amanita, Boletus, Lactarius, Russula* and *Tricholoma*; members of the Aphyllophorales are mainly wood-rotters; and those which decompose leaf and non-woody litter are found mainly in the genera *Collybia, Clitocybe, Marasmius* and *Mycena.* Further some fungi are specific to certain plant taxa, e.g. *Piptoporus betulinus* on *Betula* trunks and branches, *Amanita muscaria* in mycorrhizal association with *Betula*, *Pinus* and perhaps *Carpinus*, and *Russula mairei* with *Fagus*.

On the other hand some Basidiomycotina utilise several different resource types: *Hypholoma fasciculare* and *Tricholomopsis platyphylla* are mycelial-cord-forming fungi which decompose large woody substrata, twigs and leaf litter (Thompson: Chapter 9); some mycorrhizal formers such as *Laccaria laccata* (Hering, 1982), *Boletus* spp. and *Tricholoma* spp. (Cooke & Rayner, 1984) are also able to decompose leaf litter.

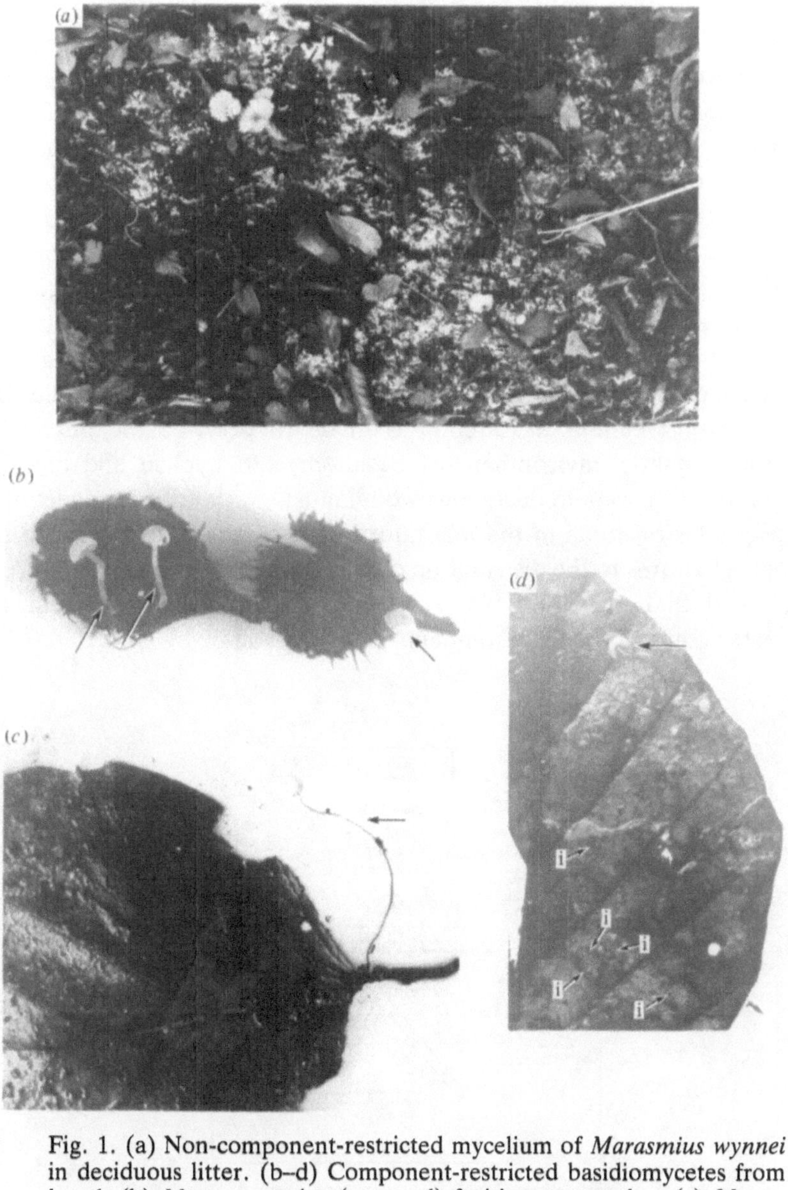

Fig. 1. (a) Non-component-restricted mycelium of *Marasmius wynnei* in deciduous litter. (b–d) Component-restricted basidiomycetes from beech (b) *Mycena* species (arrowed) fruiting on cupules. (c) *Marasmius* species (arrowed) developing from a leaf petiole. (d) *Mycena* species (arrowed) on a leaf lamina. Notice the probable interaction zone lines (i) between bleached regions. (From Cooke & Rayner, 1984.)

Evidently, then, each resource type in a deciduous woodland is characterised by rather specific associations of fungi. This is in accord with the concept of the unit community discussed by Swift (1976) and, such assemblages characteristic of a particular resource type are the nearest fungal approach to the concept of the plant association (Cook & Rayner, 1984). This raises questions of the origin of such specificity; the answer must lie in the interactions of Basidiomycotina with the abiotic and biotic components of the micro-environment associated with each resource type.

The main factors influencing basidiomycete mycelia are represented diagrammatically in modular form in Fig. 2. From this it is clear that there are numerous interactions between parameters: the components of the three modules interact with basidiomycete mycelia, with components of other modules, and with other components in a module. In the following sections an attempt is made to describe the physical and microclimatic environment of basidiomycete hyphae and mycelia in natural substrata in deciduous woodlands; to quantify the effect of such variables on fungi in the laboratory; and on the basis of this to relate such features to the distribution of basidiomycete mycelia in nature. It is hoped that principles will emerge from this discussion from which extension to other environments can be made.

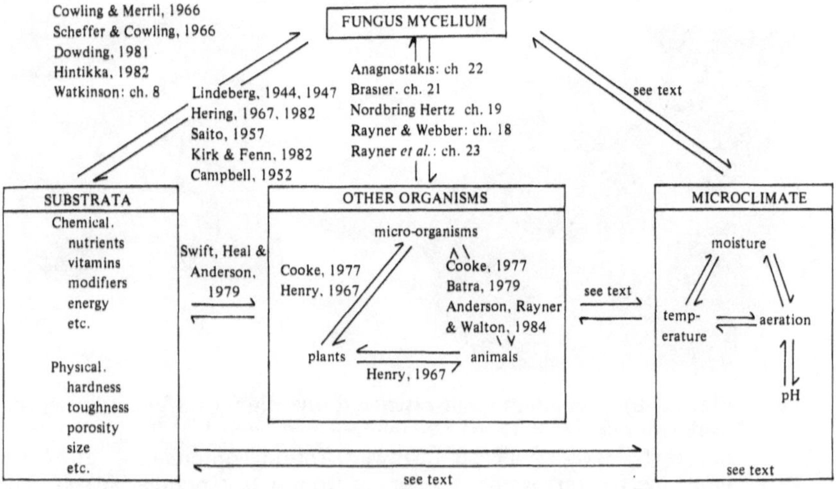

Fig. 2. Schematic representation of interactions between the fungus mycelium and the biotic and abiotic environments.

Physical and microclimatic environment

Substrata in which fungal mycelia are growing, be they soil, leaf litter, wood etc, may be regarded as matrices consisting of a solid phase comprised of variously sized organic/inorganic particles, and a system of voids. These voids occur both within organic material, e.g. vascular elements, cavities resulting from decomposition, and between the different inorganic and organic components (Fig. 3). They vary considerably in size from intermolecular spaces to cavities several millimetres or even centimetres wide. These voids are routes for entry of hyphae into the matrix, or a space which has to be crossed in order to reach another supply of organic nutrients, and are filled with a gaseous or liquid phase. Wherever hyphae are growing they will be affected by the physical and chemical conditions in the void.

Fungi interact with the environment at the hyphal level. However, the fungal hypha does not act entirely independently being only a small part of the fungal mycelium, which is a three-dimensional entity often occupying several cm^3 and sometimes even many m^3 (e.g. Thompson & Rayner, 1982, 1983; Boddy & Rayner, 1983*a,b*; Watkinson: Chapter 8; Thompson: Chapter 9; Frankland: Chapter 11). Thus, the mycelium has the unusual property of being located in numerous places at any one time and therefore may be subject to many different environmental regimes simultaneously. In terms of survival, resource capture and spread, it is the influence of biotic and abiotic variables at the mycelial level, resulting from the combined effect of these variables at the hyphal level, which is of relevance to the fungus.

Elucidation of the influence of micro-environment on the mycelium is not easy but considerably less difficult than for individual hyphae. It requires characterisation of conditions in the field along with quantification of the effects of such conditions on the mycelium. The latter can be achieved in the field but very often recourse is made to the laboratory. Extreme caution must, however, be exercised when extrapolating from artificial media, as these are necessarily relatively homogeneous whereas in nature heterogeneity is often the rule.

Temperature

Woodland and substratum microclimate reflects the local climate; thus temperature accordingly changes cyclically each day and during the year (Figs. 4, 7). It is, however, modified by various environmental features and differs from local temperature as a result of differences in aspect, degree of exposure to insolation and the buffering

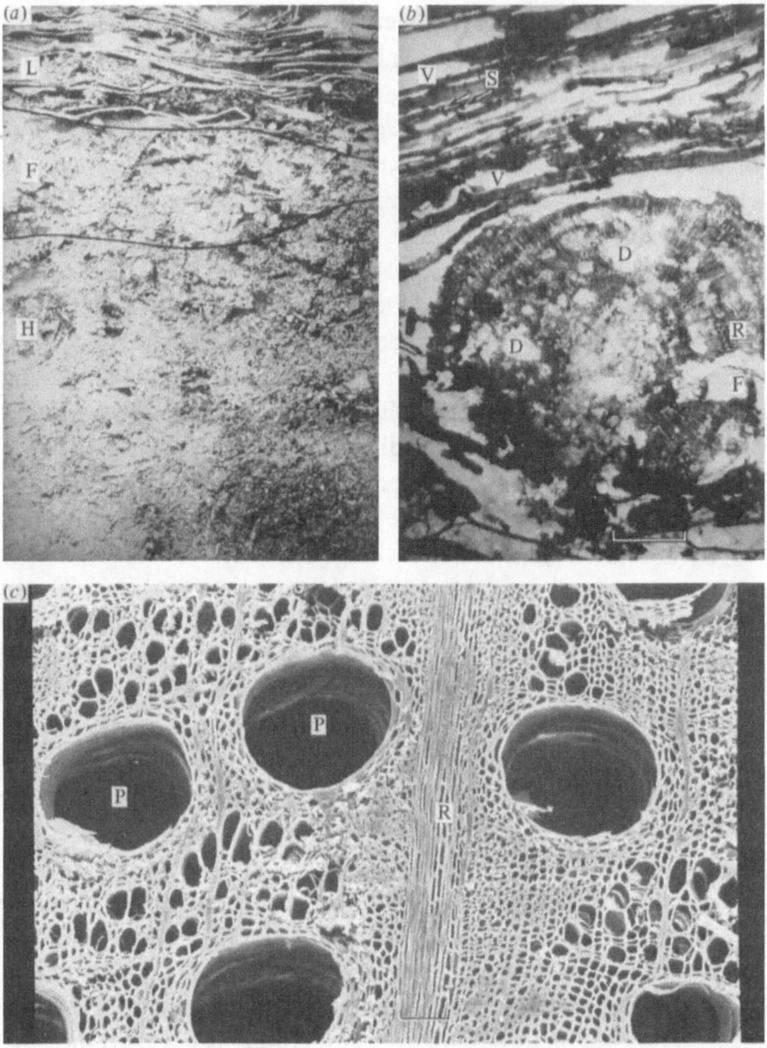

Fig. 3. Examples of microhabitats of fungal mycelia within organic substrata at different scales of resolution. (a) Detail of a gelatin-embedded vertical section through the top 5 cm of a well-developed organic soil in a *Castanea sativa* woodland showing the litter (L), fermentation (F) and humus (H) sub-horizons. (From Swift *et al.*, 1979.) (b) Detail of a section through the litter layer and a rotten twig in the same site as (a) illustrating voids between leaves (V), voids resulting from decomposition of substrates (D), and anatomical voids of the vessels and medullary rays (R). Fungal hyphae innervating twig from surrounding leaf litter (F); dark fungal stroma on leaves (S); scale bar, 1 cm (from Swift *et al.*, 1979.) (c) Scanning electron micrograph of oak sapwood showing different sized vessels (P) and medullary ray cells (R). Scale bar, 100 μm. (From M. Hale, unpublished.)

effect and different thermal capacities of the surrounding vegetation, litter, soil etc. Thus, even at the same geographical location, the temperature regimes experienced by fungal mycelia within different substrata and ecosystems can vary considerably, e.g. temperature within a woodland often has lower maxima and higher minima than in adjacent grasslands (Wilkins & Harris, 1946; Geiger, 1965).

Soil and litter are poor conductors of heat so the rate of penetration of heat from the surface into soil is slow. Thus, whereas the diurnal (and seasonal) temperature of the top centimetre or so of the litter closely follows fluctuations in air temperature, fluctuations decrease with depth and there is a lag in response, which can be as much as two hours per centimetre, so that at a depth of 15 cm the surface maximum at noon is reflected in a maximum at midnight (e.g. Swift, Heal & Anderson, 1979). At a depth of 30 cm fluctuations are often negligible. Similar phenomena occur in fallen branches and other bulky substrata (Fig. 4).

As the thermal characteristics of substrata result from the joint characteristics of the individual solid, gaseous and liquid components, and since water has a much higher thermal capacity than other components, the moisture content can significantly alter the temperature regime of a substratum. For instance Bocock, Bailey & Hornung (1982) found that the persistence of water in forest soils tended to reduce temperature fluctuations. They also found that temperature amplitude is higher in soils rich in organic matter than in mineral soils.

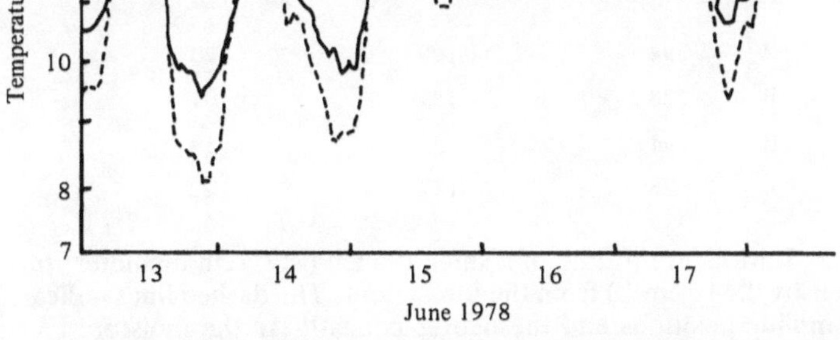

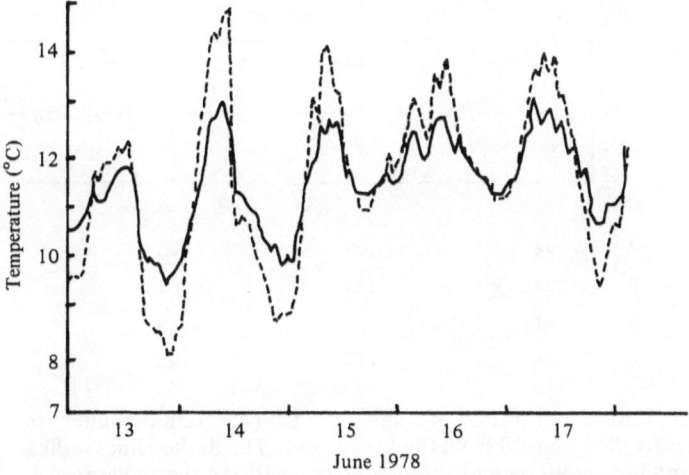

Fig. 4. Diurnal fluctuations in temperature (°C) of soil at 5 cm depth (———) and at the centre of a branch approximately 2 cm in diameter (----). (From Boddy, 1983*b*.)

Trends in temperature are thus found down the soil and litter profile and within bulky substrata, but unless large differences occur in the thermal conductivities of substrata, micro-spatial variation at the same horizontal level is very slight. This, combined with the fact that it is relatively simple to obtain accurate, continuous measurement of substratum temperature, e.g. by use of thermistor probes and data recorders, enables accurate determination of the temperature regime experienced by fungal mycelia (cf. water content and gaseous composition).

Water

Micro-spatial distribution and the expression of water content. Water content is often expressed gravimetrically as a percentage of the oven dry weight of substratum; however, this cannot be used to compare the amount of water in different substrata, e.g. between a sandy and a loam soil, between different organic materials, or between wood at different stages of decay, because these do not necessarily have the same masses in equal volumes.

Because absolute water content is an easily quantifiable and easily understood term it is frequently used, but it clearly has little biological meaning so far as micro-organisms are concerned. The volume of water *per se* required for the creation of a few centimetres length of hypha will be contained in one gramme of many apparently dry soils (Griffin, 1972).

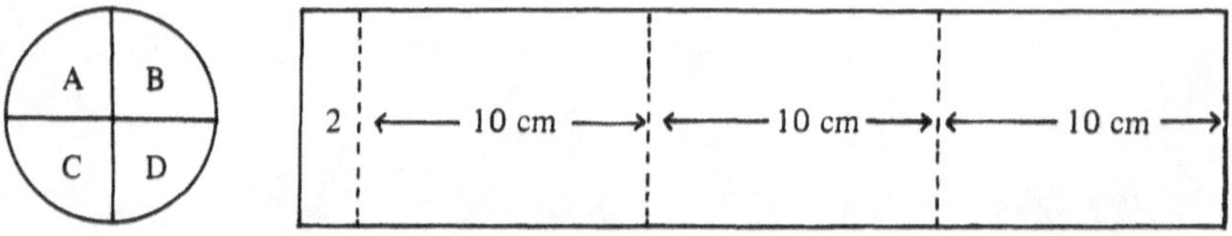

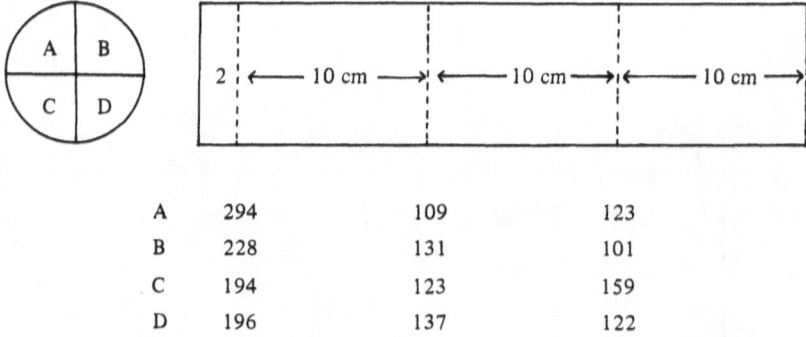

A	294	109	123
B	228	131	101
C	194	123	159
D	196	137	122

Fig. 5. Moisture profile of a small branch (<2.5 cm diameter, relative density 0.44 g cm^{-3}) from the forest floor. The dashed lines indicate the sampling positions and the figures beneath are the moisture contents (percentage oven-dry weight). Samples A and B were from the upper side and C and D from the lower side of the branch. (From Boddy, 1983*b*.)

In solid substrata water is located in the voids and the importance of water content to micro-organisms lies in the factors correlated with its presence and location. Water influences hyphal activity in two main ways: in regions of low moisture content where its lack may limit growth because of difficulties of extracellular enzymes reaching their targets and of obtaining water for other physiological activities; and, indirectly, in regions of high moisture content as a result of poor aeration.

A term which is becoming used increasingly to express the difficulty (or ease) with which water may be obtained from a substratum is that of water potential. This subject has been comprehensively treated by Griffin (1972, 1977, 1982) and will only be considered briefly here. Water potential can be regarded as having two main components: matric potential and osmotic potential. Matric potential is a result of forces associated with interfaces between water, air and the solid matrix (i.e. soil, leaf litter, wood). Osmotic potential reduces the potential energy of water by the presence of solutes within it. One of the components usually predominates to the exclusion of the other. For instance, in the sea or in syrup and preserves, the high solute content results in the osmotic component predominating. In soil and wood on the other hand, the matric component tends to predominate although dissolved nutrients will add an osmotic component and will be important in saline and heavily fertilised soils (Griffin, 1963; Williams, 1968; Boddy, 1983*c*).

In order to understand the impact of water on mycelia it is helpful to consider the distributions of mycelia and water within the substratum (e.g. those in Fig. 3). When initially saturated substrata begin to dry out water tends to be lost first from the largest voids, the smallest retaining water for the longest time (see Griffin, 1972). Concomitant with decrease in water is an increase in the gaseous phase; thus the two are intimately associated. Clearly, the void size distribution in substrata (Fig. 3) will to some extent determine the location of water and the gaseous phase and there will be considerable micro-spatial variation. Substratum matric potential will give some indication of the location of water in relation to void size and it can be measured and even maintained in experimental systems (Griffin, 1963, 1972, 1977, 1982; Rose, 1966; Holmes, Taylor & Richards, 1967; Stone & Scallen, 1967; Clarke, Jennings & Coggins, 1975). Such measurements quantify the overall moisture regime of the substratum/mycelium complex, but unless the locations of hyphae are known it reveals little at the hyphal level. Further, the distribution of water varies on a slightly larger scale within substrata according to the wetting and drying regimes experienced by different parts. This is illustrated for a small branch lying on the woodland floor (Fig. 5) and for a short stake having one end immersed in water (Fig. 6; Baines & Levy, 1979). The latter can be considered as a model for any woody resource with parts above and below ground, e.g. a cut stump, standing dead tree.

Clearly micro-spatial and other small-scale variations are likely to be of prime importance to hyphae and mycelia; however, few data of this kind are available. Meaningful gross measures of substratum moisture content are not available; even though matric potential to some extent describes micro-spatial location of water and the force required to remove it from substrata, other small-scale variation means that at present it only serves as an alternative description of moisture content and its precise meaning may be less well understood than percentage moisture content.

Macro-spatial and temporal distribution of water in woodlands. From the above discussion it is clear that it is not a simple matter to obtain meaningful information on the distribution of water in substrata in relation to the fungal mycelium. Despite this it is possible to gain some

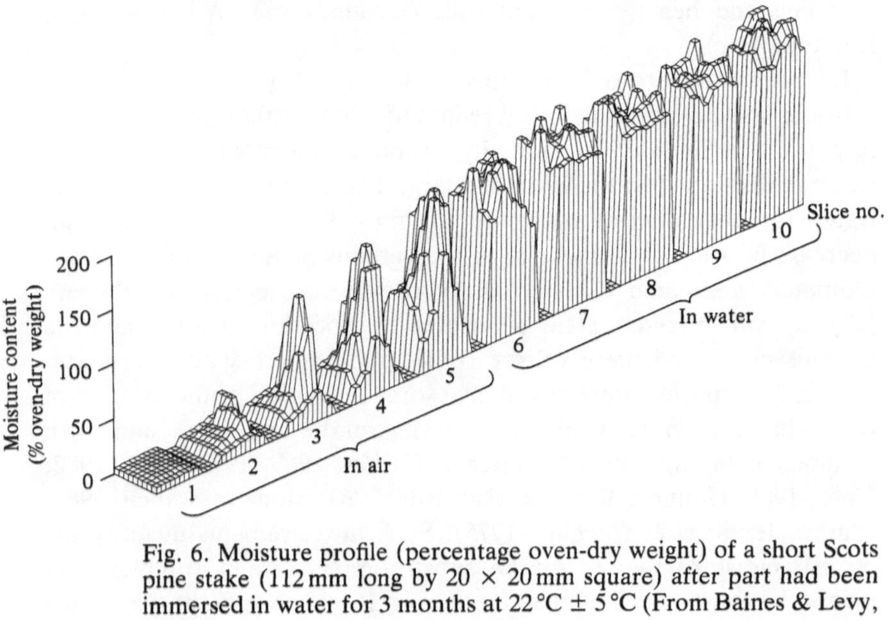

Fig. 6. Moisture profile (percentage oven-dry weight) of a short Scots pine stake (112 mm long by 20 × 20 mm square) after part had been immersed in water for 3 months at 22 °C ± 5 °C (From Baines & Levy, 1979.)

insight into the temporal and macro-spatial water regimes in woodlands from gross measurements of rainfall and percentage moisture content.

The amount and distribution of water within substrata depend upon the capacity for storage of water and its supply and loss. The capacity for storage depends on porosity which varies between substrata depending upon the anatomy, state of decay and degree of compaction. Supply of water to dead organic substrata is mainly by precipitation and loss by drainage, lateral run off and evapo-transpiration which is in turn influenced by temperature and humidity. In addition various inherent features of substrata that influence wetting and drying include: porosity; distance over which water has to travel, i.e. depth in soil, size of wood; presence or absence of bark for wood; and vegetation cover.

Temporal variation occurs as a result of local climate influencing supply and loss of water. The overall moisture relationships of the leaf litter layer have received considerable attention and accurate prediction of moisture content is now possible if precipitation and potential evapo-transpiration are known (e.g. Meentemeyer, 1974; Moore & Swank, 1975). Less information is available for other substrata but temporal variation in moisture content of small branch wood has been monitored (Fig. 7; Boddy, 1983*b*).

General statements regarding the distribution of water in woodlands can be made and large differences in water regimes would be expected to exist between substrata depending on location. For instance the quantity of water in functional sapwood (in terms of percentage water holding capacity) is always high and often considerably higher than in many dead organic substrata. Because of conditions favourable to drying, attached dead branches and leaves and standing dead trunks are generally drier than those on the forest floor, although if precipitation is light it may be intercepted in the canopy with little reaching the ground. Likewise surface litter is often drier than deeper litter, humus and other buried substrata.

Gaseous environment

The gaseous composition of substrata is the product of biological metabolic processes (thus under aerobic conditions O_2 and CO_2 are of most significance) and physical phenomena which exert their effect mainly via their influence on gaseous diffusion. The rate of diffusion within substrata is largely determined by the diffusion coefficients and the geometry of the system of voids including both the distance and shape of pathways to the external atmosphere. This will vary between different substrata and hence gaseous regimes will differ (Boynton, 1941). Any factor which increases the effective path length will decrease the rate of gaseous exchange and in soil and litter the two most significant factors are depth and water content (Fig. 8; Boynton, 1941; Yamaguchi, Flocker & Howard, 1967; Griffin, 1963). The solubility of O_2 is 35–50 times less than that of CO_2 over the range of temperatures encountered in soils (Brock, 1966); thus waterfill of voids will have a

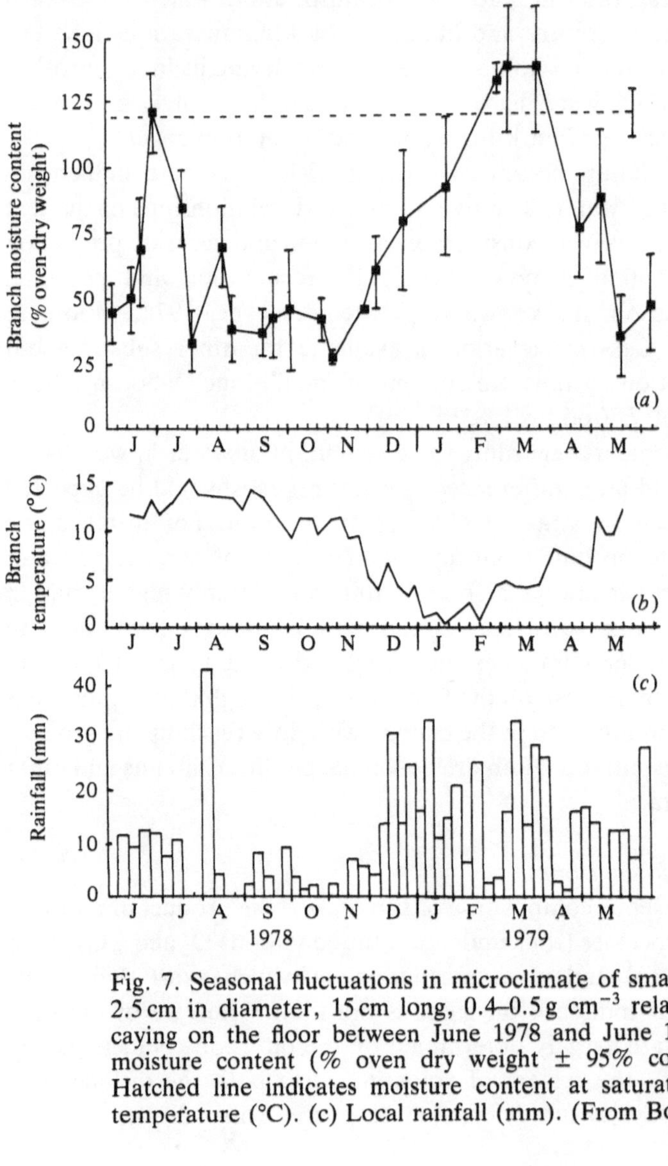

Fig. 7. Seasonal fluctuations in microclimate of small branches (1.5–2.5 cm in diameter, 15 cm long, 0.4–0.5 g cm^{-3} relative density) decaying on the floor between June 1978 and June 1979. (a) Branch moisture content (% oven dry weight ± 95% confidence limits). Hatched line indicates moisture content at saturation. (b) Branch temperature (°C). (c) Local rainfall (mm). (From Boddy, 1983*b*.)

larger effect on O_2 than CO_2. These features are illustrated for different depths in sandy loam soil columns held at different temperatures (Fig. 8; Yamaguchi *et al.*, 1967).

Similar considerations apply to other substrata such as wood, in which case path length depends on water content, anatomy, size and state of decay. In general gases penetrate undecayed wood rather slowly especially in radial and tangential directions: in *Fagus sylvatica* permeability is 65 000 times greater longitudinally than tangentially (Stamm, 1946; Smith, 1964). Permeability also differs between species, that of hardwoods usually being greater than softwoods. As decomposition proceeds the void space will increase with a concomitant decrease in path length.

As a result of these features differences in the overall gaseous composition exist between substrata: in humus and litter relatively little CO_2 accumulates under normal conditions (Romell, 1928, cited in

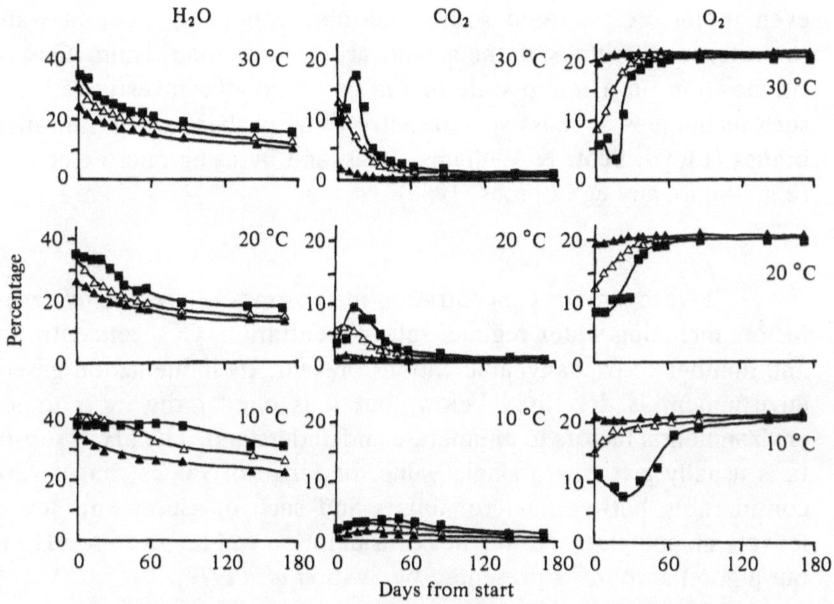

Fig. 8. Relationship between temperature, moisture content (% volume), CO_2 concentration and O_2 concentration, with time following initial saturation of sandy loam soil columns at depths of 5 cm (▲), 35 cm (△) and 65 cm (■) from the surface. (From Yamaguchi, Flocker & Howard, 1967, *Soil Science of America, Proceedings*, volume 31, 1967, pages 164–7, by permission of the Soil Science Society of America.)

Hintikka, 1982; Brierley, 1955) whereas in living or dead wood concentrations are often much higher than in air, commonly being in the region of 10–20% (Chase, 1934; Thacker & Good, 1952; Jensen, 1967, 1969*a*; Hintikka, 1982; Hintikka & Korhonen, 1970). Carrodus & Triffet (1975), using more sophisticated techniques, found that the gas in living *Acacia* stems was almost pure CO_2. In most examples increase in CO_2 is accompanied by a concomitant decrease in O_2. Seasonal fluctuations in gaseous composition often occur (Chase, 1934; Boynton, 1941; Paim & Beckel, 1963; Jensen, 1969*a*) resulting from seasonal changes in temperature, moisture and respiratory activity.

The above is based on bulk sample data of the gaseous phase. Mycelia are however, often bathed in water and in these conditions it is the quantity of dissolved gases that is significant but, as is so often the case, micro-spatial data are lacking so that conditions at the hyphal and mycelial level are unclear. That considerable micro-spatial variation is likely to exist is apparent from the discussion on water and a striking example is given by Greenwood & Goodman (1967) who showed that even in the best-aerated soils, anaerobic conditions exist in water-saturated soil crumbs if their radii are greater than 3 mm. Gaseous composition on a micro-scale *in situ* could now be investigated using such techniques as mass spectrometry using probes with Teflon membranes (Lloyd, Scott & Williams, 1983) and by using micro-electrodes (e.g. Wimpenny & Coombs, 1983).

pH

Hydrogen ion concentration in substrata is affected by several factors including water regime, salt concentration, CO_2 concentration, and number of exchangeable cations present. Its influence on gaseous environment is described below, but it is one of the most difficult environmental factors to enumerate and understand. The pH of substrata is usually given as a single value, or range of values, but it varies considerably between microhabitats and such measurements are relatively meaningless. No further consideration will be given to pH here but a good account is presented by Swift *et al.* (1979).

Interactions between abiotic variation

As has already been indicated the abiotic variables are intimately associated; thus change in one will often influence another. The following examples serve to emphasise this. Temperature affects drying of substrata, which by shortening diffusion-path length indirectly affects gaseous composition. Gaseous composition is further affected by temperature as the solubilities of both O_2 and CO_2 increase with decrease in temperature. Gaseous composition is also affected by the tendency of CO_2, but not O_2, to dissociate in water into a number of ionic forms, the relative frequency of which is markedly influenced by temperature and particularly by pH

$$CO_2 \text{ (gaseous)} \rightleftharpoons CO_2 \text{ (dissolved)} + H_2O \rightleftharpoons H_2CO_3 \rightleftharpoons H^+ + HCO_3^- \rightleftharpoons CO_3^{2-} + 2H^+$$

At or below pH 5 only dissolved CO_2 is significant but the equilibrium moves to the right with increasing pH; the bicarbonate ion is the predominant form between pH 7 and pH 10 (Brock, 1966). Thus for hyphae, which are usually bathed in a water film, measurement of gaseous CO_2 in the substratum atmosphere is not sufficient for elucidation of the conditions actually experienced. Simultaneous measurements of pH and CO_2 would remedy this but such joint measurements are rarely made. This may not be too problematic in litter as many are acidic (Williams & Gray, 1974; Swift *et al.*, 1979).

Effect of decomposer organisms on abiotic variables

That abiotic variables affect decomposer organisms is obvious, and will be discussed in detail later, but the converse is sometimes forgotten. In general, temperature is not significantly affected by the activities of micro-organisms as their output of metabolic heat is low and is usually dissipated fairly rapidly. Instances when this is not so usually arise in association with man's activities particularly in very compact substrata, a striking example being in compost heaps where temperatures greater than 60 °C may develop as a result of microbial activity.

Moisture content and gaseous composition on the other hand can be, and presumably often are, altered considerably. Water is, for instance, liberated during the breakdown of cellulose. Griffin (1977) states that complete microbial decomposition of 1.0 g yields 0.555 g water. Mycelia can translocate water to and from substrata, a prime example being that of the dry rot fungus *Serpula lacrimans* (Jennings, 1982; and Chapter 7).

Effect of microclimate on the activity of basidiomycete mycelia in culture and natural substrata

Several authors have described the influence of abiotic variables on gross decomposition rates (e.g. Bunnel, Tait, Flanagan & Van Cleve, 1977; Boddy, 1983*c*). At best these measurements give only an estimate of 'average' microbial activity in substrata within which may exist a variety of nutritional and climatic micro-environments supporting activity at widely differing levels. Few data are available for mycelial activity in the field but a preliminary picture can be built from laboratory-based studies and the few available field studies.

Temperature

Temperature influences mycelia by its effect on enzyme-catalysed reactions, and is often quantified in terms of linear extension on agar at constant temperatures. This has been reviewed by Wagener & Davidson (1954) and Cartwright & Findlay (1958) for wood-rotting basidiomycetes but not to my knowledge for litter-rotting species. Responses can be quite variable but many are mesothermic having cardinal temperatures in the region of 5°, 25° and 35 °C (Fig. 9). Linear extension rates at optimum temperature can also vary considerably between species with wood-rotting fungi often extending faster than leaf-litter decomposers (Hintikka & Korhonen, 1970). Contrary to popular belief, extension may often be quite fast, e.g. optimum extension of *Bjerkandera adusta* is 12 mm day^{-1} and *Stereum hirsutum* is almost 9 mm day^{-1} (Boddy, 1983*a*). It would be wrong to give the impression that little extension occurs at low temperatures in woodlands: *Clitocybe flaccida* and several other litter-rotting species have optima at or below 20 °C and maxima at about 25 °C (Fig. 9; Hintikka, 1964). Mitchell (unpublished) found considerable extension of fairy ring mycelia and Thompson (Chapter 9) of cord systems during winter months in Britain. Hintikka (1964) found a number of psychrophilic or psychrotolerant Basidiomycotina growing and decomposing leaf litter beneath the snow cover throughout the winter in Finnish woodlands. He observed that several of these species produced aerial mycelium, which formed dense tufts and rhizomorph-like structures when growing on agar at 5° and 10 °C but not at room temperature, and suggested that this response would aid the fungus in colonising freshly fallen litter in autumn.

Liese (1931) (cited in Cartwright & Findlay, 1958) exposed 90 different isolates of wood-rotting Basidiomycotina to temperatures of −32 °C for 14 days and found that, with the exception of some isolates of *Serpula lacrimans* (an unusual fungus in nature), all renewed vigorous growth after exposure. It seems unlikely, therefore, that mycelia of these and probably also of litter-decomposers, would be killed by low temperatures in temperate woodlands. Likewise lethal high temperatures are unlikely to occur apart from occasionally in exposed situations such as branches, where there are gaps in canopy or in felled areas.

Caution must be applied, however, when extrapolating from agar to natural substrata. Temperatures optimal for linear extension on agar may not be the same for biomass production, ability to decompose cellulose and lignin, etc. For example, it has been noted that the most rapid decay of wood often occurs 2–3 °C below the optimum for extension on agar (Gauman, 1939, cited in Cartwright & Findlay, 1958; Henningson, 1968). Further, laboratory experiments are usually performed at constant temperatures whereas fluctuations are the rule in natural substrata: Jensen (1969*b*) found an increase in growth of two wood-rotting Basidiomycotina under a fluctuating regime which was not completely accounted for by the additive effect of temperature.

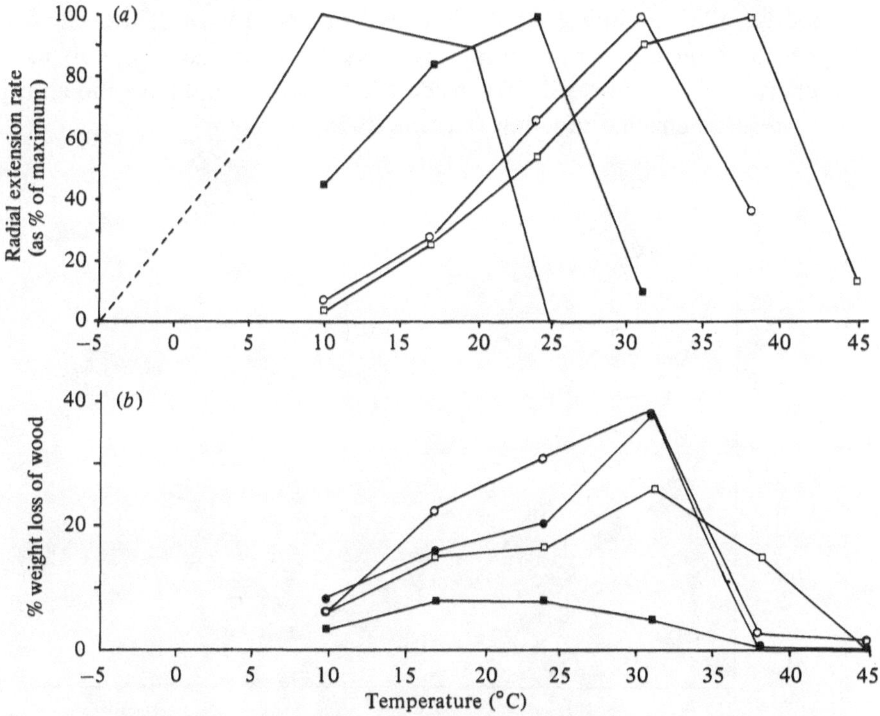

Fig. 9. Range in temperature response of selected basidiomycetes. (a) radial extension rate as a percentage of maximum, (b) decay rate of lodgepole pine wood: by a psychrophilic isolate (no symbol, From Hintikka, 1964), *Coniophora puteana* (●), *Lenzites saepiaria* (○), *Phlebia phlebioides* (□), and *Stereum sanguinolentum* (■). (From Loman, 1962, *Canadian Journal of Botany*, volume 40, pp. 1545–59, by permission of the National Research Council of Canada.)

Low water

The effect of low water on mycelia is often assessed by altering the water potential of the growth medium. Theoretically the effect of water potential on mycelia will not differ whether it is altered by osmotic or matric means (Griffin, 1972); thus either may be used. In practice, however, fungi are usually less affected by reduced osmotic, compared to matric, potentials probably as a result of solute or enzyme-diffusion problems (Griffin, 1972). When growth of mycelia is measured on agar media, the osmotic potential of which has been altered, it is essential to ensure that the effect of adding a solute is purely osmotic. Thus, several different solutes are used for comparison. Recent studies indicate that the lower limit for growth of wood- and litter-decomposing basidiomycetes is in the region of −4.0 MPa (Fig. 10; Dubé, Dodman & Flentje, 1971; Tresner & Hayes, 1971; Griffin, 1977; Wilson & Griffin, 1979; Boddy, 1983*a*, unpublished). The use of glycerol as solute allowed growth at much lower potentials probably due to the fact that polyols in general act as osmoregulators inside the cytoplasm and have the ability to preserve enzyme function (Brown, 1978).

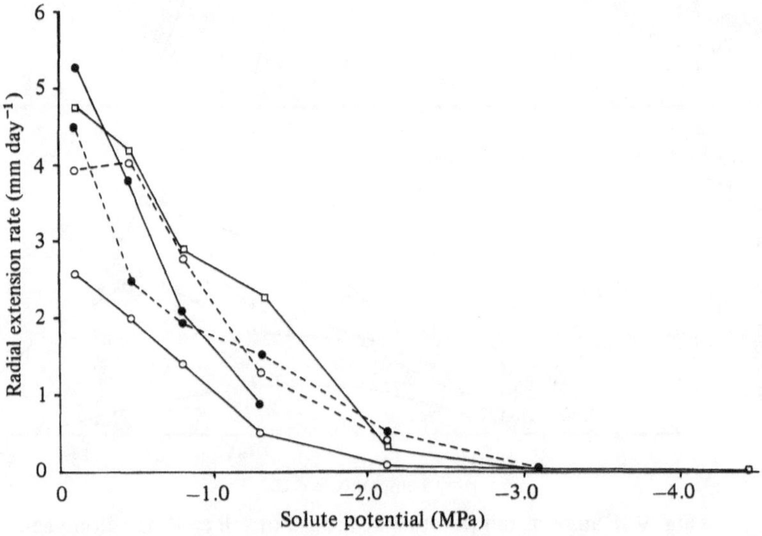

Fig. 10. Relationship between water potential (produced by KCl in malt agar) and radial extension rate (at 25 °C) of five basidiomycetes: *Exidia glandulosa* (●——●), *Hyphoderma setigerum* (●---●), *Phanerochaete velutina* (□——□), *Schizopora paradoxa* (○——○), and *Vuilleminia comedens* (○---○). (From Boddy, 1983*a*.)

By contrast with these data Wilson & Griffin (1979) found that six basidiomycetes probably from Australia (*Suillus luteus, Lycoperdon* sp., *Lactarius deliciosus, Clitocybe eucalyptorum, Geastrum* sp. and *Agaricus langei*) could withstand solute potentials below −10 MPa. Bavendamm & Reichelt (1938) found that most species that they tested grew at −14.5 MPa but these data require corroboration as the method of water-potential control was probably unsatisfactory (Griffin, 1977). Many non-basidiomycete 'soil fungi' can grow in the range −4 to −10 MPa (Griffin, 1972) and some are even able to grow below this, e.g. *Fusarium moniliforme* (Sommers, Harris, Dalton & Gardener, 1970), *Aspergillus* sp. (Pitt & Hocking, 1977).

Few data are available using other measures of activity but Griffin (unpublished, cited in Griffin, 1977) found that dry weight increase closely paralleled radial extension at various water potentials for *Gloeophyllum trabeum (= Lenzites trabea)* and *Fomes lividus.* Direct data for cellulolysis and decay rate are lacking but Griffin (1977) suggests, on the basis of work done with marine fungi (Byrne & Eaton, 1972; Meyers & Reynolds, 1959), that cellulolysis *per se* is likely to be unaffected by decrease in water potential in the range 0 to −2.8 MPa and probably even somewhat lower. Thus both growth and cellulolysis probably occur in the range 0 to −4 MPa if the physical form of the substratum in terms of void-size distribution is suitable. A critical size is that which is too small to allow diffusion of enzymes or products of degradation when holding water. Griffin (1977) suggested that in wood the limiting value for substrate availability, hyphal growth and cellulolysis might be taken as about −4 MPa. Thus, it is interesting, in evolutionary terms, that the lower limit for growth of wood-decay fungi is at about the same matric potential at which the structure of wood makes cellulose inaccessible to enzyme molecules.

Gaseous environment

Numerous studies have been performed on the effect of CO_2 on growth of basidiomycete mycelia. One of the earliest was that of Bavendamm (1928) who states that 19% CO_2 inhibits the growth of both parasitic and saprotrophic wood-rotting species but does not kill them, and also that white-rot fungi were relatively insensitive to high levels of CO_2 whereas brown-rot species were sensitive (Bavendamm, 1928, cited in Schanel, 1976). Recent studies have indicated that wood-decay species are considerably more tolerant of increase in CO_2 than litter decomposers (Hintikka & Korhonen, 1970; Schanel, 1976; Rawles,

Boddy & Rayner, unpublished). For non-wood-rotting species the maximum tolerable concentration of CO_2 was around 20–30% whereas many wood-rotting species grew well at 70% CO_2, and several that are frequently found in living or recently dead wood, were still able to grow, albeit very slowly, at approaching 100% CO_2. At concentrations around 10% the growth of many wood-rotting species was markedly stimulated as was the production of laccase and peroxidases in *Pleurotus* (Schanel, 1976). Similar effects have also been shown on the rate of mycelial biomass production (Zycha, 1937; Hintikka & Korhonen, 1970).

Hyphal and colony characteristics often alter at CO_2 concentration much above atmospheric (Hintikka & Korhonen, 1970; Schanel, 1976). Dovrtěl (1975) found changes in hyphal diameter, shape and frequency of clamp-connections at 30% CO_2, and Rawles, Boddy & Rayner (unpublished) observed the production of numerous small 'bud-like' branches on the hyphae of *Phlebia radiata, Phlebia rufa* and *Stereum gausapatum* at CO_2 concentrations above 60%.

There is some evidence that Basidiomycotina are relatively tolerant of low O_2, many being able to grow at below 1% O_2 (Bavendamm, 1928; Scheffer & Livingstone, 1937; Jensen, 1967). In the complete absence of O_2, *Serpula lacrimans* died in 2–3 days whilst 'pathogenic' fungi survived longer; *Stereum frustulatum* was the most resistant, remaining undamaged after 10 days (Bavendamm, 1928). *Heterobasidion annosum* grew well in trace amounts of O_2 (Gunderson, 1961) as did *Phlebia radiata* and *P. rufa* although 14 others failed to grow (Rawles, Boddy & Rayner, unpublished). Under such conditions anaerobic respiration may occur with the build-up of products such as ethanol, methanol, formate, acetate, lactate, and propionate. Hintikka (1969) found that wood-rotting basidiomycetes grew at slightly higher alcohol concentrations than did litter-decomposers, and there was a distinct difference in their tolerance of acetate: growth of soil- and litter-inhabiting species ceased at 0.025–0.05% whereas wood-rotters could tolerate 0.1–0.2%. Similar differences were found with formate and propionate.

Rypacek (1966) investigated aeration requirements in natural substrata by growing fungi in strips of wood having a moisture gradient. He was effectively looking at a gradient of air-filled void spaces and he estimated, by noting the wettest segment in which growth occurred, that the limit for wood-rotting Basidiomycotina was in the region of 10–20% air.

Interactive effect of abiotic variables on basidiomycete mycelia

In the field abiotic variables do not act independently upon fungal mycelia but jointly. However, relatively few studies have considered the effect of more than one variable at a time. Strong interactions between abiotic variables have been demonstrated for some fungi: for instance, in general the optimum temperature for radial growth increases by about 5 °C as water potential decreases (Ayerst, 1968; Griffin, 1978); a synergistic effect of combined lowering of O_2 and raising of CO_2 concentration has been reported in some fungi (Tabak & Cooke, 1968).

Distribution of basidiomycete mycelia in nature

The pattern of distribution of basidiomycete mycelia in woodland varies both spatially and temporally. In studies of fungal colonisation of individual resource units Basidiomycotina have often been reported as occurring relatively late in the so-called 'succession' (e.g. see Frankland: Chapter 11; Rayner & Webber: Chapter 18). These ideas have emerged largely as a result of the method of study and have, to a large extent, obscured the role of basidiomycete mycelia in the decomposition of plant and animal remains. This topic is discussed more fully by Cooke & Rayner (1984), suffice it to say here that their temporal distribution, i.e. community development, requires much further study and careful re-evaluation of ideas. Temporal patterns will be mentioned here only briefly in relation to microclimate.

The heterogeneous distribution of different substrata within woodlands and their associated microclimates results in considerable spatial variation in the distribution of basidiomycete mycelia. The distribution of these is now beginning to receive some attention particularly in bulky woody resource units (e.g. Rayner & Todd, 1979; Boddy & Rayner, 1983*a,b*; Coates, 1984) although less information is available for litter (Frankland: Chapter 11; Thompson: Chapter 9). However, there is clear indication that microclimate does to some extent influence the distribution of basidiomycete mycelia.

In temperate woodlands distribution is not usually likely to be much affected by high temperatures as these rarely rise much above those optimal for activity, although this may not be the case in exposed conditions. For instance, in a study of the distribution of decay fungi within logging slash in Canada, Loman (1962, 1965) found that the four most common basidiomycetes were remarkably consistent in their spatial location. Thus, *Lenzites saepiaria* was predominant in the central portions of the slash; *Phlebia phlebioides* was isolated mainly from upper portions; and *Stereum sanguinolentum* and *Coniophora puteana* from the lower portions. This distribution may be explained, at least in part, on the basis of their temperature tolerances, optima for growth, and relative decaying abilities (see Fig. 9). High temperatures lethal to the mycelium of *C. puteana* and *S. sanguinolentum* but not to *L. saepiaria* and *P. phlebioides* occur in the upper and central portions of slash during fine weather. Also, *P. phlebioides* had a much higher growth rate and decay rate, at 38 °C, than the other three species, and *L. saepiaria* caused more decay than the others at 31 °C. At 10 °C the greatest decay was caused by *C. puteana.* A further possibility is that the outcome of combative interactions between these fungi may vary according to temperature (cf. Rayner & Webber: Chapter 18).

Water appears to be a major determinant of both activity and distribution, either directly in the low water content region or indirectly in the high region. Low moisture content is likely to be particularly important in standing dead plant tissues and in the surface layer of fallen litter. As many Basidiomycotina do not appear to be able to grow at low water potentials those which predominate in such habitats may do so largely as a result of their ability to survive such conditions, mycelial activity only occurring under improved conditions. It is likely that this is the case for *Hyphoderma setigerum* and *Schizopora paradoxa*, two species which are usually found associated with conditions prone to desiccation in attached oak branches and trunks (Boddy, 1983*a*; Boddy & Rayner, 1983*a*; Boddy & Thompson, 1983). Many of the most active litter-degrading fungi which are not component restricted, e.g. *Collybia* spp., *Clitocybe* spp., and many cord-forming fungi, grow beneath the current year's litter where substrata are more compacted and moisture conditions are presumably more stable. Here the mycelium grows as luxuriant wefts, sheets and cords which often show maximal development during winter, extension appearing to be more dependent on moisture than temperature (Karenlampi, 1971; Thompson & Rayner, 1982, 1983; Mitchell, unpublished).

At the other extreme it is well known that waterlogging limits decay (e.g. Cartwright & Findlay, 1958; Boddy, 1983*c*). However, in temperate woodlands, although moisture contents high enough to limit decay do occur periodically (e.g. Fig. 7; Boddy, 1983*b*), it is unlikely that this would affect the distribution of mycelia, provided that it was not a permanent feature. In living tissues, on the other hand, the substratum is usually permanently saturated and Boddy & Rayner (1983*c*) have advocated that this explains the confinement of decay to non-functional sapwood in living trees.

The different tolerances of Basidiomycotina to elevated levels of CO_2 also reflect their occurrence in nature: soil- and litter-inhabiting species are generally much less tolerant than wood-rotting species (Hintikka & Korhonen, 1970), correlating with conditions found in such environments. Gaseous composition may explain why species such as *Marasmius androsaceus* and *Mycena galopus* grow on leaf litter and twigs but not on tree trunks. Even more tolerant of high CO_2 were species commonly found in attached branches (Boddy & Rayner, 1983*a*; Rawles, Boddy & Rayner, unpublished), which is probably related to their ability to colonise partially living or recently dead tissue. These basidiomycetes form extensive individuals in an apparently very short time which cannot be accounted for in terms of mycelial extension (Boddy & Rayner, 1982, 1983*a,b,c*) and it has thus been suggested that the fungi become distributed through the sapstream by modules which are only capable of very limited growth until the branch is stressed (Boddy & Rayner, 1983*c*; Cooke & Rayner, 1984). The small buds produced on hyphae of these fungi under high CO_2 concentrations may possibly serve such a purpose.

Armillaria spp. tend to become established in cut stumps and suppressed trees before cord-forming fungi (Rayner, 1975; Thompson & Boddy, 1983; Thompson: Chapter 9), a phenomenon which has been explained on the basis that *Armillaria mellea* is pathogenic and that other *Armillaria* spp. act as weak pathogens, whilst cord-forming fungi act purely saprotrophically. This ability may however be explained in terms of gaseous (in particular O_2) relations; Smith & Griffin (1971), in a comprehensive study of *Armillaria elegans,* demonstrated the importance of gaseous composition and water in growth and development of rhizomorphs. *Armillaria* rhizomorphs consist of a complex meristematic apex behind which is formed a hollow tube, surrounded by walls of closely-packed hyphal cells. They demonstrated that for optimal growth the rhizomorph required a high partial pressure of O_2 within the apex but a low one outside (otherwise melanisation took place and prevented growth). These apparently contradictory conditions are achieved by growth through moist soil and litter with the hollow rhizomorph allowing O_2 to diffuse to the tip by remaining in contact with wooden substrata in gaseous continuity with the atmosphere. Further, when crossing gas-filled voids, rhizomorphs produce short side branches which allow access of O_2. Field data did not suggest any marked reduction in growth in near-saturated soils (Griffin, 1972). Thus it might be expected that *Armillaria* spp. would be able to gain access to woody tissues before other fungi that do not have a route for direct access of O_2.

Microclimatic variables do not, by a long way, explain the distribution, and activity, of all Basidiomycotina. In nature many species are specific to certain kinds of litter whereas in pure culture they do not show strong preferences (Hering, 1982). Such restriction to a particular niche may be due to interaction with other organisms (Frankland: Chapter 11; Rayner & Webber: Chapter 18), and considerable further study is required for complete elucidation.

Conclusions

This brief account has indicated that organic substrata in woodlands consist of a mosaic of microsites, differing in terms of microclimatic parameters of temperature, moisture content, gaseous composition and pH. Field data for microclimate at this level are lacking, and information on activity of mycelia under various microclimatic regimes is largely confined to studies on relatively homogeneous artificial media in the laboratory. Further, the actual distribution of basidiomycete mycelia in nature has been little studied. Despite this, some indication of activity and distribution of mycelia in relation to microclimate has been possible. However, it is clear that the time is ripe for these gaps in our knowledge to be filled.

References

Anderson, J. M., Rayner, A. D. M. & Walton, D. W. H., eds. (1984). *Animal-Microbial Interactions.* Cambridge: Cambridge University Press.

Ayerst, G. (1968). Prevention of biodeterioration by control of environmental conditions. In *Biodeterioration of Materials*, ed. A. H. Walters & J. J. Elphick, pp. 223–41. Amsterdam, London, New York: Elsevier.

Baines, E. F. & Levy, J. F. (1979). Movement of water through wood. *Journal of the Institute of Wood Science*, **8,** 109–13.

Batra, L. R. (ed.) (1979). *Insect-Fungus Symbiosis.* New York: John Wiley & Sons.

Bavendamm, W. (1928). Neue Untersuchungen über die Lebensbedingungen holzzerstörender Pilze. Ein Beitrag zur Frage der Krankheitsepfänglichkeit unserer Holzpflazen. I. Mitteilung: Gasversuche. *Centralblatt für Bakteriologie, Parasitenkunde und Infektionskrankheiten*, **75,** 426–52, 503–33.

Bavendamm, W. & Reichelt, H. (1938). Die Abhängigkeit des Wachstums holzzerstorender Pilze vom Wassergehalt des Nährsubstrates. *Archiv für Mikrobiologie*, **9,** 486–544.

Bocock, K. L., Bailey, A. D. & Hornung, M. (1982). Variation in soil temperature with microrelief and soil depth in a newly planted forest. *Journal of Soil Science*, **33,** 55–62.
Boddy, L. (1983*a*). The effect of temperature and water potential on growth rate of wood-rotting basidiomycetes. *Transactions of the British Mycological Society*, **80,** 141–9.
Boddy, L. (1983*b*). Microclimate and moisture dynamics of wood decomposing in terrestrial ecosystems. *Soil Biology and Biochemistry*, **15,** 149–57.
Boddy, L. (1983*c*). Carbon dioxide release from decomposing wood: effect of water content and temperature. *Soil Biology and Biochemistry*, **15,** 501–10.
Boddy, L. & Rayner, A. D. M. (1982). Population structure, inter-mycelial interactions and infection biology of *Stereum gausapatum. Transactions of the British Mycological Society*, **78,** 337–51.
Boddy, L. & Rayner, A. D. M. (1983*a*). Ecological roles of basidiomycetes forming decay communities in attached oak branches. *New Phytologist*, **93,** 77–88.
Boddy, L. & Rayner, A. D. M. (1983*b*). Mycelial interactions, morphogenesis and ecology of *Phlebia radiata* and *P. rufa* from oak. *Transactions of the British Mycological Society*, **30,** 437–48.
Boddy, L. & Rayner, A. D. M. (1983*c*). Origins of decay in living deciduous trees: the role of moisture content and a re-appraisal of the expanded concept of tree decay. *New Phytologist*, **94,** 623–41.
Boddy, L. & Swift, M. J. (1983). Wood decomposition in an abandoned beech and oak coppiced woodland. *Holarctic ecology*, **6,** 320–32.
Boddy, L. & Thompson, W. (1983). Decomposition of suppressed oak trees in even-aged plantations. I. Stand characteristics and decay of aerial parts. *New Phytologist*, **93,** 261–76.
Boynton, D. (1941). Soils in relation to fruit growing: XV Seasonal and soil influences on oxygen and carbon dioxide levels of New York orchard soils. *N.Y. (Cornell) Agriculture Experiment Station Bulletin* No. 763.
Bray, J. R. & Gorham, G. (1964). Litter production in forests of the world. *Advances in Ecological Research*, **2,** 101–57.
Brierley, J. K. (1955). Seasonal fluctuations in the oxygen and carbon dioxide concentrations in beech litter with reference to the salt uptake of beech mycorrhizas. *Journal of Ecology*, **43,** 404–8.
Brock, T. D. (1966). *Principles of Microbial Ecology*. New York: Prentice-Hall.
Brown, A. D. (1978). Compatible solutes and extreme water stress in eukaryotic micro-organisms. *Advances in Microbial Physiology*, **17,** 181–242.
Bunnel, F. L., Tait, D. E. M., Flanagan, P. W. & Van Cleve, K. (1977). Microbial respiration and substrate weight loss, I. A general model of influence of abiotic variables. *Soil Biology and Biochemistry*, **9,** 33–40.
Byrne, P. J. & Eaton, R. A. (1972). Fungal attack of wood submerged in waters of different salinity. *Interactional Biodeterioration Bulletin*, **8,** 127–34.
Campbell, W. G. (1952). The biological decomposition of wood. In. *Wood Chemistry*, ed. L. E. Wise & E. C. John, 2nd edn, pp. 1061–118. New York: Reinhold Publishing Corporation.
Carrodus, B. B. & Triffett, A. C. K. (1975). Analysis of composition of respiratory gases in woody stems by mass spectrometry. *New Phytologist*, **74,** 243–6.
Cartwright, K. St. G. & Findlay, W. P. K. (1958). *Decay of Timber and Its Prevention.* London: Her Majesty's Stationery Office.
Chase, W. W. (1934). The composition, quality and physiological significance of gases in tree stems. *University of Minnesota Agricultural Experiment Station, Technical Bulletin*, No. 99, 1–51.

Clarke, R. W., Jennings, D. H. & Coggins, C. R. (1975). Growth of *Serpula lacrimans* in relation to water potential of substrate. *Transactions of the British Mycological Society*, **75**, 271–80.

Coates, D. (1984). *Biology of Intra-specific Antagonism in Wood Decay Fungi*. Ph.D. Thesis, University of Bath.

Cooke, R. (1977). *The Biology of Symbiotic Fungi*. London: John Wiley.

Cooke, R. C. & Rayner, A. D. M. (1984). *The Ecology of Saprotrophic Fungi*. London & New York: Longman.

Cowling, E. B. & Merrill, W. (1966). Nitrogen in wood and its role in wood deterioration. *Canadian Journal of Botany*, **44**, 533–44.

Dovrtěl, J. (1975). Kysličník uhličity jako ekologický factor rustu dřevokazných hub. Rig. práce. *Katedra biologie rostlin UJEP Brno.*

Dowding, P. (1976). Allocation of resources; nutrient uptake and utilisation by decomposer organisms. In *The Role of Aquatic and Terrestrial Organisms in Decomposition Processes*, ed. J. M. Anderson & A. Macfadyen, pp. 169–83. Oxford: Blackwell Scientific Publication.

Dowding, P. (1981). Nutrient uptake and allocation during substrate exploitation by fungi. In *The Fungal Community: Its Organization and Role in the Ecosystem*, ed. D. T. Wicklow & G. C. Carroll, pp. 621–35. New York: Marcel Dekker.

Dubé, A. J., Dodman, R. L. & Flentje, N. T. (1971). The influence of water activity on the growth of *Rhizoctonia solani*. *Australian Journal of Biological Sciences*, **24**, 57–65.

Frankland, J. C. (1982). Biomass and nutrient cycling by decomposer basidiomycetes. In *Decomposer Basidiomycetes: Their Biology and Ecology*, ed. J. C. Frankland, J. N. Hedger & M. J. Swift, pp 241–61. Cambridge: Cambridge University Press.

Gauman, E. (1939). Uber die Bedeutung von Wuchstoffen für des Wachstums verschierdener Pilze. *Angewandte Botanik*, **21**, 59–69.

Geiger, R. (1965). *The Climate near the Ground*, 4th edn (English translation). Harvard: Harvard University Press.

Greenwood, D. J. & Goodman, D. (1967). Direct measurements of the distribution of oxygen in soil aggregates and in columns of fine soil crumbs. *Journal of Soil Science*, **18**, 182–96.

Griffin, D. M. (1963). Soil moisture and the ecology of soil fungi. *Biological Reviews*, **38**, 141–66.

Griffin, D. M. (1972). *Ecology of Soil Fungi*. London: Chapman & Hall.

Griffin, D. M. (1977). Water potential and wood-decay fungi. *Annual review of Phytopathology*, **15**, 319–29.

Griffin, D. M. (1978). Effect of soil moisture on survival and spread of pathogens. In *Water and Plant Disease*, ed. T. T. Kozlowski, *Water Deficit and Plant Growth*, vol. 5, pp 175–97. New York: Academic Press.

Griffin, D. M. (1982). Water and microbial stress. *Advances in Microbial Ecology*, **5**, 91–136.

Gunderson, K. (1961). Growth of *Fomes annosus* under reduced oxygen pressure and the effect of carbon dioxide. *Nature*, **190**, 649.

Henningson, B. (1968). Ecology of decomposition in birch and aspen. In *Biodeterioration of Materials*, ed. A. H. Walters & J. J. Elphick, pp. 408–23. Amsterdam, London, New York: Elsevier.

Henry, S. M., ed. (1967). *Symbiosis*, vols 1, 11. New York & London: Academic Press.

Hering, T. F. (1967). Fungal Decomposition of oak leaf litter. *Transactions of the British Mycological Society*, **50**, 267–73.

Hering, T. F. (1982). Decomposing activity of basidiomycetes in forest litter. In

Decomposer Basidiomycetes: Their Biology and Ecology, ed. J. C. Frankland, J. N. Hedger & M. J. Swift, pp. 213–25. Cambridge: Cambridge University Press.

Hintikka, V. (1964). Psychrophilic basidiomycetes decomposing forest litter under winter conditions. *Communications Instituti Forestalis Fenniae*, **29** (2), 1–20.

Hintikka, V. (1969). Acetic acid tolerance in wood- and litter-decomposing Hymenomycetes. *Karstenia*, **10**, 177–83.

Hintikka, V. (1982). The colonisation of litter and wood by basidiomycetes in Finnish forests. In *Decomposer Basidiomycetes: Their Biology and Ecology*, ed. J. C. Frankland, J. N. Hedger & M. J. Swift, pp. 227–39. Cambridge: Cambridge University Press.

Hintikka, V. & Korhonen, K. (1970). Effects of carbon dioxide on the growth of lignicolous and soil-inhabiting Hymenomycetes. *Communications Instituti Forestalis Fenniae*, **62** (2), 1–22.

Holmes, J. W., Taylor, S. A. & Richards, S. J. (1967). Measurement of soil water. In *Irrigation of Agricultural Lands, Agronomy monograph* 11, ed. R. M. Hagan, H. W. Haise & T. W. Edminster, pp. 275–303, New York: Academic Press.

Jennings, D. H. (1982). The movement of *Serpula lacrimans* from substrate to substrate over nutritionally inert surfaces. In *Decomposer Basidiomycetes: Their Biology and Ecology*, ed. J. C. Frankland, J. N. Hedger & M. J. Swift, pp. 91–108. Cambridge: Cambridge University Press.

Jensen, F. K. (1967). Oxygen and carbon dioxide affect the growth of wood-decaying fungi. *Forest Science*, **13**, 384–9.

Jensen, F. K. (1969*a*). Oxygen and carbon dioxide concentrations in sound and decaying red oak trees. *Forest Science*, **15**, 246–51.

Jensen, F. K. (1969*b*). Effect of constant and fluctuating temperature on growth of four wood-decaying fungi. *Phytopathology*, **59**, 645–7.

Karenlampi, L. (1971). Weight loss of leaf litter on forest soil surface in relation to weather at Kevo Station, Finnish Lapland. *Report of the Kevo Subarctic Research Station*, **8**, 101–3.

Kirk, T. K. & Fenn, P. (1982). Formation and action of the ligninolytic system in basidiomycetes. In *Decomposer Basidiomycetes: Their Biology and Ecology*, ed. J. C. Frankland, J. N. Hedger & M. J. Swift, pp. 67–90. Cambridge: Cambridge University Press.

Liese, J. (1931). Beobachtungen uber die Biologie holzzerstorender Pilze. *Angewandte Botanik*, **13**, 138–50.

Lindeberg, G. (1944). Über die Physiologie ligninabbauender Bodenhymenomyceten. *Symbolae botanicae Upsalienses*, **8**, 1–183.

Lindeberg, G. (1947). On the decomposition of lignin and cellulose in litter caused by soil-inhabiting Hymenomycetes. *Arkiv für Botanik*, **33a**, 1–16.

Lloyd, D., Scott, R. I. & Williams, N. T. (1983). Membrane inlet mass-spectrometry measurement of dissolved gases in fermentation liquids. *Trends in Biotechnology*, **1**, 60–3.

Loman, A. A. (1962). The influence of temperature on the location and development of decay fungi in lodgepole pine logging slash. *Canadian Journal of Botany*, **40**, 1545–59.

Loman, A. A. (1965). The lethal effect of periodic high temperatures on certain lodgepole pine slash decaying basidiomycetes. *Canadian Journal of Botany*, **43**, 334–8.

Meentemeyer, V. (1974). Climatic water budget approach to forest problems. 1. The prediction of forest fire hazard through moisture budgeting. *Publications in Climatology*, **27**, no. 1. Elmer, New Jersey: C. W. Thornwaite Associates Laboratory of Climatology.

Meyers, S. P. & Reynolds, E. S. (1959). Growth and cellulolytic activity of lignicolous Deuteromycetes from marine localities. *Canadian Journal of Microbiology*, **5**, 493–503.

Moore, A. & Swank, W. T. (1975). A model of water content and evaporation for hardwood leaf litter. In *Mineral Cycling in Southeastern Ecosystems,* ed. F. G. Howell, J. B. Gentry & M. H. Smith, pp. 58–69. U.S. Energy Research & Development Administration.

Ovington, J. D. (1962). Quantitative ecology and the woodland ecosystem concept. *Advances in Ecological Research*, **1**, 103–92.

Paim, U & Beckel, W. E. (1963). Seasonal oxygen and carbon dioxide content of decaying wood as a component of the microenvironment of *Orthosoma brunneum* (Foster) (Coleoptera: Cerambycidae). *Canadian Journal of Zoology*, **41**, 1133–47.

Pitt, J. I. & Hocking, A. D. (1977). Influence of solute and hydrogen ion concentration on the water relations of some xerophilic fungi. *Journal of General Microbiology*, **101**, 35–40.

Rayner, A. D. M. (1975). *Fungal Colonization of Hardwood Tree Stumps.* Ph.D. Thesis, University of Cambridge.

Rayner, A. D. M. & Todd, N. K. (1979). Population and community structure and dynamics of fungi in decaying wood. *Advances in Botanical Research*, **7**, 333–420.

Romell, L. G. (1928). Markluftsanalyer och markluftning. *Meddelanden Statens Skogsförsöksanstalt,* **24**, 67–80.

Rose, C. W. (1966). *Agricultural Physics.* Oxford: Pergamon Press.

Rypacek, V. (1966). *Biologieholzzerstörender Pilze.* Jena: Fischer.

Saito, T. (1957). Chemical changes in beech litter under microbiological decomposition. *Ecological Review, Sendai,* **14**, 209–16.

Schanel, L. (1976). Role of carbon dioxide in growth and decaying activity of wood-rotting fungi. *Folia Facultatis Scientiarum Naturalium Universitatis Purkynianae Brunensis*, **17**, Biol. 54, op. 6, 5–54.

Scheffer, T. C. & Cowling, E. B. (1966). Natural resistance of wood to microbial deterioration. *Annual Review of Phytopathology*, **4**, 147–70.

Scheffer, T. C. & Livingstone, B. E. (1937). Relation of oxygen pressure and temperature to growth and carbon dioxide production in the fungus *Polystictus versicolor. American Journal of Botany*, **24**, 109–19.

Smith, D. N. (1964). The permeability of wood. In *Fifth World Forestry Congress Proceedings*. pp. 1546–8.

Smith, A. D. & Griffin, D. M. (1971). Oxygen and the ecology of *Armillariella elegans* Heim. *Australian Journal of Biological Sciences*, **24**, 231–62.

Sommers, L. E., Harris, R. F., Dalton, F. N. & Gardener, W. R. (1970). Water potential of three root infecting *Phytophthora* species. *Phytopathology*, **60**, 932–4.

Stamm, A. J. (1946). Passage of liquids, gases and dissolved materials through soft woods. *U.S. Department of Agriculture Technical Bulletin* No 929.

Stone, J. E. & Scallan, A. M. (1967). The effect of component removal upon the porous structure of the cell wall of wood. 11 Swelling in water and the fibre saturation point. *Tappi*, **50**, 496–501.

Swift, M. J. (1976). Species diversity and the structure of microbial communities. In *The Role of Aquatic and Terrestrial Organisms in Decomposition Processes*, ed. J. M. Anderson & A. Macfadyen, pp. 195–222. Oxford: Blackwell Scientific Publications.

Swift, M. J. (1977). The ecology of wood decomposition. *Science Progress, Oxford,* **64**, 175–99.

Swift, M. J. (1982). Basidiomycetes as components of forest ecosystems. In *Decomposer Basidiomycetes: Their Biology and Ecology*, ed. J. C. Frankland, J. N. Hedger & M. J. Swift, pp. 307–37. Cambridge: Cambridge University Press.

Swift, M. J., Boddy, L. & Healey, I. N. (1984). Wood decomposition in an abandoned beech and oak coppiced woodland in south-east England. 11. The standing crop of wood on the forest floor with particular reference to *Tipula flavolineata* and other animals. *Holarctic ecology* (in press).

Swift, M. J., Heal, O. W. & Anderson, J. M. (1979). *Decomposition in Terrestrial Ecosystems*. Oxford: Blackwell Scientific Publications.

Tabak, H. H. & Cooke, W. B. (1968). The effects of gaseous environments on the growth and metabolism of fungi. *The Botanical Review*, **34,** 126–252.

Thacker, D. G. & Good, H. M. (1952). The composition of air in trunks of sugar maple in relation to decay. *Canadian Journal of Botany,* **30,** 475–85.

Thompson, W. & Boddy, L. (1983). Decomposition of suppressed oak trees in even-aged plantations. 11. Colonisation of tree roots by cord and rhizomorph producing basidiomycetes. *New Phytologist*, **93,** 277–91.

Thompson, W. & Rayner, A. D. M. (1982). Structure and development of mycelial cord systems of *Phanerochaete laevis* in soil. *Transactions of the British Mycological Society*, **78,** 193–200.

Thompson, W. & Rayner, A. D. M. (1983). Extent, development and function of mycelial cord systems in soil. *Transactions of the British Mycological Society*, **81,** 333–45.

Tresner, H. D. & Hayes, J. A. (1971). Sodium chloride tolerance of terrestrial fungi. *Applied Microbiology*, **22,** 210–13.

Wagener, W. W. & Davidson, R. W. (1954). Heart rots in living trees. *The Botanical Review*, **20,** 61–134.

Wilkins, W. H. & Harris, G. C. M. (1946). The ecology of the larger fungi. V. An investigation into the influence of rainfall and temperature on the seasonal production of fungi in a beechwood and a pinewood. *Annals of Applied Biology*, **33,** 179–90.

Williams, J. B. (1968). Measurement of total and matric suctions of soil water using thermocouple psychrometer and pressure membrane apparatus. *Journal of Applied Ecology*, **5,** 263–72.

Williams, S. T. & Gray T. R. G. (1974). Decomposition of litter on the soil surface. In *Biology of Plant Litter Decomposition*, ed. C. H. Dickinson & G. J. F. Pugh, pp. 611–32. London: Academic Press.

Wilson, J. M. & Griffin, D. M. (1979). The effect of water potential on the growth of some soil basidiomycetes. *Soil Biology and Biochemistry*, **11,** 211–12.

Wimpenny, J. W. T. & Coombs, J. P. (1983). The penetration of oxygen into bacterial colonies. *Journal of General Microbiology*, **129,** 1239–42.

Yamaguchi, M., Flocker, W. J. & Howard, F. D. (1967). Soil atmosphere as influenced by temperature and moisture. *Soil Science Society of America, Proceedings*, **31,** 164–7.

Zycha, H. (1937). Über das Wachstum zweier holzzerstörender Pilze und ihr Verhältnis zur Kohlensaüre. *Zentralblatt für Bakteriologie, Parasitenkunde und Infektionskrankheiten,* **97,** 222–44.

13

Interrelationships between vegetative development and basidiocarp initiation

MARJATTA RAUDASKOSKI* and MERVI SALONEN†

** Department of Botany, University of Helsinki, Unioninkatu 44, SF-00170 Helsinki 17, Finland and † Department of Biology, University of Turku, SF-20500 Turku 50, Finland*

Basidiocarp formation has been studied using a variety of morphological, biochemical, genetical, and physiological approaches. In morphological and biochemical studies *Coprinus cinereus* and *Schizophyllum commune* have been used almost exclusively and the results reviewed on several occasions (Niederpruem & Wessels, 1969; Ishikawa & Uno, 1977; Schwalb, 1978; Gooday, 1979; Moore, Elhiti & Butler, 1979; Uno & Ishikawa, 1982). In genetic studies, crossing experiments with monokaryotic strains have indicated that monokaryotic fruiting in *Agrocybe aegerita* and *Polyporus ciliatus* is controlled by two genes, one of which regulates initiation and the other differentiation of the basidiocarps, and it appears that the same genes regulate dikaryotic fruiting (Esser, Semerdžieva & Stahl, 1974; Stahl & Esser, 1976; Esser & Meinhardt, 1977). In *S. commune* basidiocarp formation was originally thought to be a quantitative character regulated by several genes (Raper & Krongelb, 1958), but recent work with monokaryotic fruiters indicates regulation by genes comparable with those of *A. aegerita* and *P. ciliatus* (Esser, Saleh & Meinhardt, 1979). In studies of environmental factors, such as light, aeration, nutrients, and humidity, a larger variety of basidiomycetes have been used, including the cultivated species *Agaricus bisporus, Lentinus edodes, Pleurotus ostreatus*, and *Flammulina velutipes*. Different aspects of environmental control of fruiting in these and some other species have recently been reviewed by Tan (1978), Eger-Hummel (1980) and Manachère (1980).

Despite these studies it appears that changes in vegetative development associated with initiation of basidiocarp formation have received relatively little attention. The present review will concentrate on this feature, using, in particular, morphological and biochemical studies with *S. commune* as illustration.

Carbon dioxide as a morphogenetic factor

A volatile inhibitor of basidiocarp formation produced in cultures of *S. commune*, and identified as carbon dioxide, was first reported by Niederpruem (1963). It was suggested that the CO_2 resulted from the active metabolism of the fungus grown on a medium with high glucose content, and this was supported by initiation of basidiocarps when the glucose content of the medium was reduced or the CO_2 was trapped by potassium hydroxide (KOH). Since then, KOH has been used regularly as a CO_2 trap to ensure fruiting of monokaryotic and dikaryotic cultures of *S. commune* (Leonard & Dick, 1968; Schwalb, 1971; Sietsma, Rast & Wessels, 1977).

In a recent study of early stages of basidiocarp differentiation in dikaryotic mycelium of *S. commune* (Raudaskoski & Viitanen, 1982), it was observed that differentiation was slow or failed to take place in the light when the colonies had happened to be poorly ventilated during growth in the dark. The diameter of such colonies was regularly larger and the hyphal compartments longer than in well-aerated colonies. It was suggested that the factor changing the growth pattern and inhibiting or delaying fruiting in these colonies was CO_2 (Raudaskoski & Viitanen, 1982). To test this, a series of experiments was performed in which CO_2 and O_2 concentrations of the atmosphere in closed growth chambers were manipulated by using tablets generating H_2 and CO_2, and a CO_2 trap consisting of 1 N KOH. The diameters of colonies grown for four days in the dark in growth chambers without aeration were all significantly larger ($P < 0.001$) than those of ventilated colonies or colonies grown in the presence of KOH (Fig. 1a). The dry weights of colonies grown in closed chambers with high CO_2 concentrations were also greater, although the difference from the dry weights of aerated colonies was significant ($P < 0.001$) only in colonies grown in chambers in which the decreased O_2 tension was compensated with CO_2 (Fig. 1b, iii). The lowest dry weights were recorded for colonies grown in chambers in which CO_2 was trapped with 1 N KOH (Fig. 1b, v). These colonies also possessed hyphae with the shortest and narrowest compartments (Fig. 1c, v), whereas the longest and broadest compartments developed in growth chambers with reduced O_2 tension compensated with CO_2 (Fig. 1c, iii).

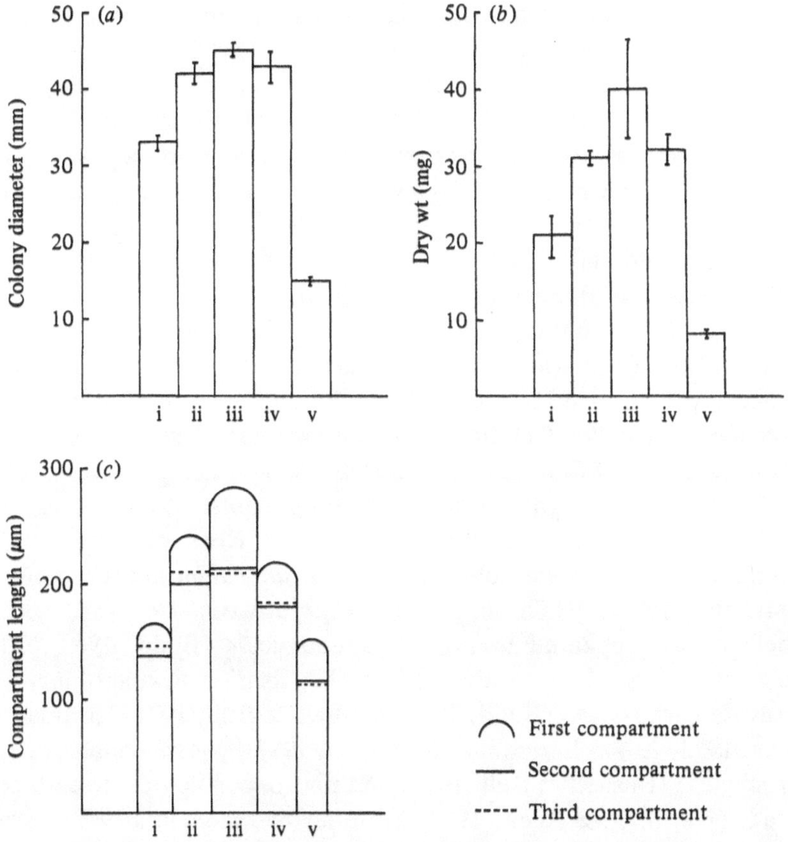

Fig. 1. The effect of CO_2 concentration on (a) diameter; (b) dry weight; and (c) hyphal compartment length and diameter of dikaryotic mycelium of *S. commune* during 4 days growth on nutritionally rich medium at 25 °C in the dark. Each column represents growth in a different atmosphere: i, normal air with ventilation; ii, CO_2 concentration increased and O_2 concentration decreased due to the respiration of the mycelium during the experiment; iii, CO_2 concentration increased and O_2 concentration decreased by 6% at the beginning of the experiment; iv, CO_2 additional to the normal air of the growth chamber at the beginning of the experiment; v, CO_2 trapped with 1 N KOH from the beginning of the experiment. The columns represent the mean ± S.E.M. of the diameter and dry weight of five colonies in one of several experiments, which all gave similar results. In (c), the mean length of 30 apical compartments is illustrated by columns, which also show the mean lengths of the subapical and third compartments. The lengths of the apical, subapical, and third compartments were always measured in the same leading hypha. The breadth of the columns represents the mean diameter of the hyphae, measured in front of the clamp-connection in 30 apical compartments.

The largest number of basidiocarps per millimetre of the colony margin occurred in aerated cultures, in which basidiocarp initials developed within 24 h after transfer to continuous light (Fig. 2*a*). Basidiocarp differentiation was also rapid in colonies grown in dark in the presence of KOH, but was slow and the number formed was lower in all colonies grown at high CO_2 levels in the dark (Fig. 2*a*). Little or no increase in diameter took place in any of the colonies after transfer to continuous light and uniform aeration (Fig. 2*b*), which is in agreement with the study by Raudaskoski & Viitanen (1982).

Effects of CO_2 on mycelial growth and differentiation have been demonstrated for a wide variety of macro- and microfungi (Tabak & Cooke, 1968; Bull & Bushell, 1976). Among microfungi, fixation of radioactively labelled CO_2 has been demonstrated. Thus in *Verticillium albo-atrum* 42% of CO_2 was fixed in proteins, 34% in nucleic acids, 17% in low-molecular-weight intracellular components, 3% in lipids and 1–2% in extracellular metabolites (Hartman, Keen & Long, 1972). Comparable results were obtained for *Fusarium culmorum* (Larmour & Marchant, 1977), while in *Neocosmospora vasinfecta* 90% of the labelled CO_2 appeared in protein amino acids (Budd, 1969). The enzymes suggested to be involved in CO_2 fixation in microfungi are pyruvate carboxylase (Budd, 1971; Bushell & Bull, 1974; Hartman & Keen, 1973, 1974*a*; Larmour & Marchant, 1977), phosphoenolpyruvate carboxylase (Bushell & Bull, 1974) and phosphoenolpyruvate carboxykinase (Hartman & Keen, 1973; Larmour & Marchant, 1977). How-

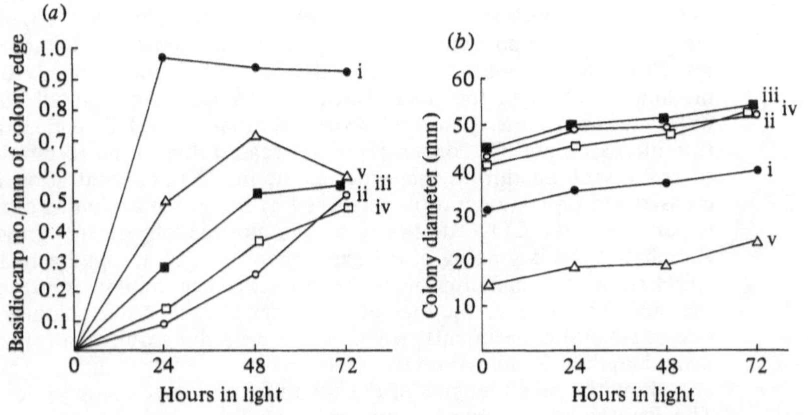

Fig. 2. The development of colonies in light after growth in the experimental conditions given in Fig. 1, i–v. (*a*) Basidiocarp production, (*b*) increase in colony diameter.

ever, in *V. albo-atrum*, phosphoenolpyruvate carboxykinase was reported to show decarboxylating activity combined with gluconeogenesis (Hartman & Keen, 1974*b*). In higher fungi CO_2 fixation has received less attention, but pyruvate carboxylase and phosphoenolpyruvate carboxylase activities have been demonstrated in cell-free extracts of hyphae of *S. commune* (Tachibana, Siode & Hanai, 1967), and pyruvate carboxylase, malic enzyme and phosphoenolpyruvate carboxykinase activities in cell-free extracts of *A. bisporus* (Rast & Bachofen, 1967; Bachofen & Rast, 1968). The primary product of CO_2 fixation by pyruvate carboxylase and phosphoenolpyruvate carboxylase is oxaloacetate (Tachibana *et al.*, 1967; Bachofen & Rast, 1968; Hartman & Keen, 1973), which thus represents anaplerotic biosynthesis of C_4 intermediates of the tricarboxylic acid cycle, and which seems to be used further mainly for the production of amino and nucleic acids and only to a limited extent for the production of other cell components.

The increase in size of the colonies and the length and diameter of the compartments in the dikaryotic mycelium of *S. commune* at high CO_2 concentrations could be due to CO_2 fixation by pyruvate carboxylase and phosphoenolpyruvate carboxylase (Tachibana *et al.*, 1967). High CO_2 concentrations, up to 16%, have also been shown to increase vegetative growth of three *Pleurotus* species (Zadražil, 1975) and some wood-rotting fungi (Thacker & Good, 1952; Hintikka & Korhonen, 1970). Due to the rich nitrogen source in the culture medium used – at least in the experiments with *S. commune* and the *Pleurotus* species (Zadražil, 1975), fixation of CO_2 may have led to increased production of amino and nucleic acids. In the microfungus *N. vasinfecta* (Budd, 1969, 1971), CO_2 fixation was shown to occur only under conditions supporting protein synthesis, suggesting a close correlation between the two processes. In *S. commune* growth at a high CO_2 level did not lead to accumulation of carbohydrate material in hyphal walls (Sietsma *et al.*, 1977), except for extracellular mucilage (Niederpruem, Marshall & Speth, 1977; Sietsma *et al.*, 1977). The lack of an increase in the activity of isocitrate lyase, one of the central enzymes of the glyoxylate cycle, in *S. commune* grown at high CO_2 concentration (Cotter, LaClave, Wegener & Niederpruem, 1970) suggests further that CO_2 does not increase carbohydrate metabolism in hyphae but operates via other mechanisms.

The above observations draw attention to the need to study the effect of CO_2 on vegetative development of dikaryotic hyphae of *S. commune* and other macrofungi by following the incorporation of radioactive ^{14}C

and identifying the products. The enzymes participating in CO_2 fixation should also be re-examined, and attention paid to carbon source and glucose content of the growth medium, both of which appear to affect CO_2 fixation in microfungi and yeasts (Ruiz-Amil *et al.*, 1965; Budd, 1971; Hartman *et al.*, 1972; Hartman & Keen, 1973; Bushell & Bull, 1974; Larmour & Marchant, 1977), as also does the nitrogen content of the medium (Gitterman & Knight, 1952; Budd, 1969).

Changes in basidiocarp-producing capacity across the colony

In the studies of *S. commune* (Raudaskoski & Viitanen, 1982; Raudaskoski & Vauras, 1982), it became evident that events connected with basidiocarp differentiation took place at the edge of well-aerated colonies. This was confirmed by following basidiocarp development from adjacent discs of agar 1 mm in diameter, excised along a radius from the edge of a dark-grown colony inwards (Raudaskoski, 1981; Raudaskoski & Yli-Mattila, in preparation). Basidiocarps never developed from discs cut and subcultured in red light, but a repeatable pattern of development took place among those excised and subcultured in white light. During four days in the dark no basidiocarps were formed from the first and second discs, that is those isolated from the 2-mm-broad marginal zone of the colony. Basidiocarp development was occasional from disc 3, whereas it occurred regularly from discs 4, 5 and 6. The capacity of the colony to form basidiocarps decreased toward the centre, and they developed only rarely from discs isolated more than 6 mm from the margin. The period of illumination (20 min) necessary for induction of basidiocarps from isolated discs was much shorter than that needed at the edge of intact colonies (3 h). This may be due either to temporary cessation of competition between reproductive and vegetative development in the discs, or to production of basidiocarp-inducing substances caused by cutting (Leonard & Dick, 1973; Leslie & Leonard, 1979*a*).

This work therefore indicated that, as suggested by Perkins (1969), basidiocarp differentiation in *S. commune* takes place in younger hyphae of dikaryotic colonies. Furthermore, in monokaryotic colonies of *S. commune*, the response to basidiocarp-inducing substances was also restricted to hyphae less than 24 h old at the time of exposure (Rusmin & Leonard, 1975), and a comparable result has been obtained in *Coprinus congregatus* when dark-grown dikaryotic colonies of different ages were illuminated; only a narrow zone corresponding to 24-h growth of the youngest hyphae responded to the light by differentiating into basidiocarp initials (Ross, 1982*a*). In the present study intercalary compartments behind the outermost apical compartments of the hyphae were more accurately defined as those capable of reacting to light by initiating basidiocarp development.

Dry weight, protein and ATP content of mycelium during basidiocarp development

In subsequent experiments dikaryotic colonies of *S. commune* were subdivided into three zones (Fig. 3), which were used for measuring dry weight, total protein and the ATP contents of hyphae during 12 h growth in light or darkness. Changes in dry weight in the three zones showed no significant difference between light and dark (Fig. 4a,b), although in light the increase is mainly due to proliferation of branches, since extension decreased gradually after transfer to light (Raudaskoski & Viitanen, 1982; see also Fig. 2), whereas in the dark the increase in the dry weight is due to both extension and branching. The latter differences in growth pattern relate especially to zones 1 and 2, and may underly the slightly lower dry weight of zone 1 in light after 4 h (Fig. 4b). Despite the lack of significant dry weight differences, higher protein and ATP contents were recorded in the hyphae of zone 2 grown in light (Fig. 4c,e). The protein content increased during the first 6 h, then fell over the next 4 h (Fig. 4c). The ATP content, which reflects changes in metabolism more sensitively than proteins do, was maximal after 4 h (Fig. 4e). The higher level of proteins and ATP in hyphae of zone 2 in light compared with the dark supports the idea that this region of the colony is indeed involved in basidiocarp initiation. In apical

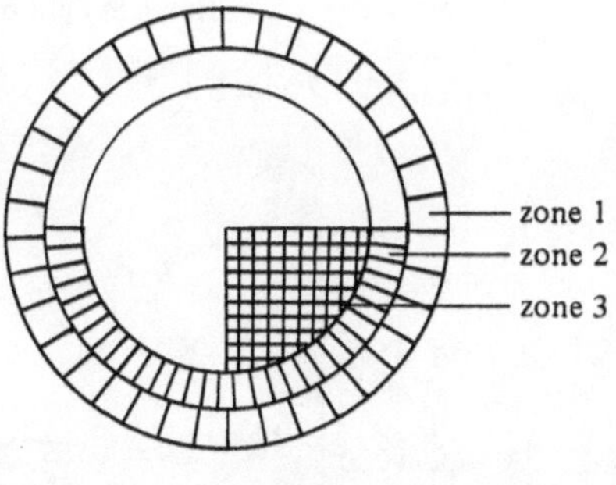

Fig. 3. The zones used in 4-day-old colonies for measuring dry weight, total protein (Bradford, 1976) and ATP (Raudaskoski & Lahti, 1983). About 7–8 colonies were needed to obtain enough fresh material from zone 1. The fresh and dry weights of the remaining mycelium of zones 2 and 3 were also determined.

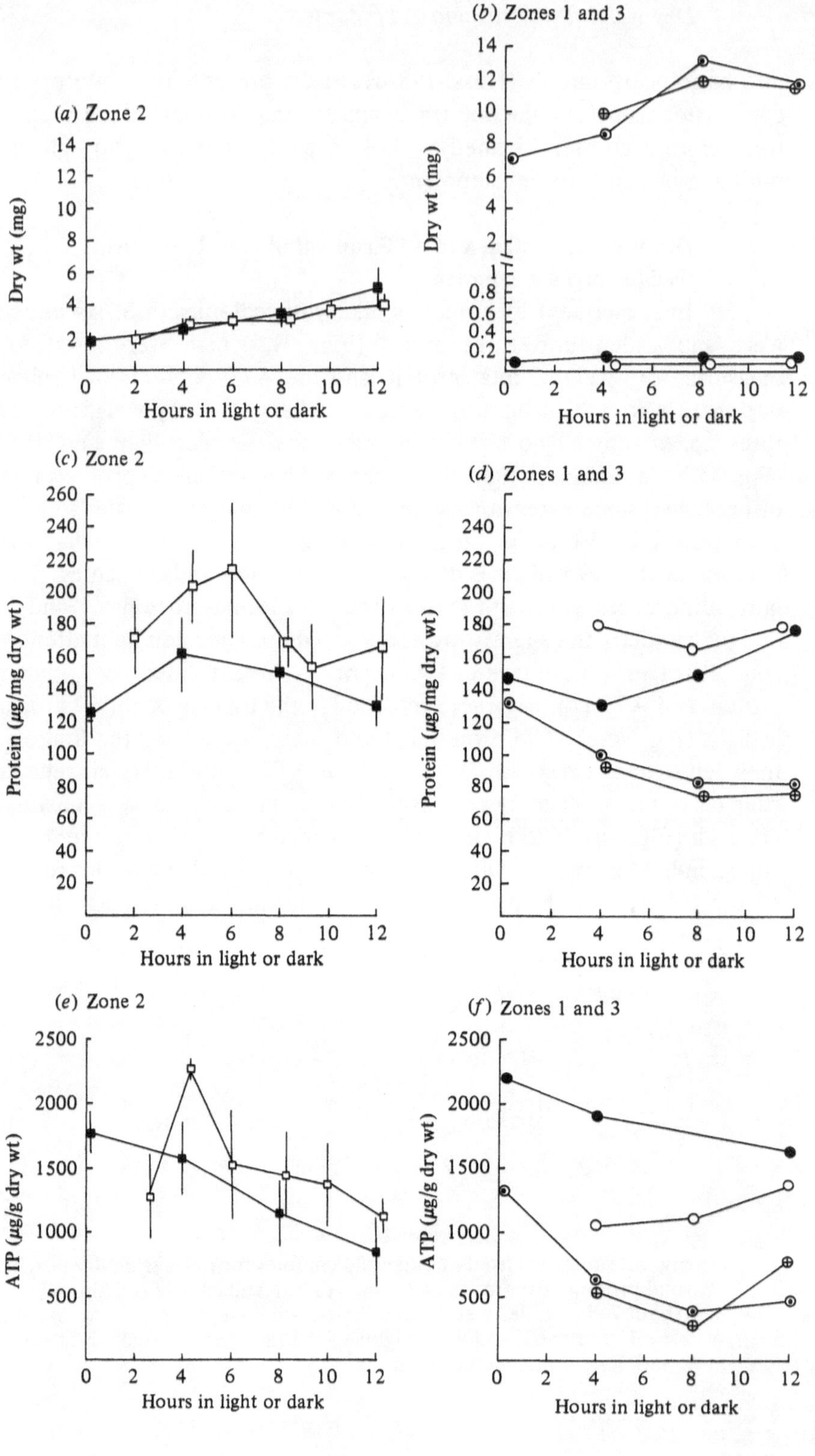
(a) Zone 2
Dry wt (mg)
Hours in light or dark
(b) Zones 1 and 3
Dry wt (mg)
Hours in light or dark
(c) Zone 2
Protein (µg/mg dry wt)
Hours in light or dark
(d) Zones 1 and 3
Protein (µg/mg dry wt)
Hours in light or dark
(e) Zone 2
ATP (µg/g dry wt)
Hours in light or dark
(f) Zones 1 and 3
ATP (µg/g dry wt)
Hours in light or dark

hyphae grown in light the protein level was higher but the ATP level lower than in those grown in the dark (Fig. 4d,f). This is in agreement with observations that in *S. commune* the expansion of the dikaryotic colony exposed to light decreases during basidiocarp formation. The total protein and ATP contents of the hyphae in the centre of the colony were similar in light and darkness (Fig. 4d,f), which indicates that this part of the colony is not significantly involved in fruit body differentiation. Although interpretation of these data are here based on the change in illumination, it must be pointed out that improved aeration in light might be partly responsible for the increase in metabolic activity, especially in zone 2.

During basidiocarp differentiation in *Volvariella volvacea* and *A. bisporus* an increase in total proteins was recorded in the former (Chang & Chan, 1973) and in soluble proteins in the latter (Paranjpe, Chen & Jong, 1979). In both cases the highest protein content was recorded at an early stage of basidiocarp formation, the 'pin' stage, which may be compared in the present study with the morphological stage occurring in *S. commune* after 8–12 h in light. In *C. cinereus* (Moore & Jirjis, 1981) and *A. bisporus* (Paranjpe *et al.*, 1979) comparisons were made of the electrophoretograms of soluble proteins in extracts from parental mycelium and from basidiocarps of different stages of development. In *C. cinereus*, considerable differences between the parental dikaryon and the 'pin' stage were detected, but not in *A. bisporus*.

Enzyme activity and basidiocarp formation

In *S. commune* special attention has been paid to phenoloxidase activity of the laccase type (Leonard, 1971; Phillips & Leonard, 1976*b*) in connection with basidiocarp formation. During basidiocarp development phenoloxidase activity, which is mainly extracellular, increased until the onset of sporulation and then declined significantly (Leonard & Phillips, 1973; Phillips & Leonard, 1976*a*). Phenoloxidase activity also appeared to be a characteristic feature of monokaryotic strains capable of monokaryotic fruiting (Leslie & Leonard, 1979*b*). High laccase activity has also been observed in dikaryotic mycelium of *C. congregatus* (Ross, 1982*b*), although the laccase of this fungus differed in certain

Fig. 4. (a, b) Dry weight, (c, d) total protein, (e, f) ATP in different zones of the colony (see Fig. 3). –□– zone 2 in light, –■– zone 2 in the dark, –○– zone 1 in light, –●– zone 1 in the dark, –⊕– zone 3 in light, –⊙– zone 3 in the dark.

respects from that of *S. commune*; it was intracellular and not able to oxidise 3-(3,4-dihydroxyphenyl)-L-alanine (DOPA) as was the phenoloxidase of *S. commune* (Leonard, 1971). When the laccase activity of *C. congregatus* in different zones of mycelia of different ages was tested, the activity was always highest at the edge of the colony in the youngest hyphae, and declined rapidly towards older parts of the colony. No differences in activity were found between light- and dark-grown colonies, although exposure of a dark-grown colony to light caused a rapid decline in activity. The zone of the colony with high laccase activity correlated well with that capable of responding to the light induction necessary for basidiocarp formation (Ross, 1982*a,b*), and it was suggested that in *C. congregatus* the laccase activity could be involved in the initial light-requiring stage of basidiocarp formation. No clear physiological role has yet been suggested for phenoloxidase activity in *S. commune*.

In a study of enzymes metabolising glucose-6-phosphate in dikaryotic mycelium and differentiating basidiocarps of *S. commune*, a decline in phosphoglucomutase activity was observed in the middle stage of basidiocarp development, while no changes in the activities of glucose-6-phosphate dehydrogenase or phosphoglucoisomerase were recorded (Schwalb, 1974, 1978). Instead, the activity of glycogen-degrading glucoamylase increased from the early to the late stages of differentiation (Schwalb & Jansons, 1973; Schwalb, 1978). In the presence of a low glucose concentration in the medium, an increase in extra- and intracellular R-glucanase activity takes place during pileus formation in *S. commune* (Wessels, 1966; Wessels & Sietsma, 1979). The increase in R-glucanase activity leads to the breakdown of extracellular water-soluble (1–3)-β,(1–6)-β-glucan and R-glucan [(1–3)-β,(1–6)-β-glucan] in the hyphal walls, which makes possible re-utilisation of the previously-formed cell constituents. In *C. cinereus* enzymes involved in the synthesis of urea were re-activated as the basidiocarp developed, while urease was repressed (Ewaze, Moore & Stewart, 1978). Urea consequently accumulated and acted as an osmoticum driving the expansion of the basidiocarp. A similar effect was induced in the mycelium by transferring it to a medium poor in carbohydrates and nitrogen. In addition, in *C. cinereus* NADP-linked glutamate dehydrogenase activity greatly increased in the developing pileus, while in the stipe and parental mycelium activity was barely detectable. The activity of this enzyme could also be increased in mycelium by transferring it to medium lacking nitrogen, with pyruvate as sole carbon source (Moore,

1981). These studies emphasise the regulatory role of the growth medium in the induction of enzyme activity, and it appears that induction of activity in a developing basidiocarp depends on lack of direct contact of the hyphae in the basidiocarp with the culture medium, allowing the hyphae in the basidiocarp to respond as would the parental mycelium on a nutritionally-deficient medium.

During basidiocarp maturation in *S. commune* the activity of inactivating proteases increased (Schwalb, 1975, 1977, 1978) and a substance inhibiting laccase activity has been reported in mature basidiocarps (Phillips & Leonard, 1976*a*). These observations have drawn attention to the possibility that some changes in enzyme activities during basidiocarp formation could possibly be due to modification of enzymes by proteases (Schwalb, 1977, 1978).

In *C. cinereus* the isozyme patterns of several common enzymes were compared in two monokaryons, in a dikaryon obtained by mating the monokaryons and in basidiocarps at different stages of maturation (Moore & Jirjis, 1981). Different isozyme patterns were produced between the monokaryons (also within them, depending on the incubation period), between the monokaryons and the dikaryon, and between the dikaryon and the developing basidiocarps. In *S. commune* no difference was observed in the isozyme patterns of several common metabolic enzymes of two otherwise co-isogenic monokaryotic strains that had different *A* and *B* incompatibility factors (Wang & Raper, 1970). The isozyme pattern of monokaryons was, however, different from that of the dikaryon formed from them, this being ascribed to the inactivating and activating effects of the incompatibility factors on combinations of genes. This result could not be repeated later, and no isozyme differences were revealed between the monokaryons and dikaryon in an experiment in which the cultural conditions and incubation period were different from those in the first experiment (Ullrich, 1977). When the mRNA sets of these monokaryons and dikaryon were analysed recently, no significant differences were revealed by saturation hybridisation of single-copy DNA with poly(A)-containing RNA (Zantinge, Dons & Wessels, 1979), polysomal RNA or total RNA (Hoge, Springer, Zantinge & Wessels, 1982) or by homologous and heterologous hybridisation of complementary DNA with poly(A)-rich RNA (Zantinge, Hoge & Wessels, 1981). Nor were any significant differences shown by polypeptides translated from monokaryon and dikaryon polysomal RNAs in a cell-free wheat germ system (Zantinge *et al.*, 1981; Hoge *et al.*, 1982). However, some differences were recorded in the proteins from the monokaryons and dikaryon (de Vries, Hoge & Wessels, 1980*a,b*). This led to the conclusion that the incompatibility factors act neither at the transcriptional level nor at the level of processing mRNA, but rather through differential modification of polypeptides during or after mRNA translation (Hoge *et al.*, 1982).

The same techniques indicated small differences between dikaryon polysomal RNAs from vegetative mycelium and those present at an early stage of basidiocarp formation. Vegetative RNA sequences occurred in somewhat lower concentration in total RNA from fruiting mycelium, and 5% of the complex RNA mass in fruiting mycelium was absent from vegetative mycelium. The translation of the RNA sequences of the fruiting mycelium in a cell-free wheat germ system indicated that the RNA sequences specific for the fruiting dikaryon coded 18 polypeptides not present in the 2-day-old dikaryon (Hoge, Springer & Wessels, 1982). It appears that overlapping of mRNA and proteins exists between vegetative mycelium and reproductive mycelium in *S. commune*. At an early stage of differentiation mainly the same proteins as in the vegetative mycelium are synthesised but for production of a new growth pattern typical to the basidiocarp initials. The few proteins specific to the differentiating basidiocarps may have a critical role in directing this process.

The ultrastructure of hyphae in different zones of dikaryotic colonies

Comparison of the ultrastructure of apical hyphal compartments grown only in the dark with those transferred after growth in dark to light for 4 h showed only small differences (Fig. 5). The apical vesicles extended somewhat deeper into the hypha and the plasmalemma was more uneven in the apices of dark-grown hyphae, perhaps indicating greater activity (Fig. 5a,b). In the subapical region fewer and longer mitochondria were recorded in dark-grown than in light-grown hyphae (Fig. 5c,d). Lipid globules were recorded in hyphae grown in both circumstances, but in light they were more closely associated with mitochondria than in dark (Fig. 5d).

In zone 2, two hyphal types were distinguished: broad hyphae with cytoplasmic contents comparable with those of hyphae in the apical zone, and smaller hyphae tightly packed with white globular material (Fig. 6a,b) similar in structure to the material reported to represent glycogen in regenerating protoplasts of *S. commune* (van der Valk & Wessels, 1976). As in the protoplasts, the globular material reacted

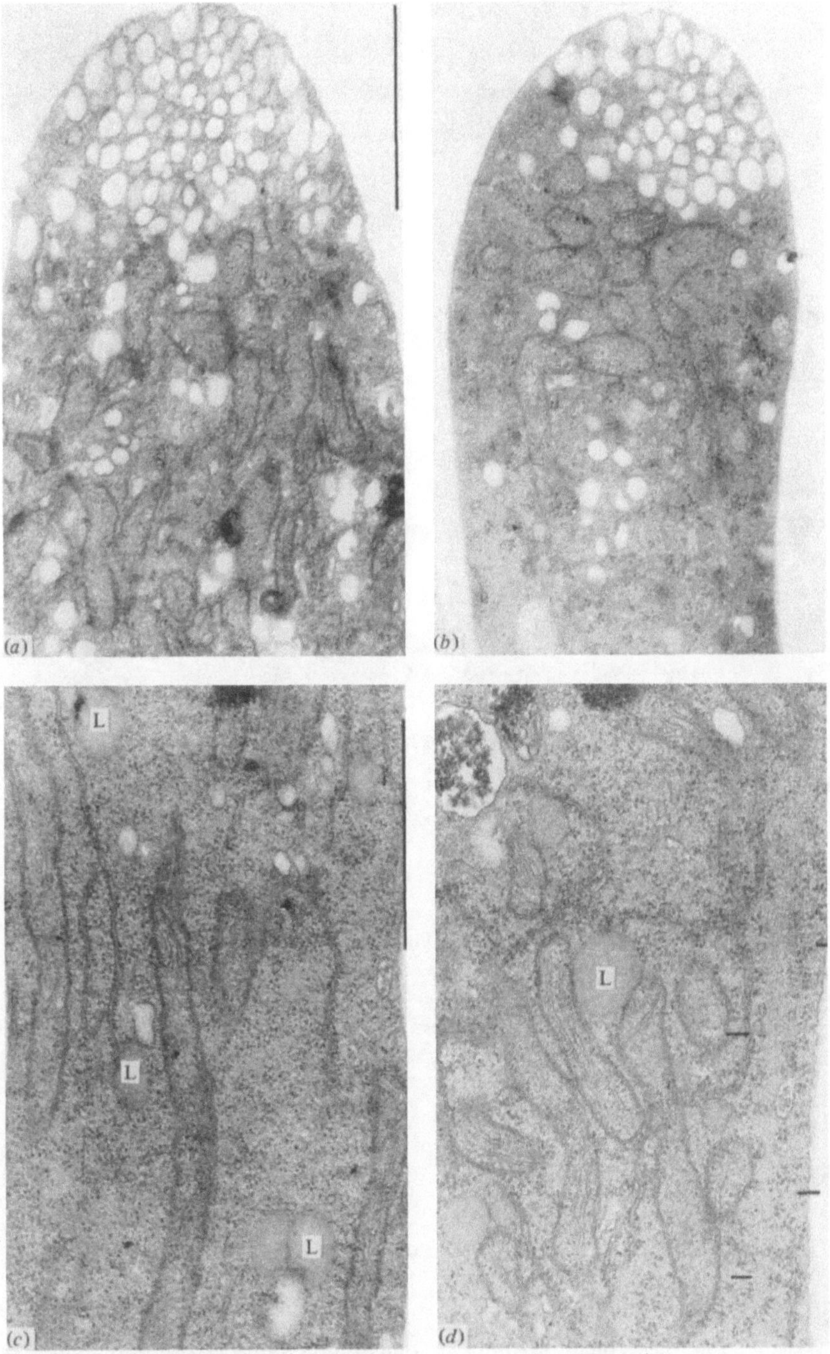

Fig. 5. Apical compartments of 4-day-old colonies grown in the dark and for 4 h in light. (a) Apex in the dark, (b) apex in the light, (c) subapical region in the dark, (d) subapical region in light. Thin filaments marked with lines. L, lipid globule. Scale bars, 1 μm.

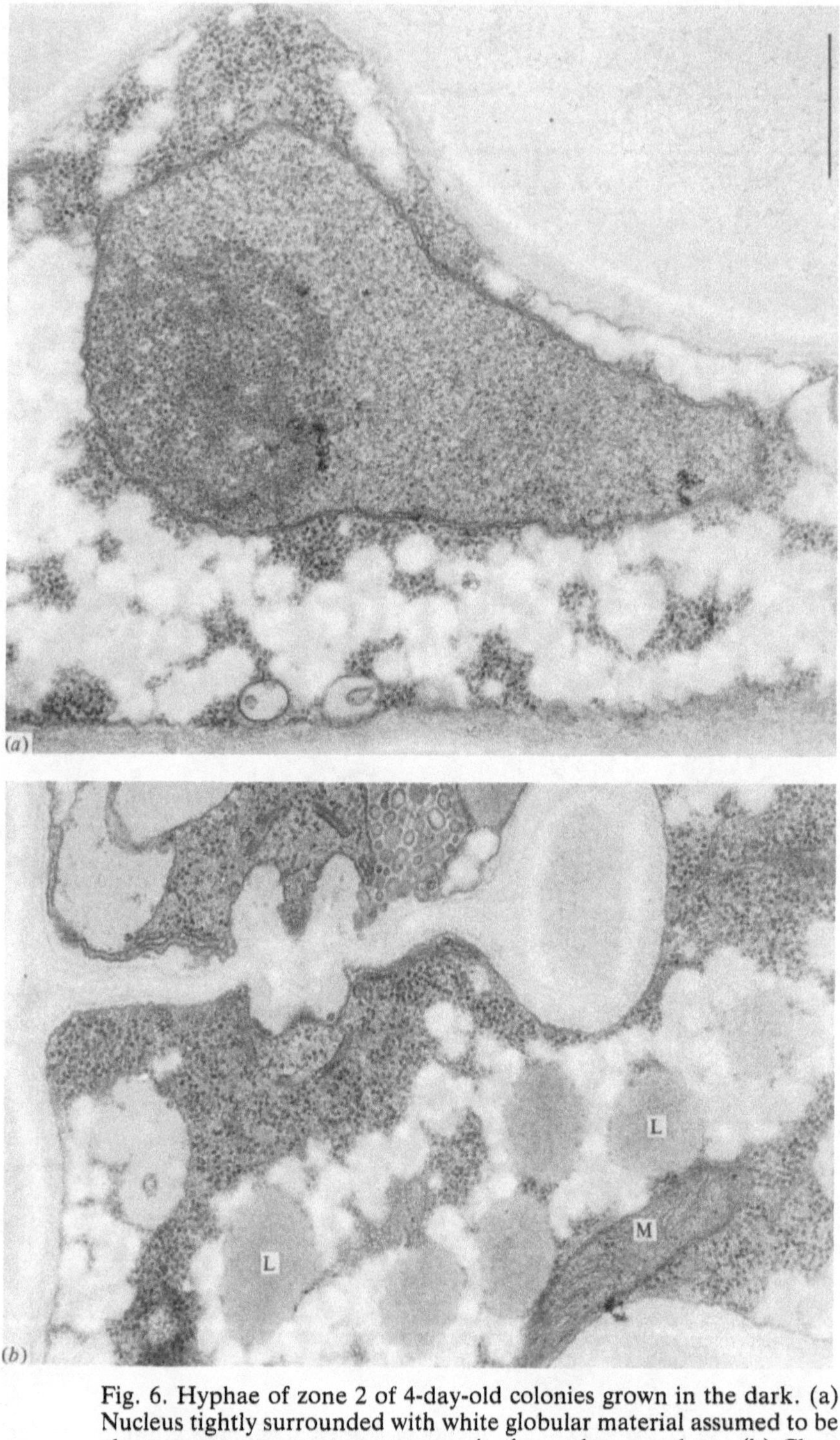

Fig. 6. Hyphae of zone 2 of 4-day-old colonies grown in the dark. (a) Nucleus tightly surrounded with white globular material assumed to be glycogen; no apparent pores occur in the nuclear envelope. (b) Close association of glycogen and lipid globules (L). M, mitochondrion. Scale bar, 0.5 μm.

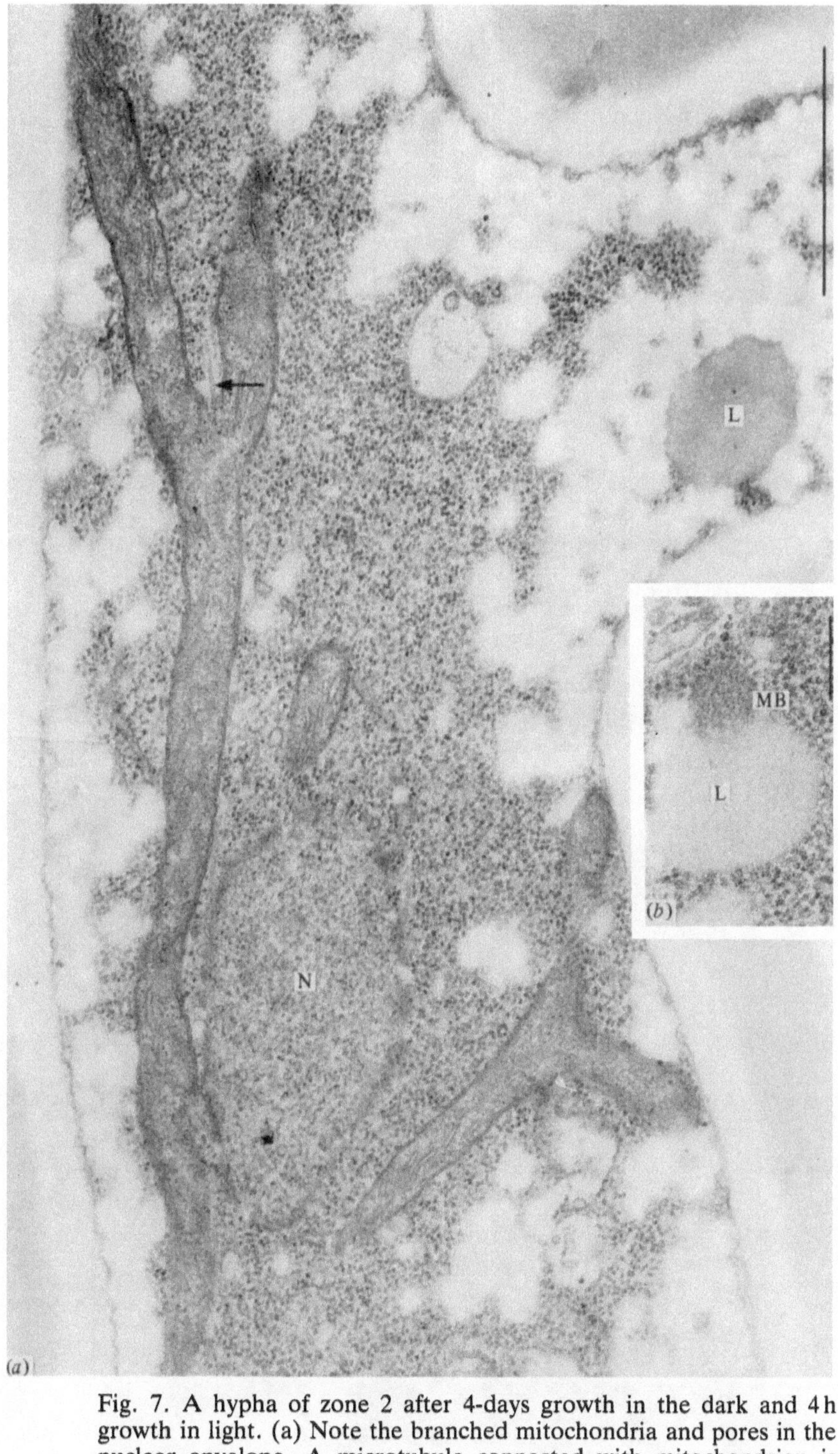

Fig. 7. A hypha of zone 2 after 4-days growth in the dark and 4 h growth in light. (a) Note the branched mitochondria and pores in the nuclear envelope. A microtubule connected with mitochondrion is marked with an arrow; scale bar, 1 μm: (b) An electron-dense microbody (MB) associated with lipid globule (L) in another part of the same hypha; scale bar, 0.2 μm. N, nucleus.

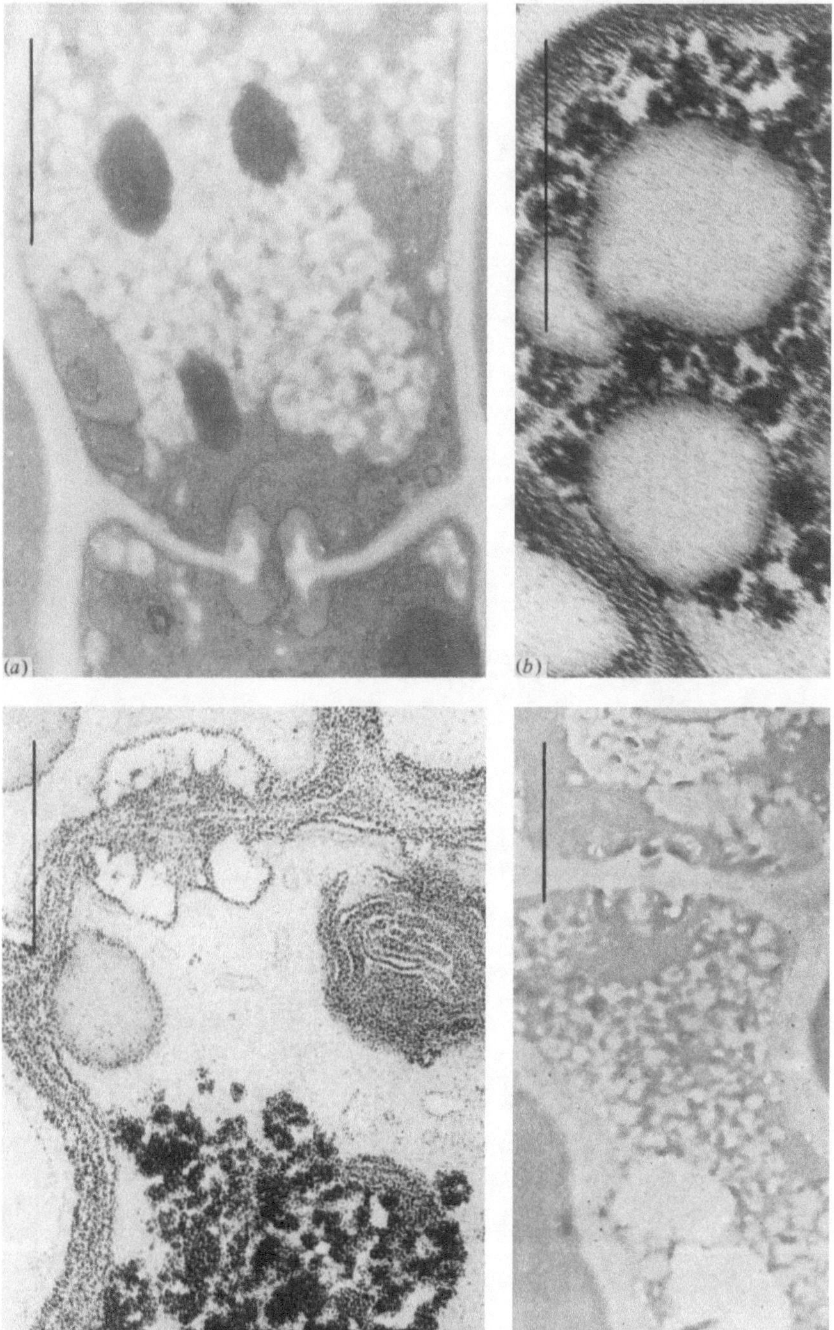
(a)
(b)
(c)
(d)

strongly when stained using the technique of Thiéry (1967, Fig. 8), while the staining pattern of the hyphal walls was comparable with that of monokaryotic hyphae of *S. commune* (van der Valk, Marchant & Wessels, 1977). On this basis the white globular material was interpreted as being glycogen (cf. Matthews & Niederpruem, 1973; Marchant, Raudaskoski & Shneyor, 1976). As in protoplasts glycogen in the hyphae was regularly associated with lipid globules (Figs. 6b, 7a,b, 8b). The structure of hyphae after 4 h in light differed slightly from that of those grown in darkness (Fig. 7). In light, mitochondria were longer and branched, there was more endoplasmic reticulum, clear pores could be distinguished in the nuclear envelope, and lipid globules were associated with electron-dense microbodies (Fig. 7b). These differences are interpreted as being due to the transfer of colonies from dark to light, and to represent initiation of growth in zone 2, which probably involves mobilisation of glycogen and lipids.

The latter view was supported by the ultrastructure of hyphae after 8 h in light (Fig. 9). Here the clamp-connections of hyphae in the apical zone produced branches (Fig. 9a) as reported by Raudaskoski & Viitanen (1982). There was now no accumulation of glycogen or lipids; instead there was accumulation of unknown, electron-dense material in vacuoles of apical hyphae (Fig. 9a) and hyphae of zone 2 (Fig. 9b). The structure of the hyphae was similar to that described for primordia of *S. commune* by van der Valk & Marchant (1978) except, in this study, for the lack of anastomoses in the hyphae. No glycogen or lipid globules were seen in the hyphae from the centre of the colony, but the hyphae were highly vacuolated and contained special mitochondria and microbodies (Fig. 9c).

The ultrastructure of apical hyphae and hyphae of zone 2 was similar when the dikaryotic colonies were grown in the dark at high CO_2 concentration. Hyphae from both zones had thin walls and septa, dense cytoplasm and no glycogen or lipid deposits (Fig. 10a). It was only in the hyphae from the centre that some accumulation of glycogen and lipids

---

Fig. 8. Cytochemical staining of the hyphae of zone 2. (a) Unstained hypha with white glycogen and black lipid globules due to interaction with osmium tetroxide. (b) Strong affinity of glycogen to Thiéry's reagent in dark-grown hypha. (c) The same reaction in a hypha kept for 4 h in light. (d) No reaction with Thiéry's reagent in a hypha kept in hydrogen peroxide instead of periodic acid. (a), (d), scale bar 1 μm; (b), (c), scale bar 0.5 μm.

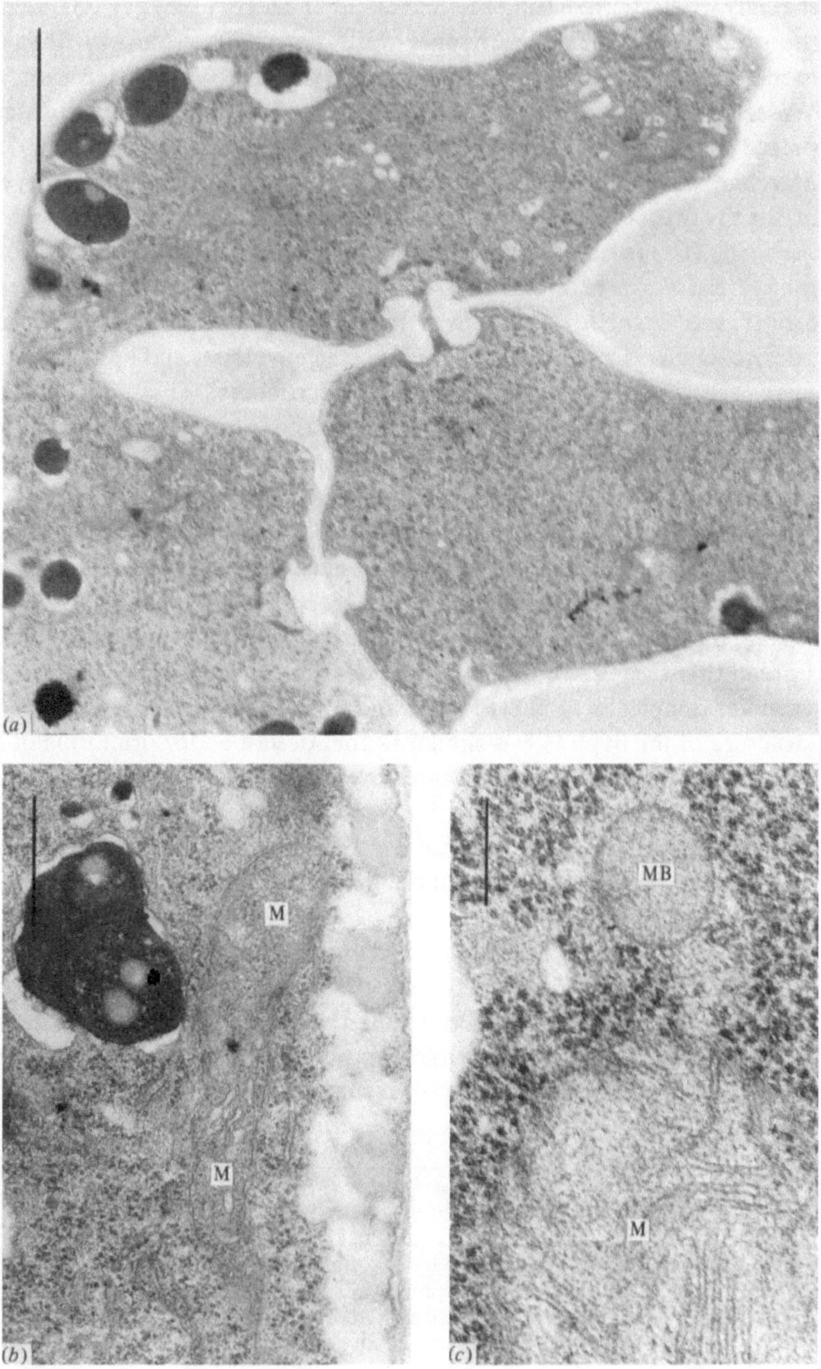
(a)
(b)
M
M
(c)
MB
M

occurred (Fig. 10c). These observations support the view that growth of a dikaryotic mycelium at high CO_2 levels on a nitrogen-rich medium may lead to enhanced accumulation of proteins and nucleic acids but to a low content of carbohydrate. The thin cell walls and septa accord with biochemical analyses of hyphal walls of *S. commune* formed in the presence of a high CO_2 level (Sietsma *et al.*, 1977), although the most prominent feature of the present study was the lack of intracellular carbohydrate and lipid reserve material in the hyphae of zone 2 and the low content in hyphae of zone 3.

In hyphae grown in the dark at a high CO_2 level and then transferred to light and normal air, protein crystals were regularly observed (Figs. 10b, 11a,b) together with electron-dense microbodies (Fig. 11c). These structures were especially abundant in apical cells sampled 100 min after transfer. Similar protein crystals have been found in developing ascospores, where they were suggested to be aggregates of spindle proteins (Zachariah & Anderson, 1975) and in the hyphae of elongating stipes of *C. cinereus*, where they are assumed to serve as reserve material for basidiocarp development (Blayney & Marchant, 1977). In *S. commune* no explanation can yet be offered for their presence.

The microbody-like structures were also frequently observed, although microbodies are usually rare in *S. commune.* In other fungi such as Chytridiomycetes, microbodies are regularly seen in the zoospores (Powell, 1976). The presence of malate synthase and isocitrate lyase activity as well as the occurrence of catalase activity and the close association of the microbodies with lipid globules have suggested that in chytridiomycete zoospores the microbodies function as glyoxysomes (Powell, 1976). In a study of fourteen species of filamentous fungi, including *Neurospora crassa* and *Phanerochaete* sp., microbodies were detected in all cases. However, their number and appearance varied both between and within species, depending on the carbon source used (Maxwell *et al.*, 1975). The lowest number of microbodies was detected when a fungus was grown in the presence of glucose (Maxwell,

---

Fig. 9. Hyphae grown for 4 days in the dark and 8 h in light. (a) Growth of the clamp-connection in a hypha of the apical zone. (b) Part of a hypha of zone 2 with some glycogen and lipid remaining. Note the accumulation of electron-dense material; M, mitochondrion. (c) Part of a hypha from the centre of the colony with round swollen mitochondrion (M) and a microbody (MB). (a), scale bar 1 μm; (b), scale bar 0.5 μm; (c), scale bar, 0.2 μm.

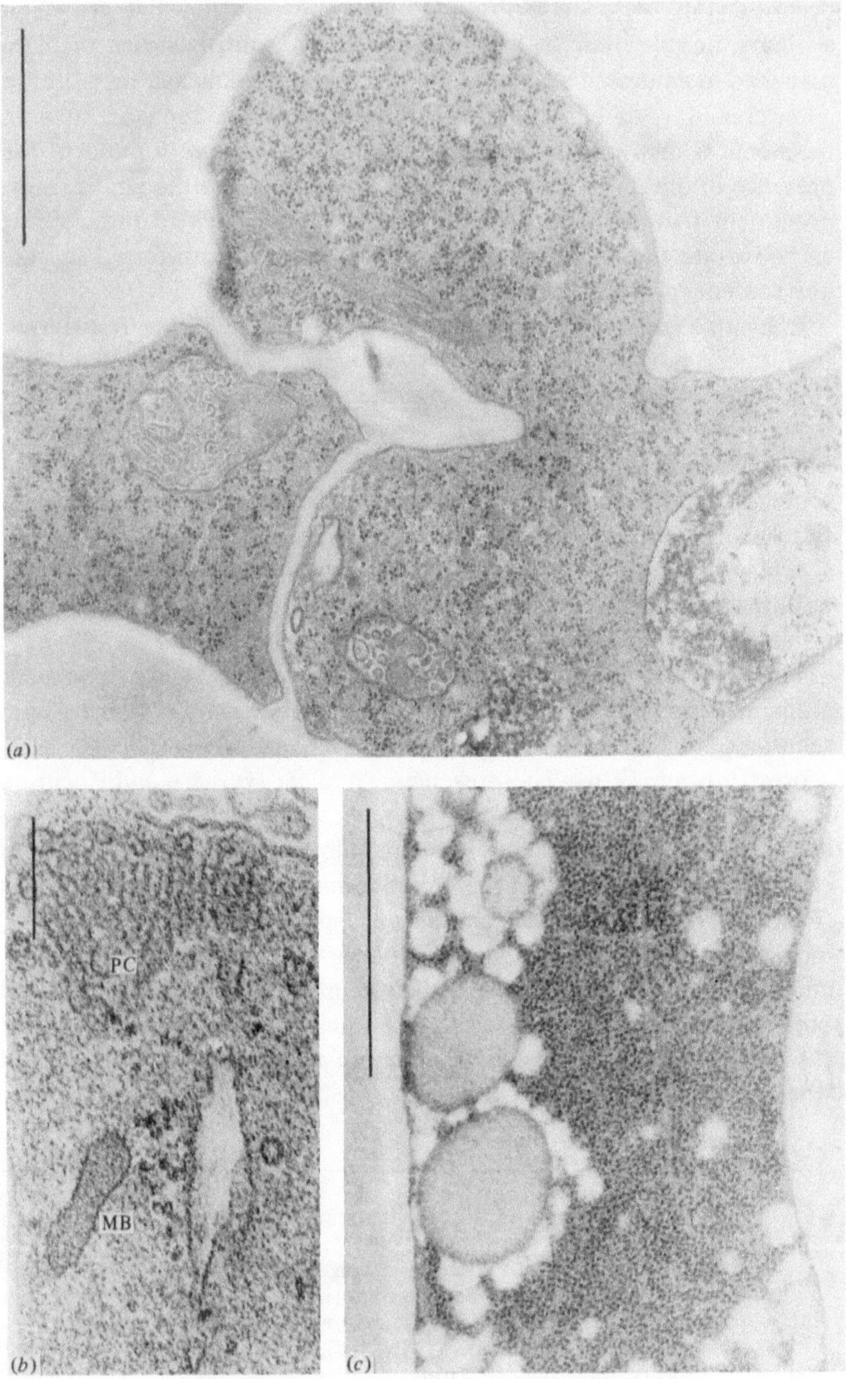
(a)
PC
MB
(b)
(c)

Armentrout & Graves, 1977), which may explain the infrequent occurrence of microbodies in hyphae of *S. commune*, which is regularly grown on medium with a relatively high glucose content. Particles with malate synthase and isocitrate lyase activities have been isolated by density-gradient centrifugation from hyphae of *C. cinereus* (O'Sullivan & Casselton, 1973; Casselton, 1976), *Aspergillus tamarii* (Graves, Armentrout & Maxwell, 1976) and *N. crassa* (Schwitzguébel, Møller & Palmer, 1981) after growth on acetate and ethanol. In the homobasidiomycete *Phanerochaete chrysosporium* microbodies were shown to be connected with hydrogen peroxide production associated with lignin degradation (Forney, Reddy & Pankratz, 1982).

The microbodies observed in association with lipid globules in the hyphae of zone 2 after transfer to light in *S. commune* could be of the glyoxysome type (Powell, 1976; Maxwell *et al.*, 1977; Kamiryo *et al.*, 1982) and involved in the breakdown and re-utilisation of the lipids, while the electron-dense microbodies in the hyphae grown at high CO_2 concentration could be of another type. They may perhaps be comparable with microbodies in the hyphae of *P. chrysosporium* (Forney *et al.*, 1982) and involved in removing the radicals produced during primary light reactions (see below). Radicals such as hydrogen peroxide (Halliwell, 1974) may especially develop, when the hyphae are incompletely prepared for light induction as they probably are after growth in the dark, at a high CO_2 level. The microbodies could also be involved in CO_2 fixation as are glycosomes in the cells of *Trypanosoma brucei* (Opperdoes & Cottem, 1982).

Conclusions

In dikaryotic colonies of *S. commune*, basidiocarp differentiation is initiated by decreased growth of apical and increased growth of subapical hyphae at the edge of the colony. The prerequisite for rapid differentiation is a mycelium with short cells in which the primary and secondary branches become close together and aggregate (Fig. 12b).

Fig. 10. Hyphae grown at high CO_2 concentration for 4 days. (a) Part of a hypha of zone 2 with no glycogen or lipid deposits. (b) Part of a hypha of zone 2 grown for 100 min in light and normal air after growth at high CO_2 concentration in the dark. Note the protein crystals (PC) and the microbody (MB). (c) Part of a hypha from the centre of the colony grown in the dark. Some glycogen and lipid has accumulated. (a), (c), scale bar, 1 μm; (b), scale bar 0.2 μm.

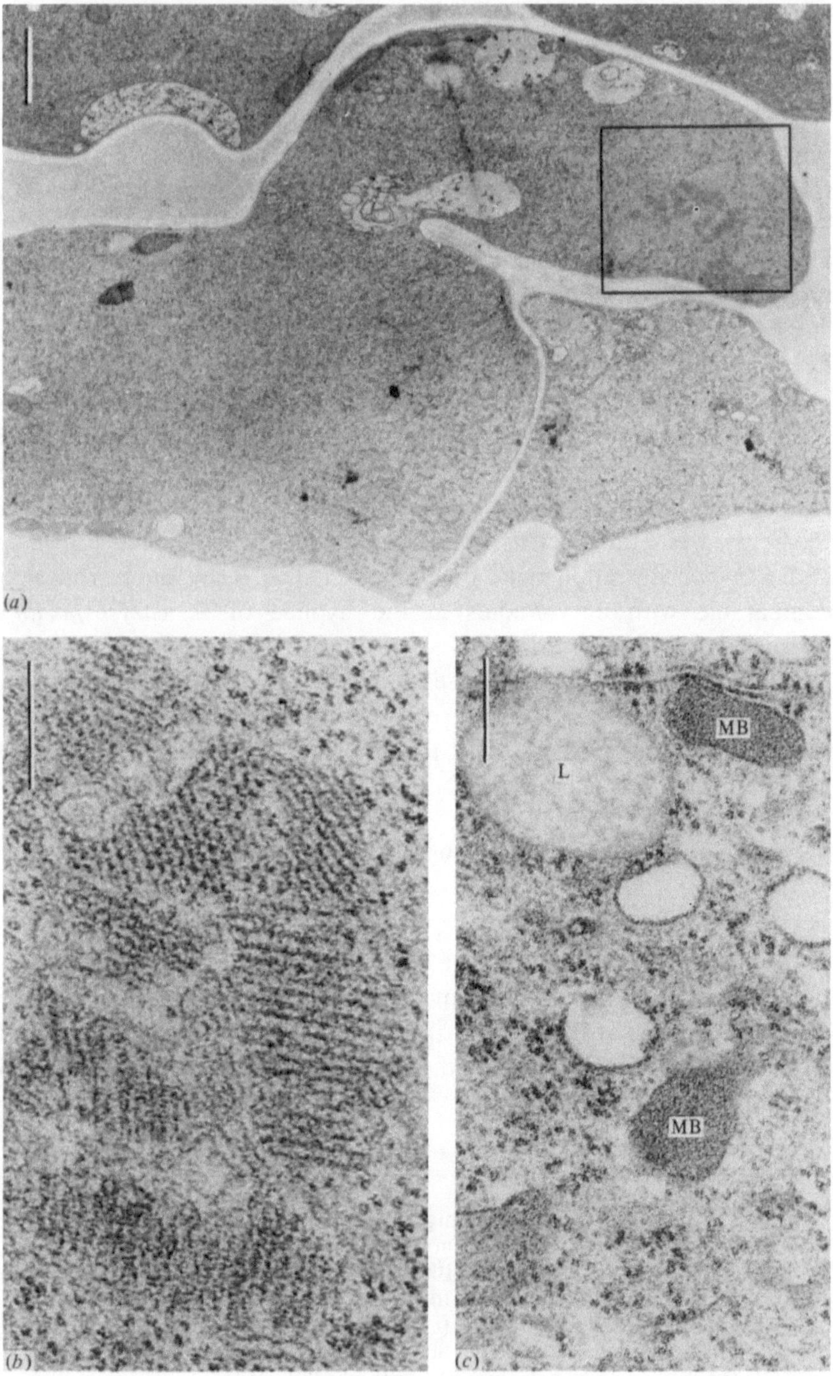
(a)
(b)
MB
L
MB
(c)

The short cells are induced by aeration (Raudaskoski & Viitanen, 1982, Fig. 1c) and appropriate temperature (Fig. 13) both in dark and light. The importance of aggregation of hyphae for basidiocarp development has already been indicated by the work with the homozygote coh-l^{-1} dikaryon of *S. commune* (Perkins & Raper, 1970), in which no hyphal aggregates or basidiocarps were found. The accumulation of glycogen and lipids in the hyphae during growth in the dark and increased aeration might also be necessary for rapid differentiation. Glycogen and lipids could serve as reserve material for growth of basidiocarps as in *C. cinereus* (Matthews & Niederpruem, 1973; Moore *et al.*, 1979). The long hyphal compartments with widely spaced branches (Fig. 12a), which develop at high CO_2 levels do not provide a good basis for rapid differentiation. The glycogen and lipid contents of the hyphae are also low, especially at the edge of the colony, where all the carbohydrates assimilated have probably been used for synthesis of the cell walls of the rapidly-extending hyphae. In these colonies a new growth pattern has to develop after transfer to light and increased aeration before differentiation can take place. In a wood-rotting fungus such as *S. commune* the ecological impact of the induction of mycelial extension by high CO_2 concentration would be an extensive distribution of the mycelium in the wood and efficient utilisation of the nutrients available (i.e. primary resource capture, cf. Rayner & Webber, Chapter 18). Close to the surface, where the CO_2 concentration is reduced to a level close to that of normal air, the growth pattern of the mycelium may change to production of short compartments, reserve material and aggregated branches. Such a mycelium would then be receptive to the light induction necessary for basidiocarp differentiation.

The as yet not very detailed action spectra for basidiocarp development in *S. commune* (Perkins & Gordon, 1969) and in other Basidiomycotina (Aschan-Åberg, 1960; Kitamoto, Suzuki & Furukawa, 1972; Badham, 1980; Durand & Jacques, 1982) indicate that blue or near-ultraviolet light stimulates basidiocarp induction, but there is no information concerning the requisite photoreceptors. In the ascomycete

Fig. 11. Hyphae grown for 100 min in light and normal air after growth for 4 days in the dark at high CO_2 concentration. (a) A hypha of the apical zone with protein crystals in the region delimited by rectangle. (b) Protein crystals at higher magnification. (c) Microbodies (MB) and lipid globules (L) typical in the apical compartments of these colonies. (a), scale bar 1 μm; (b), (c), scale bar 0.2 μm.

N. crassa the primary reactions of blue-light-controlled responses such as phase shift in circadian rhythm of conidiation and promotion of conidiation (Ninnemann & Klemm-Wolfgramm, 1980; Nakashima & Fujimura, 1982; Takahama, Shimizu-Takahama & Egashira, 1982) have received much more attention. The absorbance changes induced by blue light in the mycelium or in the mycelial extracts of *N. crassa* (Muñoz & Butler, 1975; Schmidt & Butler, 1976) suggest that a flavin is the photoreceptor pigment and that photoreduction of flavin is mediated by a b-type cytochrome. A study of the blue light reactions in an artificial system comprised of flavin mononucleotide (FMN) and cytochrome c (Schmidt & Butler, 1976) has indicated that the blue light causes the photoreduction of the flavin and that the reduced form reacts with oxygen to form superoxide anions which can further reduce cytochrome c. If no electron acceptors are present the superoxide anions may react with hydrogen to form hydrogen peroxide.

In *S. commune* the reduced growth of apical and increased growth of subapical compartments as well as the decrease in the glycogen and lipid

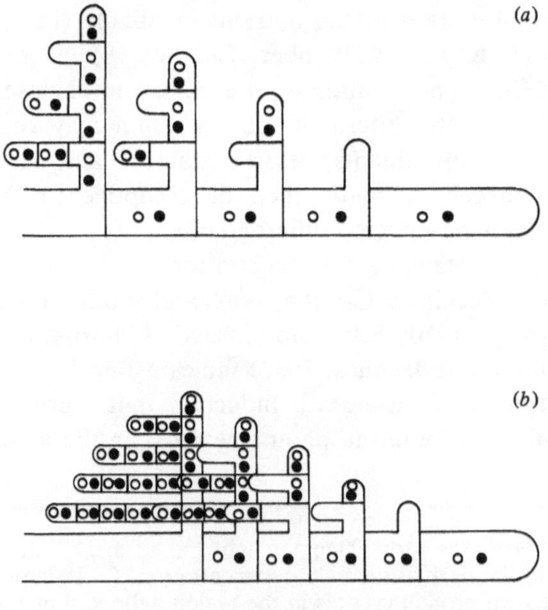

Fig. 12. A schematic drawing of the effect of compartment length on the aggregation of side branches. (a) Hyphae grown at high CO_2 concentration; (b) hyphae grown under normal aeration.

contents in the hyphae appeared to be induced by light. The decrease of glycogen in the hyphae after transfer to light is in agreement with the observations that during basidiocarp differentiation in *S. commune* the activity of glucoamylase increases and that of the phosphoglucomutase decreases (Schwalb & Jansons, 1973; Schwalb, 1974, 1978). Light is known to induce breakdown of glycogen in *C. cinereus* (Uno & Ishikawa, 1976; Moore *et al.*, 1979). Here the breakdown of glycogen was connected with an increase in glycogen phosphorylase activity and a decrease in glycogen synthetase activity. Changes in the enzyme activities could be induced by different concentrations of adenosine 3′,5′-cyclic monophosphate (cyclic AMP) in the incubation media (Uno & Ishikawa, 1976, 1978, 1982). In the unicellular green alga *Chlorella vulgaris* blue light is known to enhance breakdown of carbohydrates, which is connected to an increase in the activity of pyruvate kinase (Kowallik & Schätzle, 1980). Blue light may enhance the activity of pyruvate kinase through changing the enzyme conformation, which results in higher phosphoenolpyruvate affinity, and by increasing *de novo* synthesis of the enzyme protein (Ruyters, 1980).

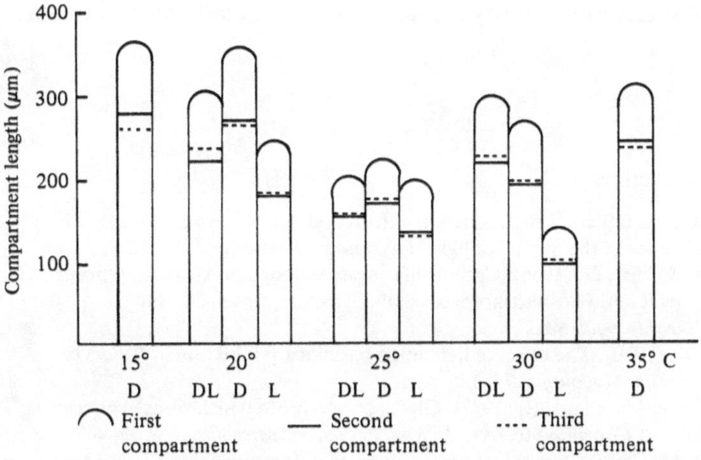

Fig. 13. The effect of temperature and illumination on hyphal compartment length in 4-day-old dikaryotic colonies. The mean length of 30 apical compartments is given by columns, which also show the mean lengths of the subapical and third compartments. The compartment lengths were measured at each temperature in colonies grown in darkness (D), in continuous light (L) and transferred from darkness to light for 4 h (DL). The lengths of the apical, subapical and third compartments were always measured in the same leading hypha.

In *S. commune*, as in *C. cinereus* and *C. vulgaris*, light may affect the carbohydrate metabolism while in *N. crassa* the key enzyme in blue-light-controlled conidiation has been suggested to be nitrate reductase (Ninnemann & Klemm-Wolfgramm, 1980). As suggested by Uno & Ishikawa (1982) the decrease and increase in activities of enzymes involved in the breakdown of carbohydrates could be regulated by levels of cyclic AMP in the hyphae. Environmental factors such as light (through a flavin receptor) and glucose concentration of the medium may regulate the level of cyclic AMP by controlling the activities of adenylate cyclase and phosphodiesterase simultaneously. The regulation could also be more direct as assumed for *C. vulgaris* (Ruyters, 1980). These suggestions would be in agreement with the results of Hoge *et al.* (1982), which indicated that the mRNA populations of vegetative and reproductive dikaryotic mycelia differed only slightly qualitatively, and that the difference expressed in the growth patterns of these mycelia could be due to modification of proteins. The synchronously and rapidly differentiating dikaryotic mycelium of *S. commune* would be good material for examining the effect of light on metabolism, while hyphae grown at high CO_2 levels could serve as the control.

This work was supported by a grant from the Academy of Finland.

References

Aschan-Åberg, K. (1960). The production of fruit bodies in *Collybia velutipes*. III. Influence of the quality of light. *Physiologia Plantarum*, **13**, 276–9.

Bachofen, R. & Rast, D. (1968). Carboxylierungsreaktionen in *Agaricus bisporus*. III. Pyruvat und Phosphoenolpyruvat als CO_2-Acceptoren. *Archiv für Mikrobiologie*, **60**, 217–34.

Badham, E. R. (1980). The effect of light upon basidiocarp initiation in *Psilocybe cubensis*. *Mycologia*, **72**, 136–42.

Blayney, G. P. & Marchant, R. (1977). Glycogen and protein inclusions in elongating stipes of *Coprinus cinereus*. *Journal of General Microbiology*, **98**, 467–76.

Bradford, M. M. (1976). A rapid and sensitive method for the quantitation of microgram quantities of protein utilizing the principle of protein-dye binding. *Analytical Biochemistry*, **72**, 248–54.

Budd, K. (1969). The assimilation of bicarbonate by *Neocosmospora vasinfecta*. *Canadian Journal of Microbiology*, **15**, 389–98.

Budd, K. (1971). Bicarbonate fixation by cell-free extracts and by mycelium of *Neocosmospora vasinfecta*. *Journal of General Microbiology*, **67**, 99–106.

Bull, A. T. & Bushell, M. E. (1976). Environmental control of fungal growth. In *The Filamentous Fungi* II, *Biosynthesis and Metabolism*, ed. J. E. Smith & D. R. Berry, pp. 1–31. London, UK: Arnold.

Bushell, M. E. & Bull, A. T. (1974). Anaplerotic carbon dioxide fixation in steady and non-steady state fungal cultures. *Proceedings of the Society for General Microbiology*, **1,** 69.

Casselton, P. J. (1976). Anaplerotic pathways. In *The Filamentous Fungi* II, *Biosynthesis and Metabolism*, ed. J. E. Smith & D. R. Berry, pp. 121–36. London, UK: Arnold.

Chang, S.-T. & Chan, K.-Y. (1973). Quantitative and qualitative changes in proteins during morphogenesis of the basidiocarp of *Volvariella volvacea. Mycologia*, **65,** 355–64.

Cotter, D. A., LaClave, A. J., Wegener, W. S. & Niederpruem, D. J. (1970). CO_2 control of fruiting in *Schizophyllum commune*: noninvolvement of sustained isocitrate lyase derepression. *Canadian Journal of Microbiology*, **16,** 605–8.

Durand, R. & Jacques, R. (1982). Action spectra for fruiting of the mushroom *Coprinus congregatus. Archives of Microbiology*, **132,** 131–4.

Eger-Hummel, G. (1980). Blue-light photomorphogenesis in mushrooms (basidiomycetes). In *The Blue Light Syndrome*, 1980, ed. H. Senger, pp. 556–62. Berlin, Heidelberg, New York: Springer-Verlag.

Esser, K. & Meinhardt, F. (1977). A common genetic control of dikaryotic and monokaryotic fruiting in basidiomycete *Agrocybe aegerita. Molecular and General Genetics*, **155,** 113–15.

Esser, K., Saleh, F. & Meinhardt, F. (1979). Genetics of fruit body production in higher basidiomycetes. II. Monokaryotic and dikaryotic fruiting in *Schizophyllum commune. Current Genetics*, **1,** 85–8.

Esser, K., Semerdžieva, M. & Stahl, U. (1974). Genetische Untersuchungen an dem Basidiomyceten *Agrocybe aegerita.* I. Eine Korrelation zwischen dem Zeitpunkt der Fruchtkörperbildung und monokaryotischem Fruchten und ihre Bedeutung für Züchtung und Morphogenese. *Theoretical and Applied Genetics*, **45,** 77–85.

Ewaze, J. O., Moore, D. & Stewart, G. R. (1978). Co-ordinate regulation of enzymes involved in ornithine metabolism and its relation to sporophore morphogenesis in *Coprinus cinereus. Journal of General Microbiology*, **107,** 343–57.

Forney, L. J., Reddy, C. A. & Pankratz, H. S. (1982). Ultrastructural localization of hydrogen peroxide production in ligninolytic *Phanerochaete chrysosporium* cells. *Applied and Environmental Microbiology*, **44,** 732–6.

Gitterman, C. O. & Knight, S. G. (1952). Carbon dioxide fixation into amino acids of *Penicillium chrysogenum. Journal of Bacteriology*, **64,** 223–31.

Gooday, G. W. (1979). Chitin synthesis and differentiation in *Coprinus cinereus.* In *Fungal Walls and Hyphal Growth*, British Mycological Society Symposium 2, 1979, ed. J. H. Burnett & A. P. J. Trinci, pp. 203–23. Cambridge, UK: Cambridge University Press.

Graves, L. B. Jr, Armentrout, V. N. & Maxwell, D. P. (1976). Distribution of glyoxylate-cycle enzymes between microbodies and mitochondria in *Aspergillus tamarii. Planta*, **132,** 143–8.

Halliwell, B. (1974). Superoxide dismutase, catalase and glutathione peroxidase: solutions to the problems of living with oxygen. *New Phytologist*, **73,** 1075–86.

Hartman, R. E. & Keen, N. T. (1973). Enzymes catalyzing anaplerotic carbon dioxide fixation in *Verticillium albo-atrum. Phytopathology*, **63,** 947–53.

Hartman, R. E. & Keen, N. T. (1974*a*). The pyruvate carboxylase of *Verticillium albo-atrum. Journal of General Microbiology*, **81,** 15–19.

Hartman, R. E. & Keen, N. T. (1974*b*). The phosphoenolpyruvate carboxykinase of *Verticillium albo-atrum. Journal of General Microbiology*, **81,** 21–6.

Hartman, R. E., Keen, N. T. & Long, M. (1972). Carbon dioxide fixation by *Verticillium albo-atrum. Journal of General Microbiology*, **73,** 29–34.

Hintikka, V. & Korhonen, K. (1970). Effects of carbon dioxide on the growth of lignicolous and soil-inhabiting Hymenomycetes. *Communicationes Instituti Forestalis Fenniae*, **69,** 1–29.

Hoge, J. H. C., Springer, J. & Wessels, J. G. H. (1982). Changes in complex RNA during fruit-body initiation in the fungus *Schizophyllum commune. Experimental Mycology*, **6,** 233–43.

Hoge, J. H. C., Springer, J., Zantinge, B. & Wessels, J. G. H. (1982). Absence of differences in polysomal RNAs from vegetative monokaryotic and dikaryotic cells of the fungus *Schizophyllum commune. Experimental Mycology*, **6,** 225–32.

Ishikawa, T. & Uno, I. (1977). A mechanism of fruiting body formation in basidiomycetes. In *Growth and Differentiation in Microorganisms* (NRI Symposia on Modern Biology, 1977), ed. T. Ishikawa, Y. Maruyama & H. Matsumiya, pp. 283–301. Baltimore, London, Tokyo: University Park Press.

Kamiryo, T., Abe, M., Okazaki, K., Kato, S. & Shimamoto, N. (1982). Absence of DNA in peroxisomes of *Candida tropicalis. Journal of Bacteriology*, **152,** 269–74.

Kitamoto, Y., Suzuki, A. & Furukawa, S. (1972). An action spectrum for light-induced primordium formation in a basidiomycete, *Favolus arcularius* (Fr.) Ames. *Plant Physiology*, **49,** 338–40.

Kowallik, W. & Schätzle, S. (1980). Enhancement of carbohydrate degradation by blue light. In *The Blue Light Syndrome* 1980, ed. H. Senger, pp. 344–60. Berlin, Heidelberg, New York: Springer-Verlag.

Larmour, R. & Marchant, R. (1977). Carbon dioxide fixation and conidiation in *Fusarium culmorum* grown in continuous culture. *Journal of General Microbiology*, **99,** 59–68.

Leonard, T. J. (1971). Phenoloxidase activity and fruiting body formation in *Schizophyllum commune. Journal of Bacteriology*, **106,** 162–7.

Leonard, T. J. & Dick, S. (1968). Chemical induction of haploid fruiting bodies in *Schizophyllum commune. Proceedings of the National Academy of Sciences of the USA*, **59,** 745–51.

Leonard, T. J. & Dick, S. (1973). Induction of haploid fruiting by mechanical injury in *Schizophyllum commune. Mycologia*, **65,** 809–22.

Leonard, T. J. & Phillips, L. E. (1973). Study of phenoloxidase activity during the reproductive cycle in *Schizophyllum commune. Journal of Bacteriology*, **114,** 7–10.

Leslie, J. F. & Leonard, T. J. (1979*a*). Monokaryotic fruiting in *Schizophyllum commune*: genetic control of the response to mechanical injury. *Molecular and General Genetics*, **175,** 5–12.

Leslie, J. F. & Leonard, T. J. (1979*b*). Monokaryotic fruiting in *Schizophyllum commune*: phenoloxidases. *Mycologia*, **71,** 1082–5.

Manachère, G. (1980). Conditions essential for controlled fruiting of macromycetes – a review. *Transactions of the British Mycological Society*, **75,** 255–70.

Marchant, R., Raudaskoski, M. & Shneyor, Y. (1976). Ultrastructure of an indigotin-producing dome mutant of *Schizophyllum commune. Journal of General Microbiology*, **96,** 333–9.

Matthews, T. R. & Niederpruem, D. J. (1973). Differentiation in *Coprinus lagopus*. II. History and ultrastructural aspects of developing primordia. *Archives of Microbiology*, **88,** 169–80.

Maxwell, D. P., Armentrout, V. N. & Graves, L. B. Jr. (1977). Microbodies in plant pathogenic fungi. *Annual Review of Phytopathology*, **15,** 119–34.

Maxwell, D. P., Maxwell, M. D., Hänssler, G., Armentrout, V. N., Murray, G. M. & Hoch, H. C. (1975). Microbodies and glyoxylate-cycle enzyme activities in filamentous fungi. *Planta*, **124,** 109–23.

Moore, D. (1981). Evidence that the NADP-linked glutamate dehydrogenase of *Coprinus cinereus* is regulated by acetyl-CoA and ammonium levels. *Biochimica et Biophysica Acta*, **661**, 247–54.

Moore, D., Elhiti, M. M. Y. & Butler, R. D. (1979). Morphogenesis of the carpophore of *Coprinus cinereus*. *New Phytologist*, **83**, 695–722.

Moore, D. & Jirjis, R. I. (1981). Electrophoretic studies of carpophore development in the basidiomycete *Coprinus cinereus*. *New Phytologist*, **87**, 101–13.

Muñoz, V. & Butler, W. L. (1975). Photoreceptor pigment for blue light in *Neurospora crassa*. *Plant Physiology*, **55**, 421–6.

Nakashima, H. & Fujimura, Y. (1982). Light-induced phase shifting of the circadian clock in *Neurospora crassa* requires ammonium salts at high pH. *Planta*, **155**, 431–6.

Niederpruem, D. J. (1963). Role of carbon dioxide in the control of fruiting of *Schizophyllum commune*. *Journal of Bacteriology*, **85**, 1300–8.

Niederpruem, D. J., Marshall, C. & Speth, L. J. (1977). Control of extracellular slime accumulation in monokaryons and resultant dikaryons of *Schizophyllum commune*. *Sabouraudia*, **15**, 283–95.

Niederpruem, D. J. & Wessels, J. G. H. (1969). Cytodifferentiation and morphogenesis in *Schizophyllum commune*. *Bacteriological Reviews*, **33**, 505–35.

Ninneman, H. & Klemm-Wolfgramm, E. (1980). Blue-light controlled conidiation and absorbance change in *Neurospora* are mediated by nitrate reductase. In *The Blue Light Syndrome* 1980, ed. H. Senger, pp. 238–43. Berlin, Heidelberg, New York: Springer-Verlag.

Opperdoes, F. R. & Cottem, D. (1982). Involvement of the glycosome of *Trypanosoma brucei* in carbon dioxide fixation. *FEBS Letters*, **143**, 60–4.

O'Sullivan, J. O. & Casselton, P. J. (1973). The subcellular localization of glyoxylate cycle enzymes in *Coprinus lagopus* (sensu Buller). *Journal of General Microbiology*, **75**, 333–7.

Paranjpe, M. S., Chen, P. K. & Jong, S. C. (1979). Morphogenesis of *Agricus bisporus*: changes in proteins and enzyme activity. *Mycologia*, **71**, 469–78.

Perkins, J. H. (1969). Morphogenesis in *Schizophyllum commune*. I. Effects of white light. *Plant Physiology*, **44**, 1706–11.

Perkins, J. H. & Gordon, S. A. (1969). Morphogenesis in *Schizophyllum commune*. II. Effects of monochromatic light. *Plant Physiology*, **44**, 1712–16.

Perkins, J. H. & Raper, J. R. (1970). Morphogenesis in *Schizophyllum commune*. III. A mutation that blocks initiation of fruiting. *Molecular and General Genetics*, **106**, 151–4.

Phillips, L. E. & Leonard, T. J. (1976*a*). Extracellular and intracellular phenoloxidase activity during growth and development in *Schizophyllum*. *Mycologia*, **68**, 267–76.

Phillips, L. E. & Leonard, T. J. (1976*b*). Benzidine as a substrate for measuring phenoloxidase activity in crude cell-free extracts of *Schizophyllum commune*. *Mycologia*, **68**, 277–85.

Powell, M. J. (1976). Ultrastructure and isolation of glyoxysomes (microbodies) in zoospores of the fungus *Entophlyctis* sp. *Protoplasma*, **89**, 1–27.

Raper, J. R. & Krongelb, G. S. (1958). Genetic and environmental aspects of fruiting in *Schizophyllum commune* Fr. *Mycologia*, **50**, 707–40.

Rast, D. & Bachofen, R. (1967). Carboxylierungsreaktionen in *Agaricus bisporus* I. Der endogene CO_2-Acceptor. *Archiv für Microbiologie*, **57**, 392–405.

Raudaskoski, M. (1981). Physiological and genetical factors affecting fruit body production in higher basidiomycetes. *Luonnon Tutkija*, **85**, 38–44.

Raudaskoski, M. & Lahti, R. (1983). Mitochondrial structure, ATP concentration and inorganic pyrophosphatase activity in a *B*-mutant strain of *Schizophyllum commune*. *Journal of General Microbiology*, **129**, 2801–8.

Raudaskoski, M. & Vauras, R. (1982). Scanning electron microscope study of fruit body differentiation in *Schizophyllum commune*. *Transactions of the British Mycological Society*, **78**, 475–81.

Raudaskoski, M. & Viitanen, H. (1982). Effect of aeration and light on fruit body induction in *Schizophyllum commune*. *Transactions of British Mycological Society*, **78**, 89–96.

Ross, I. K. (1982*a*). Localization of carpophore initiation in *Coprinus congregatus*. *Journal of General Microbiology*, **128**, 2755–62.

Ross, I. K. (1982*b*). The role of laccase in carpophore initiation in *Coprinus congregatus*. *Journal of General Microbiology*, **128**, 2763–70.

Ruiz-Amil, M., De Torrontegui, G., Palacian, E., Catalina, L. & Losada, M. (1965). Properties and function of yeast pyruvate carboxylase. *Journal of Biological Chemistry*, **240**, 3485–92.

Rusmin, S. & Leonard, T. J. (1975). Biochemical induction of fruiting bodies in *Schizophyllum commune*: a bioassay and its application. *Journal of General Microbiology*, **90**, 217–27.

Ruyters, G. (1980). Blue light effects on enzymes of the carbohydrate metabolism in *Chlorella*. 1. Pyruvate kinase. In *The Blue Light Syndrome* 1980, ed. H. Senger, pp. 361–7. Berlin, Heidelberg, New York: Springer-Verlag.

Schmidt, W. & Butler, W. L. (1976). Flavin-mediated photoreactions in artificial systems: a possible model for the blue-light photoreceptor pigment in living systems. *Photochemistry and Photobiology*, **24**, 71–5.

Schwalb, M. N. (1971). Commitment of fruiting in synchronously developing cultures of the basidiomycete *Schizophyllum commune*. *Archiv für Microbiologie*, **79**, 102–7.

Schwalb, M. N. (1974). Changes in activity of enzymes metabolizing glucose 6-phosphate during development of the basidiomycete *Schizophyllum commune*. *Developmental Biology*, **40**, 84–9.

Schwalb, M. N. (1975). Developmental control of enzyme modification during fruiting of the basidiomycete *Schizophyllum commune*. *Biochemical and Biophysical Research Communications*, **67**, 478–82.

Schwalb, M. N. (1977). Developmentally regulated proteases from the basidiomycete *Schizophyllum commune*. *Journal of Biological Chemistry*, **252**, 8435–39.

Schwalb, M. N. (1978). Regulation of fruiting. In *Genetics and Morphogenesis in the Basidiomycetes*, ed. M. N. Schwalb & P. G. Miles, pp. 135–65. London, New York: Academic Press.

Schwalb, M. N. & Jansons, V. K. (1973). Comparison of developmentally controlled glucoamylase from normal and mutant strains of the basidiomycete *Schizophyllum commune*. *Biochemical Genetics*, **9**, 359–67.

Schwitzguébel, J. P., Møller, I. M. & Palmer, J. M. (1981). Changes in density of mitochondria and glyoxysomes from *Neurospora crassa*: a re-evaluation utilizing silica sol gradient centrifugation. *Journal of General Microbiology*, **126**, 289–95.

Sietsma, J. H., Rast, D. & Wessels, J. G. H. (1977). The effect of carbon dioxide on fruiting and on the degradation of a cell-wall glucan in *Schizophyllum commune*. *Journal of General Microbiology*, **102**, 385–9.

Stahl, U. & Esser, K. (1976). Genetics of fruit body production in higher basidiomycetes I. Monokaryotic fruiting and its correlation with dikaryotic fruiting in *Polyporus ciliatus*. *Molecular and General Genetics*, **148**, 183–97.

Tabak, H. H. & Cooke, W. B. (1968). The effects of gaseous environments on the growth and metabolism of fungi. *Botanical Review*, **34**, 126–252.

Tachibana, S., Siode, J. & Hanai, T. (1967). Studies on CO_2-fixing fermentation. (XIV.) On the relationship between L-malate formation and carbohydrate metabolism of cell-free extracts of *Schizophyllum commune. Journal of Fermentation Technology*, **45,** 1130–8.

Takahama, U., Shimizu-Takahama, M. & Egashira, T. (1982). Reduction of exogenous cytochrome c by *Neurospora crassa* conidia: effects of superoxide dismutase and blue light. *Journal of Bacteriology*, **152,** 151–6.

Tan, K. K. (1978). Light-induced fungal development. In *The Filamentous Fungi* III, 1978, *Developmental Mycology*, ed. J. E. Smith & D. R. Berry, pp. 334–57. London, UK: Arnold.

Thacker, D. G. & Good, H. M. (1952). The composition of air in trunks of sugar maple in relation to decay. *Canadian Journal of Botany*, **30,** 475–85.

Thiéry, J.-P. (1967). Mise en évidence des polysaccharides sur coupes fines en microscopie électronique. *Journal de Microscopie*, **6,** 987–1018.

Ullrich, R. C. (1977). Isozyme patterns and cellular differentiation in *Schizophyllum. Molecular and General Genetics*, **156,** 157–61.

Uno, I. & Ishikawa, T. (1976). Effect of cyclic AMP on glycogen phosphorylase in *Coprinus macrorhizus. Biochimica et Biophysica Acta*, **452,** 112–20.

Uno, I. & Ishikawa, T. (1978). Effect of cyclic AMP on glycogen synthetase in *Coprinus macrorhizus. Journal of General and Applied Microbiology*, **24,** 193–7.

Uno, I. & Ishikawa, T. (1982). Biochemical and genetic studies on the initial events of fruitbody formation. In *Basidium and Basidiocarp*, ed. K. Wells & E. K. Wells, pp. 113–23. New York, Heidelberg, Berlin: Springer-Verlag.

Valk, P. van der & Marchant, R. (1978). Hyphal ultrastructure in fruit-body primordia of the basidiomycetes *Schizophyllum commune* and *Coprinus cinereus. Protoplasma*, **95,** 57–72.

Valk, P. van der, Marchant, R. & Wessels, J. G. H. (1977). Ultrastructural localization of polysaccharides in the wall and septum of the basidiomycete *Schizophyllum commune. Experimental Mycology*, **1,** 69–82.

Valk, P. van der & Wessels, J. G. H. (1976). Ultrastructure and localization of wall polymers during regeneration and reversion of protoplasts of *Schizophyllum commune. Protoplasma*, **90,** 65–87.

Vries, O. M. H. de, Hoge, J. H. C. & Wessels, J. G. H. (1980*a*). Translation of RNA from *Schizophyllum commune* in a wheat germ and rabbit reticulocyte cell-free system. Comparison of *in vitro* and *in vivo* products after two-dimensional gel electrophoresis. *Biochimica et Biophysica Acta*, **607,** 373–8.

Vries, O. M. H. de, Hoge, J. H. C. & Wessels, J. G. H. (1980*b*). Regulation of the pattern of protein synthesis in *Schizophyllum commune* by the incompatibility genes. *Developmental Biology*, **74,** 22–36.

Wang, C.-S. & Raper, J. R. (1970). Isozyme patterns and sexual morphogenesis in *Schizophyllum. Proceedings of the National Academy of Sciences of the USA*, **66,** 882–9.

Wessels, J. G. H. (1966). Control of cell-wall glucan degradation during development of *Schizophyllum commune. Antonie van Leeuwenhoek*, **32,** 341–55.

Wessels, J. G. H. & Sietsma, J. H. (1979). Wall structure and growth in *Schizophyllum commune.* In *Fungal Walls and Hyphal Growth*, British Mycological Society Symposium 2, 1979, ed. J. H. Burnett & A. P. J. Trinci, pp. 27–48. Cambridge, UK: Cambridge University Press.

Zachariah, K. & Anderson, R. H. (1975). A fibrous inclusion body in developing ascospores of *Ascobolus stercorarius. Protoplasma*, **83,** 15–26.

Zadražil, F. (1975). Influence of CO_2 concentration on the mycelium growth on three *Pleurotus* species. *European Journal of Applied Microbiology*, **1,** 327–35.

Zantinge, B., Dons, H. & Wessels, J. G. H. (1979). Comparison of poly(A)-containing RNAs in different cell types of the lower eukaryote *Schizophyllum commune. European Journal of Biochemistry*, **101,** 251–60.

Zantinge, B., Hoge, J. H. C. & Wessels, J. G. H. (1981). Frequency and diversity of RNA sequences in different cell types of the fungus *Schizophyllum commune. European Journal of Biochemistry*, **113,** 381–9.

14

Physiology and ecology of rhythmic growth and sporulation in fungi

GERNOT LYSEK

Institute for Systematic Botany and Plant Geography FU, Altensteinstrasse 6, D-1000 Berlin, F. R. Germany

Types of rhythms

Rhythms help organisms to get organised in time, that is to synchronise their activities with those abiotic or biotic environmental factors which alter periodically. Well known examples are the synchronisation of predators with the activity of their prey, the flying of pollinators to opening flowers and daily stomatal movements.

Circadian rhythms

The rhythms which regulate these and many other activities are often termed 'circadian'. These and the similar circalunar and circatidal rhythms are widely known in eukaryotes, where they are often referred to as the 'biological clock'. According to Bünning (1977) they are characterised by an endogenous, temperature-compensated period of about (but not equal to) 24 h, which is synchronised to light-dark cycles, free-running in permanent dark, and often disappearing in continuous light.

Similar rhythms are also well known in fungi, though they are obviously less frequent. The best-studied example is the circadian conidiation of the mutant *band* of *Neurospora crassa*, which has been thoroughly tested for its circadian nature (Fig. 1) (Pittendrigh, Bruce, Rosensweig & Rubin, 1959; Sargent & Kaltenborn, 1972; Feldman, 1982). Very few further examples of circadian rhythms have been demonstrated in fungi. However, for *Sclerotinia fructigena*, out of about 200 isolates tested, at least four (from exhaustive tests) showed a circadian growth rhythm (Jensen & Lysek, 1983). Other cases probably include spore discharge in Pyrenomycetes and polypores (Hodgkiss & Harvey, 1969, 1971; Nuss, 1975), but in none of these cases did the authors critically examine whether discharge was circadian, that is, endogenously regulated, or diurnal.

Diurnal rhythms

Diurnal rhythms are, according to Bünning (1977), those which are created by the alternation of light, temperature or atmospheric humidity, and correspondingly have a period of 24 h. In fungi, such rhythms are exogenously caused by environmental variables affecting growing mycelia, and are expressed as concentric growth bands, for which the terms 'rings', 'bands', 'growth bands' or 'zonations' are often used. An example is shown in Fig. 2. This growth pattern occurs widely amongst fungi and has been observed from the time they were first cultured in the laboratory (Werner, 1898). Indeed it is now known that many laboratory cultures of fungi exhibit growth rhythms in light or temperature cycles. In some cases endogenous but non-circadian rhythms can even be superimposed on diurnal ones, as found in *Sclerotinia fructigena* (Jensen & Lysek, 1983). Thus one is tempted to conclude that diurnal rhythms are most common in fungi while in other eukaryotes circadian rhythms predominate.

Endogenous rhythms uncorrelated with environmental factors – rhythmic mutants

These rhythms are also typical for fungi and characterised by their complete independence of alternating environmental factors, that is they are neither triggered by nor synchronised with such factors.

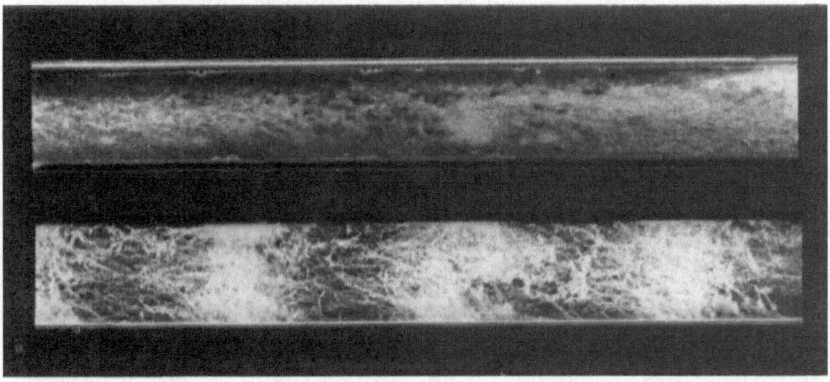

Fig. 1. Two cultures of *Neurospora crassa* in racing tubes grown in continuous dark. Above, the wild strain as control; below, the mutant band exhibiting circadian conidiation.

Typically these rhythms are also manifested as rings or bands as shown in Fig. 2. They are found mainly in the so-called 'rhythmic' or 'clock' mutants, strains which differ from wild types by exhibiting endogenous rhythms even in darkness and at constant temperatures. Such mutants

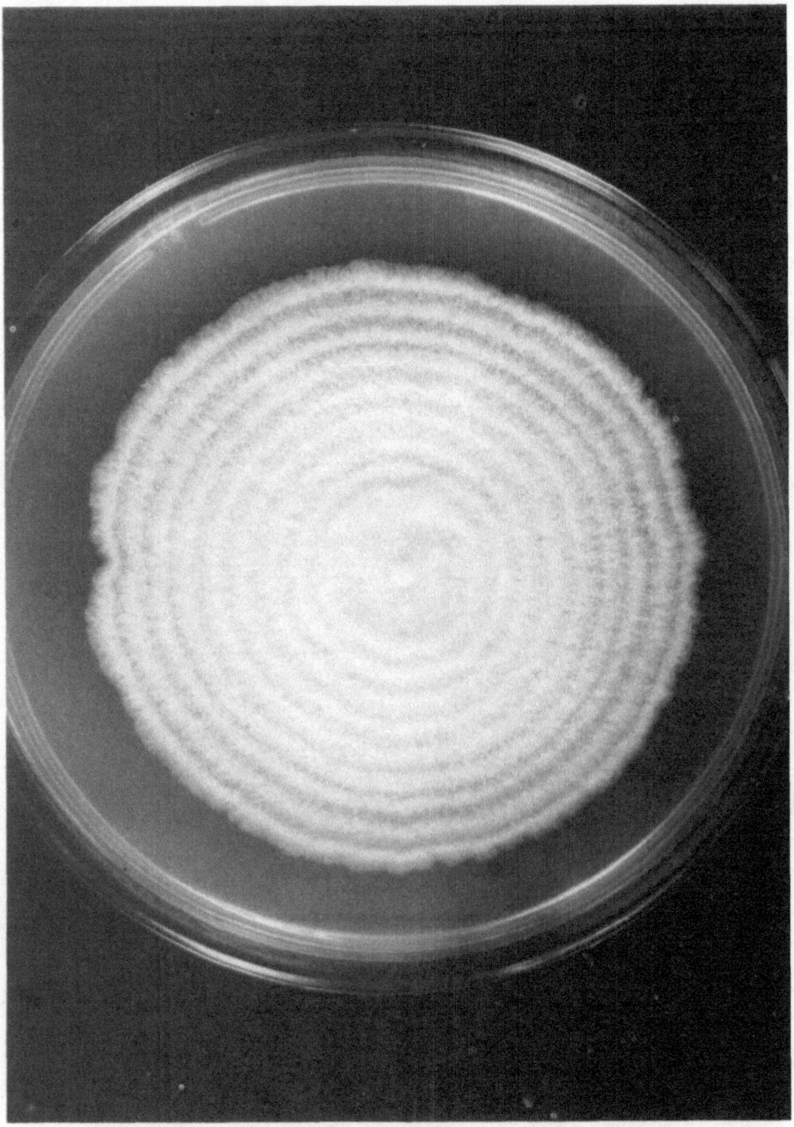

Fig. 2. Culture of *Sclerotinia fructigena* on a solid medium grown in 10:14 h light-dark cycles; the concentric bands indicating the growth rhythms are clearly seen. (From Jensen & Lysek, 1983.)

are unique to fungi and occur in, for example, *Neurospora crassa, Podospora anserina* and *Penicillium claviforme*. They are excellent as experimental objects since the wild type represents a control, allowing one to relate metabolic or other alterations to the observed rhythmic behaviour.

Endogenous, non-environmentally caused or synchronised rhythms, however, do not only occur under laboratory conditions. Lysek &

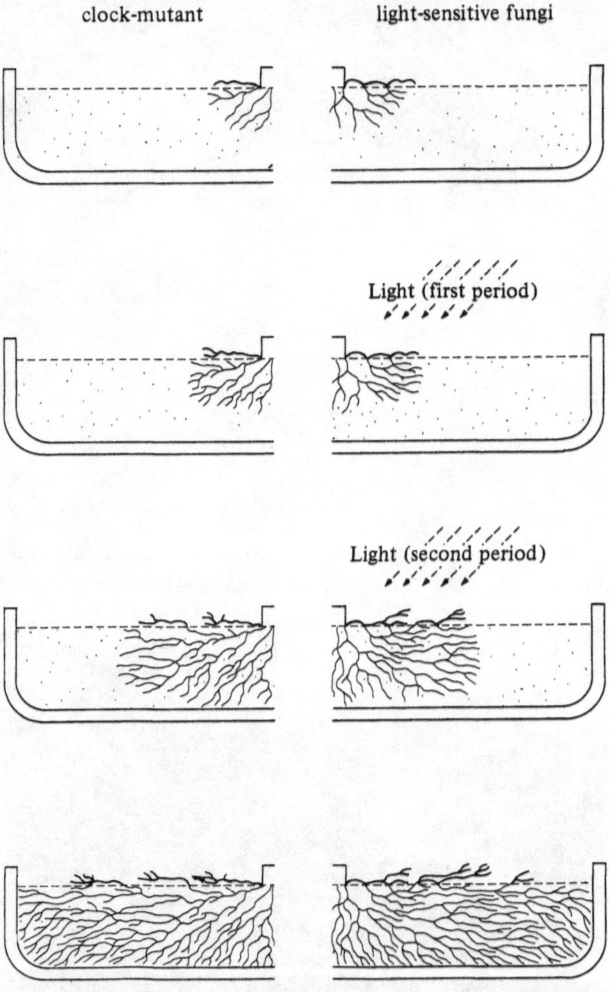

Fig. 3. The development of the mycelial bands shown in schematic cross-sections of Petri dish cultures. (After Lysek, 1972, 1978.) Left side, development due to an endogenous rhythm; right side, light-dependent growth rhythm.

Nordbring-Hertz (1981) found in *Arthrobotrys oligospora*, a nematophagous fungus, that traps were distributed in bands and hence formed periodically, but without any correlation to light- or temperature cycles.

Morphogenesis and fungal rhythms

What is happening in or to a fungus when a growth band is formed? We can consider this by reference to the clock mutant *zonata* of *Podospora anserina*, whose rhythms are well documented (Esser, 1969; Lysek, 1972, 1978; Kubicek & Lysek, 1982). The mutant exhibits well-marked growth bands, which are formed spontaneously.

The development of the bands in zonata

As is indicated in Fig. 3, the primary hyphae growing from the inoculum onto and into the medium are elongating normally. Later the hyphae close to the surface of the medium begin to stale and to develop more branches. This retards the mycelial front in the superficial layer of the colony. However, only the superficial hyphae alter their growth behaviour in this way; those within the medium continue to elongate, overtake the retarded front and penetrate the medium ahead, including the surface in front of the staling mycelium (Fig. 4). This entirely stops the mycelial front and hence gives rise to the first band. Thus bands are masses of staling hyphae replaced by new hyphae from within the medium.

This difference in growth behaviour in the colony is attributed to oxygen availability, since rhythmic growth can be associated with

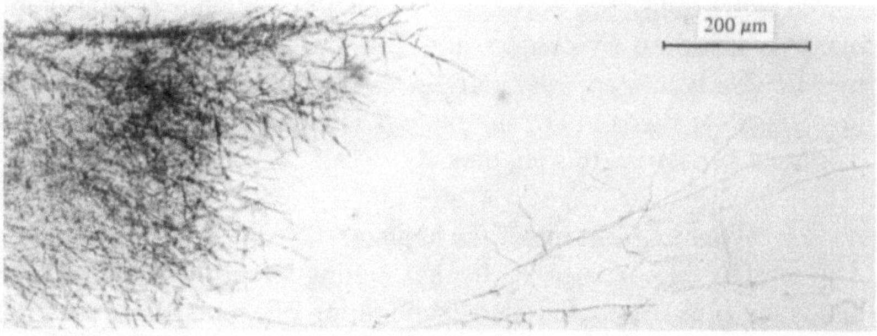

Fig. 4. Micrograph of a radial section of the edge of a band of the mutant *zonata* of *Podospora anserina*, showing the densely branched staling mycelium and the elongating hyphae. (From Kubicek & Lysek, 1982.)

enhanced oxygen consumption and/or reduced dry-weight production (Woodward & Sargent, 1973; Lysek & Esser, 1971; Lysek, 1976*a*).

Environmentally-induced rhythmic growth

As indicated above, more natural and hence more relevant are rhythms depending on or associated with alternating environmental factors, the most effective being the light-dark rhythm (Durand, 1976; Lysek, 1978). The latter may act upon a fungal colony to create a growth rhythm in the following way. As before, we can start with a colony growing from an inoculum, and primarily unaffected by light. After a period of growth, light is turned on and the most superficial hyphae, which are more exposed to light, are affected so that their elongation is reduced. They, too, consume more oxygen, since such consumption is increased with the cessation of hyphal growth. Those hyphae inside the medium, protected from the light by the medium itself, by other hyphae and subject to lack of oxygen, continue to elongate. As in the former example, these hyphae overtake the retarded mycelial front at the surface and, as soon as the inhibiting factor, light, ceases, grow ahead to give the new front which replaces the first one. Fig. 3 (right) shows these processes schematically.

With rhythmic mutants, as with light-dependent rhythmic growth, the gradient formed by the oxygen – alone or together with light – from the surface into the medium brings about different rates of elongation within the same mycelial front. Similar changes are brought about also by alterations of temperature or atmospheric humidity (Molz, 1907; Sagromsky, 1952). Thus, rhythmic growth results when different layers of a mycelial front grow under heterogeneous conditions.

The synchronisation between the individual hyphae to give complete bands is achieved by contact and anastomoses between neighbouring hyphae (Nguyen Van, 1962; Kraepelin & Francke, 1973). This effect is important for the study of the biological clock in fungi, but will not be discussed further in this chapter.

What happens inside the hyphae?

The importance of hyphal staling in producing the mycelial bands requires a more detailed examination of events inside the hyphae.

Agents inducing rhythmic growth

Table 1 lists substances known to induce rhythmic growth in fungi. The main groups are surface-active substances, glucose analogues, inhibitors of protein synthesis and inorganic cations. A common site of attack by all these, including the inhibitors of protein synthesis, may be in the hyphal membrane systems where they alter the structure and/or the permeability and hence the fluxes of ions and molecules. The importance of the hyphal membranes in rhythmic growth is indicated by the 'clock' mutant *zonata*, which is lipid-deficient (Lysek, 1976*a*); other clock mutants of *Podospora anserina* contain altered concentrations of K^+ and Na^+ (Lysek & Jennings, 1982). The effect of light in causing mycelial bands may also be explained by an effect on membrane permeability, as shown by Mani & Swamy (1981) and Rayfield (1982). The role of ionic fluxes in biological rhythms was stressed by Njus, Sulzman & Hastings (1974), who proposed ionic fluxes as the basic mechanism of the biological clock.

Hyphal growth and ionic movements

Harold (1977) and Jennings (1979) have stressed that movements of ions across the hyphal membranes are vital for polar hyphal elongation. Protons are quantitatively dominant in these movements, either being extruded from the cytoplasm into the medium surrounding the hyphae, or being taken up by the mitochondria to build up the electrochemical gradient allowing the chemi-osmotic formation of ATP (Mitchell, 1974).

The active extrusion of protons is achieved by an ATPase which transports protons and potassium and sodium ions by an electrogenic counter-transport (Slayman, 1965; Slayman & Gradmann, 1975).

The extruded protons are involved mainly in the uptake of solutes by proton symport (Komor, 1974; Komor, Rotter & Tanner, 1977), or they leak back across the plasmalemma. All these solute movements are summarised in Fig. 5.

Now, if the permeability of the plasmalemma is enhanced by one of the treatments mentioned above, the influx of protons is enhanced. Since the proton gradients are vital, the cell reacts by an increase in the turnover of the ATPase to extrude the excess protons. Hence more K^+ is accumulated, while the sodium ions, which are extruded in competition with the protons, remain inside (Lysek & Jennings, 1982). The enhanced energy consumption caused by the enhanced proton extrusion leads to the increased dependence on oxygen uptake of band formation, as mentioned above.

Table 1. *Substances which induce rhythmic growth or fructification in fungi*

Fungi	Substances	Type of rhythm	Authors
Hexose-analogues			
Neurospora crassa	L-sorbose	colonial growth	Crocken & Tatum, 1968
Podospora anserina	D-galactose, L-sorbose, 2-deoxy-D-glucose	mycelial bands	Lysek & Esser, 1971
Cations			
Aspergillus ochraceus	hydrogen ions	rings of conidiophores	Munk, 1912
Penicillium sp.			
Aspergillus niger	K^+ (+ glucose)	rings of conidiophores (endogenous)	Jerebzoff, Jerebzoff-Quintin & Lambert, 1976
Neurospora crassa	Rb^+, (K^+)	circadian conidiation	Gall & Lysek, 1981
Podospora anserina	hydrogen ions, all alkali-cations	zonated growth	Tavlitzki, 1954; Esser, 1969; Lysek & Schrüfer, 1981
Sphaerostilbe repens	Ca^{2+}	diurnal rings of sclerotia	Botton, 1978
Inhibitors of protein synthesis[a]			
Penicillium claviforme	cycloheximide, chloramphenicol, tetracyclin	endogenous bands of coremia	Sagromsky, 1976
Podospora anserina	cycloheximide, chloramphenicol	bands of aerial mycelia	Lysek, 1971
Neurospora crassa	cycloheximide, chloramphenicol	circadian conidiation	Baráthova, Betina & Nemec, 1969
Amino acids[b]			
Aspergillus niger	isoleucine	endogenous periodic sporulation	Jerebzoff & Lambert, 1969
Colletotrichum lindemuthianum	DL-serine	endogenous periodic sporulation	Jerebzoff, 1965

Sclerotium rolfsii	L-threonine	rings of sclerotia and mycelial bands	Kritzman, Chet & Henis, 1977
Membrane-affecting agents			
Podospora anserina	ionophores (monensin, nigericin, nonactin)	diurnal mycelial bands	Lysek & von Witsch, 1974
	alcohols (1-propanol, 1-butanol, 1-octanol)	zonations	Lysek & von Witsch, 1974
	detergents (Na-deoxycholate, Na-dodecylsulphate, brej-35, triton-X-100)	zonations	Lysek & von Witsch, 1974
	uncouplers (DNP)	zonations	Lysek, 1971
Penicillium claviforme	alcohols (isopropanol, tert.-amylalcohol)	endogenous periodic sporulation	Faraj-Salman, 1971
	detergents (Na-deoxycholate)	endogenous periodic sporulation	Faraj-Salman, 1971
	uncouplers (CCCP)	endogenous periodic sporulation	Sagromsky, 1976
Neurospora crassa	Na-dodecylsulphate	circadian conidiation	Cramer, Varchmin & Lysek, unpublished
Other inhibitors			
Penicillium claviforme	phenylenediamine	endogenous periodic sporulation	Sagromsky, 1976

[a] Feldman and co-workers (Nakashima, Perlman & Feldman, 1981) found an influence of cycloheximide and other substances on phase-shift behaviour and period length in the mutant band of *Neurospora crassa*.
[b] Berliner, Neurath & Yankovich (1965) found an influence of various amino acids on the rhythms of *Ascobolus* clock mutants.

Hyphal staling and loss of polarity

To understand why these altered fluxes lead to partial staling of the hyphae and hence to rhythmic growth, we have to examine further the scheme of Jennings (1979). In the apical region of the hyphae the ATPase is not yet incorporated into the membrane. Here the influx of protons and sodium ions is not balanced by their extrusion, and the potassium ions, which are moving down the hyphae, may move out into the relatively potassium-poor substrate (Fig. 5). In normally-growing hyphae this creates a marked difference between the hyphal tip and

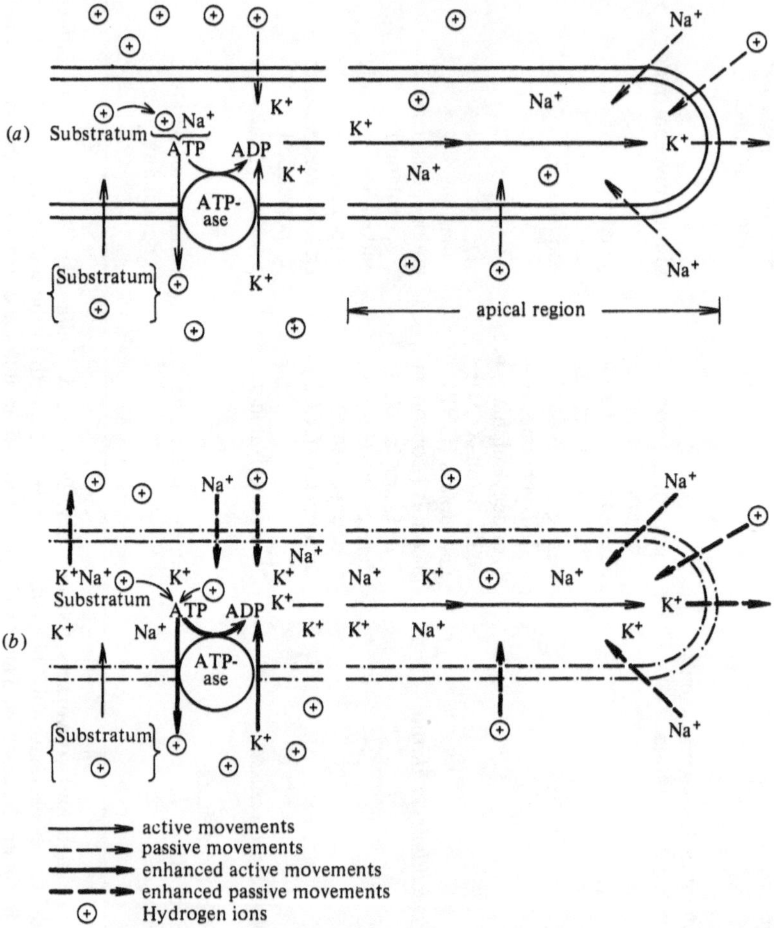

Fig. 5. Diagram showing the movements of protons, potassium and sodium ions in hyphae of a uniformly-growing (above) and of a rhythmically-growing colony (below). (Adapted from Jennings, 1979.)

those parts of the hypha further back, which results in a clear polarity along which the potassium ions and the vesicles stream into the apical region.

With greater permeability K^+ ions may leak out not only at the apex but in other regions too, while more sodium ions remain inside, as indicated above. Thus the polarity is disturbed in such a way that apical growth is affected or ceases. These alterations are summarised in Fig. 5. Experimental evidence for this scheme comes from the fact that potassium, if given in excess, leads to rhythmic growth; the polar gradient into the apical region is interrupted by the K^+ ions moving into the hyphal cytoplasm at the apical region (Lysek & Schrüfer, 1981). Marked alterations in the ion movements and contents in the clock-strains of *Podospora anserina* have been measured by Lysek & Jennings (1982).

The disappearance of the 'spitzenkörper'

Girbardt (1969) showed that the first event when hyphae stale is that the 'spitzenkörper', the organelle found in the very tip of growing hyphae of higher fungi, disappears. This is accompanied by loss of the vesicle flow into the hyphal tip. Since these vesicles transport the precursors of wall and membrane materials into the tip (Bartnicki-Garcia, 1969) an interruption of the flow also stops hyphal elongation.

Cessation of substrate uptake

Substrate uptake, for example of glucose, depends via a proton symport on the effectiveness of the ATPase system (Komor, 1974; Tarshis & Kapitanov, 1978). A hypha which is no longer elongating also loses its uptake system as new substrates are no longer available. This means that the uptake system (mainly the so-called high affinity system) is broken down (Komor, 1974). Hill (1976) found that light inhibited growth and glucose uptake of *Aspergillus ornatus* simultaneously. After such a cessation of growth a hypha now has to maintain its metabolism by using its endogenous reserves which are mainly polysaccharides.

Development of the checked hyphae

The staling of all or of part of the hyphae and the loss of polarity do not mean that these hyphae are incapable of further development. Rather they undergo a sequence of different events from the original pattern of elongation, since the staled hyphae are those which are capable of further developments or differentiation. These hyphae have terminated their elongation phase or 'trophophase' (Bu'Lock, 1967), and now 'switched off their vegetative growth' (Turian, 1975). They are now capable of differentiation which in most cases means reproduction, and hence are in their 'idiophase' (Bu'Lock, 1967). This staling is so fundamental for the further development that Durand (1976) argued that 'the morphogenetic effect of light in fungi might be often its checking of growing hyphae'. That the formation of reproductive organs is conditioned by a check to growth has also been found by Robinson (1926), Buston & Rickard (1956), Hawker (1966), Pollock (1973) and Lysek (1976*b*) (see also Raudaskoski: Chapter 13).

Further metabolic alterations in checked hyphae. As indicated, the switch from the metabolism of the growing hyphae which continuously penetrate into new medium, into the state of the resting hyphae is signalled by outward movement of substrate from the endogenous pool. Now, with the start of sporulation the metabolic flow through glycolysis, the tricarboxylic acid cycle, and respiration is enhanced. Experimental evidence for this has been found in *Chaectomium globosum* by Buston, Jarbar & Etheridge (1953) who found a correlation of mycelial sugar phosphate concentration and fruiting; in *Neurospora crassa* by Hanks & Sussman (1969) and in *Aspergillus niger* by Smith & Valenzuela-Perez (1971). These metabolic pathways produce the precursors for amino acids, and hence proteins, and for the nucleic acids, the synthesis of which is enhanced with sporulation (Dicker, Oulevy & Turian, 1969; Valenzuela-Perez & Smith, 1971; Viswanath-Reddy & Turian, 1975).

In rhythmically growing and hence periodically staling fungal colonies this correlation of metabolic alterations and sporulation has also been established. In the circadian conidiating mutant *band* of *Neurospora crassa*, Woodward & Sargent (1973) reported a simultaneous periodicity in CO_2 production; Delmer & Brody (1975) found an alternating energy charge while Martens & Sargent (1974) found synchronous rhythms in the contents of DNA and RNA. In *Podospora anserina* which forms perithecia in dikaryons only and no conidia, monokaryotic clock strains also contain enhanced concentrations of phosphorylated metabolic intermediates (Lysek & Bornefeld, 1971; Bornefeld & Lysek, 1972).

Mycelial growth rhythms and differentiation

Against this background of development from elongating to differentiating hyphae, we can ask what is the value of rhythms or mycelial growth periodicities?

Correlation of mycelial bands and rings of reproductive structures

From the accumulation of staled hyphae in the mycelial bands one might expect differentiating hyphae also to be distributed in bands. This phenomenon is well exemplified by the formation of rings of sporodochia in *Sclerotinia*, or the lines of sporodochia or perithecia of *Nectria* on wood, or the fairy rings of mushrooms and toadstools. In all these cases, however, the reproductive structures are not only formed in circles (fairy rings) but they are also formed periodically according to the mycelial growth rhythm. This is readily visible in *Sclerotinia fructigena*, where the obvious rings of sporodochia have led to the name 'ring rot'. It is, on the other hand, difficult to recognise with mushrooms in fairy rings, where only continuous observation reveals the periodic formation in concentric rings (Ingold, 1974; Dickinson, 1979).

It is reasonable to assume – though one cannot always demonstrate the fact – that these structures are formed by the mycelia and hence that their periodic formation depends on periodic events in the mycelium. These periodic events may well involve staling of hyphae, as observed in cultures. In *Sclerotinia fructigena* laboratory experiments have revealed a mycelial growth rhythm, which is mainly light-dark dependent and synchronous with conidiation (Jensen & Lysek, 1983). Figure 6 shows a direct correlation between the mycelial growth bands and the distribution of perithecia in two Ascomycetes (Lysek & von Witsch, 1974). In *Podospora anserina* the induction of a mycelial growth rhythm caused the formation of perithecia in the zonated parts of the colony, while the controls only formed perithecia near the margin. Other examples include the formation of sclerotia in mycelial rings, which was observed as early as 1908 by Reidemeister (see also Kritzman, Chet & Henis, 1977); or by Stevens & Hall (1909) with *Ascochyta chrysanthemi*. This close correlation between staling of hyphae due to mycelial rhythms and reproduction may explain the importance of rhythmic growth for fungal colonies.

Rhythms as a creation of differentiable hyphae

From the point of view of reproduction the repeatedly formed bands may be regarded as pools of hyphae initiating their idiophase, that is, which are now capable of differentiation. Thus, rhythmic growth provides the fungal colony with a large number of such hyphae and enhances the total number of reproductive units. In *Podospora anserina* cultures of a dikaryotic strain induced to rhythmic growth by low pH in the medium produced ten times more mature perithecia than the controls (G. Lysek, unpublished). Similar results were obtained with conidia of *Trichoderma viride* by K. Schrüfer (unpublished).

Rhythmic growth thus provides a mechanism for the fungal colony whereby the elongating hyphae in one part are separated from that which is staled and hence induced to initiate differentiation.

Adaptation to time

The generation of hyphae commencing differentiation starts with the very first band formed by the colony. While a uniformly growing colony needs a barrier or another external influence to generate such hyphae, a rhythmically growing colony produces hyphae which are checked and hence differentiating throughout its life span. This means, in addition to the mere enhancement of the number of propagules formed, there is a more even distribution in time, since reproduction is extended to the entire lifetime of the colony.

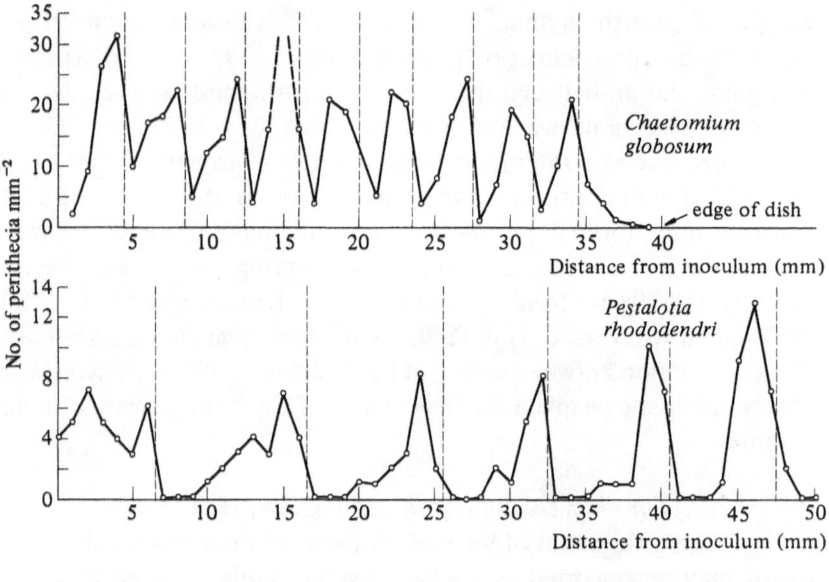

Fig. 6. Formation of mycelial bands (the edges of which are marked by the vertical dashed lines) and perithecia in *Chaetomium globosum* and *Pestalotia rhododendri*. (From Lysek & von Witsch, 1974.)

Optimising adverse conditions

As stated above, rhythmic growth results when different layers of a mycelial front grow under heterogenous conditions. Since the part of the mycelium which is affected by more unfavourable conditions forms the mycelial band and later on the differentiated structures, any unfavourable conditions are exploited by the colony in the initiation of sporulation. The examples of *Sclerotinia fructigena* or *Marasmius oreades* (Ingold, 1974) show that light or soil temperature acting differentially on the mycelia may induce rhythmic growth and hence the formation of reproductive organs. Fletcher (1976) wrote that ring production is a fundamental response to environmental change.

Other ways in which checked hyphae are produced

One may argue that a considerable number of fungi do not show rhythms but, nevertheless, form reproductive units in sufficient numbers. In these cases – wherever it is observable – the fungus uses other means to provide staled hyphae prior to reproduction. Some of these means are as follows.

Limited growth. This is the best known and most well studied; the colony is self-inhibited and thus the entire mycelium after a period of growth is converted to hyphae capable of differentiation (Gottlieb, 1971, 1975). Such a situation is well known in *Penicillium* spp. and Oomycetes.

Limitations or exhaustion of the substratum. Often the cessation of hyphal growth occurs at the edge of the substratum, when it is quite small or exhausted. Fungi, growing for example in dung, reach its surface after having penetrated and exploited it. Here they are checked and hence form reproductive structures. Also, in cases where the edge of the substratum may be less marked, the growth of the mycelium from a nutrient-rich to a nutrient-poor site may cause some of the hyphae to stale. This cessation of hyphal growth and subsequent sporulation as a result of growing out of a nutrient-rich into a nutrient-poor substratum was demonstrated by Lysek (1976*b*) with *Podospora anserina* which when grown initially on a rich medium, formed perithecia after growing for some centimetres into a water agar. Here the bulk of the mycelial front staled and formed perithecia (Fig. 7).

Barriers (including substratum/air interface). A common means of staling hyphae are barriers, where the hyphal elongation is stopped abruptly. These may be the margin of the Petri dish, where many cultured fungi sporulate. In nature these may be stones in the soil – or pieces of wood or roots. It may be a ditch, which interrupts the soil layer – and where mushrooms and toadstools are often found.

In contrast to all such cases rhythmic growth provides a means to allow differentiation even in those habitats where nutrient resources may be regarded as 'unlimited', if considered in relation to the size of the fungal colonies, as in non-component-restricted fungi of litter and soil (see Boddy: Chapter 12). Sadowska (1978), who studied the Agaricales in the Polish lowlands, reported that most of the species formed fairy rings.

Conclusions

Fungal growth rhythms should be regarded as a special example of biorhythms which are exogenously triggered and – under natural conditions – control the development of fungal colonies and thus the life cycle of many species. The entrainment to other changing factors of the environment like light or temperature, or via these factors to the activities of other organisms is of minor importance in fungi – which may explain the infrequent occurrence of endogenous rhythms. This distinguishes this group from other eukaryotes which use rhythms to start and synchronise their activities with the abiotic or biotic variables of their

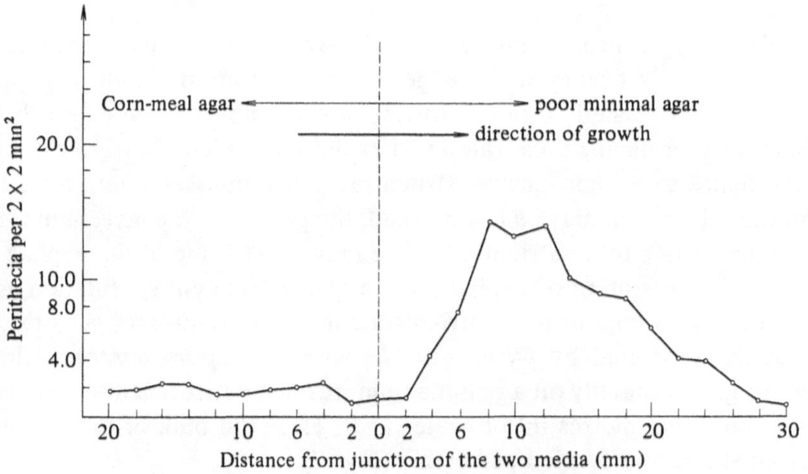

Fig. 7. Formation of perithecia in a colony growing from a rich into an empty medium. (After Lysek, 1976*b*.)

environment. This requires endogenous rhythms, which are hence universally found in animals and higher plants.

The author express his thanks to all who helped him in preparing the manuscript, especially Mrs I. Eggert, Mrs S. Salam and Mr H. Lünser.

References

Baráthova, H., Betina, V. & Nemec, P. (1969). Morphological changes induced in fungi by antibiotics. *Folia Microbiologica*, **14**, 475–83.

Bartnicki-Garcia, S. (1969). Cell wall differentiation in the phycomycetes. *Phytopathology*, **59**, 1065–72.

Berliner, M. D., Neurath, P. W. & Yankovich, B. M. (1965). The rhythms of *Ascobolus* clock mutants. *American Journal of Botany*, **52**, 635–6.

Bornefeld, T. & Lysek, G. (1972). Rhythmic mycelial growth in *Podospora anserina* V. The levels of phosphorylated intermediates. *Archiv für Mikrobiologie*, **87**, 119–28.

Botton, B. (1978). Growth of aggregated organs in *Sphaerostilbe repens*. *Canadian Journal of Microbiology*, **24**, 1039–47.

Bu'Lock, J. D. (1967). *Essays in Biosynthesis and Microbial Development*. New York, London: J. Wiley & Sons.

Bünning, E. (1977). *Die Physiologische Uhr; circadiane Rhythmik und Biochronometrie*. Berlin, Heidelberg, New York: Springer-Verlag.

Buston, H. W., Jarbar, A. & Etheridge, D. E. (1953). The influence of hexose phosphates, calcium and jute extract on the formation of perithecia by *Chaetomium globosum*. *Journal of General Microbiology*, **8**, 302–6.

Buston, H. W. & Rickard, G. (1956). The effect of a physical barrier on sporulation of *Chaetomium globosum*. *Journal of General Microbiology*, **15**, 194–7.

Crocken, B. & Tatum, E. L. (1968). The effect of sorbose on metabolism and morphology of *Neurospora*. *Biochimica et Biophysica Acta*, **156**, 1–8.

Delmer, D. P. & Brody, S. (1975). Circadian rhythms in *Neurospora crassa*: oscillation in the level of adenine nucleotide. *Journal of Bacteriology*, **121**, 548–53.

Dicker, J. W., Oulevey, N. & Turian, G. (1969). Amino acid induction of conidiation and morphological alterations in wild type and morphological mutants of *Neurospora crassa*. *Archiv für Mikrobiologie*, **65**, 241–57.

Dickinson, C. H. (1979). Fairy rings in Norfolk. *Bulletin of the British Mycological Society*, **13**, 91–4.

Durand, R. (1976). Effect of light on reproductive processes of fungi – Hypotheses concerning identity of photoreceptors. *Mycopathologia*, **60**, 3–16.

Esser, K. (1969). The influence of pH on rhythmic mycelial growth in *Podospora anserina*. *Mycologia*, **61**, 1008–11.

Faraj-Salman, A.-G. (1971). Zur Induktion einer endogenen Rhythmik bei Mutanten des Pilzes *Penicillium claviforme* Bainier. *Archiv für Protistenkunde*, **113**, 306–13.

Feldman, J. F. (1982). Genetic approaches to circadian clocks. *Annual Review of Plant Physiology*, **33**, 583–608.

Fletcher, H. J. (1976). Some conditions favouring the production of growth rings in *Graphium putredinis*. *Transactions of the British Mycological Society*, **66**, 552–4.

Gall, A. & Lysek, G. (1981). Induction of circadian conidiation by rubidium chloride. *Neurospora Newsletter*, **28**, 13.

Girbardt, M. (1969). Die Ultrastruktur der Apikalregion von Pilzhyphen. *Protoplasma (Wien)*, **67**, 413–41.
Gottlieb, D. (1971). Limited growth in fungi. *Mycologia*, **63**, 619–29.
Gottlieb, D. (1975). The ageing process in the fungus *Rhizoctonia solani. Phytophylactica (Pretoria)*, **7**, 81–90.
Hanks, D. L. & Sussmann, A. S. (1969). The relation between growth, conidiation and trehalase activity in *Neurospora crassa. American Journal of Botany*, **56**, 1152–9.
Harold, F. M. (1977). Ion currents and physiological functions in microorganisms. *Annual Review of Microbiology*, **31**, 181–203.
Hawker, L. E. (1966). *Environmental Influences on Reproduction.* In *The Fungi*, vol. II, ed. G. C. Ainsworth & A. S. Sussman, pp. 435–72. New York, London: Academic Press.
Hill, E. P. (1976). Effect of light on growth and sporulation of *Aspergillus ornatus. Journal of General Microbiology*, **95**, 39–44.
Hodgkiss, I. J. & Harvey, R. (1969). Spore discharge rhythms in pyrenomycetes. VI. The effects of climatic factors on seasonal and diurnal periodicities. *Transactions of the British Mycological Society*, **52**, 355–63.
Hodgkiss, I. J. & Harvey, R. (1971). Effects of temperature on spore discharge rhythms in pyrenomycetes. *Transactions of the British Mycological Society*, **56**, 225–34.
Ingold, C. T. (1974). Growth and death of a fairy ring. *Bulletin of the British Mycological Society*, **8**, 74–5.
Jennings, D. H. (1979). *Membrane Transport and Hyphal Growth.* In *Fungal Walls and Hyphal Growth*, ed. H. Burnett & A. P. J. Trinci, pp. 279–94. Cambridge: Cambridge University Press.
Jensen, C. & Lysek, G. (1983). Differences in the mycelial growth rhythms in a population of *Sclerotinia fructigena* (Pers.) Schröter. *Experientia*, **39**, 1401–2.
Jerebzoff, S. (1965). *Manipulation of some oscillating systems in fungi by chemicals.* In *Circadian Clocks*, ed. J. Aschoff, pp. 183–9. Amsterdam: North-Holland Publishing.
Jerebzoff, S., Jerebzoff-Quintin, S. & Lambert, E. (1976). L'induction du rhythme endogène de sporulation chez *Aspergillus niger*: rôles du rapport glucose/potassium et de micro-éléments. *Physiologia Plantarum*, **36**, 279–86.
Jerebzoff, S. & Lambert, E. (1969). Chemical induction of an endogenous zonation rhythm in *Aspergillus niger* Van Tieghem: specificity of the action of isoleucin. *Physiologie Végétale*, **7**, 181–9.
Komor, E. (1974). Proton-coupled hexose transport in *Chlorella vulgaris. FEBS Letters*, **38**, 16–18.
Komor, E., Rotter, M. & Tanner, W. (1977). Proton-cotransport system in a higher plant–sucrose transport in *Ricinus communis. Plant Science Letters*, **9**, 153–62.
Kraepelin, G. & Francke, G. (1973). Self-synchronization in yeast and other fungi. *International Journal of Chronobiology*, **1**, 163–72.
Kritzman, G., Chet, I. & Henis, Y. (1977). The relationship between rhythmic hyphal growth and circadian formation of sclerotia in *Sclerotium rolfsii. Canadian Journal of Microbiology*, **23**, 959–63.
Kubicek, R. & Lysek, G. (1982). Morphogenesis of growth bands in the clock-mutant zonata of *Podospora anserina. Transactions of the British Mycological Society*, **79**, 167–70.
Lysek, G. (1971). Rhythmic mycelial growth in *Podospora anserina.* III. Effect of metabolic inhibitors. *Archiv für Mikrobiologie*, **78**, 330–40.
Lysek, G. (1972). Rhythmic mycelial growth in *Podospora anserina.* VI. An attempt to

elucidate the growth pattern of a clock mutant. *Archiv für Mikrobiologie*, **87**, 129–37.

Lysek, G. (1976*a*). Alterations of the phospholipids in a rhythmically growing mutant of *Podospora anserina*. *Biochemie und Physiologie der Pflanzen*, **169**, 207–12.

Lysek, G. (1976*b*). Formation of perithecia in colonies of *Podospora anserina*. *Planta*, **133**, 81–3.

Lysek, G. (1978). Circadian rhythms. In *The Filamentous Fungi*, vol. 3, *Developmental Mycology*, ed. J. E. Smith & D. R. Berry, Pp. 376–88. London: Edward Arnold.

Lysek, G. & Bornefeld, T. (1971). Phosphorus content of wild type and 'clock' mutants of *Podospora anserina*. *Neurospora Newsletter*, **18**, 18.

Lysek, G. & Esser, K. (1971). Rhythmic mycelial growth in *Podospora anserina* (Ascomyc.) II. Evidence for a correlation with carbohydrate metabolism. *Archiv für Mikrobiologie*, **75**, 360–71.

Lysek, G. & Jennings, D. H. (1982). Enhanced K^+, Na^+ and Mg^{2+} contents in clock mutants of *Podospora anserina* and a role of protons in rhythmic mycelial growth. *Physiologie Végétale*, **20**, 433–41.

Lysek, G. & Nordbring-Hertz, B. (1981). An endogenous rhythm of trap formation in the nematophagous fungus *Arthrobotrys oligospora*. *Planta*, **152**, 50–3.

Lysek, G. & Schrüfer, K. (1981). Wuchsrhythmen bei *Podospora anserina* (Ascomycetes): Wirkung von Alkali-Ionen. *Berichte der Deutschen botanischen Gesellschaft*, **94**, 105–12.

Lysek, G. & Witsch, H. von (1974). Rhythmisches Mycelwachstum bei *Podospora anserina*. VII. Der Einfluß oberflächenaktiver Substanzen und Antibiotika im Dunkeln und im Licht. *Archives of Microbiologie*, **97**, 227–37.

Mani, K. & Swamy, R. N. (1981). Light-induced changes in permeability in two groups of fungi. *Experimental Mycology*, **5**, 292–4.

Martens, C. L. & Sargent, M. L. (1974). Circadian rhythms of nucleic acid metabolism in *Neurospora crassa*. *Journal of Bacteriology*, **117**, 1210–15.

Mitchell, P. (1974). A chemiosmotic molecular mechanism for proton-translocating adenosine triphosphatases. *FEBS Letters*, **43**, 189–94.

Molz, E. (1907). Über die Bedingungen der Entstehung der durch *Sclerotinia fructigena* erzeugten 'Schwarzfäule' der Äpfel. *Zentralblatt für Bakteriologie 2. Abteilung*, **27**, 175–88.

Munk, M. (1912). Bedingungen der Hexenringbildung bei Schimmelpilzen. *Zentralblatt für Bakteriologie 2. Abteilung*, **32**, 353–75.

Nakashima, H., Perlman, J. & Feldman, J. F. (1981). Cycloheximide-induced phase shifting of the circadian clock of *Neurospora*. *American Journal of Physiology*, **241**, R31–R35.

Nguyen Van, H. (1962). Rôle des facteurs internes et externes dans la manifestation de rhythms de croissance chez l'Ascomycète *Podospora anserina*. *Compte Rendu Hebdomadaire des Séances de l'Académie des Sciences, Paris*, **254**, 2646–8.

Njus, D., Sulzman, F. M. & Hastings, J. W. (1974). Membrane model for the circadian clock. *Nature*, **248**, 116–20.

Nuss, I. (1975). Zur Ökologie der Porlinge – Untersuchungen über die Sporulation einiger Porlinge und die an ihnen gefundenen Käferarten. *Bibliotheca Mycologica*, **45**, Vaduz: Cramer-Verlag FL-9490.

Pittendrigh, C. S., Bruce, V. G., Rosensweig, N. S. & Rubin, M. L. (1959). A biological clock in *Neurospora*. *Nature*, **184**, 169–70.

Pollock, R. T. (1973). Environmental factors affecting the pattern of perithecium development in *Sordaria fimicola* on agar medium. *Bulletin of the Torrey Botanical Club*, **100**, 78–83.

Rayfield, G. W. (1982). Kinetics of the light-driven proton movement in model membranes containing bacteriorhodopsin. *Biophysical Journal*, **38**, 79–84.

Reidemeister, A. (1908). Die Bedingungen der Sklerotien- und Sklerotienringbildung von *Botrytis cinerea* auf künstlichen Nährböden. Doctoral thesis Halle.

Robinson, W. (1926). The conditions of growth and development of *Pyronema confluens* Tul. (*P. omphaliodes* (Bull.) Fuckel). *Annals of Botany*, **60**, 245–72.

Sadowska, B. (1978). 'Fairy rings' on lowland meadows. *Proceedings of the 7th congress of European Mycologists*, Budapest, p. 51.

Sagromsky, H. (1952). Der Einfluß des Lichtes auf die rhythmische Konidienbildung von *Penicillium. Flora (Jena)*, **139**, 300–13.

Sagromsky, H. (1976). Induktion einer endogenen Rhythmik bei Mutanten des Pilzes *Penicillium claviforme* Bainier durch membranwirksame Stoffe. *Beiträge zur Biologie der Pflanzen*, **52**, 383–92.

Sargent, M. L. & Kaltenborn, S. (1972). Effects of medium composition and carbon dioxide on circadian conidiation in *Neurospora. Plant Physiology*, **50**, 171–5.

Slayman, C. L. (1965). Electrical properties of *Neurospora crassa*; respiration and the intracellular potential. *Journal of General Physiology*, **49**, 93–116.

Slayman, C. L. & Gradmann, D. (1975). Electrogenic proton transport in the plasma membrane of *Neurospora. Biophysical Journal*, **15**, 968–71.

Smith, J. E. & Valenzuela-Perez, J. (1971). Changes in intracellular concentrations of glycolytic intermediates and adenosine phosphates during growth cycle of *Aspergillus niger. Transactions of the British Mycological Society*, **57**, 103–10.

Stevens, F. L. & Hall, J.-G. (1909). Variation of fungi due to environment. *Botanical Gazette*, **48**, 1–30.

Tarshis, M. A. & Kapitanov, A. B. (1978). Symport H^+/carbohydrate transport into *Acholeplasma laidlawii* cells. *FEBS Letters*, **89**, 73–6.

Tavlitzki, J. (1954). Sur la croissance de *Podospora anserina* en milieu synthétique. *Compte Rendu Hebdomadaire des Séances de l'Académie des Sciences, Paris*, **238**, 2341–43.

Turian, G. (1975). Differentiation in *Allomyces* and *Neurospora. Transactions of the British Mycological Society*, **64**, 367–80.

Valenzuela-Perez, J. & Smith, J. E. (1971). Role of glycolysis in sporulation of *Aspergillus niger* in submerged cultures. *Transactions of the British Mycological Society*, **57**, 111–19.

Viswanath-Reddy, M. & Turian, G. (1975). Physiological changes during protoperithecial differentiation in *Neurospora tetrasperma. Physiologia Plantarum*, **35**, 166–74.

Werner, C. (1898). Die Bedingungen der Conidienbildung bei einigen Pilzen. Doctoral Thesis Basel and Frankfurt/M.

Woodward, D. O. & Sargent, M. L. (1973). Circadian rhythms in *Neurospora. Behaviour of Micro-Organisms, Proceedings of the 10th International Congress for Microbiology, Mexico City, 1973*, ed. A. Pérez-Miravete, pp. 282–96.

15

Senescence in *Podospora anserina* and its implication for genetic engineering

KARL ESSER, ULRICH KÜCK, ULF STAHL and PAUL TUDZYNSKI

Lehrstuhl für Allgemeine Botanik, Ruhr-Universität, Postfach 102148, D-4630 Bochum 1, Germany

Senescence or ageing is a basic biological phenomenon that concerns all living beings. With different levels of biological organisation in various organisms this syndrome has different phenotypes. But all life forms are prone via irreversible alterations of their metabolism to an endpoint of senescence, cellular death. In higher organisms the onset of senescence is marked either by a decrease in somatic growth or by a failure of either vegetative or sexual propagation or both. In single-celled organisms or even in single cells of higher organisms in culture ageing is seen as cessation of cellular division. It is most unlikely that a single principle or one causative agent can explain the versatility of ageing syndromes (for reviews see Behnke, Finch & Moment, 1979; Comfort, 1974; Curtis, 1971; Hayflick, 1975; Platt, 1976; Strehler *et al.*, 1971).

Although senescence has been investigated mainly in animals and animal cells in culture, fungi are very well suited for similar studies not only because of the ease with which they can be handled in the laboratory, but also because of their accessibility to genetic analysis. Fungi, therefore, are highly appropriate for exploration of the genetics and physiology of senescence. This is especially true for the ascomycetous fungus *Podospora anserina* for which a cytoplasmically inherited senescence was discovered more than twenty years ago by Rizet and Marcou (Marcou, 1954*a,b*, 1957, 1958, 1961; Marcou & Schecroun, 1959; Rizet, 1953*a,b*, 1957; Rizet & Marcou, 1954; and for review see Esser & Kuenen, 1967).

All strains of *Podospora anserina* freshly isolated from the wild become senescent after prolonged vegetative growth. Their growth rate decreases and aerial hyphae become slender and undulate in appearance (Fig. 1). Eventually, the hyphal tips become highly vacuolated; they swell considerably and may burst. Some days after the first appearance of such morphological changes, mycelial growth ceases and the hyphae die.

Data of Rizet and Marcou

The results of Rizet and Marcou (references cited above) may be summarised as follows:

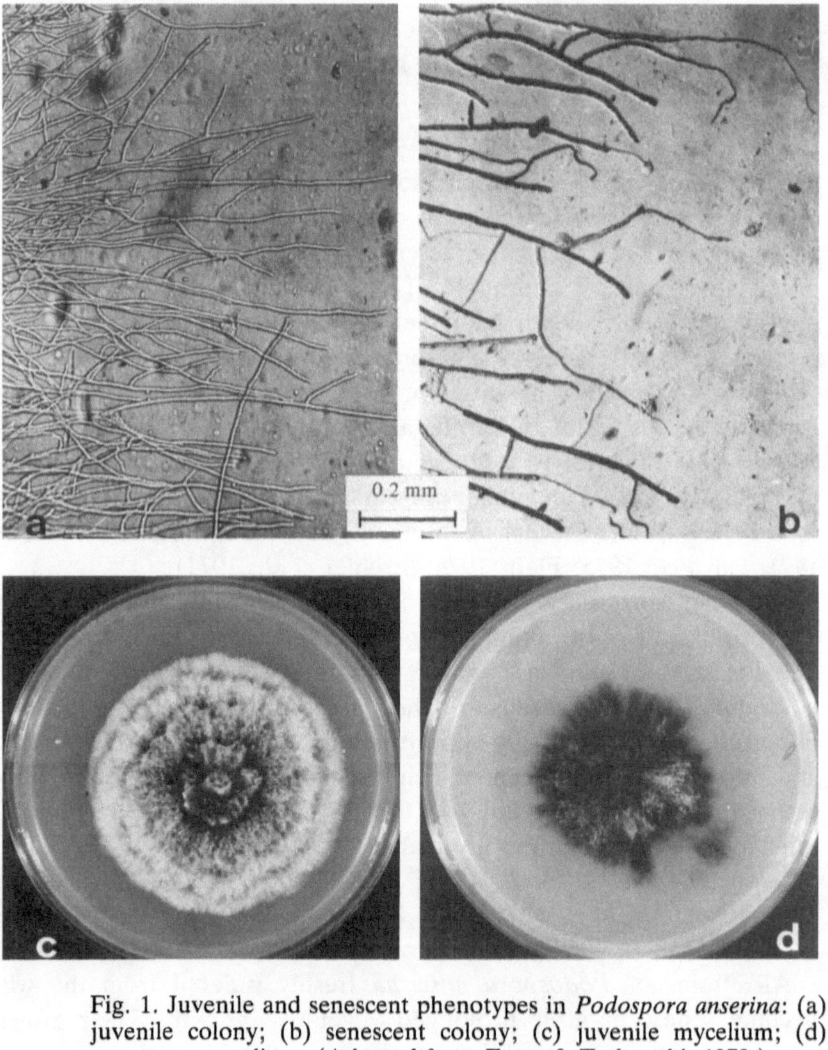

Fig. 1. Juvenile and senescent phenotypes in *Podospora anserina*: (a) juvenile colony; (b) senescent colony; (c) juvenile mycelium; (d) senescent mycelium. (Adapted from Esser & Tudzynski, 1979.)

(1) The onset of senescence depends upon environmental conditions (e.g. light, temperature) as well as genetic ones (variations in the time of onset of senescence in geographical races).
(2) Sexual reproduction is not affected by the onset of senescence since ageing strains are still able to form perithecia and ascospores with normal rates of germination.
(3) Strains may be kept juvenile indefinitely by sexual propagation at regular intervals.
(4) Senescence is maternally transmitted in reciprocal crosses between juvenile and senescent mycelia performed by spermatisation with microconidia. All ascospores formed from maternal juvenile mycelium yield juvenile clones. In reciprocal crosses, involving senescent maternal mycelia and microconidia derived from juvenile mycelia, as many as 90% of ascospores may be prematurely senescent.
(5) Senescence is caused by an extrachromosomal infective principle which may be transferred by hyphal or protoplast fusions to juvenile cells.

This early work by Rizet and Marcou did not make the nature of the infective principle for senescence clear. Following Rizet's first publications on senescence in *Podospora anserina*, similar syndromes were reported in other fungi. These include *Podospora setosa* (Rizet, 1953*a*), *Aspergillus glaucus* (Jinks, 1959; Sharpe, 1958), *Helminthosporum victoriae* (Lindberg, 1959), *Pestalotia annulata* (Chevaugeon & Digbeu, 1960), *Neurospora crassa* (Bertrand, McDougall & Pittenger, 1968; Inoue & Ishikawa, 1975; Munkres, 1976*a,b*; Munkres & Colvin, 1976; Munkres & Minssen, 1976; Sullivan & DeBusk, 1973), *Physarum polycephalum* (McCullough *et al.*, 1973) and *Phytophthora cactorum* (McIntyre & Elliot, 1974). However, none of these other fungi was submitted to a detailed genetical or physiological analysis comparable to that applied to *Podospora anserina* (see also Smith & Rubenstein, 1973*a,b*).

Genetic control of senescence

The genetic elements responsible for the nuclear cytoplasmic control of senescence were identified in our laboratory.

(1) *Nuclear genes*

In *Podospora anserina* a great variety of morphogenetic genes having a pleiotropic action on hyphal morphology and/or perithecial differentiation are also responsible for the onset of senescence. An example of the action of these genes is given in Fig. 2 showing the synergistic action of two genes, *incoloris* and *vivax*; either alone only slightly postpones the onset of senescence, while together they totally prevent its occurrence. Even seven years after these double mutants (*i viv*) were produced, the resulting strain has shown no indication of ageing. It still grows and its longevity is accompanied by an unchanging rate of growth.

Other combinations of different morphological genes can postpone the onset of senescence by several hundred days to an indefinite period (Esser & Keller, 1976; Tudzynski & Esser, 1979).

(2) *Mitochondrial plasmids*

In our search for the senescence principle it was obvious we should consider that a virus or virus-like particle (VLP) might be responsible for ageing of *Podospora anserina* mycelia, especially since some fungal diseases are viral (for review see Lemke, 1976, 1979). However, all efforts to demonstrate VLP in *Podospora anserina* by electron microscopy have so far proved negative.

However, the infective principle able to induce senescence in juvenile strains could be identified as a circular plasmid having a contour length of 0.75 μm (Stahl *et al.*, 1978). It was further shown in juvenile cells that

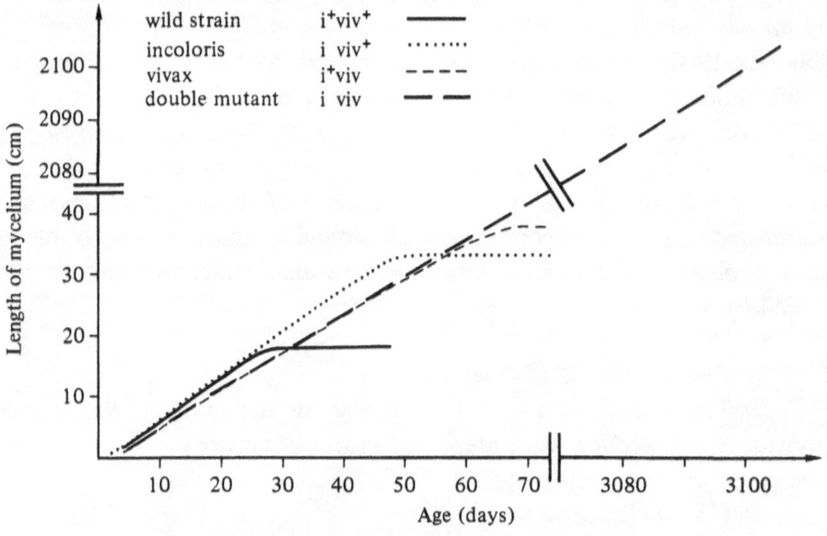

Fig. 2. Growth curves of the wild strain ($i^{+} viv^{+}$), the mutants *viv* and *i* and the double mutant *i viv* of *Podospora anserina*.

this plasmid is an integrated part of the mitochondrial chromosome. The plasmid is liberated during strain ageing and then becomes autoreplicative and effective as shown schematically in Fig. 3 (Stahl, Kück, Tudzynski & Esser, 1980; Kück, Stahl & Esser, 1981).

Other laboratories obtained similar results independently (Cummings, Belcour & Grandchamps, 1979; Jamet-Vierny, Begel & Belcour, 1980; Belcour, Begel, Mosse & Vierny, 1981).

Despite the fact that marker genes have not yet been identified on this plasmid, it carries at least one autonomously replicating sequence (ars).

The *Podospora* plasmid, as far as we know, is the first plasmid to be detected in filamentous fungi and may be manipulated like any prokaryotic plasmid. After integration of the *Podospora* plasmid into a bacterial one a shuttle vector is obtained which can replicate and be expressed in both the prokaryotic bacterium and the eukaryotic fungus (for details see Fig. 4) (Stahl *et al.*, 1982; Tudzynski, Stahl & Esser, 1980).

Thus the elucidation of nuclear-cytoplasmic control of senescence in *Podospora*, in showing that the expression of the causative agent, a mitochondrial plasmid, depends on nuclear genes, has greatly increased our understanding of the process of ageing. However, the physiological action of the senescence plasmid is still not clear. Nevertheless a hint comes from a recent observation that the *Podospora* plasmid after infecting juvenile mycelia most probably becomes integrated into the nucleus (Wright & Cummings, 1983) and may act as a transposable element in blocking essential nuclear functions.

Consequences of the discovery of the *Podospora* mitochondrial plasmid for genetic engineering

Gene cloning to date has been carried out mainly by using bacteria as host cells (Bernard & Helinsky, 1980). The systems which

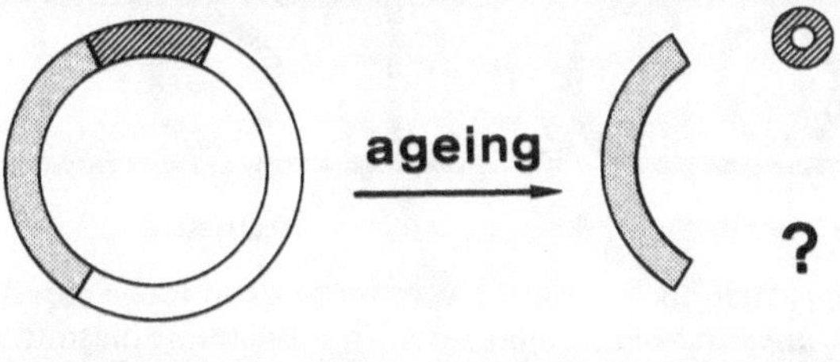

Fig. 3. Disintegration of mitochondrial DNA during ageing. While in senescent strains the majority of the circular mt DNA has disappeared, only the circular small plasmid and a linear remnant of mt DNA are found.

are suitable as vectors for recombinant DNA technology are predominantly of prokaryotic origin. Since gene cloning is becoming a central part of both fundamental (e.g. gene expression, regulation) (Losson & Lacroute, 1981) and applied research (e.g. production of therapeutants and other products relevant for biotechnology) (Johnson & Burnett, 1978), the inclusion of eukaryotes in the spectrum of host organisms would be a great step forward. One of the main reasons to use eukaryotes for such research is that when bacteria are used as hosts for

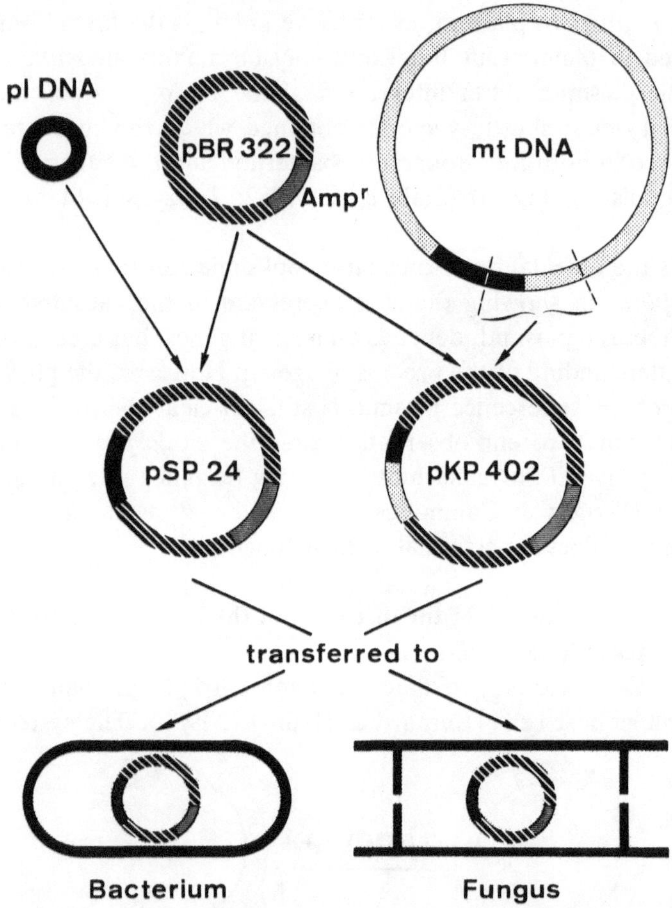

Fig. 4. Construction of shuttle vectors with mitochondrial DNA of *Podospora anserina* being integrated in a bacterial plasmid pBR 322. In using either the *Podospora* mitochondrial plasmid (above to the left) or a part of mitochondrial DNA containing an autonomously-replicating sequence (above to the right), both hybrid vectors replicate and also express equally in *E. coli* and in *Podospora*.

cloning of eukaryotic DNA, there can be failures in replication or expression of the foreign DNA, complications when extracting the product, and instability of the transformants (MacLeod, 1980).

In order to avoid complications of this kind, it is desirable that cloning of eukaryotic DNA be restricted to eukaryotic systems, i.e. eukaryotes being both the host and the vector. This situation requires: (1) a vector which guarantees replication to a high copy number and expression of the foreign DNA; (2) a selective system which allows identification of the vector in the host cell; and (3) compatibility and stability of the vector with the DNA which is to be cloned in the host.

Encouraged by the fact that we have a shuttle vector consisting of the *Podospora* plasmid, and that the bacterial plasmid pBR 322 replicates and expresses in both the prokaryotic bacterial and eukaryotic fungal host, we went a step further using sequences of the mitochondrial chromosomal DNA as vector carrying an ars replicon instead of the naturally-occurring plasmid. Figure 4 shows that the resulting hybrid plasmid can replicate and also be expressed in both the bacterium and the fungus. Since after more than one year of uninterrupted propagation all transformants still contain this hybrid plasmid, stable replication and expression for gene cloning seem assured (Stahl *et al.*, 1982).

In recent years it has become evident that mitochondrial plasmids and similar genetic elements are fairly common in eukaryotes, and we believe that the experience we have in gene cloning with *Podospora* extrachromosomal genetic traits can be extended generally for heterologous gene cloning in eukaryotes. This idea is supported by the similarity of ultrastructure and genetic organisation of mitochondria and bacteria. There is further support from the discovery of mitochondrial plasmids which may be either, as in bacteria, integrated into the mitochondrial chromosome or be present as Ti particles in the mitochondrial cytoplasm. This provides additional support for the hypothesis (Gray & Doolittle, 1982) that there is a homology between mitochondria and bacteria. Thus for eukaryotic gene cloning one should take advantage of the fact that each eukaryotic cell contains its own 'bacteria' (= mitochondria) for which either plasmids or parts of mitochondrial DNA might be used to construct vectors. By use of these vectors at least some of the above-mentioned restrictions in cloning eukaryotic genes in bacteria might be overcome:

(1) There should be no problems of incompatibility for mitochondrial DNA when used as vector to transfer genetic material in its organisms of origin.

(2) From this follows that it is highly probable that there will be no complication in replicating and expressing eukaryotic spliced genes in eukaryotes, which is not possible in many cases with prokaryotes (Esser & Lang-Hinrichs, 1983).

(3) If the proposed integration of mitochondrial plasmids into the nucleus proves to be correct, the chances of obtaining stable transformants, the main concern of large-scale production, are more likely with this system.

Initial experiments seem to confirm these ideas, it being possible to use an ars of the mitochondrial genome of *Acremonium kiliense* as universal replicon in promoting the replication and expression of chromosomal genes in other organisms, e.g. in *Saccharomyces cerevisiae* (Tudzynski & Esser, 1982).

References

Behnke, D. A., Finch, C. E. & Moment, G. B. (1979). *The Biology of Ageing*. New York: Plenum Press.

Belcour, L., Begel, O., Mosse, M. O. & Vierny, C. (1981). Mitochondrial DNA amplification in senescent cultures of *Podospora anserina*: variability between the retained, amplified sequences. *Current Genetics*, **3,** 13–21.

Bernard, H. K. & Helinsky, D. R. (1980). Bacterial plasmid cloning vehicles. In *Genetic Engineering*, vol. 2, ed. J. K. Swetlow & A. Hollaender, pp. 133–68. New York: Plenum Press.

Bertrand, H., McDougall, K. J. & Pittenger, T. H. (1968). Somatic cell variation during uninterrupted growth of *Neurospora crassa* in continuous growth tubes. *Journal of General Microbiology*, **50,** 337–50.

Chevaugeon, J. & Digbeu, S. (1960). Un second facteur cytoplasmique infectant chez le *Pestalozzia annulata. Compte rendu hebdomadaire des Séances de l'Académie, des Sciences, Paris*, **251,** 3043–61.

Comfort, A. (1974). The position of ageing studies. *Mechanisms of Ageing and Development,* **3,** 1–10.

Cummings, D. J., Belcour, L. & Grandchamps, C. (1979). Mitochondrial DNA from *Podospora anserina*. II. Properties of mutant DNA and multimeric circular DNA from senescent cultures. *Molecular and General Genetics*, **171,** 239–50.

Curtis, H. J. (1971). Genetic factors in ageing. *Advances in Genetics*, **16,** 305–24.

Esser, K. & Keller, W. (1976). Genes inhibiting senescence in the ascomycete *Podospora anserina. Molecular and General Genetics*, **144,** 107–10.

Esser, K. & Kuenen, R. (1967). *Genetics of Fungi*. Berlin, Heidelberg, New York: Springer-Verlag.

Esser, K. & Lang-Hinrichs, C. (1983). Molecular cloning in heterologous systems. *Advances in Biochemical Engineering*, **26,** 143–73.

Esser, K. & Tudzynski, P. (1979). Genetic control and expression of senescence in *Podospora anserina*. In *Viruses and Plasmids in Fungi*, ed. P. A. Lemke, pp. 595–615. New York: Marcel Dekker.

Gray, M. W. & Doolittle, W. F. (1982). Has the endosymbiont hypothesis been proven? *Microbiological Reviews,* **46,** 1–42.

Hayflick, L. (1975). Current theories of biological ageing. *Federation Proceedings*, **34**, 9–13.
Inoue, H. & Ishikawa, T. (1975). Death resulting from unbalanced growth in a temperature-sensitive mutant of *Neurospora crassa*. *Archives of Microbiology*, **104**, 1–6.
Jamet-Vierny, C., Begel, O. & Belcour, L. (1980). Senescence in *Podospora anserina*: amplification of a mt DNA sequence. *Cell*, **21**, 189–94.
Jinks, J. L. (1959). Lethal suppressive cytoplasm in aged clones of *Aspergillus glaucus*. *Journal of General Microbiology*, **21**, 397–409.
Johnson, I. S. & Burnett, J. P. (1978). Problems and potential of industrial recombinant DNA research. In *Genetic Engineering*, ed. H. W. Boyer & S. Nicosia, pp. 217–26. Amsterdam: Elsevier/North-Holland Biomedical Press.
Kück, U., Stahl, U. & Esser, K. (1981). Plasmid like DNA is part of mitochondrial DNA in *Podospora anserina*. *Current Genetics*, **3**, 151–6.
Lemke, P. A. (1976). Viruses of eukaryotic microorganisms. *Annual Review of Microbiology*, **30**, 105–45.
Lemke, P. A. (1979). *Viruses and Plasmids in Fungi*. New York, Basel: Marcel Dekker.
Lindberg, G. D. (1959). A transmissible disease of *Helminthosporium victoriae*. *Phytopathology*, **49**, 29–52.
Losson, R. & Lacroute, F. (1981). Cloning of a eukaryotic regulatory gene. *Molecular and General Genetics*, **184**, 394–9.
MacLeod, A. J. (1980). Biotechnology and the production of proteins. *Nature*, **285**, 136.
Marcou, D. (1954*a*) Sur la longévité des souches de *Podospora anserina* cultivées à divers températures. *Compte rendu hebdomadaire des Séances de l'Académie des Sciences, Paris*, **239**, 895–7.
Marcou, D. (1954*b*). Sur le réjeunissement par le froid des souches de *Podospora anserina*. *Compte rendu hebdomadaire des Séances de l'Académie des Sciences, Paris*, **239**, 1153–5.
Marcou, D. (1957). Réjeunissement et arrêt de croissance chez *Podospora anserina*. *Compte rendu hebdomadaire des Séances de l'Académie des Sciences, Paris*, **244**, 661–3.
Marcou, D. (1958). Sur la déterminisme de la sénescence observet chez l'ascomycète *Podospora anserina*. *Proceedings of the Xth International Congress of Genetics*, Vol. II, p. 179. Montreal.
Marcou, D. (1961). Notion de longévité et nature cytoplasmique du déterminant de la sénescence chez quelques champignons. *Annales des Sciences Naturelles (Botanique)*, **11**, 653–764.
Marcou, D. & Schecroun, J. (1959). La sénescence chez *Podospora anserina* pourrait être due à des particules cytoplasmiques infectantes. *Compte rendu hebdomadaire des Séances de l'Académie des Sciences, Paris*, **248**, 280–3.
McCullough, C., Cooke, D., Foxon, J., Sudberry, P. & Grant, W. (1973). Nuclear DNA-content and senescence in *Physarum polycephalum*. *Nature*, **245**, 263–5.
McIntyre, D. & Elliot, C. G. (1974). Selection for growth-rate during asexual and sexual propagation in *Phytophthora cactorum*. *Genetics Research, Cambridge*, **24**, 295–309.
Munkres, K. D. (1976*a*). Ageing of *Neurospora crassa*. III. Induction of cellular death and clonal senescence of an inositolless mutant by inositol-starvation and the protective effect of dietary antioxidants. *Mechanisms of Ageing and Development*, **5**, 163–9.
Munkres, K. D. (1976*b*). Ageing of *Neurospora crassa*. IV. Induction of senescence in wild type by dietary amino acid analogues and reversal by antioxidants and membrane stabilizers. *Mechanisms of Ageing and Development*, **5**, 171–91.
Munkres, K. D. & Colvin, H. J. (1976). Ageing of *Neurospora crassa*. II. Organic

hydroperoxide toxicity and the protective role of antioxidant and the antioxygenic enzymes. *Mechanisms of Ageing and Development*, **5,** 99–107.
Munkres, K. D. & Minssen, M. (1976). Ageing of *Neurosospora crassa*. I. Evidence for the free radical theory of ageing from studies of a natural-death mutant. *Mechanisms of Ageing and Development*, **5,** 76–98.
Platt, D. (1976). *Biologie des Alterns*. Heidelberg: Quelle & Meyer.
Rizet, G. (1953*a*). Sur l'impossibilité d'obtenir la multiplication végétative ininterrompue et illimité de l'ascomycète *Podospora anserina. Compte rendu hebdomadaire des Séances de l'Académie des Sciences, Paris*, **237,** 838–55.
Rizet, G. (1953*b*). Sur la longevité des souches de *Podospora anserina. Compte rendu hebdomadaire des Séances de l'Académie des Sciences, Paris*, **237,** 1106–9.
Rizet, G. (1957). Les modificiations qui conduisent à la sénescence chez *Podospora* sont-elles de nature cytoplasmique? *Compte rendu hebdomadaire des Séances de l'Académie des Sciences, Paris*, **244,** 663–5.
Rizet, G. & Marcou, D. (1954). Longevité et sénescence chez l'ascomycète *Podospora anserina. Compte rendu VII du Congress International (Botanique)*, vol. 10, pp. 121–8, Paris.
Sharpe, S. (1958). A closed system of cytoplasmic variation in *Aspergillus glaucus. Proceedings of the Royal Society of London*, **B148,** 355–9.
Smith, J. R. & Rubenstein, I. (1973*a*). The development of senescence in *Podospora anserina. Journal of General Microbiology*, **76,** 283–96.
Smith, J. R. & Rubenstein, I. (1973*b*). Cytoplasmic inheritance of the timing of senescence in *Podospora anserina. Journal of General Microbiology*, **76,** 297–304.
Stahl, U., Lemke, P. A., Tudzynski, P., Kück, U. & Esser, K. (1978). Evidence for plasmid-like DNA in a filamentous fungus, the ascomycete *Podospora anserina. Molecular and General Genetics*, **162,** 341–3.
Stahl, U., Kück, U., Tudzynski, P. & Esser, K. (1980). Characterization and cloning of plasmid like DNA of the ascomycete *Podospora anserina. Molecular and General Genetics*, **178,** 639–46.
Stahl, U., Tudzynski, P., Kück, U. & Esser, K. (1982). Replication and expression of a bacterial-mitochondrial hybrid plasmid in the fungus *Podospora anserina. Proceedings of the National Academy of Sciences of the USA*, **79,** 3641–5.
Strehler, B. L., Hirsch, G., Gussek, D., Johnson, R. & Bick, M. (1971). Codon-restriction theory of ageing and development. *Journal of Theoretical Biology*, **33,** 429–35.
Sullivan, J. L. & DeBusk, A. G. (1973). Inositol-less death in *Neurospora* and cellular ageing. *Nature New Biologist*, **243,** 72–4.
Tudzynski, P. & Esser, K. (1979). Chromosomal and extrachromosomal control of senescence in the ascomycete *Podospora anserina. Molecular and General Genetics,* **173,** 71–84.
Tudzynski, P. & Esser, K. (1982). Extrachromosomal genetics of *Cephalosporium acremonium*. II. Development of a mitochondrial DNA hybrid vector replicating in *Saccharomyces cerevisiae. Current Genetics*, **6,** 153–8.
Tudzynski, P., Stahl, U. & Esser, K. (1980). Transformation to senescence with plasmid like DNA in the ascomycete *Podospora anserina. Current Genetics*, **2,** 181–4.
Wright, R. M. & Cummings, D. J. (1983). Integration of mitochondrial gene sequences within the nuclear genome during senescence in a fungus. *Nature*, **302,** 86–8.

16

The mycelial biology of *Endothia parasitica*. I. Nuclear and cytoplasmic genes that determine morphology and virulence

SANDRA L. ANAGNOSTAKIS
Department of Plant Pathology and Botany, The Connecticut Agricultural Experiment Station, P.O. Box 1106, New Haven, CT 06504, USA

When the stately American chestnut trees (*Castanea dentata*, Fig. 1) lining the avenues of Bronx Zoo in New York City began to wilt and die, the cause was traced to the mycelium of a fungus (Merkel, 1905). The mycelium grew in and under the bark of the trees disrupting xylem and phloem, and the girdling cankers quickly led to death of twigs, branches, and trunks of the mighty trees. In 1906 Murrill reported that this was a new species of *Diaporthe* (Murrill, 1906). A discussion then started among American mycologists regarding the taxonomic position of the fungus because it was very similar in appearance to two American species of *Sphaeria* found as saprophytes on chestnut. *S. gyrosa* had very different ascospores and was soon ruled out as the progenitor, but specimens of *Sphaeria radicalis* from Europe and the United States were morphologically very similar in their cultural characteristics, pycnidia, perithecia, and ascospores to the new pathogen. The main distinguishing characteristic was pathogenicity on American chestnut. Was this, then, a new mutant form of *S. radicalis?* That question was presumed settled when a plant explorer named Meyer, working in China and Japan, found a fungus which produced slowly-expanding cankers on oriental chestnut trees that were rarely lethal (Shear & Stevens, 1913; Fairchild, 1913; Shear & Stevens, 1916). American mycologists used isolates from China to inoculate American chestnut trees, and found that they produced killing cankers. This was, of course, before plant quarantine regulations in the USA. It is now assumed that oriental chestnut trees imported for ornamental plantings brought the blight along, but curious mycologists may have done their share in introducing the new fungus!

The pathogen proceeded to spread out from its sites of introduction until its range included all of the natural range of the American chestnut tree. By 1950 an estimated 3.6 million hectares (9 million acres) of these

Fig. 1. An American chestnut (*Castanea dentata*) photographed in Connecticut in 1905. This tree was 23 inches dbh and 83 feet tall.

valuable trees were dead or dying (Anonymous, 1954). Efforts to seek a cure continue because the trees are still producing sprouts from the roots, which are not invaded by the fungus. The sprouts in turn become infected and rarely survive to sexual maturity.

Paul J. and H. W. Anderson named the new fungus *Endothia parasitica* (Murr.) And. (Anderson & Anderson, 1912). Recently, Barr (1978) has suggested that the name should be *Cryphonectria parasitica* (Murr.) Barr, but the older name will be retained here.

The task of taxonomy in enabling identification of a taxon is to examine many specimens and produce descriptions which contain the major, easily noticed features that are common to all. However, beyond these common characteristics there is often a wide range of variation within species populations. With many fungi, for example, it is often possible to recognise individual strains by the appearance of their mycelia in laboratory culture. In this chapter, the basis for some of this variation will be examined, and its significance in relation to the population structure of *E. parasitica*, its pathogenicity, and possibilities for biological control will be considered.

Environmental effects

Laboratory cultures of *E. parasitica* have been described as having mycelia that were at first white and then yellow (Anderson, 1914; Shear, Stevens & Tiller, 1917). The pigments have been well studied and appear from yellow, in acid solution, to magenta, in basic solution (Anderson, 1914; Roane & Stripes, 1978). Uninucleate conidia are formed in orange pycnidia and extruded in sticky orange ribbons called spore horns.

Light has been reported to induce asexual sporulation in *E. parasitica* (Anderson, 1914). When the mycelium is grown in an alternating light/dark regime pycnidia form on the part of the mycelium young enough to be induced by the light. This results in concentric rings of pycnidia separated by sterile mycelium of hyphae which were too old for pycnidial induction when the light period began. In continuous light, pycnidia are fewer and scattered. In continuous darkness the pycnidia appear much later, even fewer are formed, and they are scattered over the mycelium (cf. Raudaskoski & Salonen: Chapter 13; Lysek: Chapter 14). Pycnidia also form in large numbers along the edge of a knife-cut through mycelium, and along barrage lines between vegetatively incompatible strains (see Chapter 22).

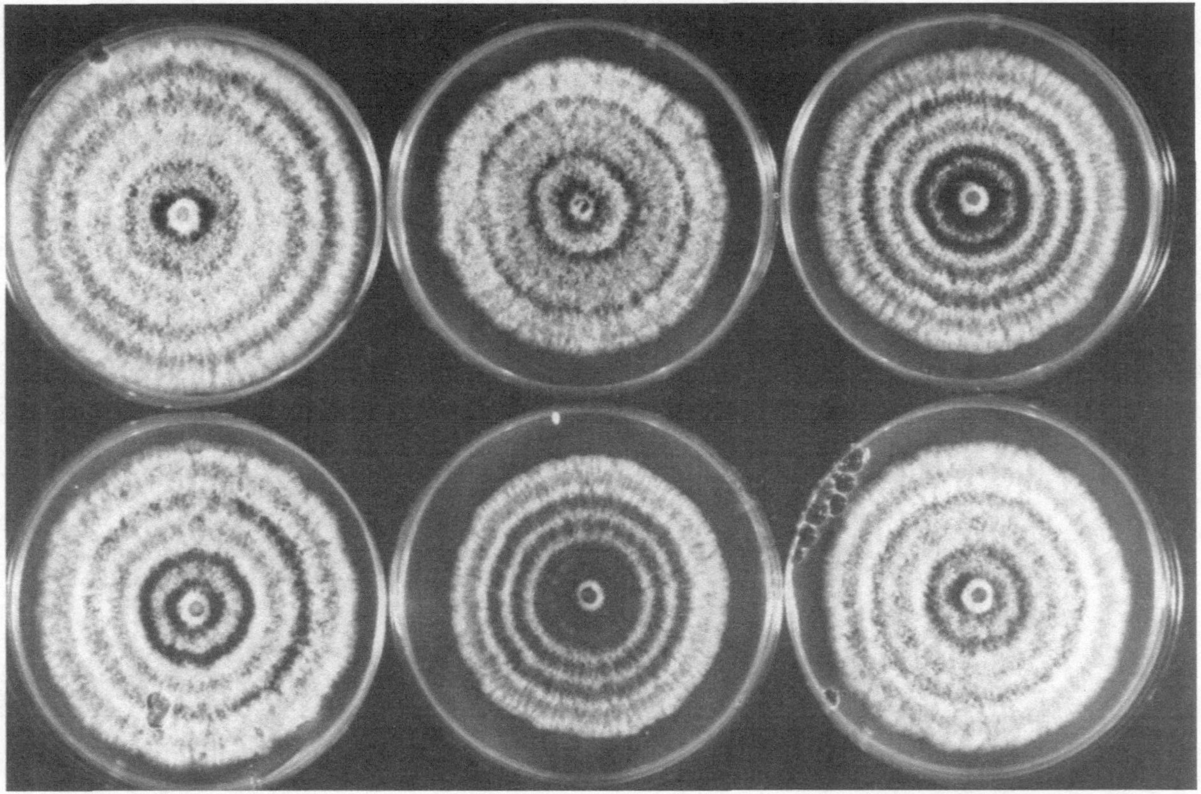

Fig. 2. Six virulent strains of *Endothia parasitica* grown under the same conditions. The gross morphology of individuals remains the same after repeated mass-mycelial transfer.

Although this general description covers the behaviour of many of the strains in our culture collection, there are many exceptions.

Nuclear genes

The pycnidia vary in size and distribution when different strains are grown under the same conditions (Fig. 2). These characteristics seem to be controlled by several genes because crosses yield progeny which segregate in a normal distribution for these features.

Certain characteristics, however, segregate as if determined by single nuclear genes. These include forms in which the pigment, which is normally yellow, is either totally lacking or dark brown, with off-white or dark-brown pycnidia and conidial spore horns; forms with slow colony extension but with dense branching (typical 'colonial' morphology) (Fig. 3); and forms with appressed submerged hyphae with little or no aerial mycelium (Fig. 4) (Puhalla & Anagnostakis, 1971; Anagnostakis, 1980; Anagnostakis, 1982). Mutations to these anamorphs are common after UV irradiation, which is not surprising. However, it *is* surprising that some of these anamorphs have been recovered from forest-grown host trees. It is worth noting that the appearance of the teleomorph is not changed by the single gene mutations that make mycelium, pycnidia, and conidial masses white or brown. The perithecia on white and brown strains appear identical to those on 'normal' yellow strains.

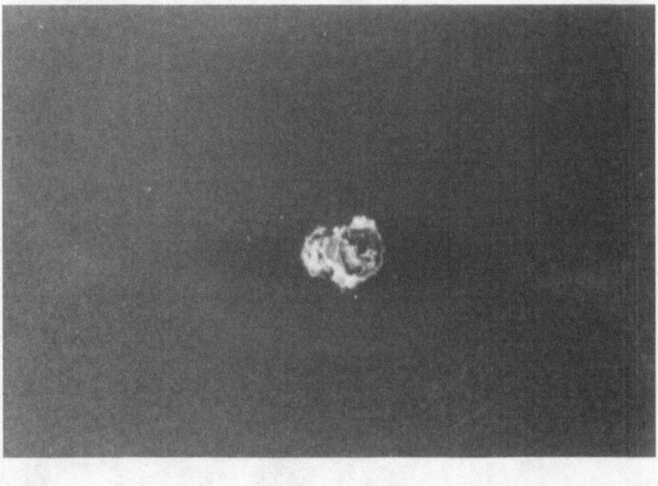

Fig. 3. A single nuclear gene mutation (*col-l*) leads to the abnormal growth form called 'colonial'. This *Endothia parasitica* culture was grown under the same conditions, and is of the same age as those in Fig. 2.

Perhaps the most interesting of the single gene mutations that I have studied in *E. parasitica* is one which can change the anamorph from a member of the *Sphaeropsidales*, with conidia in pycnidia, to a member of the *Moniliales*, with conidiophores on the vegetative mycelium. This kind of mutation has been recovered from many field isolates after isolating single (uninucleate) conidial cultures. Five such strains were used as males in crosses with genetically marked *E. parasitica* strains to prove that the morphology was indeed determined by a single nuclear gene. I have called this genotype *flat*. The mycelium of *flat* strains is dense, dark orange, and rarely produces aerial hyphae. As has already been implied, however, its most striking feature is a lack of pycnidia; instead conidia are formed along vegetative hyphae on clusters of short, flask-shaped conidiogenous cells with curved necks. This is a very

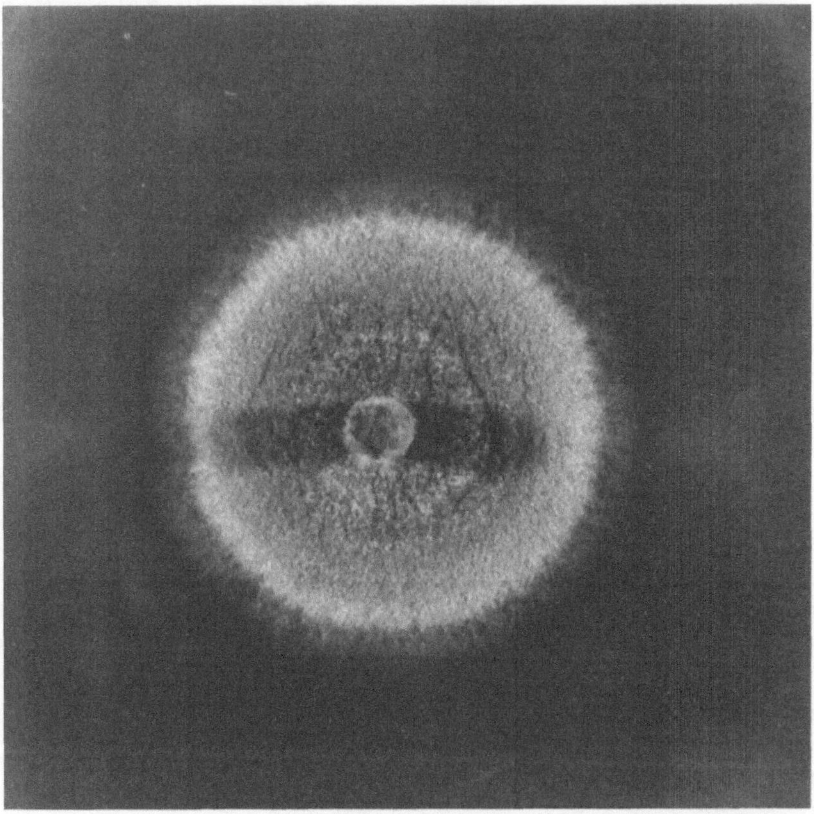

Fig. 4. *Endothia parasitica* with a single mutant nuclear gene (*flat-1*) which nearly eliminates the production of aerial hyphae.

dramatic mutation, and should make taxonomists of Deuteromycotina a little nervous!

Cytoplasmic genes

Chestnut trees were also important in Europe where *Castanea sativa* had been extensively planted by the Romans. *E. parasitica* was found in northern Italy in 1938 and proceeded to devastate chestnut populations there as it had in the USA (Woodruff, 1946; Pavari, 1949). The death of large stands of *C. sativa* caused great concern since the Italian country people relied on the nuts for food. Then, in 1951, an Italian plant pathologist, Biraghi, noticed a chestnut coppice near Genoa that seemed unusually healthy. He found that this coppice 'was once severely damaged by *Endothia parasitica* and . . . it was impossible then to find any living shoot older than four or five years . . . (However, in 1951) about 85% of the shoots were infected by *Endothia parasitica*, but only a few showed the usual symptoms characteristic of the blight . . .' (Biraghi, 1953). He found cankers that were callusing and that the fungus was restricted to the outer layer of bark on these trees. His persistent claims attracted the attention of a French mycologist, J. Grente, who visited Italy in the early 1960s and took bark from callusing cankers to his laboratory in Clermont-Ferrand. From these samples, he isolated morphologically distinctive forms of the blight fungus that had reduced virulence, and called them exclusive hypovirulent strains (Grente, 1965). These hypovirulent forms cured existing blight when they were inoculated into the bark around the margins of killing cankers. Later, in co-operation with Sauret, he published several reports on these curative hypovirulent strains (Grente & Sauret, 1969*a,b*; Grente, 1975; Grente, 1981).

Once a canker had been successfully cured by treatment with a hypovirulent (H) strain, the virulent (V) mycelium that had caused the canker was converted to hypovirulent. Grente and Sauret described the behaviour of their H strains in culture and suggested that in the host, hyphae of the V strain causing the canker anastomosed with hyphae of the introduced H strain which allowed genetic determinants in the cytoplasm to be transferred, converting the V strain to H.

European curative strains are now reported from Italy, Switzerland, Spain, and France. A variety of these strains have generously been sent to us by Grente, Turchetti, and others, and a great deal of work has now been done on them at The Connecticut Agricultural Experiment Station. We have found that all of the European curative hypovirulent

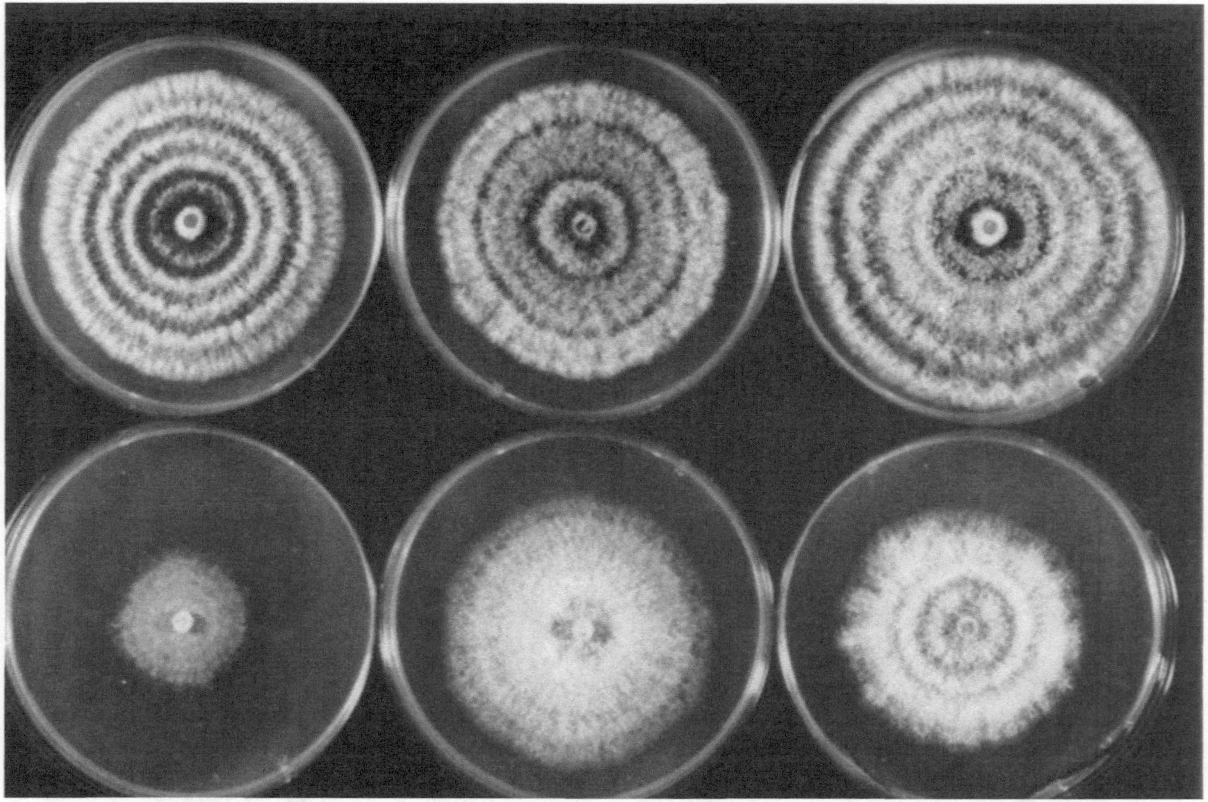

Fig. 5. Three virulent (V) strains of *Endothia parasitica* (top row) and their hypovirulent (H) converts (bottom row). Each H strain is directly under its V progenitor and has the same nuclear genotype. The difference in morphology is caused by cytoplasmic genes from Italian curative strains.

strains have cytoplasmic determinants that drastically alter mycelial morphology (Fig. 5).

Since I now intend to discuss these cytoplasmic determinants, I will use the convention of italicising nuclear gene symbols and enclosing in square brackets symbols for genes in the cytoplasm (whether they are viral, plasmid, mitochondrial, or whatever).

When [H] genomes are transferred into the cytoplasm of strains which lacked them, the morphology and virulence of the V strains are changed. In Fig. 6, V and H strains have been paired on Cellophane over agar medium and the change that has occurred in the morphology of the virulent strains is easily visible. The point at which this change takes place presumably marks the area where successful anastomoses have occurred, and where cytoplasm with [H] genes was transferred. Of course, from a practical point of view the most important characteristic of H strains is their reduced virulence. Important here is the fact that the genes for this reduced virulence are genetically as dominant as those for the morphology, so that when virulent [V] and hypovirulent [H] cytoplasm are mixed, it is the [H] virulence genotype which is expressed. This is why these H strains are so effectively curative: the [H] genes completely dominate in their new fungal environment.

We have found no curative strains that do not contain double-stranded RNA of very high molecular weight (6×10^6 daltons) (Day *et al.*, 1977; Dodds, 1980). This provides strong circumstantial evidence that the [H] genes are on or are associated with these dsRNA molecules. It is worth noting that most of the fungal viruses that have been studied have lower-molecular-weight dsRNA molecules as their genomes. None of these have any noticeable effect on the morphology or virulence of the fungi they inhabit. The three examples of fungi containing high-molecular-weight dsRNA molecules are all cases in which the fungi *are* affected in some noticeable way by the presence of the dsRNA: *E. parasitica, Ustilago maydis* (Day, 1981), and *Rhizoctonia solani* (Castanho, Butler & Shepherd, 1978).

Mycelial morphology of European H strains shows the same kind of variation that we see among 'normal' American and European V strains. The H type studied the most (Fig. 5) grows a little more slowly than V strains with the same nuclear genome. It produces less pigment, and therefore, is usually referred to as 'white'. These strains also produce fewer pycnidia and conidia. I have never succeeded in getting H strains to serve as females in controlled crosses in the laboratory, although the conidia function perfectly well as spermatia; that is, as males.

The only biochemically defined characteristic that I have been able to link with the [H] genome is reduced production of oxalate on laboratory media. We are studying this phenomenon to see whether it can be associated with any of the other phenotypic expressions of [H] (Havir &

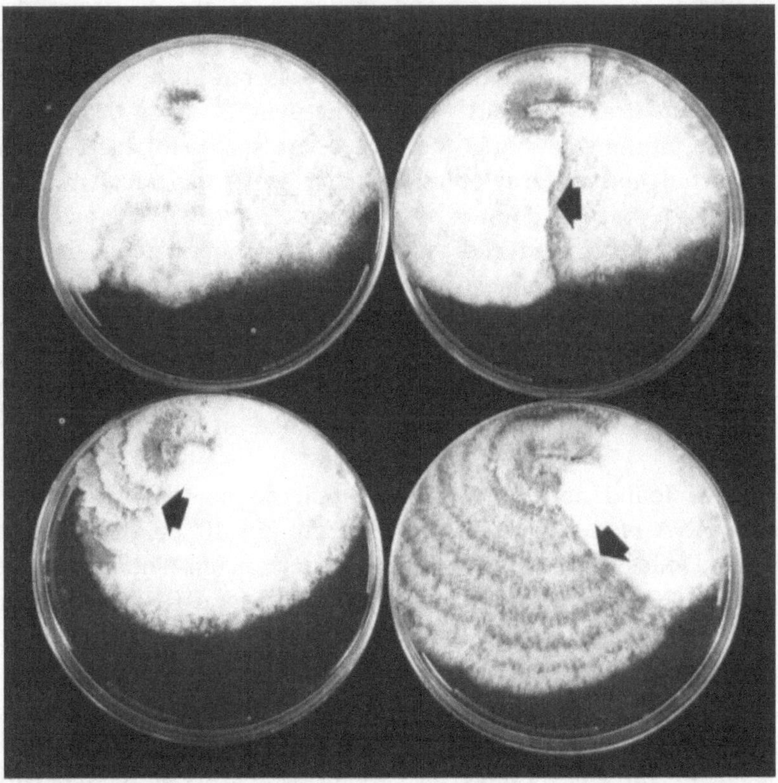

Fig. 6. Four different virulent strains of *Endothia parasitica* (left inoculum on each plate) paired on Cellophane over agar medium with a hypovirulent strain (right inoculum on each plate). The strains in the top left plate were in the same vegetative compatibility group (see Chapter 22) and their mycelia merged together as they grew. The virulent member of this pair was converted to hypovirulent very early in growth and no mycelium with virulent morphology resulted. In the lower right plate the two strains were in very different vegetative compatibility groups and no conversion has taken place. The barrage line between the strains is indicated with an arrow. The other two pairs (top right and bottom left) have formed weak barrage lines (at arrows), and the virulent strains have been converted to hypovirulent.

Anagnostakis, 1983). Oxalate production by other fungi has been linked to their pathogenicity (Noyes & Hancock, 1981; Maxwell, 1973), and McCarroll & Thor (1978*b*) suggested that oxalic acid produced by *E. parasitica* might chelate and remove calcium from cell-wall and membrane components in host bark, and reduce the ambient pH. This would then allow the polygalacturonase produced by the fungus to function. The action of this and other enzymes could produce the gelatinous zone seen in advance of the mycelium in cankers (McCarroll & Thor, 1978*a*). Further possibilities relating to the role of oxalate production in fungi are discussed by Jennings (Chapter 7).

The dsRNA molecules are usually present in H strains in rather low titre, and strains derived from single uninucleate conidia often include some strains lacking dsRNA and behaving as normal V strains. Conditions of culture of the H strains affect frequency of V recovery. Grente (1981) reported that culture of an H strain in continuous light yielded about 80% V strains from conidia taken from superficial and submerged pycnidia. The highest percentage of white H strains from single conidial clones of the H strain came from the superficial pycnidia formed under alternating light/dark growth conditions. This variation probably reflects differences in dsRNA replication rates in hyphae growing under different conditions.

The lack of pigment associated with [H] genomes is not found in strains with the genotype *flat* [H]. When strains with *flat* phenotypes were first recovered as single conidial clones from curative field isolates, the morphology (called Jaune Régénéré or Pigmentato) was assumed to be a different expression of an [H] genome (Grente & Sauret, 1969*a,b*; Bonifacio & Turchetti, 1973). The stability of the morphology when subsequent single conidial clones were made from *flat* strains belied cytoplasmic control. Although the original *flat* [H] strains were curative (and contained dsRNA), *flat* progeny from matings of five different *flat* [H] × *flat*$^{+}$ [V] were not curative. This was not surprising since dsRNA viruses have never been reported to be transmitted to ascospore progeny of infected Ascomycetes (Rawlinson, Hornby, Pearson & Carpenter, 1973; Bozarth, 1977), although they are transmitted to the basidiospore progeny of infected Basidiomycetes (Day, 1981; Castanho & Butler, 1978). Virulence of the *flat* strains was low. When such strains were paired with standard white H strains, dsRNA molecules were transferred into the *flat* strains, and their morphology was not changed. Thus, these natural single-gene mutants somehow resist morphological expression of the [H] genes.

Frequency of recovery of *flat* clones from conidia of white H strains is influenced by conditions of culture and isolation. Grente (1981) reported that when one white H strain was grown on malt agar in continuous light, conidia from superficial pycnidia yielded 2% *flat* strains, but in total darkness, conidia from pycnidia produced on the surface all yielded strains with *flat* phenotypes. The latter recovery rate is too high to be explainable by a sudden increase in mutation rate, and is probably due to selection favouring rare mutant nuclear types in the mycelium growing in the darkness. There is another possibility that provides rather interesting speculation. The *flat* genotype might be an integration of a dsRNA gene or its DNA copy into the nuclear genome of the fungus. If this were the case, stress studies with *flat* [V] strains should yield *flat*$^{+}$ [H] or *flat* [H] strains.

Conclusion

We clearly need more information about the nature and control of virulence in the genus *Endothia* before we can use avirulence as the sole feature distinguishing between species. The ultimate test of species in fungi must be the ability of a new strain to cross with known strains with the production of viable progeny. This leads me to wonder: what is, or was, *Endothia* (*Sphaeria*) *radicalis*? I have not been able to get a recent isolate of this 'species' from Italian or US mycologists, even though it was once reported as common in both countries (Shear *et al.*, 1917). Has its habitat been taken over by its more aggressive relative? Did they hybridise in areas where *E. parasitica* was introduced, resulting in a disappearance of the *radicalis* type? Does *E. radicalis* still exist in England where it was first described (from the New Forest), and where *E. parasitica* is still unknown?

The major source of variation in *Endothia* is genetic recombination during sexual reproduction. The integrity of individual strains is controlled, if not always maintained, in the vegetative phase by an elaborate system of vegetative incompatibility (cf. Chapter 22). We need a better understanding of the population structure and range of variation possible in the species. This would help us develop logical strategies for promoting a stable biological control of chestnut blight using curative strains.

References

Anagnostakis, S. L. (1980). Notes on the genetics of *Endothia parasitica*. *Neurospora Newsletter*, **27**, 36.

Anagnostakis, S. L. (1982). Genetic analyses of *Endothia parasitica*: linkage data for four single genes and three vegetative compatibility types. *Genetics*, **102**, 25–8.

Anderson, P. J. (1914). *The Morphology and Life History of the Chestnut Blight Fungus*. The Commission for the Investigation and Control of the Chestnut Tree Blight Disease in Pennsylvania Bulletin **7**. Harrisburg, Pennsylvania: State Printer. 44 pp. + XIX.

Anderson, P. J. & Anderson, H. W. (1912). *Endothia virginiana*. *Phytopathology*, **2**, 261–2.

Anonymous. (1954). *The US Department of Agriculture Farmers Bulletin*, **2068**, p. 21.

Barr, M. E. (1978). *The Diaporthales in North America*, Mycologia Memoir #7. Lehre, Germany: J. Cramer. 232 pp.

Biraghi, A. (1953). Possible active resistance to *Endothia parasitica* in *Castanea sativa*. In *Reports of the 11th Congress of the International Union of Forest Research Organizations, Rome*, pp. 643–5.

Bonifacio, A. & Turchetti, T. (1973). Differenze morphologiche e fisiologiche in isolati di *Endothia parasitica* (Murr.) And. *Annali Accademia Italiana di Scienze Forestali*, **2**, 111–31.

Bozarth, R. F. (1977). Biophysical and biochemical characterization of virus-like particles containing a high molecular weight dsRNA from *Helminthosporium maydis*. *Virology*, **80**, 149–57.

Castanho, B. & Butler, E. E. (1978). *Rhizoctonia* decline: a degenerative disease of *Rhizoctonia solani*. *Phytopathology*, **68**, 1505–10.

Castanho, B., Butler, E. E. & Shepherd, R. J. (1978). The association of double-stranded RNA with *Rhizoctonia* decline. *Phytopathology*, **68**, 1515–19.

Day, P. R. (1981). Fungal virus populations in corn smut from Connecticut. *Mycologia*, **73**, 379–91.

Day, P. R., Dodds, J. A., Elliston, J. E., Jaynes, R. A. & Anagnostakis, S. L. (1977). Double-stranded RNA in *Endothia parasitica*. *Phytopathology*, **67**, 1393–6.

Dodds, J. A. (1980). Revised estimates of the molecular weights of dsRNA segments in hypovirulent strains of *Endothia parasitica*. *Phytopathology*, **70**, 1217–20.

Fairchild, D. (1913). The discovery of the chestnut bark disease in China. *Science*, **38**, 297–9.

Grente, J. (1965). Les formes Hypovirulentes d'*Endothia parasitica* et les espoirs de lutte contre le chancre du châtaignier. *Comptes Rendus Hebdomadaires des Séances de l'Académie d'Agriculture de France*, **51**, 1033–7.

Grente, J. (1975). La lutte biologique contre le chancre du châtaigner par "Hypovirulence contagieuse." *Annales de Phytopathologie*, **7**, 216–18.

Grente, J. (1981). Les variants hypovirulents de l'*Endothia parasitica* et la lutte biologique contre le chancre du châtaignier. Ph.D. Thesis, Université de Bretagne Occidentale, Brest, France. 195 pp.

Grente, J. & Sauret, S. (1969*a*). L'hypovirulence exclusive, phénomène original in pathologie végétale. *Compte Rendu Hebdomadaire des Séances de l'Académie des Sciences, Paris, Série D*, **268**, 2347–50.

Grente, J. & Sauret, S. (1969*b*). L'"hypovirulence exclusive" est-elle contrôlée par des déterminants cytoplasmiques? *Compte Rendu Hebdomadaire des Séances de l'Académie des Sciences, Paris, Série D*, **268**, 3173–6.

Havir, E. A. & Anagnostakis, S. L. (1983). Oxalate production by virulent but not by hypovirulent strains of *Endothia parasitica*. *Physiological Plant Pathology*, **23**, 369–76.

Maxwell, D. P. (1973). Oxalate formation in *Whetzelinia sclerotiorum* by oxaloacetate acetylhydrolase. *Physiological Plant Pathology*, **3,** 279–88.

McCarroll, D. R. & Thor, E. (1978*a*). Death of a chestnut: the host pathogen interaction. In *American Chestnut Symposium Proceedings*, ed. W. MacDonald. West Virginia Agricultural Experiment Station and US Department of Agriculture.

McCarroll, D. R. & Thor, E. (1978*b*). The role of oxalic acid in the pathogenesis of *Endothia parasitica.* In *American Chestnut Symposium Proceedings*, ed. W. MacDonald. West Virginia Agricultural Experiment Station and US Department of Agriculture.

Merkel, H. W. (1905). A deadly fungus on the American Chestnut. In *New York Zoological Society, 10th Annual Report*, pp. 97–103.

Murrill, W. A. (1906). A new chestnut disease. *Torreya*, **6,** 186–9.

Noyes, R. D. & Hancock, J. G. (1981). Role of oxalic acid in the *Sclerotinia* wilt of sunflower. *Physiological Plant Pathology*, **18,** 123–32.

Pavari, A. (1949). Chestnut blight in Europe. *Unisilva*, **3,** 8–13.

Puhalla, J. E. & Anagnostakis, S. L. (1971). Genetics and nutritional requirements of *Endothia parasitica. Phytopathology*, **61,** 169–73.

Rawlinson, C. J., Hornby, D., Pearson, V. & Carpenter, J. M. (1973). Virus-like particles in the take-all fungus *Gaeumannomyces graminis. Annals of Applied Biology*, **74,** 197–209.

Roane, M. K. & Stripes, R. J. (1978). Pigments in the fungal genus *Endothia. Virginia Journal of Science*, **29,** 137–41.

Shear, C. L. & Stevens, N. E. (1913). The chestnut-blight parasite (*Endothia parasitica*) from China. *Science*, **38,** 295–7.

Shear, C. L. & Stevens, N. E. (1916). The discovery of the chestnut-blight parasite (*Endothia parasitica*) and other chestnut fungi in Japan. *Science*, **43,** 173–6.

Shear, C. L., Stevens, N. E. & Tiller, R. J. (1917). *Endothia parasitica and Related Species. US Department of Agriculture Bulletin* **380.** Washington, D.C. 82 pp.

Woodruff, J. B. (1946). Chestnut blight in Italy. *Trees* (the journal of the U.S. National Arborist Association), **April,** 8–9, 16.

17

Variation and heterokaryosis in *Rhizoctonia solani**

NEIL A. ANDERSON

Department of Plant Pathology, University of Minnesota, St Paul, MN 55108, USA

Rhizoctonia solani is a cosmopolitan plant pathogen of considerable economic importance. Parmeter & Whitney (1970) listed the characteristics of the mycelium as having multinucleate compartments, dolipore septa, branch hyphae with a septum and constriction near the main hypha, and some shade of brown colour. The teleomorph is *Thanatephorus cucumeris*. Talbot (1970) has termed it a collective species because it includes at least five formerly described species, and indicated that while he did not expect the nomenclature of *T. cucumeris* to remain stable he did expect that the use of the name *Thanatephorus* would persist. The purpose of this paper is to review the role which homogenic and heterogenic incompatibility systems play in relation to variation and heterokaryosis in this collective species.

Heterogenic incompatibility

Rhizoctonia solani is composed of a number of reproductively isolated populations, each termed an anastomosis group (AG). The AG concept was first proposed by Schultz (1937). Confirmation and refinements were made by Richter & Schneider (1953); Watanabe & Matsuda (1966); Parmeter, Sherwood & Platt (1969), and by Sherwood (1969). Ogoshi (1972), proposed (1976) subsequently two subgroups in both AG1 and AG2. The two subgroups for AG1 are species previously known as *Corticium sasakii* and *Corticium microsclerotia*. Those for AG2 are based on pathogenicity: type 1 are the crucifer pathogens or as Ogoshi called them, the 'Rush type pathogens'; type 2 isolates are root canker pathogens of sugar beet and carrot. Kuninaga, Yokosawa &

* Paper No. 1870, Scientific Journal Series, Minnesota Agricultural Experiment Station, St Paul, MN 55108, USA.

Ogoshi (1979) described two additional AGs bringing the total to seven. Isolates of AGs 1, 2, 3 and 4 have been exchanged and anastomosis tests have confirmed their presence in Europe, Japan, and the US. Ogoshi (1975) indicated that his AG5 is probably similar to group B of Richter & Schneider (1953) but that anastomosis tests had not been made. AG5 isolates are present in Minnesota soils and these isolates do anastomose with Ogoshi's AG5 tester strains (N. A. Anderson, unpublished). It appears that the AG phenomenon is associated with speciation in *R. solani* in all parts of the world.

Hyphae of field isolates within each AG anastomose. Those in different AGs fail to recognise each other and fusion does not occur. Following fusion between field isolates of the same AG on nutrient- or water-agar, five to six compartments on either side of the fusion compartment become vacuolated and die. A cleared zone of dead compartments was first noted in paired AG 2-1 isolates and called the killing reaction by Flentje, Stretton & McKenzie (1967). It is due to heterogenic incompatibility (see also Brasier: Chapter 21; Anagnostakis: Chapter 22; Rayner *et al.*; Chapter 23) and has now been noted in all AGs. In fact its presence between two isolates can be used as a taxonomic tool, indicating that they belong to the same AG.

In AG2, some isolates have been found that anastomose with both type 1 and type 2 isolates (Parmeter *et al.*, 1969). Serological tests made by Adams & Butler (1979) using sixty-four field isolates representing all AGs of *R. solani*, indicated that each AG was distinctive serologically but that type 1 and type 2 isolates of AG2 and the sasakii and microsclerotia types of AG1 were not distinguishable by serological tests. It appears that these subgroups are in the process of diverging and that the micro-evolutionary changes necessary to negate hyphal recognition and fusion have not been completed.

In AG1, microsclerotial forms, Castanho, Butler & Shepherd (1978) found that dsRNA was associated with a degenerative disease of one isolate. Attempts by Castanho & Butler (1978) to transfer the disease agent to four other AG1 microsclerotial strains via hyphal anastomosis were unsuccessful and the authors concluded that the dsRNA was strain specific. It is also likely that heterogenic incompatibility impedes the spread of the disease agent. An analogous situation in *Endothia parasitica* is discussed in detail by Anagnostakis (Chapters 16, 22). Hashiba & Yamada (1982) have, however, developed a procedure to obtain protoplasts from AG-1 (Sasakii) isolates and this may facilitate transfer of nuclei and cytoplasm between isolates and be useful in genetical and biological control studies (cf. Croft & Dales: Chapter 20).

No studies have yet been made on the genetic control of heterogenic incompatibility in any AG of *R. solani*. However, the intensity of the reaction varies in different pairings and indications are that it is probably controlled by several genes. Within an AG, the killing reaction is evident in pairings between different field isolates, in pairings between field isolates and homokaryons, and in pairings between different homokaryons (Fig. 1), but does not occur when an isolate is paired with itself. In pairings between homokaryons obtained from field isolates from widely separated geographic areas, the killing reaction is often so intense that only a few small tufts of heterokaryotic hyphae are formed (Fig. 2). This was the case when Anderson, Stretton, Groth & Flentje (1972) paired sexually compatible AG4 homokaryons from field isolates from widely separated geographic areas. On the other hand, when selected homokaryons from the same field isolate are paired, large tufts occur. The largest tufts that I have observed formed at the junction of compatible homokaryons that had been in-bred and selected for large tufts for three generations (N. A. Anderson, unpublished). Thus it appears, in AG4 isolates, that formation of tuft hyphae which initiates the heterokaryon and is part of the homogenic incompatibility system

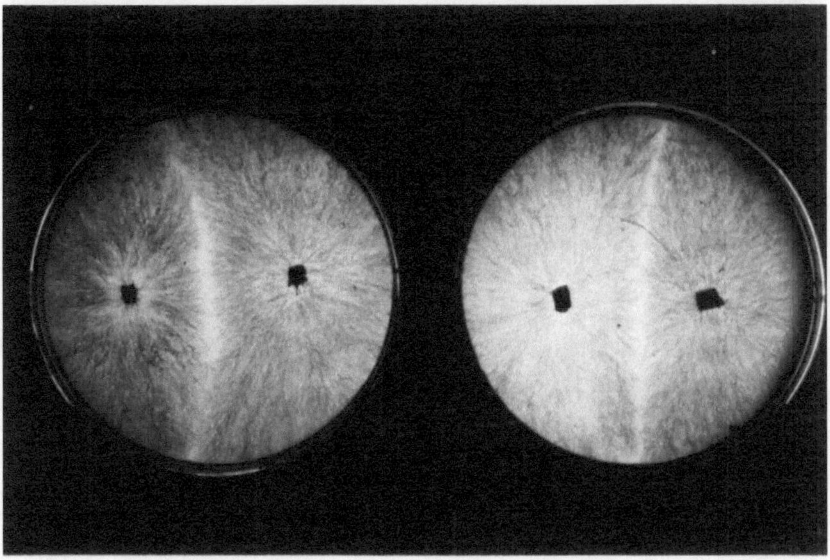

Fig. 1. Heterogenic incompatibility reaction between AG4 homokaryons with similar H factors.

(see below) is maximised when the heterogenic incompatibility system is minimised. The interaction between homogenic and heterogenic incompatibility in Basidiomycotina is discussed by Rayner *et al.* in Chapter 23.

Homogenic incompatibility

As previously indicated, homogenic incompatibility in *R. solani* results in heterokaryotic tufts of mycelium between certain pairs of homokaryons (Fig. 3). It was first detected in AG1 microsclerotial isolates by Whitney & Parmeter (1963). In AG4 it was first reported by Anderson *et al.* (1972) who found it to be controlled by two closely linked genes collectively termed the H-factor which regulates outbreeding and heterokaryon initiation via the formation of tuft hyphae when homokaryons with different H-factors are paired on migration complete agar (Fig. 3). Apparently the H-factor is a master switch that allows many genes to function once it is operative. Puhalla & Carter (1976) were able to show using heterokaryons forced between auxotrophic mutants with the same H-factor that similar H-factor heterokaryons could be synthesised but did not form tuft hyphae. They were, however, not as stable as dissimilar H-factor heterokaryons, and it was concluded that while the H-factor promotes outcrossing, as previously

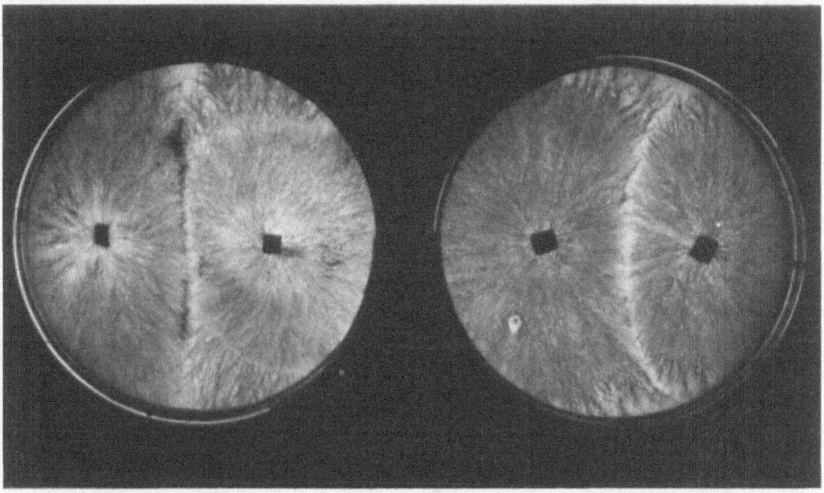

Fig. 2. Simultaneous heterogenic and homogenic reactions. Petri dish on left, note line of darkened tuft hyphae between H11 (left) and H12 (right) homokaryons from field isolate 42. Petri dish on right, very small heterokaryotic tuft hyphae forming between H11 (left) and H23 (right) homokaryons from field isolates from Britain and Canada respectively. Note strong killing reactions.

reported, it also regulates nuclear pairing and heterokaryon stability and hence is an incompatibility factor. Adams & Butler (1982) reached the same conclusion that the H-factor was a homogenic incompatibility system from their studies on homokaryotic fruiting in AG4 isolates.

In culture, the rapidly growing heterokaryotic tuft hyphae are not in contact with the agar surface. Small knots of hyphae form down in the agar at the base of the tuft hyphae and appear to function as micro-pumps supplying nutrients to the growing tips. In soil, the apical compartments of the heterokaryotic mycelium may be capable of independent absorption of nutrients.

Evidence that in AG4 the H-factor is composed of two genes up to 2.2 cross-over units apart (Anderson *et al.*, 1972) is based on detection of recombinant H-factors in homokaryons that form a mycelial tuft when paired with both parental H-factor tester strains. In an experiment with AG4 field isolate 42 which contains H factors H11 and H12, a total of 500 single basidiospore cultures were established; 251 formed the tuft reaction with the H11 and 221 with the H12 parental cultures. A total of eleven homokaryons (2.2%) formed tuft hyphae with both parent cultures. On the basis of their pairing interactions, the eleven homo-

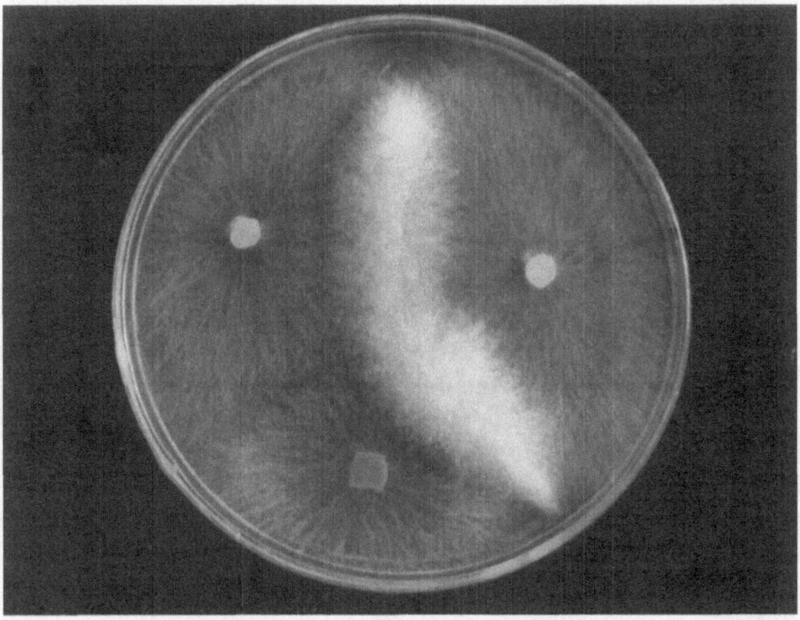

Fig. 3. Homogenic incompatibility reactions between three homokaryons from the same AG4 field isolate.

karyons were placed into two groups. Tufts formed in inter-group pairings but not in intra-group pairings.

One homokaryon from each of the two recombinant classes was selected. These were paired on migration complete agar and small bits of the tuft hyphae removed, and placed on water agar for 24 h; then a hyphal tip was excised and placed on potato dextrose agar supplemented with 0.5% yeast extract and allowed to grow for 3 days at 25 °C; 3-mm discs were cut from the periphery of the colony and transferred to soil extract agar at 20 °C for sporulation. A total of 502 single basidiospore cultures were established from this heterokaryon and 248 formed a tuft with one recombinant parent and 237 with the other. Eight cultures formed a tuft reaction with both recombinant parent cultures. Again these recombinant cultures could be placed into two classes on the basis that tufts formed only in inter-group pairings, none in intra-group pairings. Four of these second generation recombinant H-factors had the H11, the other four the H12 H-factor. Thus the original H-factors of isolate 42 had been reconstructed. The fact that tuft hyphae formed when the recombinant H-factor cultures were paired with both parental H-factors indicated that only one of the two genes of the H-factor must be heterozygous for the tuft hyphae to form and for the homogenic incompatibility system to function. The heterokaryotic tuft hyphae that formed at the junction of recombinant H-factor cultures (sib pairings) were usually much larger and more vigorous than those formed when homokaryons from different field isolates were paired. When these large mycelial tufts formed, no killing reaction was noted. This accords with the idea of simultaneous and reciprocal functioning of homogenic and heterogenic incompatibility mentioned previously.

Flentje, Stretton & Hawn (1963) have shown in AG4 isolates that nuclei in the apical compartments of hyphae divide synchronously. Genetic evidence from AG1 and AG4 field isolates is that they contain two kinds of nuclei, and all pathogenic AG1 and AG4 field isolates that have sporulated in culture to the present time have been found to contain only two H-factors. Data from Anderson *et al.* (1972) with AG4 field isolates and synthesised heterokaryons indicate there was nuclear pairing and strict outcrossing between the two nuclear components of the heterokaryons. No selfing was observed and 1 : 1 : 1 : 1 segregation ratios were noted for the two parental and two non-parental gene pairs studied. More than two different H-factor nuclei have apparently been introduced into a heterokaryon by Bolkan & Butler (1974) but these isolates were non-pathogenic. Such heterokaryons will be discussed in more detail below.

In both AG1 and AG4, small tufts occasionally form when a field isolate is paired with one of its homokaryons or with a homokaryon from a different field isolate with unlike H-factors. However, it should be noted that without the addition of 1% charcoal to the medium (see discussion of heterokaryosis below) this is a rare event. Among the 500 single spore cultures that were established from field isolate 42 mentioned earlier, 17 did not usually form a tuft with either the H11 or H12 parents, producing the killing reaction instead. However, 4 of these 17 cultures did occasionally form very small tufts with one or both tester isolates, and when these sporulated, it was found that they contained both the H11 and H12 factors. It appears that an important adjunct of heterogenic incompatibility is the limitation of the number of H-factors that exist in a heterokaryon to two, thus maintaining the fitness as a pathogen.

To obtain an estimate of the number of H-factors in AG4 isolates in nature, Anderson *et al.* (1972) obtained two different H-factor homokaryons from each of 15 field isolates from Australia, Britain, Canada, and the United States and paired them in all possible combinations. Two South Australian isolates from the same field had similar H-factors so only one pair of H-factor homokaryons was retained in the sample. Pairings that resulted in hyphal tufts at the junction of the colonies indicated that dissimilar H-factors were involved. Pairings that did not result in tuft hyphae being formed were repeated up to eight times and indicated that similar H-factors were involved. A total of 14 distinct H-factors were found in the sample. Two recombinant H-factors obtained from isolate 42 from Great Britain and two recombinant H-factor cultures from isolate 146 from Minnesota were then added to the sample. One recombinant H-factor from isolate 42 was identical with the H2 factor from Minnesota, the other recombinant H-factor was similar to H19 from Saskatchewan, Canada. One recombinant H-factor from isolate 146 was similar to H4 from western Nebraska, US, the other recombinant H-factor was unique and formed a tuft with all H-factor cultures in our sample. Thus a total of 15 H-factors was detected. Using the formula of Dobzhansky & Wright (1941), a total of 17 H-factors was estimated to exist in the population as a whole. Burnett (1964) has indicated that once 20 alleles are available in a bipolar mating system that the potential outbreeding bias (non-sister : sister matings) becomes asymptotic. The outbreeding efficiency of a bipolar system with 17 alleles is 94%.

Multiple alleles exist at both genes of the H-factor. So far it has not been possible to get the necessary recombinant H-factors to assign the number of alleles to each of the two genes. However, using the four recombinant H-factors plus the eight repeated H-factors found in the sample, it appears there are three alleles at each gene (N. A. Anderson, unpublished). The allele designation of seven H-factors is yet to be determined.

In *Schizophyllum commune* suppressor genes have been found to affect recombination sites and control recombination within the two locus B-factor (Koltin & Stamberg, 1973). A similar mechanism may be important in AG1 and AG4 isolates of *R. solani.* Increased recombination of the homogenic incompatibility factors has the potential to increase inbreeding and could be important to an organism introduced into a new ecological niche. One mechanism that does influence the inbreeding : outbreeding ratio in AG1 and AG4 is production of 2-, 3-, 4- and 5-spored basidia. Counts were made on the number of sterigmata per basidium for three AG4 and one AG1 microsclerotial isolates that sporulated on water agar. The proportions of binucleate spores produced by these four isolates were estimated by Anderson (1982) to be 17.8, 32.9, 33.3 and 11.4% respectively. The proportions of the above spores that would contain two different H-factors were estimated to be 11.7, 21.5, 21.9 and 7.5%. Genetic data from sporulating AG4 field isolates indicate the proportions of spores with two different H-factors (they did not form a tuft when paired with either H-factor tester parent) as determined by Anderson *et al.* (1972) were 0.6, 3.4 and 6.4%. It might be possible to detect shifts in the number of 2-, 3- and 4-spored basidia and thus a shift in their inbreeding : outbreeding ratios by altering the medium they are grown on in the laboratory prior to sporulation. Alternatively, observations could be made on the ratio of 2-, 3-, 4-spored basidia of AG4 isolates during the several years they are pathogens on a heterozygous crop like alfalfa and then note the ratio the following year when they are pathogens of pure line cultivars like flax or peas.

No analysis has been made on the fine structure of the H-factor of AG1 microsclerotial isolates. The following analysis was made using AG1 field isolate 53 from Minnesota (N. A. Anderson, unpublished) from which 490 single spore cultures were grown and paired with the two parental H-factor cultures. A total of 246 homokaryons formed a tuft with H-factor 101, 233 formed a tuft with H-factor 102, four homokaryons did not form a tuft with either parental H-factor culture, and seven homokaryons formed a tuft with both H-factor cultures. The seven homokaryons that formed a tuft with both parental H-factor cultures were paired in all combinations and two classes were detected. Intra-class pairings did not produce tuft hyphae but inter-class pairings did. Recombinant H-factor cultures were paired and heterokaryons established but so far they have failed to sporulate, and hence to allow completion of the fine-structure analysis. However, because the pairing reaction of the seven recombinant homokaryons indicated two classes of recombinants, the evidence suggests that two genes are involved in the AG1 H-factor.

Homokaryotic fruiting has been studied in AG4 isolates by Adams & Butler (1982). They found that all progeny from one homokaryon were self-sterile and that 96% of the homokaryotic progeny from six other homokaryons were self-sterile. In their studies, heterokaryosis enhanced basidiospore formation. They concluded that AG4 isolates of *R. solani* were not homothallic and that the H-factor was a homogenic incompatibility system as Puhalla & Carter had indicated (1976). Adams & Butler (1983*a,b*) indicated that the study on homokaryotic fruiting was enhanced by their previous work on the nutritional and environmental conditions that favour sporulation of AG1 and AG4 isolates.

Heterokaryosis

There appear to be several kinds of heterokaryons among the AGs of *R. solani.* In AG1, microsclerotial types, and AG4, the H-factor is involved in the initiation and stability of the heterokaryon as well as maintaining the ratio of the two H-factor components in its multinucleate hyphal tip compartments. No such H-factor mechanism has been found in AG2 and AG3 isolates. These isolates are found at deeper levels in the soil and there are indications that they have a more-limited pathogenic host range. The AG1 and AG4 isolates reside near the soil surface and often are foliar pathogens of many host species.

The tuft hyphae that form at the junction of paired AG1 and AG4 homokaryons with dissimilar H-factors may be regarded as a special mechanism by which the heterokaryon is initiated, allowing it to grow away from staling products. This phenomenon is apparently rather rare in higher fungi but similar heterokaryotic tufts have been reported by Raper (1974) in *Agaricus bitorquis*, by Sanchez, Leary & Endo (1976) in

Fusarium oxysporum f. sp. *lycopersici*, by Fatemi & Nelson (1977) in *Pyricularia oryzae*, and by Puhalla (1968) in *Ustilago maydis.*

Proof of heterokaryosis in the tuft hyphae of AG1 microsclerotial isolates was presented by Whitney & Parmeter (1963). In AG4 isolates, Anderson *et al.* (1972) used auxotrophic mutants and dissimilar H-factors to synthesise heterokaryons. Small bits of the tuft hyphae were placed on water agar; after 24 h growth, hyphal tips were excised and placed on potato dextrose agar supplemented with 0.5% yeast extract. After 3 days, transfers were made to soil extract agar and the plates incubated at 20 °C. Upon sporulation strict outcrossing ratios of the two parental and two non-parental genes were obtained.

Puhalla & Carter (1976) studied the stability of AG4 heterokaryons with similar and dissimilar H-factors, using hyphal tips excised from forced heterokaryons growing on minimal medium. The frequency of prototrophic: auxotrophic hyphal tips obtained from the two kinds of heterokaryons was used to measure their stability. Hyphal tip cultures not prototrophic on minimal medium had their growth requirement determined by growing them on minimal medium plus the required growth substance. Heterokaryons with dissimilar H-factors had fewer auxotrophic tips and more prototrophic tip compartments when grown on minimal medium than similar H-factor heterokaryons. The authors concluded that heterozygous H-factors contributed to the stability of the heterokaryon.

Hyphal anastomosis has been shown to occur between similar AG1 H-factor homokaryons by Garza-Chapa & Anderson (1966). The H-factor does not control anastomosis but is involved in the formation of heterokaryotic tuft hyphae and other post-fusion events as a homogenic incompatibility system. The role of similar H-factor heterokaryons in nature is not known. Although they might contribute to genetic variation within an AG, it does not appear likely that they would be important plant pathogens in nature due to their instability and would be selected against. As mentioned earlier, all pathogenic AG1 and AG4 field isolates have been found to contain only two H-factors.

Hyphae of field isolates and synthesised heterokaryons of AG1 and AG4 have multinucleate tip compartments. Nuclear division in these compartments is synchronised and most genetic studies have indicated that with dissimilar H-factors there is a 1:1 ratio of the two kinds of nuclei. As mentioned in the section on heterogenic incompatibility, it appears that the killing reaction is a means of maintaining the genetic integrity of the heterokaryon when it encounters other heterokaryons or homokaryons, as probably occurs generally in Basidiomycotina (Rayner *et al.*, Chapter 23). In a recent test (N. A. Anderson, unpublished), field isolate 42 from Great Britain, with H11 and H12 H-factors, was paired on migration complete agar with the H1 and H2 homokaryons from isolate 127 from Minnesota. Sixteen of each of the above pairings were made and no tuft hyphae were observed. However, when field isolate 42 was paired on the same medium with compatible recombinant H-factor cultures, 20 and 69, from isolate 42, nine of fourteen pairings involving culture 20, and twelve of sixteen pairings involving culture 69 formed hyphal tufts. It appears that nuclear transfer via the above tuft hyphae is comparable to di-mon pairings in other Basidiomycotina and occurs most readily when the strains are closely related and the heterogenic incompatibility effects are minimised. In work with forced heterokaryons there was an indication that a choline mutant somehow affected the 1:1 nuclear ratio in the hyphal tip compartments (Puhalla & Carter, 1976). In cases where the tip compartment contained only one kind of nucleus one would expect tuft hyphae to form quite readily when paired with a homokaryon.

Butler & Bolkan (1973) found that AG4 heterokaryons containing three or four different H-factors could be synthesised on agar medium containing 1% charcoal. Their data indicate that tuft hyphae and heterokaryons formed in the following pairings that did not contain a common H-factor: field plus field isolate, field isolate plus a homokaryon, heterokaryon plus heterokaryon, heterokaryon plus homokaryon. However, all heterokaryons with more than two H-factors were non-pathogenic. Bolkan & Butler also obtained fifteen AG4 isolates from soil and in tests on germinating cabbage seedlings in Petri dish tests, found that nine isolates were non-pathogenic, two had low virulence and four were highly virulent. None of these isolates sporulated so the number of H-factors could not be determined. The authors stated that the nine non-pathogenic isolates might be heterokaryons with more than two H-factors. Anderson (1982) has warned plant pathologists not to mix AG4 isolates when growing inoculum for disease tests. Additional research is needed on the nuclear condition and number of H-factors in AG4 field isolates that are nonpathogenic.

In AG2, two subgroups have been identified by Ogoshi (1976). Anastomosis usually occurs only within the groups listed as type 1 or type 2, but some isolates have been found that will fuse with both types. Although no H-factor mechanism was detected in AG2 isolates by Flentje & Stretton (1964) it appears that field isolates are heterokaryotic. Dodman (1972) used auxotrophic mutants to synthesise forced AG2-1 heterokaryons. Segregation ratios from these sporulating heterokaryons indicated strict outcrossing. In another study on AG2-1 isolates from crucifers, heterokaryons were not forced and Stretton & Flentje (1972*a*) noted some selfing. The existence of a homogenic incompatibility system in AG2 isolates is yet to be determined.

In Minnesota soils, Grisham (1978) isolated AG1 microsclerotial forms, AG2 types 1 and 2, AG3, AG4 and AG5 isolates from canker lesions on carrot roots. Only the AG2-2 isolates caused a canker and crown rot disease of carrot. In the autumn of the growing season just prior to harvest, hymenia of *R. solani* were noted on carrot petioles. However these were AG2-1 isolates, were radish pathogens from the radish crop the previous year, and did not infect carrot. When basidiospores of these AG2-1 isolates were cast onto nutrient agar the spores germinated but colonies developed only when several germ tubes anastomosed. With these isolates it appears that some mechanism is operating that enforces heterokaryosis soon after spore germination.

Both Ogoshi (1975) and Parmeter *et al.* (1969) reported AG2 isolates that caused a web blight disease of plants as well as isolates that mainly caused disease of hypocotyls and roots of crucifers. Based on anastomosis studies, Ogoshi placed both of these different isolates in AG2-1. Stretton & Flentje (1972*b*) attempted to make heterokaryons between the above two AG2-1 forms by using homokaryons originating from *Phaseolus* leaves and crucifer stems. From approximately 100 pairings involving *Phaseolus* plus crucifer mutants only four wild type heterokaryons were recovered, and only one of these sporulated. Since Stretton & Flentje (1972*a*) had already reported that heterokaryons formed readily between homokaryons from four different crucifer isolates, it appears that the aerial and soil forms of AG2-1 are diverging. From a pathologist's point of view they should therefore be regarded as different evolutionary units when testing for disease resistance.

AG3 isolates of *R. solani* are primarily pathogens of potato. They cause stem and stolon lesions and form sclerotia on tubers. Murray (1981) reported that AG3 isolates caused a stunt disorder of barley in Scotland. Studies on heterokaryosis have been hindered because AG3 isolates do not sporulate readily in culture. However, Murray (1981) was able to induce sporulation in sixteen AG3 isolates in the laboratory, and later by refining his techniques was able to get AG2-1, AG3 and AG5 isolates to sporulate (Murray, 1982). Our experience in Minnesota is that basidiospore germination on nutrient agar is usually less than 5%.

In nature basidiospores form at the base of potato stems and are common once the foliage is well developed and a high humidity can be maintained. It would be interesting to determine the breeding biology of AG3 in parts of the world where the potato is indigenous. Kuninaga *et al.* (1979) reported that AG3 and AG4 isolates were found only in cultivated soils. Perhaps AG3 isolates that are so common wherever potatoes are grown were introduced into these areas as sclerotia on tubers.

Zachman (1972) paired 13 single AG3 basidiospore cultures in all combinations and only one pairing readily gave him a heterokaryon based on morphology and pathogenicity. Hill (1980) paired single basidiospore cultures and obtained 'heterokaryons' that differed in morphology from the parent cultures but was unable to get the putative heterokaryons to sporulate to prove heterokaryosis. No tuft reaction was reported in Hill's pairings of single basidiospore cultures and he concluded there was no H-factor mechanism in the AG3 isolates he studied.

The AG3 isolates of potato found in Minnesota most closely resemble AG2-2 isolates that cause a canker of carrots, sugar and table beets. Kuninaga *et al.* (1979) reported that isolates in AG B-1 anastomosed with field isolates in AG2-1, AG2-2, AG3 and AG6, but not with field isolates in AG1 and AG4. From their work it appears that events leading to speciation are more advanced in AG1 and AG4, and that AG2 and AG3 isolates are more closely related to each other than to the aerial forms.

The role of heterokaryosis in pathogenicity and virulence

The role of heterokaryosis in pathogenicity of AG2-1 and AG4 isolates has been studied. A method to study the infection process was first defined in AG2-1 isolates by Flentje *et al.* (1967). The six steps that were determined to be controlled by single genes are: (1) inhibition, no fungus growth on hypocotyls; (2) growth but no attachment to the hypocotyl; (3) growth and attachment of hyphae to the hypocotyl; (4) infection cushions formed but no penetration of host cell walls; (5) hypersensitive reaction; and (6) pathogenic reaction.

Stretton & Flentje (1972*a,b*) used this technique to study the role of heterokaryosis between AG2-1 isolates of similar and different pathogenicity respectively. As mentioned earlier, homokaryons from crucifer isolates from widely separated geographic areas formed heterokaryons readily. The aerial and soil-inhabiting isolates of AG2-1 did anastomose but heterokaryosis between homokaryons from the two forms was rare. This was significant as Stretton & Flentje used a macerating technique to obtain heterokaryons and this method may decrease the heterogenic incompatibility reaction. Therefore it appears that within AG2-1, an isolating mechanism limits heterokaryosis and outcrossing between the aerial and soil-inhabiting forms.

In studies with AG4 isolates, Anderson & Stretton (1978) studied two different heterokaryons comprised of homokaryons with mutations affecting step 2 and step 4 of the infection cycle. The heterokaryons were pathogenic and highly virulent. The mutant factors segregated as single genes and were independent of the H-factors. A third heterokaryon composed of a step 4 mutant and a wild type homokaryon was pathogenic and segregated in a 1:1 ratio of pathogenic to non-pathogenic. A fourth heterokaryon was synthesised between two step-2 mutant homokaryons with different H-factors. This heterokaryon was pathogenic and it appears that the mutations in these two homokaryons were in different cistrons. Stretton & Flentje (1972*b*) reported on a heterokaryon composed of two step-3 mutants that apparently were alleles as this heterokaryon was non-pathogenic.

Vest & Anderson (1968) studied the rôle of heterokaryosis on virulence in AG1 microsclerotial isolates using homokaryons with low-virulence ratings. A total of 62 heterokaryons were synthesised between homokaryons of low virulence and 51 were significantly more virulent than either parent on at least one of the two flax cultivars on which they were tested. Virulence was thought to be conditioned by genes having overdominance or epistatic effects.

The auxotrophic, infection process, and virulence mutants used in studies at Minnesota were all naturally occurring. This indicates that AG1 and AG4 field isolates do shelter mutations and are a storage mechanism for genetic variation. Heterokaryosis is important in the infection process and the amount of disease an isolate causes. It promotes sporulation in an isolate and in dissimilar H-factor heterokaryons promotes outbreeding and heterokaryon stability.

References

Adams, G. C. & Butler, E. E. (1979). Serological relationships among anastomosis groups of *Rhizoctonia solani. Phytopathology*, **69,** 629–33.

Adams, G. C. & Butler, E. E. (1982). A re-interpretation of the sexuality of *Thanatephorus cuumeris* anastomosis group four. *Mycologia*, **74,** 793–800.

Adams, G. C. & Butler, E. E. (1983*a*). Influence of nutrition on the formation of basidia and basidiospores in *Thanatephorus cucumeris. Phytopathology*, **73,** 147–51.

Adams, G. C. & Butler, E. E. (1983*b*). Environmental factors influencing the formation of basidia and basidiospores in *Thanatephorus cucumeris. Phytopathology*, **73,** 152–5.

Anderson, N. A. (1982). The genetics and pathology of *Rhizoctonia solani. Annual Review of Phytopathology*, **20,** 329–47.

Anderson, N. A. & Stretton, H. M. (1978). Genetic control of pathogenicity in *Rhizoctonia solani. Phytopathology*, **68,** 1314–17.

Anderson, N. A., Stretton, H. M., Groth, J. V. & Flentje, N. T. (1972). Genetics of heterokaryosis in *Thanatephorus cucumeris. Phytopathology*, **62,** 1057–65.

Bolkan, H. A. & Butler, E. E. (1974). Studies on heterokaryosis and virulence of *Rhizoctonia solani. Phytopathology*, **64,** 513–22.

Burnett, J. H. (1964). The natural history of recombination systems. In *Incompatibility in Fungi*, ed. K. Esser & J. R. Raper. New York: Springer-Verlag, 124 pp.

Butler, E. E. & Bolkan, H. (1973). A medium for heterokaryon formation in *Rhizoctonia solani. Phytopathology*, **63,** 542–3.

Castanho, B. & Butler, E. E. (1978). *Rhizoctonia* decline: studies on hypovirulence and potential use in biological control. *Phytopathology*, **68,** 1511–14.

Castanho, B., Butler, E. E. & Shepherd, R. J. (1978). The association of double-stranded RNA with *Rhizoctonia* decline. *Phytopathology*, **68,** 1515–19.

Dobzhansky, T. & Wright, S. (1941). Genetics of natural populations. V. Relations between mutation rates and accumulation of lethals in populations of *Drosophila pseudoobscura. Genetics*, **26,** 23–52.

Dodman, R. L. (1972). Heterokaryon formation and genetic recombination between auxotrophic and morphological mutants of *Thanatephorus cucumeris. Australian Journal of Biological Science*, **25,** 739–48.

Fatemi, J. & Nelson, R. R. (1977). Intra-isolate heterokaryosis in *Pyricularia oryzae. Phytopathology*, **67,** 1523–5.

Flentje, N. T., Stretton, H. M. & Hawn, E. J. (1963). Nuclear distribution and behaviour throughout the life cycles of *Thanatephorus*, *Waitea* and *Ceratobasidium* species. *Australian Journal of Biological Science*, **16,** 450–67.

Flentje, N. T. & Stretton, H. M. (1964). Mechanisms of variation in *Thanatephorus cucumeris* and *T. praticolus. Australian Journal of Biological Science*, **17,** 686–704.

Flentje, N. T., Stretton, H. M. & McKenzie, A. R. (1967). Mutation in *Thanatephorus cucumeris. Australian Journal of Biological Science*, **20,** 1173–80.

Garza-Chapa, R. & Anderson, N. A. (1966). Behavior of single-basidiospore isolates and heterokaryons of *Rhizoctonia solani* from flax. *Phytopathology*, **56,** 1260–8.

Grisham, M. (1978). Variations in the pathogenicity and host specificity of isolates of *Rhizoctonia solani* associated with carrots. Ph.D. Thesis, University of Minnesota, St Paul. 75 pp.

Hashiba, T. & Yamada, M. (1982). Formation and purification of protoplasts from *Rhizoctonia solani. Phytopathology*, **72,** 849–53.

Hill, C. B. (1980). The biology and host specificity of anastomosis group 3 isolates of *Rhizoctonia solani* Kuhn from potato. M.S. Thesis, University of Minnesota, St Paul. 175 pp.

Koltin, Y. & Stamberg, J. (1973). Genetic control of recombination in *Schizophyllum commune*: location of a gene controlling B-factor recombination. *Genetics*, **74,** 55–62.

Kuninaga, S., Yokosawa, R. & Ogoshi, A. (1979). Some properties of Anastomosis Group 6 and B1 in *Rhizoctonia solani* Kuhn. *Annals of Phytopathology Society of Japan*, **45,** 207–14.

Murray, D. I. L. (1981). *Rhizoctonia solani* causing barley stunt disorder. *Transactions of the British Mycological Society*, **76**, 383–95.

Murray, D. I. L. (1982). A modified procedure for fruiting *Rhizoctonia solani* on agar. *Transactions of the British Mycological Society*, **79**, 129–35.

Ogoshi, A. (1972). Some characters of hyphal anastomosis groups in *Rhizoctonia solani* Kuhn. *Annals of the Phytopathology Society of Japan*, **38**, 123–9.

Ogoshi, A. (1975). Grouping of *Rhizoctonia solani* Kuhn and their perfect stages. *Review of Plant Protection Research*, **8**, 93–103.

Ogoshi, A. (1976). Studies on the grouping of *Rhizoctonia solani* Kuhn with hyphal anastomosis and on the perfect stages of groups. *Bulletin of the National Institute of Agricultural Science Section C*, **30**, 1–63.

Parmeter, J. R. Jr., Sherwood, R. T. & Platt, W. D. (1969). Anastomosis grouping among isolates of *Thanatephorus cucumeris*. *Phytopathology*, **59**, 1270–8.

Parmeter, J. R. Jr. & Whitney, H. S. (1970). Taxonomy and nomenclature of the imperfect state. In *Rhizoctonia solani*: *Biology and Pathology*, ed. J. R. Parmeter Jr, pp. 7–19. Berkeley: The University of California Press.

Puhalla, J. E. (1968). Compatibility reactions on solid medium and interstrain inhibition in *Ustilago maydis*. *Genetics*, **60**, 461–74.

Puhalla, J. E. & Carter, W. W. (1976). The role of the H locus in heterokaryosis in *Rhizoctonia solani*. *Phytopathology*, **66**, 1348–53.

Raper, C. A. (1974). The biology and breeding potential of *Agaricus bitorquis*. *Mushroom Science*, **9**, 1–10.

Richter, H. & Schneider, R. (1953). Untersuchungen zur morphologischen und biologischen Differenzierung von *Rhizoctonia solani* K. *Phytopathologische Zeitschrift*, **20**, 167–226.

Sanchez, L. E., Leary, J. V. & Endo, R. M. (1976). Heterokaryosis in *Fusarium oxysporum* f. sp. *lycopersici*. *Journal of General Microbiology*, **93**, 219–26.

Schultz, H. (1937). Vergleichende Untersuchungen zur Okologie, Morphologie, und Systematik des 'Vermehrung pilzes'. *Arbeiten Biologischen Reichanstalt fur Land- und Forstwirtschaft, Berlin*, **22**, 1–41.

Sherwood, R. T. (1969). Morphology and physiology in four anastomosis groups of *Thanatephorus cucumeris*. *Phytopathology*, **59**, 1924–9.

Stretton, H. M. & Flentje, N. T. (1972*a*). Inter-isolate heterokaryosis in *Thanatephorus cucumeris*. I. Between isolates of similar pathogenicity. *Australian Journal of Biological Science*, **25**, 293–303.

Stretton, H. M. & Flentje, N. T. (1972*b*). Inter-isolate heterokaryosis in *Thanatephorus cucumeris*. II. Between isolates of different pathogenicity. *Australian Journal of Biological Science*, **25**, 305–18.

Talbot, P. H. B. (1970). Taxonomy of the perfect state of *Rhizoctonia solani*. In *Rhizoctonia solani: Biology and Pathology*, ed. J. R. Parmeter Jr., pp. 20–31. Berkeley: The University of California Press.

Vest, G. & Anderson, N. A. (1968). Studies on heterokaryosis and virulence of *Rhizoctonia solani* isolates from flax. *Phytopathology*, **58**, 802–7.

Watanabe, B. & Matsuda, A. (1966). Studies on the grouping of *Rhizoctonia solani* Kuhn pathogenic to upland crops. *Designated Experiment (Plant diseases and Insect pests) No. 7, Agriculture, Forestry and Fisheries Research Council and Ibaraki Agricultural Experiment Station* (In Japanese with English summary).

Whitney, H. S. & Parmeter, J. R. Jr. (1963). Synthesis of heterokaryons in *Rhizoctonia solani* Kuhn. *Canadian Journal of Botany*, **41**, 879–86.

Zachman, R. (1972). Untersuchungen uber die Variabilitat von *Rhizoctonia* im Hinblick auf die Resistenzzuchtung der Kartoffel. Ph.D. Thesis. Stuttgart: University of Hanover. 189 pp.

18

Interspecific mycelial interactions - an overview

A. D. M. RAYNER† and JOAN F. WEBBER*
*†School of Biological Sciences, University of Bath, Claverton Down, Bath, BA2 7AY, UK and *Forest Research Station, Alice Holt Lodge, Wrecclesham, Farnham, Surrey, GU10 4LH, UK*

The value of interaction studies

Two organisms may be said to interact when the presence of one in some way affects the performance of another. Such interactions are a salient feature of the fungal way of life, being an inevitable consequence of proximity or contact between individuals of the same or different species. They may occur quite simply because of common exploitation of a resource, or, as in the more complex example of mycoparasitism, because one organism serves directly as a nutrient source for another. Interspecific interactions in the former category are the particular concern of this chapter.

For organisms with a determinate body form, interspecific interactions are often regarded as becoming minimised by partitioning and specialism into separate niches. However, in mycelial fungi the indeterminate body form, combined with their versatility in requirements for growth and reproduction, increase the probability of interactions being both more obviously manifest and more frequent determinants of their distribution at any one time. Studies of interactions are therefore important if the development of natural fungal communities is to be understood. By the same token, they can also provide an understanding of the context within which the activities of fungi can be exploited in the biological control of pests and pathogens, under natural or semi-natural conditions. More fundamentally, studies of interaction patterns may help to provide an insight of processes involved in the functioning of mycelia and their capacity to co-ordinate their activities.

Approaches to interaction biology

Previous approaches

Interactions of fungi amongst themselves and with other types of organisms have long been studied, the diversity in approaches perhaps equalling the range of the interactions themselves. Attempts to clarify the interactions have, as with much of fungal ecology, been derived from ideas developed from other fields (Burkholder, 1952; Clark, 1954; Odum, 1954). This introduces the danger of applying concepts and principles appropriate for some organisms, but inappropriate for fungi. Further, confusion may arise if classifications of interactions between fundamentally different groups of organisms (as occurs for example, in the formation of mycorrhizas and leguminous root nodules) are applied to closely similar organisms. Another difficulty arises from the emphasis which has been traditionally placed upon soil fungi, which may have distorted the development of principles, overemphasising some facets, whilst underemphasising others.

In an early attempt at synthesis, Park (1960) described harmless or beneficial associations as symbiosis, but those detrimental to at least one of the interacting fungal partners as antagonistic (Fig. 1). Since then, it has become widely accepted that the mechanisms of antagonistic action include antibiosis, parasitism and competition – the latter usually involving rivalry for nutrients and growth factors, rarely for space and oxygen (Park, 1960, 1968; Clark, 1965; Baker & Cook, 1974). Whilst the boundaries between these three forms of antagonism were recog-

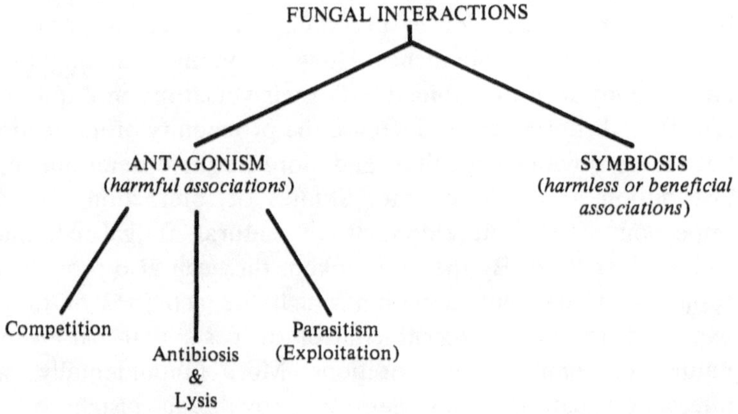

Fig. 1. Previous schema of fungal interactions. (After Park, 1960, 1968; Clark, 1965; Baker & Cook, 1974.)

nised to be indistinct, this has not prevented sometimes protracted debate concerning which mechanism is the most widespread or important. Early work on fungistasis in soils (Dobbs & Hinson, 1953; Dobbs, Hinson & Bywater, 1960) suggested to many that non-specific antibiosis was the primary form of antagonistic action. Characterisation of specific antibiotic substances, and the pairing of fungal isolates in culture, which often demonstrated inhibition in growth of one 'target' organism (mutual inhibition perhaps being often overlooked), tended to reinforce this concept (Brian, 1957).

However, the primary division shown in Fig. 1 into antagonism and symbiosis, whilst arguably applicable to interactions between dissimilar groups of organisms, is not a particularly helpful dichotomy when applied to mycelial interactions. It is difficult to reconcile the few known examples of harmless or beneficial associations between fungi with the widely accepted view of symbiosis as a persistent intimate association between individuals. In addition, a return to the original view of de Bary (1887) that all such persistent, intimate associations are symbioses regardless of their harmful, harmless or beneficial consequences (Lewis, 1973; Cooke, 1977), make present use of the term symbiosis ambiguous.

The subdivision of antagonism into three mechanisms may also confuse. Firstly, the term competition is used only in a narrow sense, out of keeping with the broader concepts of general ecology and evolution, and subservient in hierarchical terms to antagonism. The paradox is apparent in the exclusion from the scheme of Garrett's important concept of *competitive saprophytic ability*, because it encompases both competition for nutrients and antibiosis (Garrett, 1956, 1970). Secondly, whilst antibiosis has tended to be given prominence, other mechanisms, such as hyphal interference, have not been incorporated into the scheme. As already suggested this may be a legacy of the early emphasis on soil fungi. Soil being an open, heterogeneous, finely divided, spore-filled environment, observation of any active fungal mycelia within this system must be difficult. Consequently, experimental approaches to investigating interactions in soil have of necessity often been *in vitro*. As many soil fungi are typically those which produce abundant spores which germinate and grow rapidly on culture plates, notably many Fungi Imperfecti, these and their interactions will be given prominence.

A final and perhaps most important difficulty with a terminology based solely on consideration of mechanism rather than the outcome of interactions creates difficulty is that it introduces the constraint that before classification can be made, mechanism has to be decided upon, often without appropriate data to hand. Directly observable outcomes of interactions therefore provide a much more flexible basis for classification, particularly since a single outcome may result from a combination of mechanisms.

Garrett placed outcome before mechanism in his approach to interaction studies, when considering competitive saprophytic ability. Experimentally, this involved the exposure of an uncolonised resource (such as an agar plate, a wheat straw) to a mixed microbial inoculum (usually in soil) and analysis of the resultant colonisation patterns. Organisms achieving dominance in the early stages of colonisation were regarded as possessing high competitive ability, and only secondarily were the mechanisms of their success considered. However, consideration of what might follow once a resource had become fully occupied was omitted. Interaction outcomes during both early and late phases of colonisation were subsequently recognised by Gibbs & Smith (1978) and termed 'primary' and 'secondary' antagonism respectively. For reasons which will become apparent, this categorisation may be misleading when used in a general context, since it implies that events occurring during primary colonisation should necessarily be interpreted as 'antagonism'.

More recently, an attempt has been made to apply the concepts of *exploitation competition* and *interference competition* to fungi, these terms having initially been developed for other organisms (Lockwood, 1981; Wicklow, 1981*a*). Exploitation competition is restricted to the depletion of resources by one organism, or population, without reducing the access of another organism, or population, to the same resource pool. Interference competition involves behavioural or chemical mechanisms by which access to a resource is influenced by the presence of a competitor (McNaughton & Wolf, 1973). But the application of these terms to mycelial fungi is both inadvisable and misleading (Cooke & Rayner, 1984). During resource depletion many, and possibly most, mycelial fungi do restrict access of others, either by efficient nutrient uptake or through dense branching hyphal systems or other mechanisms (see below). Thus the term exploitation competition divorces competition for nutrients from a consideration of the space in which they are contained. The outcome of using the term is the conclusion that competition for nutrients is of prime importance, that for space being negligible (Lockwood, 1981). Whilst correct within context, the conclusion can be misleading since with most mycelial fungi occupation of space is inseparably linked with the utilisation of resources.

In view of these varied approaches and mixed philosophies we believe there is a need for a rationalisation of outlook commensurate with the mycelial way of life, that also takes account of the full range of fungal activities in natural habitats. A first step towards this is to examine how the types of interaction are related to the life strategies of mycelial fungi.

Interactions and life strategies

As will be apparent from other contributions to this Symposium, the fungal mycelium in its natural habitat is a heterogeneous complex, dynamic entity in which phases of establishment, exploration and exploitation of resources and reproduction occur in overlapping sequence. The timing and duration of these phases crucially affects the interactive properties of a mycelium and hence its role in population and community development.

Thus in some fungal species, rapid germination and extension growth are associated with essentially explorative mycelium, utilising only readily assimilable substrates with early commitment to reproduction. In other cases, the exploitative phase is prolonged, involving utilisation of more refractory substrates and delayed commitment to reproduction. These alternatives are the extremes of the spectrum between *r*- and *K*-selection in which organisms are visualised as being fitted either for an ephemeral or sustained existence respectively (e.g. MacArthur & Wilson, 1967; Harper & Ogden, 1970; Andrews & Rouse, 1982). Whilst an ephemeral existence obviates the requirement for maintenance of activity in the presence of other organisms, a sustained existence demands either the ability to withstand (or even exploit) the presence of other organisms or of conditions (stress) which militate against their survival. These considerations have led Grime (1979) to suggest for higher plants that three major ecological strategies exist: ruderal, competitive and stress-tolerant, and these same strategies can also be applied to fungi (Pugh, 1980). Using the term combative rather than competitive, the former being perhaps more appropriate to fungi, Cooke & Rayner (1984) have then attempted to rationalise the diversity of fungal activities within these three primary strategies.

Proposed schema

Following the above, mycelial interactions are described on the basis of proposals made by Cooke & Rayner (1984) and illustrated in Fig. 2. The primary division is into competitive, neutralistic and mutualistic interactions between two participants, depending respectively upon whether the outcome is detrimental to either or both, detrimental to neither but not beneficial to both, or beneficial to both.

One feature of the schema is that the hierachical position of competition and antagonism is reversed from that in Fig. 1, competition being used in its more usual broad sense, and antagonism more specifically as the mechanism underlying a particular type of competition (i.e. combat). We have followed the simple definition of competition between organisms, namely that of an active demand by two or more individuals of the same or different species for the same resource. Following from the earlier discussion of exploitation and interference competition and primary and secondary antagonism it is evident that during colonisation of an initially vacant resource, two basically different aspects of competition must be distinguished, which are termed primary resource capture and combat.

Primary resource capture describes the process of gaining initial access to, and influence over, an available resource. It is an apt term for mycelial fungi in which the indeterminate thallus facilitates potentially continuous expansion of territory. Since primary resource capture is dependent solely on priority of arrival and sequestration of resources before the occupied domains become contiguous, it does not involve direct challenge between individuals in close proximity except perhaps during initial establishment at the surface of the resource. Success in

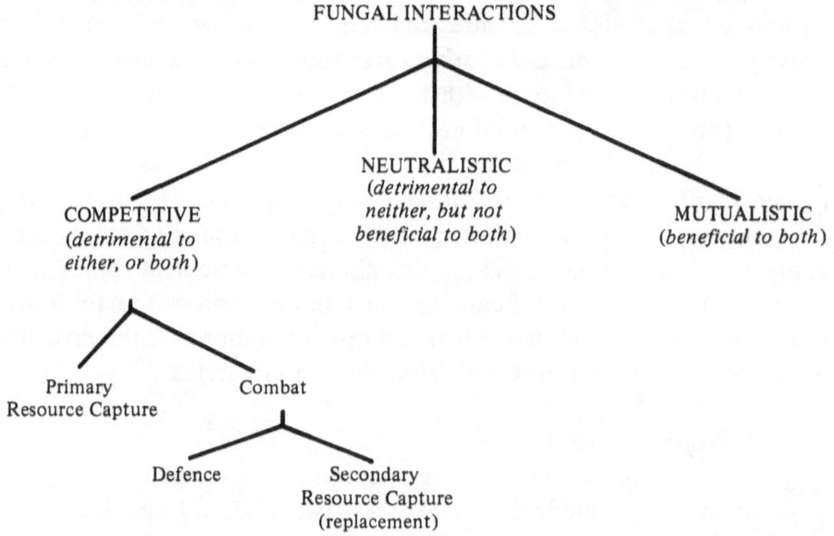

Fig. 2. Proposed schema of fungal interactions. (After Cooke & Rayner, 1984.)

primary resource capture is therefore determined by such factors as effective dispersal mechanisms, spore germination, mycelial extension rates, possession of suitable enzymes to utilise available substrates, and tolerance of adverse conditions (stress) associated with the resource. Thus fungi with at least partially ruderal strategies are favoured in primary capture of resources relatively free from stress by powers of early arrival and rapid exploration, whilst stress-tolerant fungi are favoured by being adapted to physicochemical or nutritional stress associated with the resource.

As primary resource capture proceeds so the domains of different individuals increasingly come into contact. When this happens three events may follow. There may be unilateral or bilateral intermingling via neutralistic or mutualistic interactions (see below), mutual exclusion, or replacement of one individual by another. The latter two events may under some circumstances result from purely nutritional factors; thus effective uptake of nutrients by one individual may be sufficient to deny another individual access to its domain. Alternatively, one individual which has exhausted supplies of nutrients which are assimilable to it may then be replaced by another individual capable of utilising the residue. However, mutual exclusion and replacement also occur as a result of more direct physiological challenge between individuals such that access to one individual's domain may be prevented by active defence, or achieved via active secondary resource capture mechanisms. This is described as combat, and those fungi which exhibit it as combative, or as having a combative ecological strategy. Collectively the mechanisms underlying combat are encompassed by the term antagonism. Of course the combative ability of a fungus is determined partly by its physiological state, so that changes in physicochemical or nutritional conditions associated with resource utilisation will affect the outcome of interaction between two combatants.

As has been implied, defence and secondary resource capture represent two different aspects of combat. In defence, access to resources gained by primary capture is denied to another individual. Hence success of a fungus able principally in defence will rest on its ability for effective primary capture. In secondary resource capture, access is gained to an already occupied domain, so that fungi able in this aspect of combat need not necessarily be effective in primary resource capture.

Although emphasis is normally placed on the competitive interactions just described, there are two obvious ways in which fungi may associate non-competitively in nature, via neutralism and mutualism.

There are two possible kinds of neutralistic associations between fungi. In the first, association results in no discernible benefit or harm to the associated species, interaction being either absent or so minor as to produce no detectable effects. This type of essentially passive coexistence, though it seems to be common between fungi and bacteria, is probably rare between fungi. In the second kind of neutralistic association, one of the associates benefits in some way from the activities of the other without conferring any benefit or harm in return.

Mutualistic associations are those in which each associate benefits from the activities of the other. In view of the wide range of habitats open to fungi, and the great variety of fungi commonly found within them, it is possible that mutualism and the second kind of neutralistic interaction commonly occur by chance. However, as will become apparent, too few examples are presently available to assess the extent and significance of neutralism and mutualism and their physiological bases.

Interactions in culture

Neutralistic and mutualistic interactions

The most direct initial evidence for a neutralistic or mutualistic interaction in culture is intermingling of mycelia. However, it must be noted that intermingling may be facilitated in sparse mycelial systems because interdigitating hyphal branches are allowed access to uncolonised substratum. Hence the nutritional conditions must be taken into account before a decision is made to the effect that a particular combination of individuals show an intermingling response. Low nutrient availability may allow apparent intermingling between correspondingly sparse mycelial systems, which under conditions of higher nutrient concentrations might be mutually exclusive. Here, intermingling at low concentrations is really due to continued exploration and primary resource capture. By the same token, it must be accepted that intermingling mycelia may still be depleting resources from the same resource pool and hence be in competition (cf. earlier mention of exploitation competition). For the association to be truly neutralistic different portions of the resource pool should be utilised. One way in which this could be achieved in culture is where 'staled' parts of a colony of one individual become occupied by another. Similarly, some fungi may survive on waste products or exudates from hyphae of another, especially if the latter are autolysing. *Penicillium* and *Cladosporium* contaminants of basidiomycete cultures probably behave in this way.

Under conditions which favour dense mycelial growth, intermingling between mycelia appears an exceptional event. However, fungi which have constitutively sparse, explorative mycelia such as certain coenocytic Mucorales, may invade domain already occupied by another fungus without necessarily deleteriously affecting the latter. In a study of the interactions of the Dutch Elm disease pathogen, *Ceratocystis ulmi*, with a wide range of bark-inhabiting fungi, only two instances of an apparent intermingling reaction were recognised, namely with *Cladosporium herbarum* and *Graphium penicillioides* (Webber, 1979; Fig. 3). Here, intermingling was manifested by stimulated production of synnemata of *C. ulmi* within the colonies of *C. herbarum* or *G. penicillioides*. Axenic cultures of *C. ulmi* normally lack synnemata.

Stimulation of sporulation has been very widely reported as a consequence of mycelial interactions, and represents one way in which an individual may benefit without necessarily bringing about deleterious effects in the other. It may therefore be regarded as one basis for neutralism and for mutualism. However, it is more likely that stimulation of the reproductive mode results from deleterious action on the vegetative mode, and that any advantage accrued must be viewed against the basic life strategy and energetics of the fungus concerned.

One example of the type of interaction just described is provided by what has been termed the '*Trichoderma* effect', observed amongst many heterothallic species of *Phytophthora*. The effect is confined to the *A2* compatibility type and, perhaps significantly, is especially marked amongst species invading roots of woody hosts, such as *P. cinnamomi*, *P. cambivora* and *P. palmivora*, where contact with *Trichoderma* species is likely to occur (Brasier, 1975*a*,*b*, 1978). The effect involves stimulation of the production of oospores via self-fertilisation in mycelial cultures exposed to volatile antibiotics from *Trichoderma* strains, variability between *Trichoderma* strains in eliciting the effect being correlated with their production of antibiotics. Since the effect is due to antibiotics, it is evident that it may at least partially function via deleterious effects on the vegetative phase, and indeed *Trichoderma* eventually replaces the *Phytophthora* species. However, the deleterious effect of replacement might be viewed as being counteracted partly by the production of survival structures (the oospores), but also by the stimulation of sexual reproduction in the absence of the *A1* compatibility type. Typically there seems to be a marked imbalance between the compatibility types in *Phytophthora* populations, for example amongst a world-wide sample of *P. cinnamomi* isolates, 28 were *A1* and 632 *A2*.

Such an imbalance may be seen either as leading to a requirement for, or being a consequence of, the *Trichoderma* effect.

Another example of a change of mode brought about by interactions occurs in stimulation of rhizomorph and mycelial cord formation. Thus rhizomorph development in *Armillaria* is stimulated by the presence of

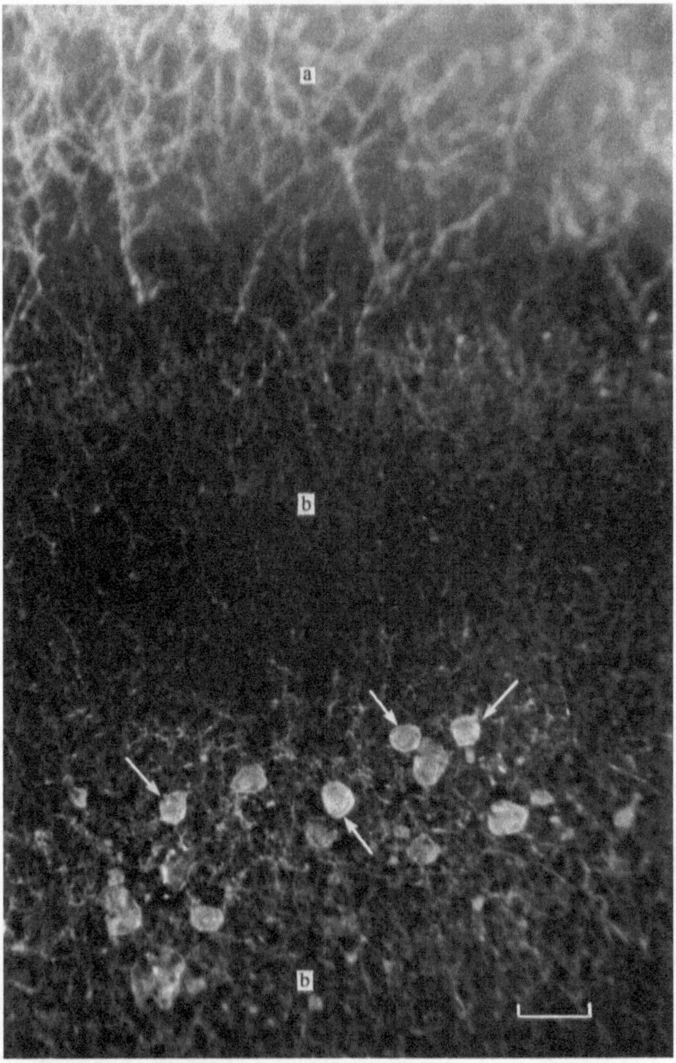

Fig. 3. Intermingling reaction between *Ceratocystis ulmi* (a) and *Cladosporium herbarum* (b) on 2% malt agar. The heads of the synnemata (arrowed) produced by *C. ulmi* are clearly visible within the *C. herbarum* colony. Scale bar, 2 mm.

Aureobasidium pullulans, apparently via ethanol production (Pentland, 1965, 1967) and in *Sphaerostilbe repens* it is stimulated by a number of fungi including *Penicillium thomii, Aspergillus niger* and *A. amstelodami* (Botton & El-Khouri, 1978). Mycelial cords of fungi such as *Phallus impudicus* and *Phanerochaete velutina* appear to be induced to form at a distance by *Penicillium* species and then may show directed growth towards the latter (L. Boddy & A. D. M. Rayner unpublished; Fig. 4). As is indicated by Thompson (Chapter 9) mycelial cord formation in several Basidiomycotina is favoured by growth under non-sterile conditions.

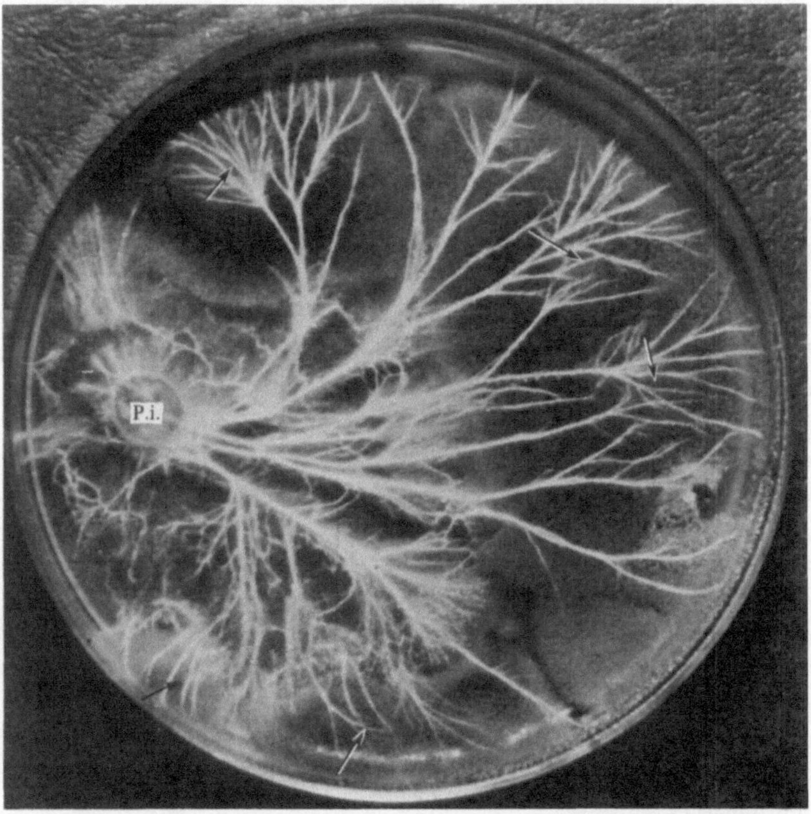

Fig. 4. Interaction between *Phallus impudicus* (P.i.) inoculated on one side of a 2% malt agar plate, and a series of colonies of *Penicillium* (the positions of the centres of some of which are arrowed) which developed around the periphery of the plate. Cords of *P. impudicus* have grown towards the *Penicillium* colonies prior to fanning out across them. (Photograph by Mr S. J. Payne.)

Neutralistic interactions may occur when readily assimilable substrates released by extracellular enzyme action of one fungus on a more complex and hence refractory substrate, are made available to another, associated species, not itself capable of breaking down the more refractory substrate. Thus *Melanospora destruens,* unable to use inulin, has been reported to grow well on a medium containing no other carbon source in mixed culture with fungi which can break down this substrate. Similarly, it grows more luxuriantly on sucrose-containing media in combination with fungi with greater powers of inversion (Hawker, 1947). Garrett had this situation in mind when he postulated the occurrence of so-called 'secondary sugar fungi' associating with cellulolytic and lignolytic species in decomposing plant residues. Whilst the existence of such fungi has been widely assumed (Hudson, 1968) there is little direct evidence for it. Tribe (1966) suggested that *Pythium oligandrum* associating with *Botryotrichum piluliferum* and *Fusarium* species on cellulose films was behaving as a secondary sugar fungus, but subsequent studies indicated that it may be in fact behaving as a mycoparasite (Deacon, 1976). Hedger & Hudson (1974) obtained evidence that *Thermomyces lanuginosus* growing in association with *Chaetomium thermophile* on filter-paper cellulose or wheat straw benefited from reducing sugars released by the cellulolytic activity of the latter fungus. Whilst mycoparasitism of *C. thermophile* cannot be ruled out they found no direct evidence for it.

It has been pointed out that removal of reducing sugars by an associated sugar fungus may limit catabolite repression, and hence enhance cellulase production by a cellulolytic fungus (Hulme & Stranks, 1970). Also since cellulase is an enzyme complex, mixtures of cellulases from different fungi growing together may be more efficient than those of any individual (Wood, 1969; Hulme & Shields, 1975). Whilst these examples illustrate synergism in cellulose breakdown, they should not automatically be regarded as mutualistic associations; enhanced cellulose breakdown may not necessarily be in keeping with the particular strategies of the participants. Nevertheless, such examples do illustrate how complementary physiological activities can provide a basis for mutualism.

Complementarity can also be found in culture by growing fungi on media lacking vital growth substances. Thus *Nematospora gossypii* and *Bjerkandera adusta* can grow in mixed culture on a synthetic medium lacking biotin, inositol and thiamin, but not individually. Whilst *B. adusta* can synthesise biotin and inositol, it cannot synthesise thiamin.

N. gossypii can synthesise thiamin, but not biotin and inositol (Kogl & Fries, 1937).

Competitive interactions

It is seldom possible to predict from cultural experiments the effectiveness of a fungus in primary resource capture, since this is frequently associated with a specificity to the resource which can never be reproduced under cultural conditions. Studies of mycelial extension rates and enzymic capacity can provide some indirect measure of ability in primary resource capture, but these methods may have questionable relevance under field conditions. Therefore, studies of competitive interactions in culture have usually concentrated on the outcome of combative confrontation between mycelia (usually two paired opposite one another on agar medium) rather than to the preliminary events of resource capture as the colonies become established. Where sufficient nutrients are available to allow dense mycelial development two outcomes of confrontation between mycelia are usually observed: deadlock in which neither individual invades territory occupied by the other, and partial or total replacement, in which one individual destructively invades another's territory. Whilst as indicated previously, these outcomes may sometimes result from purely nutritional factors, in others they result from direct physiological challenge (combat/antagonism).

Interactions of the latter type may be classified in terms of their underlying mechanisms (Fig. 5). Two main types are recognised. These are firstly, interactions which are mediated at a distance through diffusible or volatile substances (antibiotics) which result in inhibition or

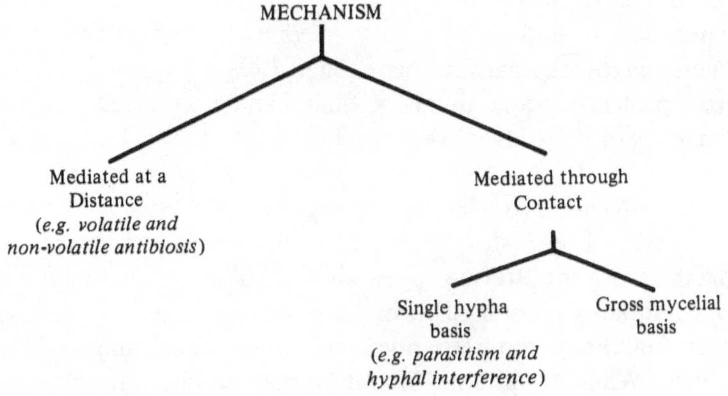

Fig. 5. Mechanisms of combat.

cessation of growth of one or both mycelia. Unilateral inhibition may be a prelude to replacement, whilst bilateral inhibition inevitably results in deadlock. The second type of interaction occurs only following contact, and may again result in deadlock or replacement. Contact between individuals, with its consequent varied outcome, may be effected either at the level of individual hyphae, as in hyphal interference and necrotrophic mycoparasitism, or by gross contact between mycelial systems.

There appears to be some correlation between the types of interaction in which a fungus becomes involved and its mycelial organisation and life strategy. As a generalisation, fungi with coenocytic, rapidly extending mycelia tend not to have antagonistic properties; septate Ascomycotina, Deuteromycotina and Mucorales (*Mortierella* species) often operate antagonistically via antibiosis; whilst Basidiomycotina usually operate via contact antagonism. That some of these differences in interactive properties are based on functional qualities associated with mycelial organisation, and that this transcends taxonomic divisions, is emphasised by the behaviour of certain Basidiomycotina. *Phlebia radiata* and *Phlebia rufa* exhibit what is known as astatocoenocytic behaviour (Boidin, 1971) and both primary and secondary mycelia develop behind a rapidly extending broad front of appressed coenocytic hyphae. This front is followed by septate mycelium, with extensive aerial proliferation and, in secondary mycelia, clamp-connections (Boddy & Rayner, 1983*a*; Fig. 6). The coenocytic front is non-combative, and is readily replaced by many fungi paired against it. However, once contact with septate mycelium of the *Phlebia* species is established this typically develops a dense zone of aerial mycelium, which advances across opposed mycelia via successive waves of lysis (Fig. 7*a*). It is as though exploration, consolidation and combat are mediated through morphologically diverse mycelial stages in these fungi. Less obvious but similar transitions probably occur in many fungi, and may greatly influence the dynamics of their interaction responses.

Mechanisms of combat (antagonism)

Antibiosis. The historical accident of the discovery of penicillin, together with the striking appearance of colony inhibition at a distance, have probably combined to make antibiosis in its broadest sense the most frequently reported phenomenon between fungi interacting in culture. Whilst long-range inhibition may be brought about by a wide variety of mechanisms, including the production and accumulation of general waste products, changes in pH etc., the accent has usually been on single volatile or diffusible substances produced by particular strains of fungi. This accent follows from the attempts to discover practically useful antibiotics, and their efficacy towards particular 'target' organisms considered for one reason or another as pests. There is no doubt that this approach has been useful in leading to the isolation and characterisation of particular antibiotic substances. Included amongst these is griseofulvin, isolated originally by Brian, Curtis & Hemming (1946) from *Penicillium janczewskyi* and described as a 'curling factor' because of its effect on germ tubes of *Botrytis allii* and other fungi with chitin walls.

Successful as the search for antibiotics may have been (see also Moss: Chapter 6), it has provided only limited insight into their role in the dynamics of mycelial interactions themselves, perhaps due to the

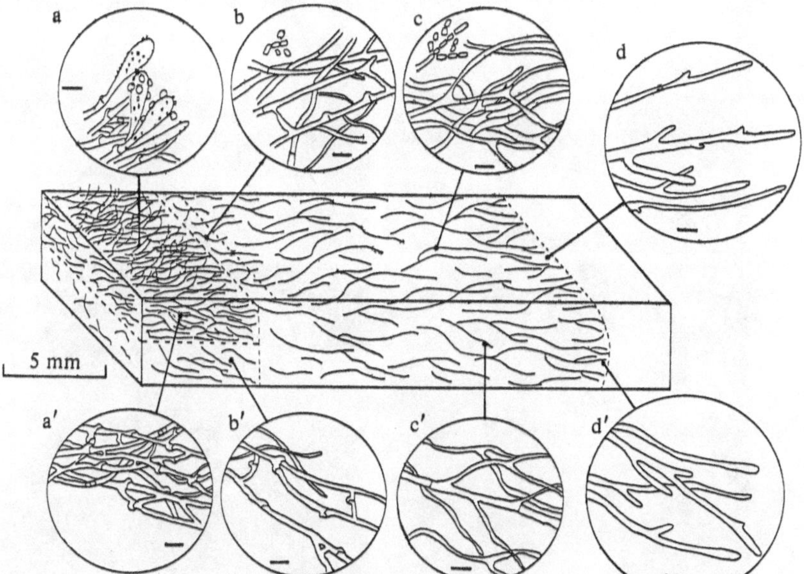

Fig. 6. Colony characteristics of a dikaryotic culture of *Phlebia radiata* growing through malt agar. Direction of growth left to right; a–d, surface mycelium; a′–d′, submerged mycelium; scale bar, 10 μm. a, hyphae with clamp-connections and cystidioles; b, heterogeneous mycelium with septa, anastomoses, pseudoclamps and oidia; c, heterogeneous mycelium, mostly coenocytic with oidia; d, wide coenocytic marginal hyphae. a′, closely packed mycelium with anastomoses and clamp-connections; b′, as in a′ but loosely packed; c′, heterogeneous mycelium, mostly coenocytic; d′, wide, coenocytic marginal hyphae. (After Boddy & Rayner, 1983*b*.)

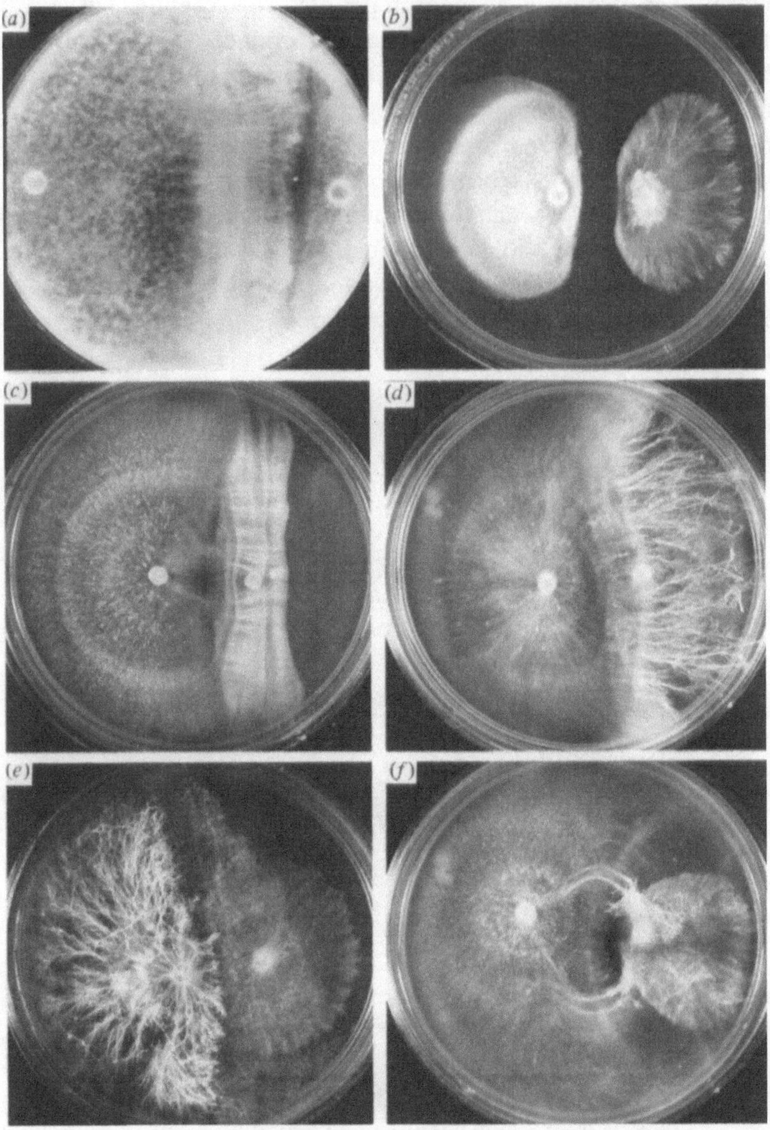

Fig. 7. Combative interactions involving various wood- and litter-inhabiting Basidiomycotina on malt agar. (a) *Phlebia radiata* (left) versus *Bjerkandera adusta*. Successive fronts of *P. radiata* are in the process of replacing *B. adusta* preceded by lysis of the latter. (b) Mutual inhibition at a distance between *Collybia confluens* (left) and *Clitocybe flaccida*. (c) Replacement of *Collybia butyracea* (right) by a dense mycelial front of *Collybia peronata*. (d) Replacement of *Clitocybe nebularis* (right) by *Collybia peronata*, with the normally appressed mycelium of the latter producing cords. (e) Replacement of *Clitocybe*

overemphasis of unilateral effects on a particular target organism, and on single chemical substances. Thus mutual inhibition, which is an extremely common feature of antibiotic interactions, has received little attention. In addition, the fact that a particular isolate may show antibiotic action only against certain fungi but not others, has been interpreted in relation to variation in sensitivity to the antibiotic, rather than differences in elicitation of its production.

These features of mutual inhibition, and apparent specificity of action, however, imply a more complex situation, possibly involving a reciprocal exchange of chemical signals between participants. For example, amongst certain litter-decomposing Agaricales, *Collybia confluens* and *Clitocybe flaccida* show a strong mutual inhibition when paired against each other in culture on 3% malt agar but are not inhibited in pairings against certain other species (Fig. 7*b*). They must therefore be responding in a very specific manner to each others' metabolic products, of which at least two specific substances must be involved. This type of situation suggests that some 'antibiotic' reactions involve a process of reciprocal exchange of chemical signals and recognition at a distance.

Hyphal contact. Whilst antibiosis was long regarded as the principle mechanism underlying many interactions, it has become clear that growth inhibition may often not occur prior to actual contact between hyphae or mycelia. One important mechanism resulting in contact inhibition has been termed hyphal interference (Ikediugwu, Dennis & Webster, 1970). This characteristically involves mutual or unilateral death of hyphae, or hyphal compartments as a result of intimate contact with one another. The death of cytoplasm in contacted compartments follows changes in membrane permeability, increased refractivity, vacuolation and eventual lysis. The process between *Coprinus heptemerus* and *Ascobolus crenulatus, Heterobasidion annosum* and *Phlebia gigantea* has been studied with the electron microscope. (Ikediugwu, 1976*a,b*). The affected compartments show a variety of abnormal

Caption for Fig. 7 (*cont.*)

flaccida (right) by *Collybia dryophila*, with the normal cord-forming growth of the latter reverting to a diffuse, appressed mycelium. (f) Replacement of *Clitocybe flaccida* (right) by cords of *Collybia peronata* with the morphogenetic stimulus associated with cord-formation being transmitted far back into the colony of the latter. (Photographs b–f by A. G. Mitchell, from Cooke & Rayner, 1984.)

features. They contain lipid globules, vesicles of various types, and organelles such as mitochondria and nuclei become swollen. The swollen mitochondria subsequently become vacuolate as eventually does the whole cell. At the point of contact with the antagonising hypha a characteristic extraplasmalemmal zone develops with various types of vesicles radiating from it.

Hyphal interference appears to be a particular feature of higher fungi, especially Basidiomycotina. It also appears to involve a form of surface recognition and the process shows some similarities with hyphal fusion which has aborted at an early stage of development. Somatic incompatibility within many Basidiomycotina is typically expressed as a post-fusion event at the intra-specific level, the interval between fusion and lysis varying with different species. Very closely related species, such as *Phlebia radiata* and *Phlebia rufa* appear to exhibit fusion followed by virtually immediate lysis (Boddy & Rayner, 1983*a*), whilst *Stereum rameale* and *Stereum rugosum* produce very marked hyphal interference (Rayner & Turton, 1982). A connection between the ability for hyphal fusion and hyphal interference might help to explain the apparent predominance of the latter in the Basidiomycotina. Interestingly, the advancing non-combative coenocytic phase of *Phlebia* species totally lacks hyphal fusions (Boddy & Rayner, 1983*a*).

It may be only a relatively short step from hyphal interference to the point at which a hypha of one fungus, having killed that of another, is able to derive nutrients from it. The latter is probably always associated with the destructive action of *Gliocladium roseum*, which can attack a wide range of host species (Barnett & Lilly, 1962). However, parasitism may also be related to secondary capture of the resource in which the parasitised fungus is growing, as for example in the apparent parasitism observed in *Pseudotrametes gibbosa* and *Bjerkandera adusta* (Fig. 8). Both cause white rot of deciduous wood, but *P. gibbosa* selectively colonizes wood previously occupied by *Bjerkandera adusta* (Rayner, 1978; Rayner & Todd, 1979).

The possible close connection between hyphal fusion, hyphal interference and parasitism is further suggested by observations of coiling reactions in di-mon mating of *Schizophyllum commune* (Nguyen & Niederpruem: Chapter 4). Such coiling is a characteristic feature of necrotrophic parasites, but biotrophic parasites which often obtain their nutrients from living cells may show a wide range of intercellular reactions (Barnett & Binder, 1973). That Buller should have regarded parasitic fusions as being a third kind of hyphal fusion (Gregory:

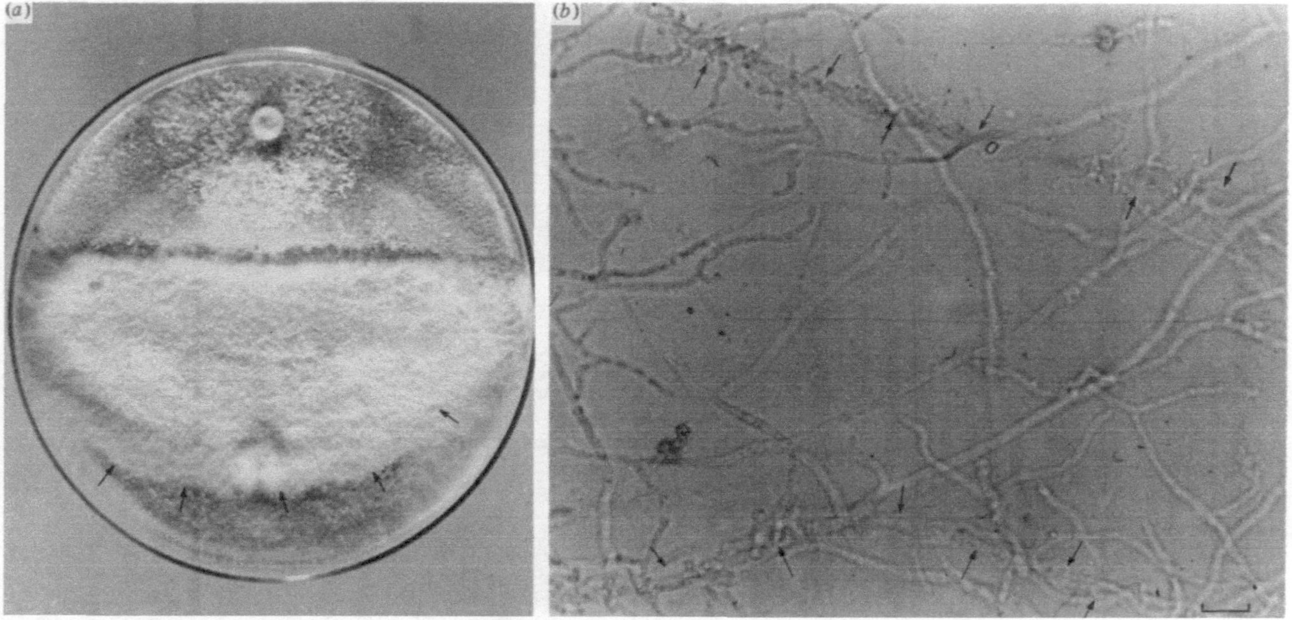

Fig. 8. Necrotrophic mycoparasitism of *Bjerkandera adusta* by *Pseudotrametes gibbosa*. (a) Interaction on 3% malt agar showing replacement by a dense mycelial front (arrowed) of *P. gibbosa*. (After Rayner & Todd, 1979.) (b) Aggregation of branches of *P. gibbosa* around two dead hyphae of *B. adusta* (arrowed). Scale bar, 10 μm.

Chapter 1) might have significance here. However, biotrophic mycoparasitism does not lead to secondary resource capture – the parasitised fungus is the resource, and is therefore beyond the scope of this discussion.

Mycelial contact. Observation of basidiomycete interactions in particular, suggest that mechanisms involving gross mycelial contact between colonies are important determinants of deadlock and replacement. Very often a dense zone of mycelium is produced at the interaction interface; this may be mutual in deadlock, or unilateral in replacement interactions where a dense wave of mycelium progressively advances across the opposing colony, often preceeded by lysis (Fig. 7c). Similarly, aggregated structures, such as cords, are often differentiated prior to replacement. Although such replacement may be brought about by lysis and parasitism on a huge scale, the underlying mechanisms of such gross interactions are still largely unknown. In some cases it may be that they simply have a 'smothering' effect on the mycelium replaced. Nonetheless, the specificity with which such interactions occur in different combinations, together with the elicitation of dense mycelium, suggest that some form of recognition response is involved.

Mycelial co-ordination during interactions. Combative interactions with another mycelium are likely to test the powers of co-ordination of mycelial activities, including effectiveness of mobilising resources from a non-interacting region to a site of confrontation. This feature is particularly evident amongst Basidiomycotina many of which exhibit marked combative ecological strategies. Little work has been done in this connection, but it certainly provides a potentially useful approach to studying the physiological functioning of mycelia.

Evidence for such co-ordination of activities and the dynamic nature of combative interactions is seen in Fig. 7 d–f, where for example 'lines of communication' to an interacting front of mycelial cords is evident (Fig. 7f). Similarly, a mycelium which is actively replacing another (for example *Pseudotrametes gibbosa* replacing *Bjerkandera adusta*) often remains relatively sparse behind the replacement front as compared with control plates or some deadlock interactions (Fig. 8).

The dynamic equilibrium behind two combative mycelial fronts is sometimes evident when bi-directional replacement occurs. This is commonly seen in interactions between *Piptoporus betulinus* and *Phlebia radiata* for example, where replacement occurs through 'access' points or 'bridgeheads' in the interaction fronts (Fig. 9a). With *P. radiata* and *Coriolus versicolor* the normal interaction, following replacement of the coenocytic phase of *P. radiata*, is a deadlock between vigorous interaction fronts. However if a strip inoculum of *P. radiata* subcultured from an established mycelium is placed against *C. versicolor* the balance of power is changed, so that lysis of *C. versicolor* occurs opposite the strip inoculum, whilst replacement of explorative *P. radiata* mycelium occurs at the edges of the interaction front (Fig. 9b).

Interactions in natural substrata

Problems of identification

The great value of cultural studies of interactions is their convenience and the fact that underlying mechanisms can often be ascertained simply by observation. However, doubt always exists concerning the relevance of such interactions under natural conditions, and so an attempt must be made to ascertain what events may be occurring under these conditions. These attempts usually start with the use of 'semi-natural' substrata, under laboratory conditions, and culminate in the study of the dynamics of truly natural communities. An important feature is that whilst intermingling, replacement and deadlock interactions

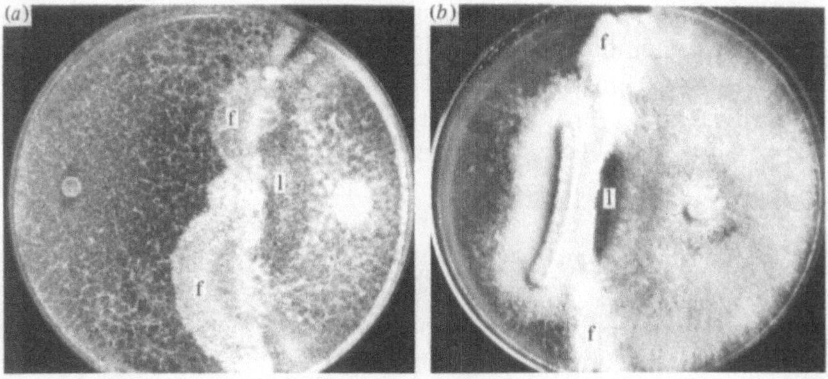

Fig. 9. Bi-directional replacement interactions on malt agar involving *Phlebia radiata* (left) versus: (a) *Piptoporus betulinus*, which has produced dense fan-like fronts through gaps in the *P. radiata* reaction front, but has undergone lysis elsewhere; (b) *Coriolus versicolor* – here a strip inoculum from established mycelium of *P. radiata* has been used. *C. versicolor* has undergone lysis immediately opposite the strip, but has formed replacement fronts over the sparse mycelium on either side of it. (Photograph by Nicola Kemp.) l, lysis; f, replacement fronts.

may be observed in natural substrata; interaction mechanisms must remain obscure unless supported by cultural work.

Because of the many difficulties inherent in studying interactions in the field, much of the evidence for them is indirect, based on the studies of succession on decomposing substrata. Unfortunately, interpretation of interactions in this way has been clouded by a lack of rigour and false assumptions in successional approaches to fungal ecology. In part these relate to insufficient consideration of the development of natural communities in relation to both time and space, as opposed to observations of purely temporal changes in floristic composition, sometimes based on the presence of reproductive structures as opposed to mycelia (Rayner & Todd, 1979; Cooke & Rayner, 1984). One particularly important consideration which is often neglected, is the fact that extension of indeterminate mycelial thalli is an integral part of community development. It is therefore always difficult to be certain of the time at which fungi first colonise a substratum, as opposed to when they become dominant.

Approaches to identification

Direct inoculation of natural or semi-natural substrata in the laboratory. This usually involves pairings of pure cultures of fungi in 'appropriate' substrata such as pine needles, leaf discs or wood-blocks. These are often sterilised to eliminate contaminants – the first stage of distancing from field conditions. Secondly, it is always difficult to maintain such natural substrata under appropriate micro-environmental conditions, especially with regard to moisture availability and aeration, so that although the methods are convenient, the results obtained are frequently disappointing adding little or nothing to those obtained from cultural experiments.

One unusual example of how this type of approach can be very effective, however, assesses the interactions of coprophilous fungi. Previously-sterilised food can be inoculated with spores of fungi prior to being fed to an animal, and the effects of interaction on fungal development within the dung then studied (Wicklow, 1981*b*).

Primary capture of a vacant resource. Another approach, applicable in the laboratory, is to expose a unit of uncolonised resource, such as a wheat straw or leaf disc, to a natural or semi-natural inoculum such as unsterile soil, or soil sterilised and inoculated with several different fungi. This type of approach formed the basis of the 'Cambridge' method, used by Garrett to study competitive saprophytic colonisation referred to previously, and more recently, has been adapted by Fisher (1977, 1979) in a study of aero-aquatic hyphomycetes.

As already emphasised, this type of approach really takes into account only competition for primary resource capture, unless the experiment is maintained over a prolonged period. The difficulties of maintaining conditions appropriate to those of the field, mentioned for direct inoculation, may still apply. Furthermore, the approach may not be consistent with natural patterns of colonisation, in that sudden exposure of a vacant resource unit is possibly a rare phenomenon in nature. It is therefore likely to give undue prominence to fungi with ruderal characteristics.

Direct inoculation into substrata in the field. This approach obviates the difficulties of maintaining appropriate environmental conditions and has proved useful, for example, in elucidating the interactions that occur between bark-inhabiting fungi (Webber, 1979; Lonsdale & Sherriff, 1983). In such a study, Webber (1979) inoculated a variety of elm bark saprotrophs into uncolonised inner bark (phloem) and into already-colonised inner bark of recently cut elm logs which were left in the field. By later estimating the extent of colonisation by fungi introduced into the fresh bark, as opposed to the extent of colonisation by fungi introduced into already-occupied bark, it was possible to assess which of the saprotrophs tended to be effective in primary resource capture and which in secondary resource capture respectively.

However, using such methods the timing of inoculation may be inappropriate in relation to the natural processes of community establishment and development, or inocula may be unrepresentative of the natural methods of arrival and establishment. The latter problem was encountered in the study of elm bark saprotrophs already mentioned. One particular species, *Phomopsis oblonga* only colonised inner bark extensively when inoculated in the form of a precolonised bark plug, failing to do so when introduced as a spore suspension. Later observations have suggested that inner-bark colonisation by this fungus would not normally occur from a spore inoculum as it apparently exists in a mycelial form in the outer bark of healthy elms, and invades the inner bark from this inoculum source with the onset of the tree's death (Webber & Gibbs, 1983).

Following processes of natural community development. Ultimately, the approach most likely to provide us with an understanding of mycelial

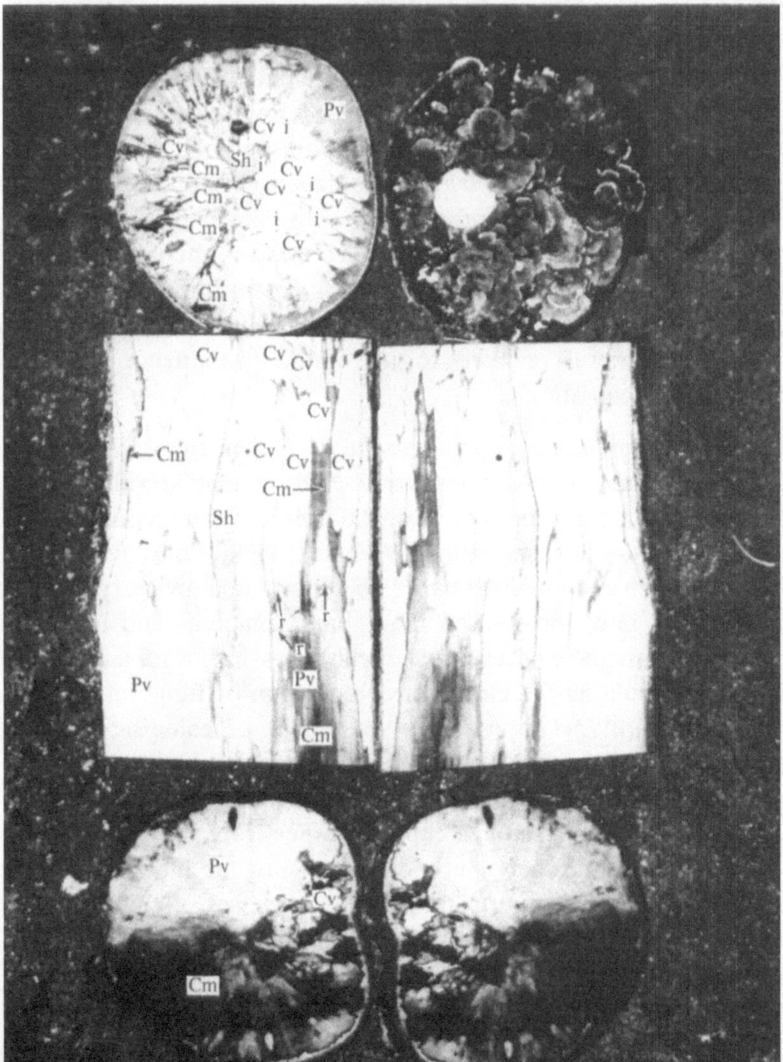

Fig. 10. Community organisation in a beech log which had been placed upright with its base buried in soil and litter two years previously at a deciduous woodland site. The log has been cut transversely near the top and bottom surfaces, and longitudinally for the intervening length, so that the three-dimensional distribution of columns of decay and discolouration can be ascertained. The upper surface is covered by basidiocarps of *Coriolus versicolor*. The distribution of some of the fungi present is indicated by the following symbols: Ab, *Armillaria bulbosa*; Cm, *Chaetosphaeria myriocarpa* (conidial stage) occupying interaction zones and other regions not occupied by active decay fungi;

interactions as they occur in nature, is to study them within the context of natural colonisation processes. However, this demands that sufficient account be taken of spatio-temporal characteristics of mycelia, and, as already indicated this has not always been a feature of succession studies. In turn, this requirement means that spatially defined resource units are the most suitable for study, and that analysis of interactions involves necessarily destructive methods. Wood represents such a spatially defined resource, and the ease with which populations and communities of mycelial individuals within it can be observed and their distribution mapped by direct methods makes it arguably the best example of the successful application of this approach. However, this should not be taken to imply that the approach is unlikely to be successful in other, perhaps less bulky substrata, although the techniques required to study mycelial distribution on a micro- rather than a macro-scale may need to be more sophisticated in these venues (Frankland, 1981). As a general principle, as the size of a resource unit is reduced, so the available surface for its colonisation, in relation to volume, increases proportionately. The result is that similar numbers of mycelial individuals may often be present in substrata of markedly different sizes (Cooke & Rayner, 1984; Boddy: Chapter 12; Rayner *et al.*: Chapter 23).

Two approaches are available for the study of natural colonisation processes in defined resource units. In the first, the community patterns present in a resource unit at any one time may be analysed, from which the events that have given rise to that pattern may be extrapolated. Comparison of such resource units at various stages of colonisation may provide further, indirect evidence of community development processes, but doubt must always remain concerning the inevitably speculative conclusions drawn. In wood, it is often quite easy, by analysis of the community pattern at any one time, to infer interaction patterns. For example, the presence of zone lines between different individuals, or of a series of relict zone lines left behind by a retreating individual can provide direct evidence of deadlock and replacement interactions respectively (Rayner & Todd, 1979; Fig. 10). As an example of the

Caption for Fig. 10 (*cont.*)

Cv, *Coriolus versicolor* individuals; Pv, *Phanerochaete velutina* invading from the base; Sh, *Stereum hirsutum*. i, interaction zone lines between individuals of the same and different species; r, relic zone lines left behind during replacement by *P. velutina*. (After Coates, 1984.)

potential usefulness of this approach, Boddy & Rayner (1983*b*) have recently applied it in a study of colonised attached oak branches, and been able to draw conclusions concerning the ecological roles of twelve different Basidiomycotina involved.

The other approach which allows more certainty about sequential processes, is simultaneously to expose large numbers of resource units to colonisation, then to sample these over a period of time and to analyse the community patterns at each sampling. Figure 11 illustrates community development processes observed in partly buried cut beech logs during such an experiment. The main disadvantage of this approach is the same as that described with the 'Cambridge' method; that it may not reflect the natural colonisation processes where large volumes of colonisable resource are not typically made suddenly available in this way.

Interactions and predictive ecology

In examining the role of mycelial interactions in natural community development processes, two issues are paramount. One con-

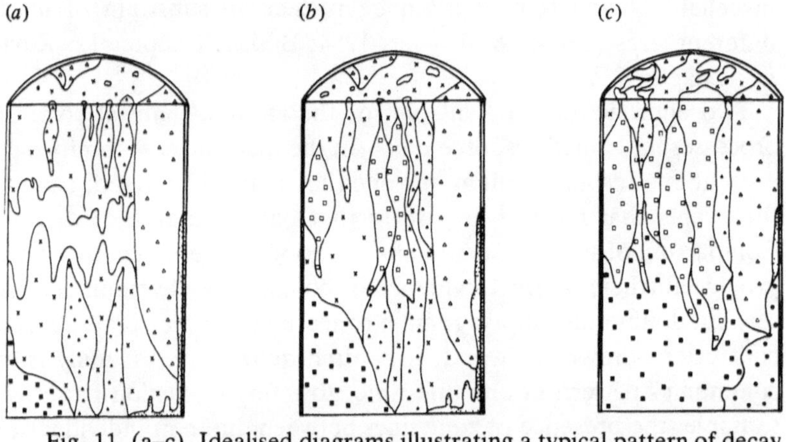

Fig. 11. (a–c). Idealised diagrams illustrating a typical pattern of decay community development in a cut beech log from the same experiment as that shown in Fig. 10. The community patterns at 6, 12 and 18 months respectively are shown. x, stained or discoloured wood containing microfungi and/or the Basidiomycotina *Chondrostereum purpureum* and *Corticium evolvens*; ▲, *Xylaria hypoxylon*; △, *Hypoxylon nummularium*; □, combative air-borne Basidiomycotina, e.g. *Coriolus versicolor, Bjerkandera adusta* and *Stereum hirsutum*; ■, combative early-arriving cord-formers, e.g. *Phallus impudicus* and *Tricholomopsis platyphylla*; ●, combative late-arriving cord-formers, e.g. *Phanerochaete velutina*; stipple, *Armillaria bulbosa*. (From Coates, 1984.)

cerns the extent to which events occurring under laboratory conditions are likely to change under different circumstances. Related to this, there is the need to consider the particular circumstances under which fungal colonisation occurs in nature, and hence the extent to which mycelial interactions are likely to be crucial determinants of the process.

Relationship between laboratory and field events

The most obvious way of ascertaining the relationship between laboratory and field events is to compare them directly and determine whether they are essentially the same, or, if they differ, in what ways. Such comparisons may ultimately yield understanding of which types of interactions observed in culture are likely to be repeated in nature.

Observations with wood decay fungi suggest that interactions observed under cultural conditions favouring dense mycelial growth (as on 2–3% malt agar) and mediated via hyphal or mycelial contact mechanisms will be broadly similar to those in naturally decomposing wood. Furthermore, there is evidence of strong correlation between the combative ability of these fungi in culture and their ecological roles (Rypacek, 1966; Henningson, 1967; Rayner, 1978; Thompson, 1982; Boddy & Rayner, 1983a,b; Coates, 1983). Primary colonisers of standing or recently felled trees are typically non-combative, being readily replaced. These include true heartrot fungi such as *Laetiporus sulphureus* and *Rigidoporus ulmarius*, together with species colonising recently functional sapwood, such as *Chondrostereum purpureum, Corticium evolvens, Flammulina velutipes* and *Peniophora quercina.* Airborne colonists of exposed non-living woody surfaces, such as the ubiquitous *Coriolus versicolor* and *Stereum hirsutum* are capable of replacement of pioneers, and of deadlock against one another. Mycelial cord-forming species, such as *Tricholomopsis platyphylla, Phallus impudicus, Phanerochaete velutina* and *Hypholoma fasciculare* invading from the soil and litter are highly combative, readily replacing many other species, in spite of their typically moderate or slow mycelial extension rates in culture. This corresponds with the fact that many seem to become established in wood via secondary resource capture. The closely related species pair, *Phlebia radiata* and *Phlebia rufa* provide further demonstration of the principles. *P. rufa* appears to be a pioneer invader of attached, light-suppressed branches and trunks especially of oak (*Quercus* species), whilst *P. radiata* is a ubiquitous invader of felled, fallen or already decomposing deciduous timber. *P.*

rufa is less combative, and fruits more readily in culture than *P. radiata* (Boddy & Rayner, 1983*a*).

These examples should not be taken to imply that Basidiomycotina are the only fungi in which a combative hierarchy may be present. Certain Deuteromycotina and Ascomycotina such as *Trichoderma* and *Scytalidium* species show a strong capacity to replace even some of the most combative wood-decaying fungi such as *Phanerochaete velutina*. These non-Basidiomycotina appear to have an essentially scavenging role in nature, and often develop in wood which has undergone decomposition or been subjected to disturbance. Their combative characteristics may assist establishment at later stages of community development, but much interest has also been raised by the possibility of introducing them directly into undecayed wood as biological control agents of decay (Hulme & Shields, 1972). Furthermore, in locations where Basidiomycotina do not readily become established, as in bark, similar patterns of behaviour may occur amongst other fungi.

These observations apply to fungal colonisation of essentially uncomminuted resources, that is before they become subject to extensive invasion by fauna. It is in such resources that purely mycelial interactions are likely to play a significant rôle in community development processes. Within these terms, the emphasis on soil as a habitat in early studies of fungal interactions, with its focus on antibiosis as a major mechanism of antagonism, is seen as being somewhat unfortunate. Very probably, the essential life styles of many soil fungi consists of long periods of inactivity under starvation conditions, alternating with sudden bursts of activity associated with localised nutrient enrichment, as from root exudates. Such fungi would be expected to have predominantly stress-tolerant ruderal characteristics, with an absence of combative mechanisms. However it may be that in comminuted resources such as soil (where diffusion is facilitated), antagonistic action in the form of antibiosis can be effective, although it has only occasionally been possible to prove that these substances are produced *in vivo*. The *Trichoderma* effect outlined earlier, is one such example of diffusible antibiosis that apparently occurs in soil at root surfaces (Brasier, 1978). Animal invasion of woody substrata with accompanying comminution, which introduces a major destabilising influence via direct effects both on fungi and on the resource (Anderson, Rayner & Walton, 1983), may also allow antibiotic producing fungi to exert a significant effect under natural conditions.

Interactions and community development patterns

It will now be evident that the significance of any interaction will need to be qualified by the circumstances under which fungal development takes place. In addition, the stage at which colonisation occurs will be an important determinant of the actual contribution (if any), which interactions make to community processes. For example, the conditions under which colonisation is initiated may be sufficiently selective, that is impose sufficient stress, to exclude all but one or a few fungi with consequent reduction in the number of potential interactions. Also, during early stages of colonisation, whilst the community is still 'open', the emphasis will be on primary resource capture, whilst at later stages, as the community 'closes', combative interactions are likely to assume increasing importance.

In this context, there is a need to understand the range of pathways along which natural community development may flow. Cooke & Rayner (1984) have attempted to provide a framework for such an understanding, and this is illustrated by the schema in Fig. 12. They suppose that colonisation processes may be initiated under two distinctive circumstances, either under conditions of some form of stress (probably the predominant circumstance in nature) or in the relative absence of such conditions (predominant in many 'experimental' successions). Hereafter, the sequence of events is determined by stress-alleviation, such as humidification of orginally relatively dry material as a result of metabolic activity, or stress-aggravation, involving imposition of new or additional stress, such as elevation of temperature also due to metabolic activity, and disturbance through the provision of new colonisable substrata by partial or total destruction of the resident biomass, or enrichment with additional resources.

Practical exploitation. The most obvious practical outlet for interaction studies is in the development of effective biological control for use against a pathogen or 'pest' fungus. In addition, there is the general ecological need to understand the consequences of any form of interference with ecosystem functioning, and the probability of exploiting 'favourable' interactions in such processes as waste disposal and food production.

Two basic methods are available in relation to exploitation of fungal interactions. The first is directly to introduce the favoured organism(s) into the system. The second is by manipulation of environmental conditions to alter the balance in favour of selected organisms. In all

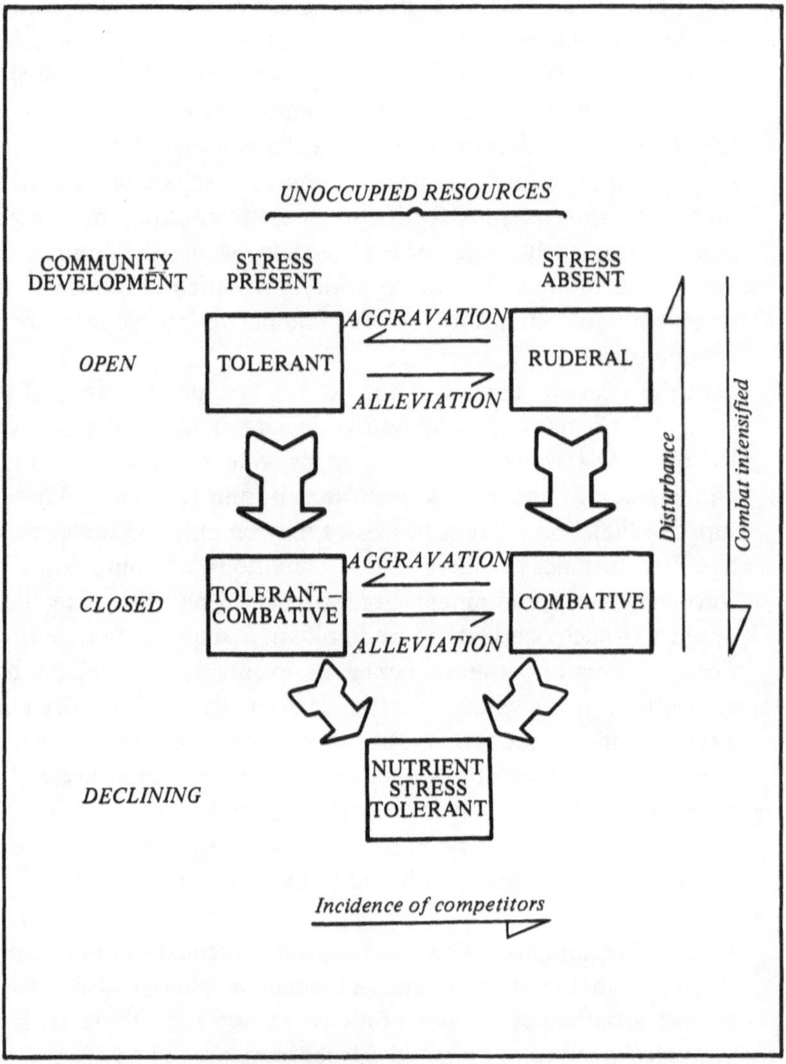

Fig. 12. Diagram of possible community development pathways from colonisation of a totally unoccupied resource, through an open community stage with still unoccupied resources available for primary capture, to a closed community with all initial primary capture completed. Culmination is a declining community stage characterised by severe nutrient stress. In the absence of competitors, developing tolerant communities may progress directly to declining tolerant communities without an intermediate combative stage. (Simplified from Cooke & Rayner, 1984.)

cases the successful application of knowledge of interactions is dependent on understanding the circumstances under which they are likely to operate.

Thus in relation to biological control there is little to be gained from attempting to introduce a suitable agent into a system in which the conditions do not favour establishment. Related to this, the principles underlying preventative as opposed to curative biological control measures are likely to be fundamentally different, in that preventative measures will usually involve primary resource capture competition, whilst curative measures are likely to involve secondary resource capture or direct parasitism. Adhering to the principle that prevention is better than cure, first considerations should be paid to abilities of a potential control agent in primary capture. From this point of view, its interaction performance in culture may be a secondary consideration, as selection of potentially useful agents should not necessarily be based on observations of replacement or antibiotic reactions. It is only necessary for the biotic agent to prevent establishment of the target organism, not to replace or kill the latter. Furthermore, since community development processes have been initiated in the absence of the target organism, it is unlikely to establish at a later stage even if the primary control agents are themselves subsequently replaced. In part, this statement reflects the fact that many pest fungi are by their very nature, primary colonisers.

It follows from this that a useful first approach to investigating possibilities of biological control is to study processes of primary capture. Thence a selected species may be given an advantage in primary capture, via direct inoculation and/or manipulation of conditions, which deflect community establishment processes away from the target organism. Consciously or subconciously, this approach has underlain several examples of biological control. Thus direct inoculation of *Phlebia gigantea* to freshly cut conifer stumps has been widely used to prevent establishment of *Heterobasidion annosum.* The essential feature here is that *P. gigantea* was initially selected as a primary coloniser of stump surfaces, rather than a direct 'antagonist' of *H. annosum.* Following inoculation there is extensive primary capture by *P. gigantea* which limits establishment of *H. annosum*, perhaps partly associated with hyphal interference mechanisms which prevent replacement by the latter (Rishbeth, 1963; Ikediugwu *et al.*, 1970). Another similar example is the treatment of pruning wounds on plum trees with *Trichoderma viride*, which prevents the establishment of

Chondrostereum purpureum (Grosclaude, Ricard & Dubos, 1973). A classical indirect mechanism is the ring barking of tropical trees prior to felling and establishment of tea plantations, which reduces problems associated with *Armillaria* due to shift towards other root-colonising fungi in treated trees (Leach, 1937). Similarly ammonium sulphamate treatment of hardwood stumps may favour root colonisation by fungi other than *Armillaria* species (Rayner, 1975, 1977).

Within the context of biological control via exploitation of primary resource capture, the emphasis is on the manipulation of an endogenous community to alter the balance away from a target organism. Nonetheless, it has sometimes been suggested that a successful biological control agent is unlikely to be found within an existing system where the target organism is already in balance with its potential competitors, and hence that emphasis should be on the introduction of exotic agents (Baker & Cook, 1974). This is clearly inappropriate for primary capture mechanisms of control, but may be more pertinent when considering curative measures employing secondary resource capture or parasitism. Whilst applicable in certain instances of control of animal pests, it remains to be seen whether this has application to fungi. In all events establishment under the prevailing natural conditions remains of overriding importance.

References

Anderson, J. M., Rayner, A. D. M. & Walton, D. W. H. (ed.) (1983). *Invertebrate–Microbial Interactions*. Cambridge, UK: Cambridge University Press.

Andrews, J. H. & Rouse, D. I. (1982). Plant pathogens and the theory of r- and k-selection. *The American Naturalist*, **120**, 283–96.

Baker, K. F. & Cook, R. J. (1974). *Biological Control of Plant Pathogens*. San Francisco, USA: Freeman & Co.

Barnett, H. L. & Binder, F. L. (1973). The fungal host–parasite relationship. *Annual Review of Phytopathology*, **11**, 273–92.

Barnett, H. L. & Lilly, V. G. (1962). A destructive mycoparasite *Gliocladium roseum*. *Mycologia*, **54**, 72–7.

de Bary, A. (1887). *Comparative Morphology and Biology of the Fungi, Mycetozoa and Bacteria*. Oxford, UK: Clarendon Press.

Boddy, L. & Rayner, A. D. M. (1983*a*). Mycelial interactions, morphogenesis and ecology of *Phlebia radiata* and *P. rufa* from oak. *Transactions of the British Mycological Society*, **80**, 437–48.

Boddy, L. & Rayner, A. D. M. (1983*b*). Ecological roles of basidiomycetes forming decay columns in attached oak branches. *New Phytologist*, **93**, 77–88.

Boidin, J. (1971). Nuclear behaviour in the mycelium and the evolution of the Basidiomycetes. In *Evolution in the Higher Basidiomycetes*, ed. R. H. Petersen, pp. 129–48. Knoxville, USA: University of Tennessee Press.

Botton, B. & El-Khouri, M. (1978). Synnema and rhizomorph production in *Sphaerostilbe repens* under the influence of other fungi. *Transactions of the British Mycological Society*, **70**, 131–6.
Brasier, C. M. (1975*a*). Stimulation of sex organ formation in *Phytophthora* by antagonistic species of *Trichoderma.* I. The effect *in vitro. New Phytologist*, **74**, 183–94.
Brasier, C. M. (1975*b*). Stimulation of sex organ formation in *Phytophthora* by antagonistic species of *Trichoderma.* II. Ecological implications. *New Phytologist*, **74**, 195–8.
Brasier, C. M. (1978). Stimulation of oospore formation in *Phytophthora* by antagonistic species of *Trichoderma* and its ecological implications. *Annals of Applied Biology*, **89**, 135–9.
Brian, P. W. (1957). The ecological significance of antibiotic production. *Microbial Ecology*, 7th Symposium of the Society of General Microbiology, pp. 168–88. Cambridge, UK: Cambridge University Press.
Brian, P. W., Curtis, P. J. & Hemming, H. G. (1946). A substance produced by *Penicillium janczewskii* Zal. I. Biological assay, production and isolation of 'curling factor'. *Transactions of the British Mycological Society*, **29**, 173–87.
Burkholder, P. R. (1952). Cooperation and conflict among primitive organisms. *American Scientist,* **40,** 601–31.
Clark, F. E. (1965). The concept of competition in microbial ecology. In *Ecology of Soil Borne Plant Pathogens*, ed. K. F. Baker & W. C. Snyder, pp. 339–45. London, UK: John Murray.
Clark, G. L. (1954). *Elements of Ecology*. London, UK: John Wiley.
Coates, D. (1984). Biology of intraspecific antagonism in wood decay fungi. Ph.D. Thesis, University of Bath, UK.
Cooke, R. C. (1977). *The Biology of Symbiotic Fungi*. London, UK: John Wiley.
Cooke, R. C. & Rayner, A. D. M. (1984). *The Ecology of Saprotrophic Fungi*. London, UK: Longman.
Deacon, J. W. (1976). Studies on *Pythium oligandrum*, an aggressive parasite of other fungi. *Transactions of the British Mycological Society*, **66**, 383–91.
Dobbs, C. G. & Hinson, W. H. (1953). A widespread fungistasis in soils. *Nature*, **172**, 197.
Dobbs, C. G., Hinson, W. H. & Bywater, J. (1960). Inhibition of fungal growth in soils. In *Ecology of Soil Borne Plant Pathogens*, ed. K. F. Baker & W. C. Snyder, pp. 130–47. London, UK: John Murray.
Fisher, P. J. (1977). New methods of detecting and studying saprophytic behaviour of aereo-aquatic hyphomycetes from stagnant water. *Transactions of the British Mycological Society*, **68**, 407–11.
Fisher, P. J. (1979). Colonisation of freshly abscissed and decaying leaves by aero-aquatic hyphomycetes. *Transactions of the British Mycological Society*, **73**, 99–102.
Frankland, J. C. (1981). Mechanisms in fungal successions. In *The Fungal Community: Its Organisation and Role in the Ecosystem*, ed. D. T. Wicklow & G. C. Carroll, pp. 403–26. New York, USA: Marcel Dekker.
Garrett, S. D. (1956). *Biology of Root Infecting Fungi.* Cambridge, UK: Cambridge University Press.
Garrett, S. D. (1970). *Pathogenic Root Infecting Fungi.* Cambridge, UK: Cambridge University Press.
Gibbs, J. N. & Smith, M. E. (1978). Antagonism during the saprophytic phase of the life cycle of *Heterobasidion annosum* and *Ceratocystis ulmi. Annals of Applied Biology*, **89**, 125–8.
Grime, J. P. (1979). *Plant Strategies and Vegetation Processes*. Chichester, UK: John Wiley.

Grosclaude, C., Ricard, J. & Dubos, B. (1973). Inoculation of *Trichoderma viride* spores via pruning shears for biological control of *Stereum purpureum* on plum tree wounds. *Plant Disease Reporter*, **57**, 25–8.

Harper, J. L. & Ogden, J. (1970). The reproductive strategy of higher plants. I. The concept of strategy with special reference to *Senecio vulgaris* L. *Journal of Ecology*, **58**, 681–9.

Hawker, L. E. (1947). Further experiments on growth and fruiting of *Melanospora destruens* Shear in the presence of various carbohydrates, with special reference to the effects of glucose and of sucrose. *Annals of Botany*, **11**, 245–9.

Hedger, J. N. & Hudson, H. J. (1974). Nutritional studies of *Thermomyces lanuginosus* from wheat straw compost. *Transactions of the British Mycological Society*, **62**, 129–43.

Henningsson, B. (1967). Interactions between microorganisms found in birch and aspen pulpwood. *Studia Forestalia Suecica*, **53**, 1–31.

Hudson, H. J. (1968). The ecology of fungi on plant remains above the soil. *New Phytologist*, **67**, 837–74.

Hulme, M. A. & Shields, J. K. (1972). Effects of primary fungal infection upon secondary colonisation of birch bolts. *Material und Organismen*, **7**, 177–88.

Hulme, M. A. & Shields, J. K. (1975). Antagonistic and synergistic effects for biological control of decay. In *Biological Transformation of Wood by Microorganisms*, ed. W. Liese, pp. 52–63, New York, USA: Springer-Verlag.

Hulme, M. A. & Stranks, D. W. (1970). Induction and regulation of cellulase by fungi. *Nature*, **226**, 469–71.

Ikediugwu, F. E. O. (1976*a*). Ultrastructure of hyphal interference between *Coprinus heptemerus* and *Ascobolus crenulatus*. *Transactions of the British Mycological Society*, **66**, 281–90.

Ikediugwu, F. E. O. (1976*b*). The interface in hyphal interference by *Peniophora gigantea* against *Heterobasidion annosum*. *Transactions of the British Mycological Society*, **66**, 291–6.

Ikediugwu, F. E. O., Dennis, C. & Webster, J. (1970). Hyphal interference by *Peniophora gigantea* against *Heterobasidion annosum*. *Transactions of the British Mycological Society*, **54**, 307–9.

Kogl, F. & Fries, N. (1937). Uber den Einfluss von Biotin, Aneurin und Meso-inosit auf das Wachstum verschieden Pilzarlen. *Zeitschrift für physiologische Chemie*, **249**, 23–110.

Leach, R. (1937). Observations on the parasitism and control of *Armillaria mellea*. *Proceedings of the Royal Society, Series* B, **121**, 561–73.

Lewis, D. H. (1973). Concepts in fungal nutrition and the origin of biotrophy. *Biological Reviews*, **48**, 262–78.

Lockwood, J. L. (1981). Exploitation competition. In *The Fungal Community: Its Organisation and Role in the Ecosystem*, ed. D. T. Wicklow & G. C. Carroll, pp. 319–50. New York, USA: Marcel Dekker.

Lonsdale, D. & Sherriff, C. (1983). Some aspects of the ecology of *Nectria* on beech. In *Proceedings of the 1982 IUFRO Beech Bark Disease Conference*, Hamdon, Connecticut, USA, ed. D. R. Houston. (In press.)

MacArthur, R. H. & Wilson, E. D. (1967). *The Theory of Island Biogeography*. Princetown, N.J., USA: Princetown University Press.

McNaughton S. J. & Wolf, L. L. (1973). *General Ecology*. New York, USA: Holt, Rinehart & Winston.

Odum, E. P. (1954). *Fundamentals of Ecology*. London, UK: W. B. Saunders Co.

Park, D. (1960). Antagonism – the background to soil fungi. In *The Ecology of Soil Fungi*, ed. D. Parkinson & J. S. Waid, pp. 148–59. Liverpool, UK: University Press.

Park, D. (1968). The ecology of terrestrial fungi. In *The Fungi*, **vol. 3,** ed. G. C. Ainsworth & A. S. Sussman, pp. 1–39. London, UK: Academic Press.

Pentland, G. D. (1965). Stimulation of rhizomorph development of *Armillaria mellea* by *Aureobasidium pullulans* in artificial culture. *Canadian Journal of Microbiology*, **11,** 345–50.

Pentland, G. D. (1967). Ethanol produced by *Aureobasidium pullulans* and its effect on the growth of *Armillaria mellea. Canadian Journal of Microbiology*, **13,** 1631–9.

Pugh, G. J. F. (1980). Strategies in fungal ecology. *Transactions of the British Mycological Society*, **75,** 1–14.

Rayner, A. D. M. (1975). Fungal colonization of hardwood tree stumps. Ph.D. Thesis, University of Cambridge, UK.

Rayner, A. D. M. (1977). Fungal colonization of hardwood stumps from natural sources. II. Basidiomycetes. *Transactions of the British Mycological Society*, **69,** 303–12.

Rayner, A. D. M. (1978). Interactions between fungi colonizing hardwood stumps and their possible role in determining patterns of colonization and succession. *Annals of Applied Biology*, **89,** 131–4.

Rayner, A. D. M. & Todd, N. K. (1979). Population and community structure and dynamics of fungi in decaying wood. *Advances in Botanical Research*, **7,** 333–420.

Rayner, A. D. M. & Turton, M. N. (1982). Mycelial interactions and population structure in the genus *Stereum*: *S. rugosum, S. sanguinolentum* and *S. rameale. Transactions of the British Mycological Society*, **78,** 483–94.

Rishbeth, J. (1963). Stump protection against *Fomes annosus*. III. Inoculation with *Peniophora gigantea. Annals of Applied Biology*, **52,** 63–77.

Rypacek, V. (1966). *Biologie holzzerstorender Pilze.* Jena: Fischer.

Shigo, A. L. (1958). Fungi isolated from oak wilt trees and their effects on *Ceratocystis fagacearum. Mycologia*, **50,** 757–69.

Thompson, W. (1982). Biology and ecology of mycelial cord-forming basidiomycetes in deciduous woodlands. Ph.D. Thesis, University of Bath, UK.

Tribe, H. T. (1966). Interactions of soil fungi on cellulose film. *Transactions of the British Mycological Society*, **49,** 457–66.

Webber, J. F. (1979). Interactions between bark saprophytes and the Dutch elm disease pathogen, *Ceratocystis ulmi.* Ph.D. Thesis, University College of Wales, Aberystwyth, UK.

Webber, J. F. & Gibbs, J. N. (1984). Colonisation of elm bark by *Phomopsis oblonga. Transactions of the British Mycological Society*, **82,** 348–52.

Wicklow, D. T. (1981*a*). Interference competition and the organisation of fungal communities. In *The Fungal Community: Its Organisation and Role in the Ecosystem,* ed. D. T. Wicklow & G. C. Carroll, pp. 351–78. New York, USA: Marcel Dekker.

Wicklow, D. T. (1981*b*). The coprophilous fungal community: a mycological system for examining ecological ideas. In *The Fungal Community: Its Organisation and Role in the Ecosystem,* ed. D. T. Wicklow & G. C. Carroll, pp. 47–76. New York, USA: Marcel Dekker.

Wood, T. M. (1969). Relationship between cellulolytic and pseudocellulolytic microorganisms. *Biochimica et Biophysica Acta,* **192,** 531–4.

19

Mycelial development and lectin–carbohydrate interactions in nematode-trapping fungi

BIRGIT NORDBRING-HERTZ

Department of Microbial Ecology, University of Lund, Ecology Building, Helgonavägen 5, S-223 62 Lund, Sweden

Nematophagous fungi comprise the group of fungi that are natural enemies of nematodes. More than 150 species are known today differing very much in taxonomic position, morphology, physiology and ecology. All share the unique feature of attacking living nematodes and consuming them. During recent years several reviews of these fungi have appeared covering their biology and ecology (Barron, 1977, 1981; Lysek & Nordbring-Hertz, 1983), the experimental methods used in handling them (Barron, 1982) as well as their use in biological control (Mankau, 1980).

In our laboratory, we have studied how these fungi recognise their prey. With few exceptions, the ability to attack living nematodes is connected with a specific developmental phase of the fungal mycelium. Many factors can affect or regulate this differentiation. The search for a recognition mechanism in this predator–prey relationship, therefore, is intimately connected with the study of the factors governing morphogenesis of the fungi. In this paper I review our results from studies on three different fungi with different saprophytic/parasitic ability with respect to their ability to recognise and capture nematodes. *Arthrobotrys oligospora* captures nematodes with hyphal structures of the adhesive network type (Fig. 1a,b,c). In *Dactylaria candida* adhesive knobs are formed on the mycelium and nematodes adhere to one or a few of these (Fig. 1d,e). *Meria coniospora*, finally, is an endoparasitic fungus with adhesive conidia as infective structures (Fig. 1f).

Factors affecting hyphal differentiation in *A. oligospora*

A. oligospora is one of the most common nematode-trapping fungi and also one that provides great opportunities to study fungal development as it produces a saprophytic mycelial phase and a predacious phase when traps are formed. The tendency to form traps varies greatly from strain to strain, but generally the transition from the saprophytic to the predacious phase is influenced by a series of biotic

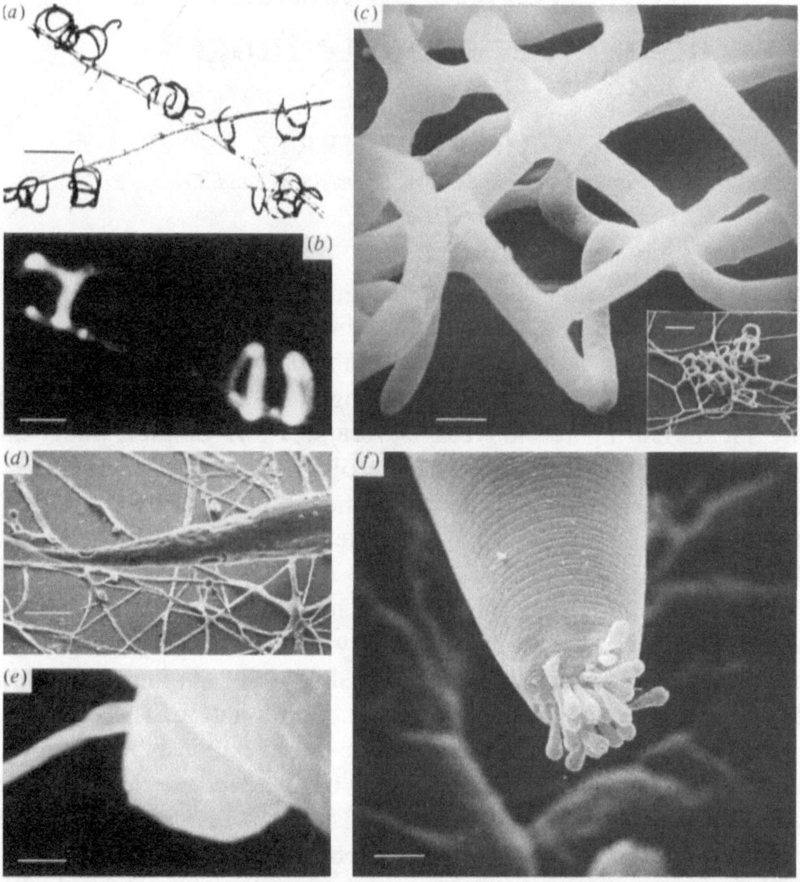

Fig. 1. Light, fluorescence and SEM micrographs of nematode-trapping devices in *A. oligospora* (a–c), *D. candida* (d, e) and *M. coniospora* (f). (a) Lactophenol-blue stained adhesive traps. Bar, 50 μm. (From Nordbring-Hertz, 1977*a*.) (b) Fluorescein-diacetate (FDA) stained material showing high metabolic activity in traps. (Courtesy of Dr B. Söderström.) Bar, 25 μm. (From Nordbring-Hertz, 1977*a*.) (c) Fully developed adhesive traps, detail and (inset) entire trap. Bars, 5 μm and 25 μm, respectively. (d), (e) Fungal colony with adhesive knobs in vicinity of nematode, and knob adhered to nematode cuticle. Bars, 20 μm and 1 μm, respectively. (f) Adhesive conidia attached to sensory organs of nematode. Bar, 5 μm. (From Jansson, 1982*c*.)

and abiotic factors. It is well known that nematodes or proteinaceous material induce the formation of traps (e.g. Pramer & Kuyama, 1963). In *A. oligospora* (ATCC 24927), small peptides with a high proportion of non-polar and aromatic amino acids in combination with a low nutrient status bring about a large number of traps (Nordbring-Hertz, 1973). This peptide-induced trap formation has provided a valuable laboratory system for studying differences between hyphae and traps.

However, both peptide-induced and nematode-induced trap formation are also influenced by a series of environmental factors, especially in the initial phases. Important factors in these phases of trap formation are: nutrient level and colony age, moisture content, pH, light, O_2/CO_2, and volatile compounds (Nordbring-Hertz, 1977*a*; Nordbring-Hertz & Odham, 1980). This became especially evident in a study (Nordbring-Hertz, 1977*a*) where single living nematodes as inducers of trap formation exerted a rapid effect, which could not be explained solely as an effect of addition of proteinaceous material. There are indications that the transport of ions across the plasma membrane might be important in the initial phases of hyphal differentiation in *A. oligospora* (B. Nordbring-Hertz, unpublished). Furthermore, in some cases the induction of trap formation shows an endogenous regulation (Lysek & Nordbring-Hertz, 1981; Czesny & Lysek, personal communication).

Thus, there is a very delicate balance in the transition from a saprophytic to a predacious phase and many factors can affect or regulate this differentiation. However, the actual mechanism remains unknown.

Irrespective of whether the traps are nematode-induced or formed as a result of other factors they are fully functional in capturing living nematodes as soon as they are formed and stay so as long as they keep intact their three-dimensional structure. The trap has a large number of other characteristics, summarised in Table 1.

Morphogenesis in relation to saprophytic/parasitic ability

The ability of nematophagous fungi to destroy nematodes (predacity) is intimately connected with their ability to form trapping structures. Cooke (1963) showed that the more saprophytic nematode-trapping fungi were weaker predators than the more predacious species. Jansson (1982*a*) divided the fungi into three ecological groups based on their saprophytic/predacious ability and their tendency to form trapping structures, respectively (Table 2). Further, in a soil microcosm investigation Jansson (1982*b*) showed that along with an increasing attraction

Table 1. *Comparison between characteristics of traps and hyphae in* Arthrobotrys oligospora

Traps	Hyphae	Reference
Capture of nematodes	No capture	1
Penetration of nematode cuticle by traps	No penetration by hyphae	2
Mucilaginous coat	No mucilaginous coat	2, 3
Electron-dense organelles, thick cell wall	No organelles, thin cell wall	2, 3
High K^+-content	Lower K^+-content	4
High metabolic activity	Lower metabolic activity	5
Attraction of nematodes strong	Attraction weaker	6
Presence of lectin(s)	No lectin detected	7, 8

1 Barron, 1977.
2 Nordbring-Hertz & Stålhammar-Carlemalm, 1978.
3 Veenhuis, Nordbring-Hertz & Harder, unpublished.
4 Nordbring-Hertz, 1977*b*.
5 Nordbring-Hertz, 1977*a*.
6 Jansson, 1982*b*.
7 Nordbring-Hertz & Mattiasson, 1979.
8 Borrebaeck, Mattiasson & Nordbring-Hertz, 1984.

of nematodes from group 1 to group 3 there was also an increased predacity.

A. oligospora, D. candida and *M. coniospora* can be considered as type species of each of the groups. As mentioned before, *A. oligospora* is a relatively good saprophyte; nevertheless, it is possible to manipulate this fungus to form network traps, thus increasing its predacious ability. *D. candida* forms adhesive knobs spontaneously and, probably because of this ability, is more predacious. *M. coniospora* does not form a mycelium outside its host and is considered as an obligate parasite.

Recognition in the nematode–nematophagous fungus system

In studying recognition in host–parasite relationships the question of host specificity is of major importance. Specificity in the choice of prey is little understood in the nematophagous fungi. The more saprophytic species seem to capture plant-parasitic, fungal-feeding and bacterial-feeding nematodes with equal ease (Jansson & Nordbring-Hertz, 1980). One approach to the question of host specificity in this predator–prey system is to study the pattern of attraction or repulsion of

Table 2. *Ecological groups of nematophagous fungi (From Jansson, 1982a)*

	Group 1	Group 2	Group 3
	Nematode-trapping fungi		Endoparasitic fungi
Trapping organs	adhesive networks; inducible by nematodes or with chemicals	adhesive knobs, adhesive branches, (non)-constricting rings; spontaneously produced	conidia, either adhering to nematode cuticles, or ingested by nematodes
Growth pattern	fast-growing and relatively good saprophytes; weak predacious ability	relatively slow-growing; weak saprophytes; great predacious ability	mostly obligate parasites; very slow-growing
Type species	*Arthrobotrys oligospora* *A. conoides*	*Dactylaria candida* *Monacrosporium cionopagum* *D. gracilis*	*Meria coniospora* *Harposporium anguillulae*
	——— increasing predacity ———→		
	——— increasing attraction ———→		

nematodes in the vicinity of the fungi. Another and more direct approach is to study capture of nematodes and the molecular mechanism behind adhesion to the trap surface.

Attraction of nematodes

Using an agar plate attraction assay (Jansson & Nordbring-Hertz, 1979) we showed that the attraction of different plant-parasitic and bacterial-feeding nematodes towards different fungi did not show any general pattern (Jansson & Nordbring-Hertz, 1980). However, the attraction intensity of a fungus differed with the developmental phase of the fungus. Jansson (1982*b*) showed that mycelia of *A. oligospora* with adhesive traps were twice as potent as mycelia without traps in attracting the bacterial-feeding nematode *Panagrellus redivivus* (Fig. 2). Similarly, adhesive conidia of endoparasitic nematophagous fungi attracted the nematodes whereas non-adhesive ones did not (Jansson, 1982*c*). Thus, there seems to be a connection between the adhesiveness of the fungal structures and the ability to attract nematodes.

Lectin-mediated capture of nematodes

Lectins have been used for some time for detecting surface structures of a carbohydrate nature in a large number of systems. But the role of lectins in nature is still obscure, although some systems have

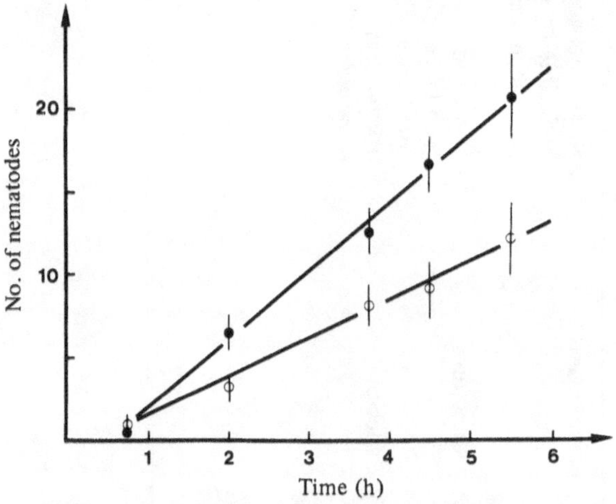

Fig. 2. Attraction of nematodes to pure mycelium (hollow circle) and to trap-containing mycelium (filled circle) of *A. oligospora*. The method used was the attraction-intensity assay of Jansson & Nordbring-Hertz (1979). (From Jansson, 1982*b*.)

Table 3. *Evidence for a trap lectin in* A. oligospora

Property	Reference
Inhibition of capture by *N*-acetyl-D-galactosamine (GalNAc)	1
Binding of red blood cells	1
Demonstration of GalNAc on nematode surface	1
Inhibition of capture by trypsin and glutaraldehyde	2
Binding of (^{125}I)-labelled fungal homogenate to GalNAc-Sepharose and to nematodes	3
Inhibition of binding of homogenate by trypsin treatment	3
Isolation and characterisation of a carbohydrate-binding, GalNAc-specific, Ca^{2+}-dependent protein (subunit mol. wt ~20 000)	4

1 Nordbring-Hertz & Mattiasson, 1979.
2 Nordbring-Hertz, Friman & Mattiasson, 1982.
3 Mattiasson, Johansson & Nordbring-Hertz, 1980.
4 Borrebaeck, Mattiasson & Nordbring-Hertz, unpublished.

been studied in great detail, e.g. the legume–*Rhizobium* association and the slime mould lectins (for review see Bauer, 1981; Barondes, 1981). Apart from the slime mould lectins very few fungal lectins have been isolated.

A. oligospora. Based on our experience of the development and characteristics of traps (Table 1) and of their great ability to trap nematodes we postulated that there must be an interaction between surface macromolecules of the trap and nematode, respectively. We now know that the trap of *A. oligospora* carries a carbohydrate-binding protein (a lectin) binding specifically to *N*-acetyl-D-galactosamine (GalNAc) on the nematode surface. Table 3 shows the significant items of evidence which have been accumulated.

We suggested that if a lectin–carbohydrate interaction is involved in adhesion, then pre-exposure of the lectin-carrying surfaces to specific carbohydrates would prevent attachment of the nematodes to the traps (Nordbring-Hertz & Mattiasson, 1979). In these inhibition experiments, as in many other interaction studies, we used a dialysis membrane technique recently described in detail (Nordbring-Hertz, 1983). Pretreatment of a trap-containing colony with carbohydrates and subsequent addition of nematodes showed that only *N*-acetyl-D-galactosamine at 20 mM substantially inhibited capture. This result suggested that a fungal lectin is involved in the interaction. The result was further substantiated by demonstration of the capture of a model prey (red blood cells, blood group A with GalNAc terminally) and by demonstration of the presence of GalNAc residues on the surface of the nematode by the aid of a peroxidase-lectin-conjugate (Nordbring-Hertz & Mattiasson, 1979). The protein nature of the trap molecule responsible was evident from further *in vivo* experiments, where traps pretreated with trypsin or glutaraldehyde lost the ability to capture nematodes (Nordbring-Hertz, Friman & Mattiasson, 1982). Intact fungal material was also radiolabelled using (^{125}I)-iodosulphanilic acid which only labels surface-bound protein structures. A (^{125}I)-labelled fungal homogenate was prepared and this material bound both to GalNAc-Sepharose and to nematodes. Addition of GalNAc to the homogenate in both cases reduced binding. Trypsin treatment of the homogenate abolished binding to GalNAc-Sepharose in these *in vitro* experiments giving further evidence for a carbohydrate-binding protein (Mattiasson, Johansson & Nordbring-Hertz, 1980).

The conclusive evidence of a carbohydrate-binding protein, however, was demonstrated recently. Since there is a very delicate balance between trap and hyphal formation in this fungus (see above), there is a major difficulty in producing enough trap-containing mycelium for biochemical experiments. However, it is now possible to produce liquid culture-grown fungal material with a high traps-to-hyphae ratio. Labelled fungal homogenate was purified by affinity chromatography on GalNAc-Sepharose. The carbohydrate-binding protein, eluted in one single peak, bound to Ca^{2+} as shown by metal chelate affinity chromatography (Borrebaeck, Lönnerdahl & Etzler, 1981) and had a protomeric molecular weight of 20 000 as demonstrated by SDS-gel electrophoresis and confirmed by autoradiography (Borrebaeck, Mattiasson & Nordbring-Hertz, unpublished). Interestingly, presence of calcium is indicated by X-ray microanalyses in the fungal trap cell wall as well as in trap organelles (Nordbring-Hertz & von Hofsten, unpublished), further suggesting the dependence on this element of the carbohydrate-binding protein.

In *A. oligospora* lectin–carbohydrate binding seems to be the signal which initiates those interactions leading to the penetration of the nematode cuticle and digestion of the nematode. These latter events need further study at ultrastructural, cytochemical and biochemical levels. Previous ultrastructural studies have demonstrated the differences between traps and hyphae in different nematode-trapping fungi

(Heintz & Pramer, 1972; Nordbring-Hertz & Stålhammar-Carlemalm, 1978; Table 1). These differences have been more clearly delineated in recent studies (Fig. 3a,b) which have also increased our awareness of the possible roles of some of the organelles in adhesion, penetration and digestion (Veenhuis, Nordbring-Hertz & Harder, unpublished). In this connection, the study of lytic extracellular enzymes, which have previously not been investigated in this system, would be of major importance.

D. candida. D. candida (CBS 220.54) which forms adhesive knobs spontaneously on the hyphae is an extremely efficient nematode-trapping fungus in the laboratory. There are many indications of a similar recognition mechanism in this fungus to that in *A. oligospora.* Thus, the protein nature of the receptor molecule of the traps has been indicated by the inhibition of capture by trypsin and glutaraldehyde (Nordbring-Hertz *et al.*, 1982). Further, out of 20 carbohydrates only 2-deoxy-D-glucose totally inhibited capture. However, further evidence for the presence of a lectin–carbohydrate interaction in this fungus–nematode relationship is still lacking.

Neither *A. oligospora* nor *D. candida* shows any host specificity as

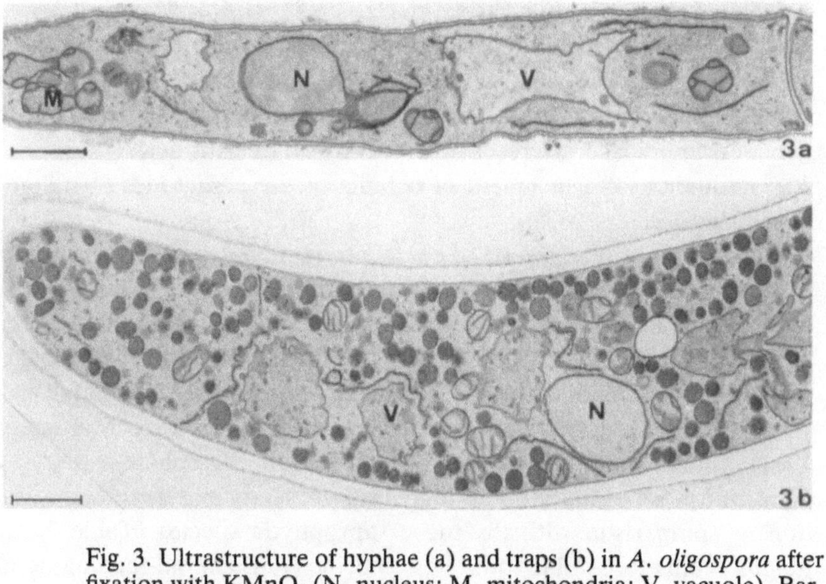

Fig. 3. Ultrastructure of hyphae (a) and traps (b) in *A. oligospora* after fixation with $KMnO_4$ (N, nucleus; M, mitochondria; V, vacuole). Bar, 1 μm. (Courtesy of M. Veenhuis, Biological Centre, University of Groningen, Netherlands.)

judged by ability to capture different types of nematodes (Jansson & Nordbring-Hertz, 1980). However, the attraction of nematodes as well as the ability to destroy them is stronger in mycelia of *D. candida* than in that of *A. oligospora* (Table 2; Jansson & Nordbring-Hertz, 1979). Indeed, there is an increased attraction of nematodes with an increased requirement for nematodes as nutrients. It is reasonable to assume that the increased attraction is connected with the abundance of adhesive knobs on the mycelia, as both adhesive networks in *A. oligospora* and also adhesive conidia in *M. coniospora* increase the ability to attract nematodes to these fungi (Jansson, 1982*b,c*).

M. coniospora. In *M. coniospora* (CBS 615.82) the connection between the attraction of nematodes and the adhesion of conidia as possible recognition mechanisms is more obvious. Here the infective structures, conidia with an adhesive knob, adhere preferentially to the nematode sensory organs (Fig. 1f) which are sites of chemoreception. Furthermore, nematodes lose their ability to be attracted when the chemoreceptors are blocked by adhered conidia. To elucidate whether a recognition mechanism similar to that of *A. oligospora* is present in *M. coniospora*, the adhesive conidia have been treated with solutions of carbohydrates before placing in the presence of the nematode *P. redivivus* (Jansson & Nordbring-Hertz, 1983). Out of 21 carbohydrates *N*-acetylneuraminic acid, a sialic acid, caused the strongest inhibition of adhesion. These experiments would indicate the presence of sialic acid at the chemoreceptors of the nematodes. When nematodes were treated with neuraminidase, which causes splitting off of sialic acids, there was a 20% reduction of attachment of conidia to the nematodes (Jansson & Nordbring-Hertz, 1983).

The fact that the adhesion of conidia to nematodes can be suppressed by pre-incubating conidia with specific carbohydrates led us to conclude that a fungal lectin could be involved also in this interaction. Thus, although not strictly a mycelial structure, the conidia probably bear a carbohydrate-binding protein which is developmentally regulated as in *A. oligospora* (Nordbring-Hertz, Friman, Johansson & Mattiasson, 1981) and associated with formation of the adhesive phase.

Furthermore, in addition to increased predacity and increased attraction in comparison with the more saprophytic species (Table 2) this fungus seems to exhibit some host specificity. Bacterial-feeding nematodes are infected exclusively at the sensory organs whereas plant-parasitic ones are infected over the whole cuticle (Jansson & Nordbring-Hertz, 1983).

Fungal–fungal interactions

As shown above, nematophagous fungi offer a large variety of hyphal structures, most of which are functional nematode-trapping devices (c.f. Fig. 1). In the genus *Arthrobotrys* intermediates between hyphal coils, traps and conidial initials are fairly common (e.g. Jansson & Nordbring-Hertz, 1981), and the development of either structure to fully developed traps or conidia is highly dependent on environmental conditions. Since there is such a broad spectrum of hyphal developmental structures, it would not be surprising if interactions with other fungi also occur. Tzean & Estey (1978) were the first to report on nematophagous fungi as mycopathogens. In their study, hyphal coils were formed preferentially around hyphae of plant-parasitic fungi suggesting a specific interaction. Species of *Trichoderma* which have been used successfully in biological control of *Rhizoctonia solani* and *Sclerotium rolfsii* in field and greenhouse experiments (Elad, Chet & Katan, 1980) attach themselves to their host by hyphal coils similar to those of the nematode-trapping species (Elad, Chet, Boyle & Henis, 1983). Lectin activity in a host–mycoparasite relationship was recently demonstrated with *R. solani* and *T. harzianum* (Elad, Barak & Chet, 1983).

To elucidate whether surface recognition of the same type as that in

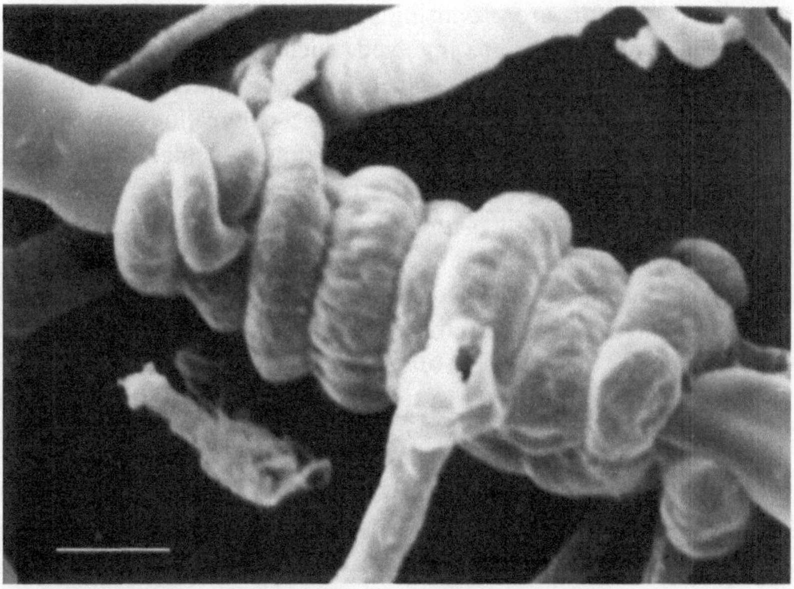

Fig. 4. SEM of the interaction between *A. oligospora* and *R. solani*. *A. oligospora* encircles hyphae of *R. solani* with trap-like structures. Bar, 5 μm.

the nematode–fungus relationships was effective also in a nematode-trapping fungus–fungus interaction we used the same dialysis membrane technique as above (Nordbring-Hertz, 1983). Both *A. oligospora* and *A. superba* attacked several fungi with the aid of hyphal coils. Figure 4 shows *A. oligospora* coils surrounding hyphae of *R. solani.* The contents of the hyphae of *R. solani* gradually disappeared. The hyphal coils, although of similar appearance to traps, differ from traps in that they are not able to capture nematodes at any stage of development. Furthermore, they clearly differ from traps at the ultrastructural level, for instance they lack the typical trap organelles (M. Veenhuis, personal communication). In preliminary inhibition experiments, we were unable to detect a specific surface interaction. Thus, so far there is no indication that *A. oligospora* recognises its fungal host in the same way as its nematode prey.

Concluding remarks

The nematode–nematophagous fungus system offers an excellent system to study both fungal development as influenced by environmental factors and host–parasite relationships of different levels of complexity. Although host specificity is not pronounced in this system, surface recognition on the molecular level has been detected in several species. The results are promising for studies of more specific associations where nematodes are involved, such as fungi which attack cyst nematodes and which have gained increased interest during recent years. Furthermore, the fact that there is a connection between attraction (chemotaxis) and adhesion in one fungus–nematode relationship points to the possibility of a similar connection in other systems. One likely candidate for study is fungal propagules (zoospores) which move chemotactically to their plant hosts. Another example is those mycoparasitic relationships in which directed growth by the antagonist might initiate an interaction. In all these cases a certain phase of the mycelial development is involved. Much more work, however, is needed to explain these intimate associations and information has to come from all aspects of mycelial studies – ecological, physiological and molecular.

The experimental work reported here was supported by grants from the Swedish Natural Science Research Council. I am grateful to Dr Hans-Börje Jansson for discussions during the preparation of this review.

References

Barondes, S. H. (1981). Lectins: their multiple endogenous cellular functions. *Annual Review of Biochemistry*, **50**, 207–31.

Barron, G. L. (1977). *The Nematode-Destroying Fungi.* Canadian Biological Publications Ltd, Box 214, Guelph, Ontario N1H 6J9, Canada.

Barron, G. L. (1981). Predators and parasites of microscopic animals . In *Biology of Conidial Fungi*, vol. 2, ed. G. T. Cole & W. B. Kendrick, pp. 167–200. New York: Academic Press.

Barron, G. L. (1982). Nematode-destroying fungi. In *Experimental Microbial Ecology*, ed. R. G. Burns & J. H. Slater, pp. 533–52. London: Blackwell Scientific Publications.

Bauer, W. D. (1981). Infection of legumes by Rhizobia. *Annual Review of Plant Physiology*, **32**, 407–49.

Borrebaeck, C. A. K., Lönnerdahl, B. & Etzler, M. E. (1981). Metal chelate affinity chromatography of the *Dolichos biflorus* seed lectin and its subunits. *FEBS Letters*, **130**, 194–6.

Borrebaeck, C. A. K., Mattiasson, B. & Nordbring-Hertz, B. (1984). Isolation and partial characterization of a carbohydrate-binding protein from a nematode-trapping fungus. *Journal of Bacteriology* (in press).

Cooke, R. C. (1963). Ecological characteristics of nematode-trapping hyphomycetes. I. Preliminary studies. *Annals of Applied Biology*, **52**, 431–7.

Elad, Y., Barak, R. & Chet, I. (1983). Role of lectins in mycoparasitism. *Journal of Bacteriology*, **154**, 1431–51.

Elad, Y., Chet, I., Boyle, P. & Henis, Y. (1983). Parasitism of *Trichoderma* spp. on *Rhizoctonia solani* and *Sclerotium rolfsii* – scanning electron microscopy and fluorescence microscopy. *Phytopathology*, **73**, 85–8.

Elad, Y., Chet, I. & Katan, J. (1980). *Trichoderma harzianum*: a biological agent effective against *Sclerotium rolfsii* and *Rhizoctonia solani. Phytopathology*, **70**, 119–21.

Heintz, C. E. & Pramer, D. (1972). Ultrastructure of nematode-trapping fungi. *Journal of Bacteriology*, **110**, 1163–70.

Jansson, H. B. (1982*a*). Attraction of nematodes to nematophagous fungi. Ph.D. Thesis, Lund University.

Jansson, H. B. (1982*b*). Predacity by nematophagous fungi and its relation to the attraction of nematodes. *Microbial Ecology*, **8**, 233–40.

Jansson, H. B. (1982*c*). Attraction of nematodes to endoparasitic nematophagous fungi. *Transactions of the British Mycological Society*, **79**, 25–9.

Jansson, H. B. & Nordbring-Hertz, B. (1979). Attraction of nematodes to living mycelium of nematophagous fungi. *Journal of General Microbiology*, **112**, 89–93.

Jansson, H. B. & Nordbring-Hertz, B. (1980). Interactions between nematophagous fungi and plant-parasitic nematodes: attraction, induction of trap formation and capture. *Nematologica*, **26**, 383–9.

Jansson, H. B. & Nordbring-Hertz, B. (1981). Trap and conidiophore formation in *Arthrobotrys superba. Transactions of the British Mycological Society*, **77**, 205–7.

Jansson, H. B. & Nordbring-Hertz, B. (1983). The endoparasitic nematophagous fungus *Meria coniospora* infects nematodes specifically at the chemosensory organs. *Journal of General Microbiology*, **129**, 1121–6.

Lysek, G. & Nordbring-Hertz, B. (1981). An endogenous rhythm of trap formation in the nematophagous fungus *Arthrobotrys oligospora. Planta*, **152**, 50–3.

Lysek, G. & Nordbring-Hertz, B. (1983). Die Biologie nematodenfangender Pilze. *Forum Mikrobiologie*, **6**, 201–8.

Mankau, R. (1980). Biological control of nematode pests by natural enemies. *Annual Review of Phytopathology*, **18,** 415–40.

Mattiasson, B., Johansson, P. A. & Nordbring-Hertz, B. (1980). Host-microorganism interaction: studies on the molecular mechanisms behind the capture of nematodes by nematophagous fungi. *Acta Chemica Scandinavica*, **B34,** 539–40.

Nordbring-Hertz, B. (1973). Peptide-induced morphogenesis in the nematode-trapping fungus *Arthrobotrys oligospora. Physiologia Plantarum*, **29,** 223–33.

Nordbring-Hertz, B. (1977*a*). Nematode-induced morphogenesis in the predacious fungus *Arthrobotrys oligospora. Nematologica*, **23,** 443–51.

Nordbring-Hertz, B. (1977*b*). X-ray microanalysis of the nematode-trapping organs in *Arthrobotrys oligospora. Transactions of the British Mycological Society*, **68,** 53–7.

Nordbring-Hertz, B. (1983). Dialysis membrane technique for studying microbial interactions. *Applied and Environmental Microbiology*, **45,** 290–3.

Nordbring-Hertz, B., Friman, E., Johansson, P. A. & Mattiasson, B. (1981). Host-microorganism interaction: developmentally regulated lectin-mediated capture of nematodes by nematode-trapping fungi. In *Lectins, Biology, Biochemistry and Clinical Biochemistry*, vol. I, ed. T. C. Bøg-Hansen, pp. 43–50. Berlin: W. de Gruyter.

Nordbring-Hertz, B., Friman, E. & Mattiasson, B. (1982). A recognition mechanism in the adhesion of nematodes to nematode-trapping fungi. In *Lectins, Biology, Biochemistry and Clinical Biochemistry*, vol. II, ed. T. C. Bøg-Hansen, pp. 83–90. Berlin: W. de Gruyter.

Nordbring-Hertz, B. & Mattiasson, B. (1979). Action of a nematode-trapping fungus shows lectin-mediated host–microorganism interaction. *Nature*, **281,** 477–9.

Nordbring-Hertz, B. & Odham, G. (1980). Determination of volatile nematode exudates and their effects on a nematode-trapping fungus. *Microbial Ecology*, **6,** 241–51.

Nordbring-Hertz, B. & Stålhammar-Carlemalm, M. (1978). Capture of nematodes by *Arthrobotrys oligospora*: an electron microscope study. *Canadian Journal of Botany*, **56,** 1297–1307.

Pramer, D. & Kuyama, S. (1963). Symposium on biochemical bases of morphogenesis in fungi. II. Nemin and the nematode-trapping fungi. *Bacteriological Reviews*, **27,** 282–92.

Tzean, S. S. & Estey, R. H. (1978). Nematode-trapping fungi as mycopathogens. *Phytopathology*, **68,** 1266–70.

20

Mycelial interactions and mitochondrial inheritance in *Aspergillus*

J. H. CROFT and R. B. G. DALES
Department of Genetics, University of Birmingham, P.O. Box 363, Birmingham, B15 2TT, UK

The range of mycelial interactions in *Aspergillus* and their possible consequences

It has now become clear that heterogenically controlled vegetative or somatic incompatibility is of widespread occurrence among mycelial fungi (Rayner *et al.*: Chapter 23). The natural functions of vegetative incompatibility within a species remain uncertain but two possible biological roles are evident for *Aspergillus*.

Firstly, vegetative incompatibility could act as a genetic isolating mechanism. This has been discussed, in the case of several species of basidiomycetes, in terms of strain 'individualism' (Todd & Rayner, 1980) and the consequences of this mechanism for the development and maintenance of population structure and diversity are reviewed by Rayner *et al.* (Chapter 23). In *Aspergillus nidulans* a similar role seems possible. A sample of natural isolates taken from locations at widely scattered sites in England and Wales have been shown to fall into several heterokaryon compatibility (h-c) groups (Grindle, 1963*b*). A study of genetic variation found for several characters both within and between h-c groups showed clearly that members of any one h-c group were closely related to each other and less closely related to members of any other h-c group (Jinks, Caten, Simchen & Croft, 1966; Butcher, Croft & Grindle, 1972). It can be concluded that the natural population of *A. nidulans* consists of a mixture of asexually reproducing clones, the h-c groups, dispersed over a wide geographical area. The h-c groups are thought to represent divergently evolving lines genetically isolated from each other by vegetative incompatibility (Croft & Jinks, 1977).

Secondly, vegetative incompatibility may act as an effective barrier to cytoplasmic exchange and thus reduce the spread of deleterious cytoplasmic mutations and viral infection. This was first illustrated by Caten (1972) who showed that the cytoplasmically inherited mutant character 'vegetative death' in *A. amstelodami* transferred freely between vegetatively compatible pairs of strains but not at all, or with a very reduced frequency, between incompatible pairs of strains. This mutant character is phenotypically similar to the group of 'ragged' mutant strains which have been shown to be due to a suppressive mitochondrial mutation (Lazarus, Earl, Turner & Küntzel, 1980*a*). Vegetative incompatibility also prevents the inter-strain transfer of some metabolic intermediates (R. M. Bloomfield, A. C. Newton & C. E. Caten, personal communication). In these cases pairs of mutants affecting cleistothecial wall colour in *A. amstelodami* and nitrate utilisation in *A. amstelodami* and in *Septoria nodorum* can be seen to complement along the junction of the two colonies only if the strains are compatible. The effectiveness of vegetative incompatibility in preventing inter-strain transfer of cytoplasmic genetic determinants has also been demonstrated for other mitochondrial mutations such as oligomycin and chloramphenicol resistance in *A. nidulans* (Rowlands & Turner, 1976), and for genes, probably carried on dsRNA molecules, which can alter both the morphology and virulence of *Endothia parasitica* (Anagnostakis & Waggoner, 1981; Anagnostakis: Chapters 16 and 22) and possibly also in *Ceratocystis ulmi* (Brasier: Chapter 21).

Sexual crosses with little, if any, reduction in fertility can be made between vegetatively incompatible strains of *A. nidulans* (Butcher, 1968). Thus it is possible that such crosses could have played a part in generating the present population structure of this species. However, in *A. amstelodami*, vegetative incompatibility can impede sexual crossing (Caten, 1973) and it has been argued that sexual crosses between h-c groups in *A. nidulans* are likely to be rare in nature with the majority of progeny strains having poorly adapted genotypes which are unlikely to compete well with the established strains (Jinks *et al.*, 1966; Butcher, 1969; Croft & Jinks, 1977). Also, it should be noted that the population structures of asexual species of *Aspergillus* appear to be similar to that found in *A. nidulans* (Caten, 1971), as do those of certain non-outcrossing Basidiomycotina such as *Stereum sanguinolentum* (Rayner & Turton, 1982; Rayner *et al.*: Chapter 23).

Mechanisms underlying many interspecific mycelial interactions are not well understood at the present time. In the natural ecosystem it is probable that competition between a wide range of unrelated species is of considerable ecological importance (see Rayner & Webber: Chapter

18). Very few interspecific mycelial interactions are likely to result in the production of novel genotypes via genetic exchange since various kinds of functional incompatibility between distantly related genomes can be expected. These will include chromosomal structural differences, and various poorly adapted genetic combinations both at the nuclear and nucleocytoplasmic levels, as well as sytems of vegetative incompatibility which may act to prevent or reduce effective hyphal fusion. A programme of interspecific hybridisation in the genus *Aspergillus* has been carried out in order to study the possibility of gene flow between species and to assess the role that this might play in evolutionary processes. Both somatic and sexual hybridisation have been considered and in this account particular attention will be paid to the evidence obtained from the study of the inheritance of the mitochondrial genome.

Interspecific somatic hybridisation

Effects of vegetative incompatibility

The taxonomy of *Aspergillus* has been described in considerable detail by Raper & Fennell (1965). The 200 or more species have been classified into some 18 species-groups, each of which is divided further into individual species or varieties. This taxonomic classification is summarised in Fig. 1. The degree to which it reflects genetic affinity is open to both speculation and experimentation. It should be noted that it is a hierarchical classification and that the h-c groups form a further level in the hierarchy below that of the species.

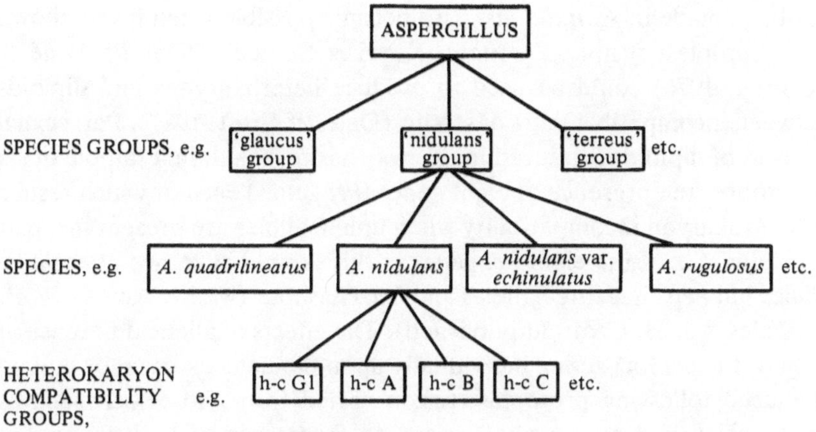

Fig. 1. Taxonomic classification of *Aspergillus*.

In *A. nidulans* vegetative incompatibility has mostly been considered strictly in terms of heterokaryon incompatibility, that is the inability to produce neutral heterokaryons (Caten & Jinks, 1966) as revealed by the presence of striped heterokaryotic conidial heads in mixed cultures of pairs of strains differing in conidial colour (Grindle, 1963*a*). The results obtained by this test have been found to correlate exactly with attempts to produce heterokaryons by nutritional selection using pairs of complementary auxotrophic mutations (Dales, Moorhouse & Croft, 1983). Evidence for vegetative incompatibility in other species often depends on observations of barrage or antagonistic reactions between pairs of mycelia and in most cases the genetic basis of these reactions is not known and is probably complex (cf. Brasier: Chapter 21; Anagnostakis: Chapter 22; Rayner *et al.*: Chapter 23). In *A. nidulans* similar reactions are often seen. The genetic control of these may be complex but both nuclear and cytoplasmic control of some of them has been demonstrated. In all cases they are independent of the strictly defined heterokaryon incompatibility system. Thus, presence or absence of a barrage and the ability or inability to form heterokaryotic conidial heads can segregate independently among the progeny of a cross, and a pair of strains which produces a barrage can be heterokaryon-compatible, and likewise a pair of strains which grow into each other without any visible reaction can be fully heterokaryon-incompatible. However, in practice, the various barrage and antagonistic reactions will tend to reduce heterokaryon formation in heterokaryon-compatible pairs of strains.

A detailed analysis of the genetics of heterokaryon incompatibility has been made in *A. nidulans*. This became possible when it was shown that protoplast fusion (Ferenczy, Kevei & Szegedi, 1975*a,b*; Anné & Peberdy, 1976) could be used to produce heterokaryons and diploids between incompatible pairs of strains (Dales & Croft, 1977). Parasexual analysis of diploids produced in this way has revealed, in a sample of six h-c groups, the presence of eight genes (*het* genes) each of which results in heterokaryon incompatibility when unlike alleles are present in a pair of strains. Six of the eight *het* genes in this sample have two alternative alleles but *hetC* has three alleles and *hetB* has four (M. M. Anwar, R. B. G. Dales & J. H. Croft, unpublished). The effects of allelic difference at each of the *het* loci acting individually upon the heterokaryon or diploid produced following protoplast fusion varied from those loci, such as *hetA*, which had only a small effect, to those such as *hetB* where the heterokaryon or diploid was poorly growing, with a deep red or brown pigmentation and very few small conidial heads (Dales *et al.*, 1983). The effects of allelic differences at more than one *het* gene are in part cumulative so that the diploids produced by protoplast fusion between most pairs of natural isolates from different h-c groups grow very poorly and are highly pigmented.

Protoplast fusion between *A. nidulans* and members of other species groups such as *A. amstelodami, A. versicolor, A. niger* and *A. terreus* have failed to give any evidence of interspecific hybridisation, and the same is true between *A. nidulans* and certain members of the 'nidulans' group, such as *A. stellatus, A. unguis* and *A. heterothallicus*. However, fusion between several other species combinations in the 'nidulans' group has resulted in the recovery of colonies which have a similar morphology to the diploids produced between heterokaryon-incompatible strains within *A. nidulans*. This has been achieved with *A. nidulans, A. quadrilineatus, A. rugulosus, A. violaceus* and *A. nidulans* var. *echinulatus* (F. Kevei, personal communication).

Genetic analysis of benomyl-induced haploid segregants derived from these hybrid colonies show clearly that they represent allodiploids (Clausen, Keck & Hiesey, 1945) between the parental species. In the case of *A. nidulans* plus *A. rugulosus* the segregants showed free recombination of markers representing all eight linkage groups of *A. nidulans* (Kevei & Peberdy, 1977, 1979) suggesting that the overall genome organisation of these two species is similar. However, in the case of *A. nidulans* and *A. nidulans* var. *echinulatus* (K. L. Evans & J. H. Croft, unpublished) five of the eight *A. nidulans* linkage groups showed free segregation but linkage groups III, V and VIII gave only parental combinations of markers among the segregants. Thus *A. nidulans* var. *echinulatus* probably carries complex translocations involving the chromosomes equivalent to *A. nidulans* linkage groups III, V and VIII. It should be noted that *A. nidulans* var. *echinulatus* and *A. nidulans* var. *latus* referred to later, are not separate species in the strict taxonomic sense. However, the studies described in this chapter indicate that, genetically, they do represent distinct species on an equivalent rank to the taxonomic species described in the 'nidulans' group (Raper & Fennell, 1965). Consequently, they will be referred to as species in this chapter.

The presence of genes controlling heterokaryon incompatibility between *A. nidulans* and *A. nidulans* var. *echinulatus*, similar to the *het* genes within *A. nidulans*, has been demonstrated by the analysis of the segregation of heterokaryon incompatibility among segregants derived from the allodiploid. The information gained also indicated that there was considerable genetic heterogeneity between these two species as can be seen from sexual backcrosses of segregants to the parental species. The only backcrosses which gave normal progenies were those involving segregants which carried the full appropriate parental combination of markers. All other backcrosses, or crosses among the segregants themselves, gave abnormal progenies. Particularly high frequencies of aneuploids (up to 50%) were found which indicates the presence of chromosome structural heterozygosity between the parents of these crosses (Upshall & Käfer, 1974). However, it was possible to demonstrate the absence of any *het* genes on *A. nidulans* linkage groups I and IV and their equivalent *A. nidulans* var. *echinulatus* homologues and the presence of one *het* gene on each of the chromosome pairs equivalent to *A. nidulans* linkage groups II, VI and VII and at least one on the III-V-VIII complex. New combinations of the alleles at these *het* loci will mean that few of the segregants (8 out of the 64 possible linkage group combinations) will be vegetatively compatible with either parental strain.

Thus it appears that the species which will hybridise somatically to produce allodiploids are closely related and will produce viable recombinant somatic segregants at the haploid level even though chromosomal genetic divergence can be demonstrated. Some of these recombinant segregants are sexually infertile and there is evidence that some may carry duplications. However, many are sexually fully self-fertile. It is interesting to note that the ascospore wall sculpturing, which is a characteristic taxonomic feature in the 'nidulans' group (Raper & Fennell, 1965), is intermediate between that of the two parental species in most of these self-fertile interspecific recombinants. These strains, while appearing as typical members of the 'nidulans' group, are therefore likely to be vegetatively incompatible and not sexually fully fertile with both parental species and hence should strictly be regarded as taxonomically distinct.

Polymorphism of the mitochondrial genome in Aspergillus *species*

Restriction enzyme analysis of mitochondrial (mt) DNA may be useful in taxonomic studies of *Aspergillus* (Kozlowski, Bartnik & Stępień, 1982). Very full restriction-site maps of mtDNA of *A. nidulans* (Stępień, Bernard, Cooke & Küntzel, 1978; Lazarus *et al.*, 1980*b*) and *A. amstelodami* (Lazarus *et al.*, 1980*a*) have been published. In *A.*

Table 1. *Variation in size of the mitochondrial genomes of some species of Aspergillus classified in the 'nidulans' group*

Species	Approximate size (kb)
A. nidulans var. *echinulatus*	39
A. rugulosus	37.5
Unidentified species	37.5
A. nidulans	32.5
A. nidulans var. *latus*[a]	30.5
A. quadrilineatus	30.5

[a] Identification of this species requires expert confirmation.

nidulans the location of several genes on the mitochondrial genome has been demonstrated (Macino *et al.*, 1980) and the nucleotide sequence of the mtDNA of this species is almost completely known (Waring *et al.*, 1981; Grisi *et al.*, 1982; Köchel & Küntzel, 1982; Netzler, Köcher, Basak & Küntzel, 1982). When the mtDNA of certain species belonging to the 'nidulans' group is examined it is found that although the overall physical map is similar for the majority of restriction sites, there being a few species-specific sites in most cases, a major difference between species is in the size of the genome. This varies from about 30 kilobases (kb) to about 39 kb (Table 1). Detailed comparisons of the restriction-site maps of *A. nidulans* and *A. nidulans* var. *echinulatus* have shown that the size differences are due to inserts at certain positions in the mitochondrial genome (Earl *et al.*, 1981). The major inserts are of about 1 to 2 kb in size. Smaller inserts or deletions may also be present but these will not be considered here. The inserts are clustered in two regions, one of which is in the region of the gene coding for subunit 1 of cytochrome oxidase and the other is in the region of the gene coding for apo-cytochrome b, and it is likely that these inserts represent introns similar to those described by Davies *et al.* (1982). The mitochondrial genome of *A. quadrilineatus* has not been studied in as much detail, but it differs in size from that of *A. nidulans* by the lack of two further inserts, one in each of the two regions described above (A. J. Earl, R. A. Spooner & G. Turner, unpublished). Thus the mitochondrial genomes of these three species differ from each other by the presence or absence of variable combinations of seven inserts (Fig. 2). The evidence for the species specificity of the structure of the mitochondrial genome is of necessity based on small samples, but no differences in the mtDNA have been found in four independent isolates of *A. nidulans*, three from the UK and one from the USA. Similarly three independent isolates of

A. nidulans var. *echinulatus* have each been shown to contain mtDNA carrying all of the inserts. A fourth pattern of mitochondrial genome structure is seen in *A. rugulosus* which is intermediate in size between *A. nidulans* and *A. nidulans* var. *echinulatus*. This appears to be due to the presence of the three inserts in the cytochrome oxidase region and the absence of the two inserts in the apo-cytochrome b region. More than one species may have a similar mitochondrial genome structure based on restriction endonuclease mapping data. Thus an unidentified species of African origin has similar mtDNA to that found in *A. rugulosus* and two independent UK isolates which have been provisionally identified as *A. nidulans* var. *latus* have mtDNA similar to that found in *A. quadrilineatus*. It has also been shown that *A. tamarii* and *A. wentii* possess mtDNA which have indistinguishable restriction

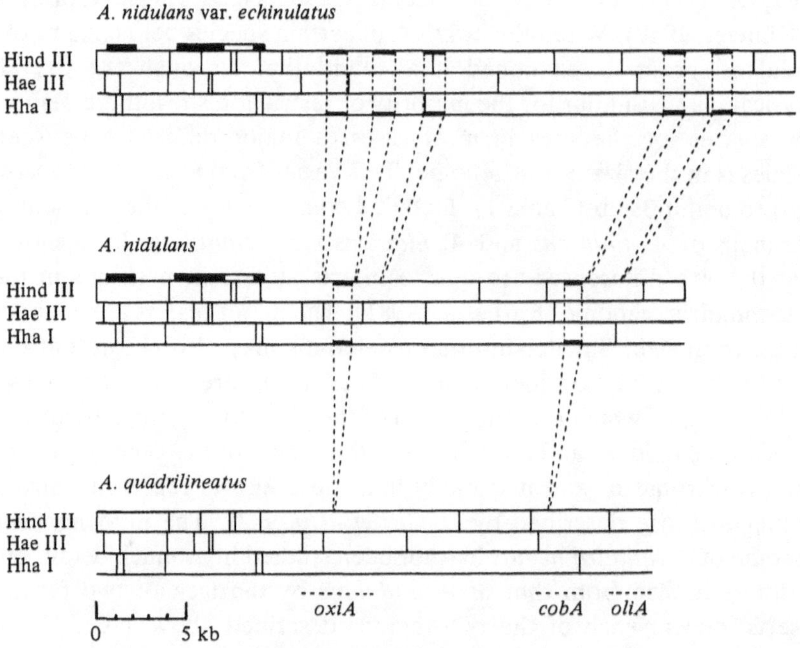

Fig. 2. Linearised restriction-site maps of the circular mitochondrial genomes of three 'nidulans' group species of *Aspergillus*. The black bars above the Hind III lines indicate positions of the small and large ribosomal RNA genes where determined. Approximate locations of three other genes are shown (*oxiA* codes for cytochrome oxidase subunit 1; *cobA* for apo-cytochrome b; *oliA* for an altered ATPase subunit 6 which confers oligomycin resistance). Positions of the seven variable inserts are connected by vertical broken lines.

patterns in Eco RI and Hind III double digests (Kozlowski *et al.*, 1982). However, this information in itself does not necessarily indicate an extremely close phylogenetic relationship between these two species.

Inheritance of the mitochondrial genome in somatic hybrids

Two experimental approaches to the study of the inheritance of the mitochondrial genome in interspecific somatic hybrids have been used. In the first, mitochondrial markers such as oligomycin, chloramphenicol or mucidin resistance have been used to select for the transfer of mitochondrial genetic material from one species to another (Croft *et al.*, 1980). This approach became possible when it was demonstrated that mitochondrial markers could be transferred between vegetatively incompatible strains of *A. nidulans* by protoplast fusion. The second approach was to analyse the mtDNA contained in the allodiploid produced by the fusion of two species. In this case no selective mitochondrial markers were used.

When attempts are made to select for transfer of mitochondrial markers from one species to another several results are possible. In transfer experiments involving *A. nidulans* with distantly related species or with those 'nidulans' group species which do not form allodiploids there is currently no evidence for any genetic interaction and selective marker transfer does not take place. However, in those 'nidulans' group species which will form allodiploids, marker transfer occurs readily. In these cases there are two possible results – either the mitochondrial genome transfers unchanged to the recipient species, or only part of the donor mitochondrial genome transfers and a recombinant genome is established. There appear to be simple general rules for this process. Thus selection for the transfer of the smaller *A. nidulans* mitochondrial genome to the *A. nidulans* var. *echinulatus* nuclear background has always resulted in the establishment of a recombinant genome (Earl *et al.*, 1981). By contrast, in the reciprocal experiment the larger *A. nidulans* var. *echinulatus* mitochondrial genome is transferred apparently unchanged into *A. nidulans*. In a similar experiment the whole *A. nidulans* var. *echinulatus* genome can be transferred to *A. quadrilineatus* thus providing this species with an mtDNA molecule some 30% larger than that which it normally contains, yet with no obvious deleterious effects. This is consistent with the possibility that these inserts carry no essential genetic information and are concerned only with their own excision at the time of mRNA processing (Davies *et al.*, 1982).

The genome which is regularly produced by the transfer of oligomycin or chloramphenicol resistance from *A. nidulans* to *A. nidulans* var. *echinulatus* has been characterised by restriction-site analysis (Earl *et al.*, 1981). It is found to be intermediate in size between that of the two parental species and contains the group of three inserts carried by *A. nidulans* var. *echinulatus* in the cytochrome oxidase region but does not contain the two inserts in the apo-cytochrome b region. It can be readily explained as a recombinant molecule (Fig. 3) and it is interesting to note that it has a similar structure to the mitochondrial genome found naturally in *A. rugulosus* and in one other unidentified species (Table 1). A few different recombinants have also been recovered. These include cases where the genome structure is similar to the common class of recombinant, but where the recombination event to the left of the group of three inserts has taken place at a variety of positions as can be determined from the several species-specific restriction sites in that

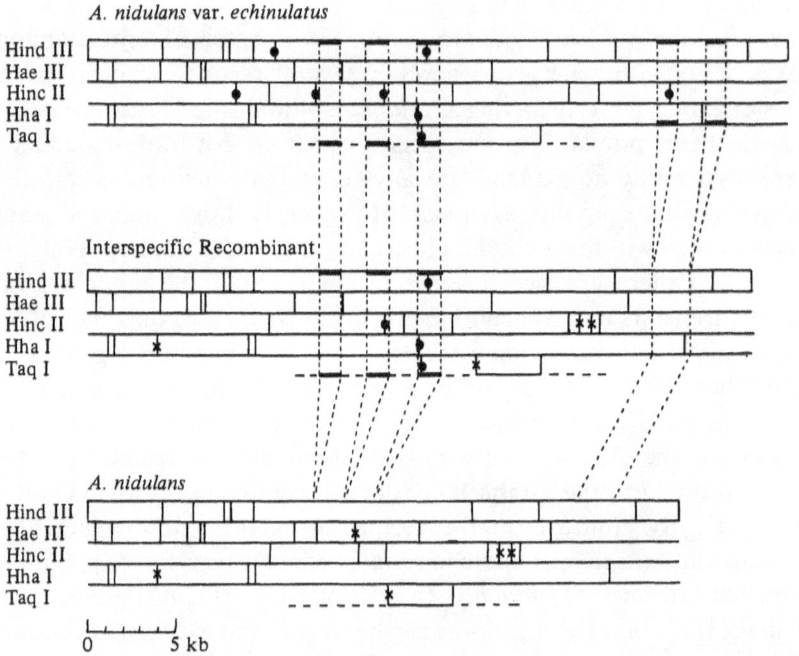

Fig. 3. Linearised restriction-site maps of the mitochondrial genomes of two 'nidulans' group species of *Aspergillus* and of the most common recombinant between them. Sites present in *A. nidulans* var. *echinulatus* not present in *A. nidulans* are shown ♦, and sites present in *A. nidulans* not present in *A. nidulans* var. *echinulatus* are shown X. Only three of the many Taq I sites are given. The positions of the five inserts are connected by vertical broken lines.

region. Two structurally different recombinant genomes have also been produced. One of these had become oligomycin resistant but had retained all five inserts whereas the other contained only the first two of the group of three inserts in the cytochrome oxidase region. Although all of these recombinants can be explained by conventional double-recombination events, the actual mechanism by which these recombinants arose remains unknown.

Mitochondrial DNA has been extracted from an allodiploid produced between *A. nidulans* and *A. nidulans* var. *echinulatus* and also from haploid progeny derived from a second independently produced allodiploid between these two species. No mitochondrial selective markers were involved in either case. In both cases the mitochondrial genomes were recombinant. In one this was the common type of recombinant (Fig. 3) whereas the other retained only the first two of the group of three inserts. Thus it would appear that there is a high probability that mitochondrial recombination will occur spontaneously in the formation of the allodiploid between two species and that this will result in the establishment of a recombinant mitochondrial genome in the haploid interspecific recombinants derived from that allodiploid.

In all cases discussed in this section all of the strains derived from the various protoplast fusion experiments contain only one molecular type of mtDNA. Although a mixed population of mitochondrial genomes must exist immediately after protoplast fusion there is no evidence that this may persist for any detectable period. The mechanism by which recombinant or reassorted mitochondrial genomes become established in the colonies which regenerate following protoplast fusion remains unknown.

Interspecific sexual hybridisation

A tentative report of sexual hybridisation between *A. nidulans* and *A. rugulosus* was made fifteen years ago (Tyc, 1968). Subsequent attempts to make interspecific sexual crosses between *A. nidulans* and other 'nidulans' group species have given similar results (J. H. Croft, unpublished). Hybrid cleistothecia have been produced in crosses with *A. quadrilineatus, A. nidulans* var. *latus* and one unidentified species. It has not proved possible to cross *A. nidulans* with *A. nidulans* var. *echinulatus* but it is thought that the reason for this is physical rather than primarily genetic since these two species show strong barrage reactions with each other when grown in mixed culture.

In all cases these interspecific crosses show a very low fertility and the hybrid cleistothecia contain very few ascospores, only a proportion of which may be viable. Those which do germinate develop into colonies which fall into three morphological classes. The first class contains normal vigorous colonies which are apparently of one or other parental type. The second class consists of small unstable colonies which are probably representative of various types of aneuploid. The third class is represented by stable morphologically-abnormal colonies which are similar in appearance to the allodiploid colonies described previously. It is thought that these represent some form of polyploid hybrid and that the extreme morphology is due to the presence of allelic differences at *het* genes. These interspecific progenies have not yet been studied intensively but it is possible that the aneuploids and morphologically abnormal colonies could both give rise to stable haploid colonies which carry nuclear genetic information derived from both parental species.

Inheritance of the mitochondrial genome in sexual hybrids

Clear evidence concerning the mode of inheritance of mtDNA in interspecific sexual crosses is not yet available. The maternal inheritance of mitochondrial markers has previously been demonstrated in *A. nidulans* (Rowlands & Turner, 1976). From their results and the earlier work of Apirion (1963) it can be concluded that hybrid cleistothecia in a mixed culture will fall into two types according to which of the two parental strains makes the maternal contribution and it would appear, even in this homothallic fungus, that a process analogous to fertilisation in heterothallic species takes place. When the experimental procedure of Rowlands and Turner is followed using a suitable pair of *A. nidulans* strains, one of which is carrying the large mitochondrial genome of *A. nidulans* var. *echinulatus*, it has been shown that two classes of hybrid cleistothecia are formed in a mixed culture (D. M. Jadayel & J. H. Croft, unpublished). In one of these all of the progeny contain the small *A. nidulans* mitochondrial genome while the other gives progeny all of which contain the large *A. nidulans* var. *echinulatus* genome. No recombinant mtDNA has been found in any progeny of this sexual cross. Although this point must be verified by the analysis of progeny derived from an actual interspecific cross it appears likely that the mitochondrial genomes of different species will inherit maternally and without recombination during sexual interspecific hybridisation.

Discussion

There is now a considerable volume of evidence which shows that the systems of vegetative incompatibility which operate within species of fungi do act to restrict gene flow within those species. In *A. nidulans* it has been proposed that the h-c groups are evolving divergently by conventional mutation and selection and that this might be a process by which a species will evolve into a group of sibling species (Mather, 1965) which is what many of the taxonomic species or varieties belonging to the 'nidulans' group clearly represent (Croft & Jinks, 1977). In *A. nidulans* sexual crossing is possible and such an event taking place between members of different h-c groups would bring about recombination both of the *het* genes and of the general genetic background. This could result in the origin of new individuals which might compete successfully in the natural ecosystem along with the established members of the species and thus produce a new or modified population of h-c groups. It should be noted that sexual crossing in *A. nidulans* is thought to be rare in nature and competitively successful individuals among the progeny of a cross are also thought to be rare (Butcher, 1969). There is clear evidence for stabilising selection in *A. nidulans* leading to the selection of balanced intermediate phenotypes (Croft & Jinks, 1977). It is probable that any strain produced by the sexual crossing of existing strains would still be recognisably *A. nidulans* in its characteristics. At the level of the mitochondrion no variation would be expected since, on current evidence, no variation in the mitochondrial genome exists within the species. However, investigation of a much larger sample of individuals is required in order to confirm this observation.

Asexual species of *Aspergillus* appear to have a similar population structure to that of *A. nidulans* (Caten, 1971). In these cases, and in the sexual species, a similar result to that produced by sexual crossing would be achieved if occasional somatic hybridisation were possible between vegetatively incompatible strains. However, the amount of genetic recombination would be expected to vary since it has been demonstrated that the frequencies of spontaneous chromosome non-disjunction and of intra-chromosomal mitotic recombination vary widely from species to species (see Caten, 1981).

The experimental results discussed in this chapter show that interspecific hybridisation between established species of *Aspergillus* is possible in the laboratory. Indeed this can be achieved with no more difficulty than between incompatible strains within *A. nidulans*. Such hybridisation is possible only between species which are closely related and which may be thought of as sibling species. Possible genetic reasons for the failure to hybridise certain pairs of species, even some classified in the 'nidulans' taxonomic group, are currently under investigation.

Somatic hybridisation following protoplast fusion of closely related species gives rise to the full allodiploid formed by the nuclear fusion of the two parental haploid nuclei. The initial barriers to the formation of such a hybrid are, in the main part, due to the vegetative incompatibility system and the extreme morphologies of the allodiploids are the result of the effects of the *het* genes. The range of variability among the haploid segregants derived from allodiploids will depend on the balance of the degree of genetic difference and the degree of chromosomal structural heterozygosity present between the two parental species. Thus the example of the *A. nidulans* plus *A. rugulosus* hybrid reported by Kevei & Peberdy (1977, 1979) shows little evidence for structural heterozygosity, whereas that of the *A. nidulans* plus *A. nidulans* var. *echinulatus* hybrid reported here shows considerable evidence for the presence of translocations and other genetic differences between the two species.

Sexual hybridisation is also possible between many if not all of these closely related species. These crosses are of only a low viability and give rise to many abnormal colonies such as aneuploids. Nevertheless interspecific recombination is possible by this route, though this has not yet received a full investigation.

When the mitochondrial genome is considered there is circumstantial evidence that sexual hybridisation between species will yield progenies with one or other parental type of mtDNA since it is likely that the mitochondrial genome will inherit maternally. By contrast, interspecific somatic hybridisation gives rise to progenies in which the recombination of the mitochondrial genome may even be the most likely outcome. Thus hybridisation of *A. nidulans* and *A. nidulans* var. *echinulatus* has given rise to interspecific recombinants which contain a mitochondrial genome unlike that of either parent, but similar to those found in other naturally occurring species. In fact, in the laboratory, interspecific recombinant strains have been produced which appear, on nuclear and mitochondrial characteristics, to classify best as new species, or at least as the progenitors of new species upon which mutation and selection may act.

Thus, in contrast to the thesis put forward initially in this chapter, it is now proposed as a working hypothesis that rare interspecific hybridisation may play an important role in the evolution of these genetically-related species groups. On the basis of the variation found in the mitochondrial genome and of the ease with which mitochondrial recombination takes place during somatic hybridisation it is thought that this process may play a more significant role than sexual hybridisation. Exactly how this would take place remains obscure. Several models have been suggested for the process of mitochondrial recombination. These include conventional recombination, a process analogous to transposition, and a process of unidirectional gene conversion similar to that described for the *omega* and *var-1* regions of the yeast mitochondrial genome (Heyting & Menke, 1979; Strausberg, Vincent, Perlman & Butow, 1978). However, in the absence of evidence for mitochondrial transformation in *Aspergillus*, it would appear that plasmogamy of the two parental species is necessary before mitochondrial recombination can take place. For one explanation of how this might be achieved in nature it is not necessary to look further than the process of protoplast fusion. It has been shown that ageing cultures of *A. nidulans* autolyse and during this process release lytic enzymes (Isaac & Gokhale, 1982) and this is currently the major source of lytic enzymes used for protoplast production in our laboratory. It is thus suggested that two species of *Aspergillus* growing together on a source of nutrients would completely or partly lyse and, in the high osmotic environment which the nutrient source might present, the resultant protoplasts or partially digested hyphae could fuse naturally. This is clearly a speculative proposal, but one which it should not be impossible to test.

We are grateful to Mrs K. L. Evans, Miss D. M. Jadayel and Drs M. M. Anwar, R. M. Bloomfield, C. E. Caten, A. J. Earl, F. Kevei, A. C. Newton, R. A. Spooner and G. Turner for permission to refer to their unpublished results. We wish to acknowledge the financial support of the Science and Engineering Research Council and the Agricultural Research Council.

References

Anagnostakis, S. L. & Waggoner, P. E. (1981). Hypovirulence, vegetative incompatibility, and the growth of cankers of chestnut blight. *Phytopathology*, **71,** 1198–1202.

Anné, J. & Peberdy, J. F. (1976). Induced fusion of fungal protoplasts following treatment with polyethylene glycol. *Journal of General Microbiology*, **92,** 413–17.

Apirion, D. (1963). Formal and physiological genetics of ascospore colour in *Aspergillus nidulans. Genetical Research*, **4,** 276–83.

Butcher, A. C. (1968). The relationship between sexual outcrossing and heterokaryon incompatibility in *Aspergillus nidulans. Heredity*, **23,** 443–52.

Butcher, A. C. (1969). Non-allelic interactions and genetic isolation in wild populations of *Aspergillus nidulans. Heredity*, **24,** 621–31.

Butcher, A. C., Croft, J. H. & Grindle, M. (1972). Use of genotype-environmental interaction analysis in the study of natural populations of *Aspergillus nidulans. Heredity*, **29,** 263–83.

Caten, C. E. (1971). Heterokaryon incompatibility in imperfect species of *Aspergillus. Heredity*, **26,** 299–312.

Caten, C. E. (1972). Vegetative incompatibility and cytoplasmic infection in fungi. *Journal of General Microbiology*, **72,** 221–9.

Caten, C. E. (1973). Genetic control of heterokaryon incompatibility and its effect on crossing and cytoplasmic exchange in *Aspergillus amstelodami. Genetics*, **74,** 540.

Caten, C. E. (1981). Parasexual processes in fungi. In *The Fungal Nucleus*, ed. K. Gull & S. G. Oliver, pp. 191–214. Cambridge: Cambridge University Press.

Caten, C. E. & Jinks, J. L. (1966). Heterokaryosis: its significance in wild homothallic Ascomycetes and Fungi Imperfecti. *Transactions of the British Mycological Society*, **49,** 81–93.

Clausen, J., Keck, D. D. & Hiesey, W. M. (1945). *Experimental Studies on the Nature of Species.* II. *Plant Evolution through Amphiploidy and Autoploidy with Examples from the Madiinae. Carnegie Institution of Washington Publication 564.*

Croft, J. H., Dales, R. B. G., Turner, G. & Earl, A. J. (1980). The transfer of mitochondria between species of *Aspergillus*. In *Advances in Protoplast Research*, ed. L. Ferenczy & G. L. Farkas, pp. 85–92. Budapest: Akadémiai Kiadó; and Oxford: Pergamon Press.

Croft, J. H. & Jinks, J. L. (1977). Aspects of the population genetics of *Aspergillus nidulans*. In *Genetics and Physiology of Aspergillus*, ed. J. E. Smith & J. A. Pateman, pp. 339–60. New York and London: Academic Press.

Dales, R. B. G. & Croft, J. H. (1977). Protoplast fusion and the isolation of heterokaryons and diploids from vegetatively incompatible strains of *Aspergillus nidulans. FEMS Microbiology Letters*, **1,** 201–3.

Dales, R. B. G., Moorhouse, J. & Croft, J. H. (1983). The location and analysis of two heterokaryon incompatibility (*het*) loci in strains of *Aspergillus nidulans. Journal of General Microbiology*, **129,** 3637–42.

Davies, R. W., Waring, R. B., Ray, J. A., Brown, T. A. & Scazzocchio, C. (1982). Making ends meet: a model for RNA splicing in fungal mitochondria. *Nature*, **300,** 719–24.

Earl, A. J., Turner, G., Croft, J. H., Dales, R. B. G., Lazarus, C. M., Lünsdorf, H. & Küntzel, H. (1981). High frequency transfer of species specific mitochondrial DNA sequences between members of the Aspergillaceae. *Current Genetics*, **3,** 221–8.

Ferenczy, L., Kevei, F. & Szegedi, M. (1975*a*). Increased fusion frequency of *Aspergillus nidulans* protoplasts. *Experientia*, **31,** 50–2.

Ferenczy, L., Kevei, F. & Szegedi, M. (1975*b*). High-frequency fusion of fungal protoplasts. *Experientia*, **31,** 1028–30.

Grindle, M. (1963*a*). Heterokaryon compatibility of unrelated strains in the *Aspergillus nidulans* group. *Heredity*, **18,** 191–204.

Grindle, M. (1963*b*). Heterokaryon compatibility of closely related wild isolates of *Aspergillus nidulans. Heredity*, **18,** 397–405.

Grisi, E., Brown, T. A., Waring, R. B., Scazzocchio, C. & Davies, R. W. (1982). Nucelotide sequence of a region of the mitochondrial genome of *Aspergillus nidulans* including the gene for ATPase subunit 6. *Nucleic Acids Research*, **10,** 3531–9.

Heyting, C. & Menke, H. (1979). Fine structure of the 21S ribosomal RNA region on yeast mitochondrial DNA. III. Physical location of mitochondrial genetic markers and the molecular nature of *omega*. *Molecular and General Genetics*, **168**, 279–91.

Isaac, S. & Gokhale, A. V. (1982). Autolysis: a tool for protoplast production from *Aspergillus nidulans*. *Transactions of the British Mycological Society*, **78**, 389–94.

Jinks, J. L., Caten, C. E., Simchen, G. & Croft, J. H. (1966). Heterokaryon incompatibility and variation in wild populations of *Aspergillus nidulans*. *Heredity*, **21**, 227–39.

Kevei, F. & Peberdy, J. F. (1977). Interspecific hybridisation between *Aspergillus nidulans* and *Aspergillus rugulosus* by fusion of somatic protoplasts. *Journal of General Microbiology*, **102**, 255–62.

Kevei, F. & Peberdy, J. F. (1979). Induced segregation in interspecific hybrids of *Aspergillus nidulans* and *Aspergillus rugulosus* obtained by protoplast fusion. *Molecular and General Genetics*, **170**, 213–18.

Köchel, H. G. & Küntzel, H. (1982). Mitochondrial L-rRNA from *Aspergillus nidulans*: potential secondary structure and evolution. *Nucleic Acids Research*, **10**, 4795–801.

Kozlowski, M., Bartnik, E. & Stępień, P. P. (1982). Restriction enzyme analysis of mitochondrial DNA of members of the genus *Aspergillus* as an aid in taxonomy. *Journal of General Microbiology*, **128**, 471–6.

Lazarus, C. M., Earl, A. J., Turner, G. & Küntzel, H. (1980*a*). Amplification of a mitochondrial DNA sequence in the cytoplasmically inherited 'ragged' mutant of *Aspergillus amstelodami*. *European Journal of Biochemistry*, **106**, 633–41.

Lazarus, C. M., Lünsdorf, H., Hahn, U., Stępień, P. P. & Küntzel, H. (1980*b*). Physical map of *Aspergillus nidulans* mitochondrial genes coding for ribosomal RNA: an intervening sequence in the large rRNA cistron. *Molecular and General Genetics*, **177**, 389–97.

Macino, G., Scazzocchio, C., Waring, R. B., McPhail Berks, M. & Davies, R. W. (1980). Conservation and rearrangement of mitochondrial structural gene sequences. *Nature*, **288**, 404–6.

Mather, K. (1965). The genetical interest of incompatibility in fungi. In *Incompatibility in Fungi*, ed. K. Esser & J. R. Raper, pp. 113–17. Berlin: Springer-Verlag.

Netzler, R., Köchel, H. G., Basak, N. & Küntzel, H. (1982). Nucleotide sequence of *Aspergillus nidulans* mitochondrial genes coding for ATPase subunit 6, cytochrome oxidase subunit 3, seven unidentified proteins, four tRNAs and L-rRNA. *Nucleic Acids Research*, **10**, 4783–94.

Raper, K. B. & Fennell, D. I. (1965). *The Genus Aspergillus*. Baltimore: Williams & Wilkins.

Rayner, A. D. M. & Turton, M. N. (1982). Mycelial interactions and population structure in the genus *Stereum*: *S. rugosum*, *S. sanguinolentum* and *S. rameale*. *Transactions of the British Mycological Society*, **78**, 483–93.

Rowlands, R. T. & Turner, G. (1976). Maternal inheritance of mitochondrial markers in *Aspergillus nidulans*. *Genetical Research*, **28**, 281–90.

Stępień, P. P., Bernard, U., Cooke, H. & Küntzel, H. (1978). Restriction endonuclease cleavage map of mitochondrial DNA from *Aspergillus nidulans*. *Nucleic Acids Research*, **5**, 317–30.

Strausberg, R. L., Vincent, R. D., Perlman, P. S. & Butow, R. A. (1978). Asymmetric gene conversion at inserted segments on yeast mitochondrial DNA. *Nature*, **276**, 577–83.

Todd, N. K. & Rayner, A. D. M. (1980). Fungal individualism. *Science Progress, Oxford*, **66**, 331–54.

Tyc, M. (1968). An attempt to produce interspecific hybrids between *Aspergillus nidulans* and *Aspergillus rugulosus*. *Aspergillus News Letter*, **9**, 20–1.

Upshall, A. & Käfer, E. (1974). Detection and identification of translocations by increased specific non-disjunction in *Aspergillus nidulans*. *Genetics*, **76**, 19–31.

Waring, R. B., Davies, R. W., Lee, S., Grisi, E., McPhail Berks, M. & Scazzocchio, C. (1981). The mosaic organisation of the apocytochrome b gene of *Aspergillus nidulans* revealed by DNA sequencing. *Cell*, **27**, 4–11.

21

Inter-mycelial recognition systems in *Ceratocystis ulmi*: their physiological properties and ecological importance

C. M. BRASIER

Forest Research Station, Alice Holt Lodge, Farnham, Surrey, GU10 4LH, UK

Inter-mycelial recognition systems are physiological mechanisms promoting 'self-non-self' discrimination between individual mycelia or spores. An alternative term for these phenomena is compatibility/incompatibility systems, since they ultimately control different forms of genetic isolation. In plant pathogens, a principal reason for studying such systems, apart from any intrinsic mycological or genetical interest, is because of their likely role in regulating pathogenicity and population behaviour. In the Dutch elm disease fungus, *Ceratocystis ulmi,* in which three such systems are known, periods of active research on this aspect have inevitably tended to be encouraged by the occurrence of epidemics. Considerable early progress was made during what would now be considered the first epidemic of Dutch elm disease in the 1920s–40s, including the identification of the causal fungus, *Graphium ulmi* by Schwarz (1922); the discovery of the sexual stage, *Ceratocystis* (*Ophiostoma*) *ulmi* and the demonstration of heterothallism (Buisman, 1932; Swingle, 1936; Schafer & Lyming, 1950). A second wave of extremely destructive epidemics across much of the Northern Hemisphere in recent years has led to further intensification of research activity. The present work will review current knowledge of the mycelial population structure and recognition systems of *C. ulmi* drawing mainly on much previously unpublished research by the author. It will also consider the significance of the recognition systems for the ecological and pathological behaviour of the fungus, and for the outcome of the present epidemics.

The status of *C. ulmi* below the species level

One of the first facts to emerge from investigations into the causes of the current epidemics was that *C. ulmi* existed not, as had been generally assumed, as a continuum of variation within one population, but as three discrete reproductively isolated sub-populations or sub-groups each with its own characteristic range of variation. These have been termed the non-aggressive strain, and the North American (NAN) and Eurasian (EAN) races of the aggressive strain (Fig. 1). The two major sub-groups, the aggressive and non-aggressive strains, differ in most important physiological respects and are considered equivalent to sub-species. The EAN and NAN sub-groups within the aggressive strain also differ in many respects but their similarities broadly outweigh their differences. They have been designated 'races' in the broader zoological sense (sensu Stebbins, 1950) and not in the rather narrow sense of 'physiologic race' as often used by plant pathologists.

A major feature of the sub-groups is their differences in pathogenicity. The non-aggressive strain is relatively weakly pathogenic. The two races of the aggressive strain are highly pathogenic, the NAN tending to be slightly more so. These differences also have considerable historical significance, since it is the non-aggressive strain which is now believed to have been responsible for the first epidemics in Europe and North

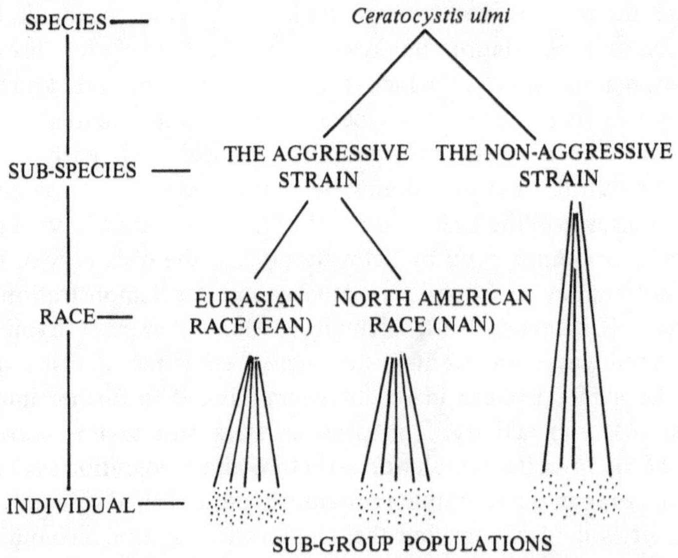

Fig. 1. The division of the *C. ulmi* population into three reproductively isolated sub-groups or sub-populations.

America in the 1920s and 1930s, whereas the NAN and EAN are responsible for the present epidemics. In North America both the NAN aggressive and non-aggressive sub-groups now occur together. In Europe and South West Asia the situation is more complex: the NAN aggressive has recently been imported from North America and is spreading eastwards; the EAN aggressive is migrating overland from the East; and all three sub-groups, non-aggressive, NAN and EAN are now intermixing in many locations (for a review, see Brasier, 1983*a*). The outcome of this intermixing, and more specifically the future behaviour of the fungus and hence of the disease, is likely to be decided to a considerable extent by the mycelial interactions controlled by the fungus' recognition systems. Before examining these, it is necessary to consider where such interactions will take place.

Where mycelial interaction occurs

There are two points in the annual cycle of Dutch elm disease where mycelial interactions can occur. The first is in the feeding grooves cut by the vector scolytid beetles in the twig crotches of healthy elms (the point at which infection of healthy trees by *C. ulmi* occurs). Here, a short mycelial phase may be established (Webber & Brasier, 1984). Very little is known, however, about the ecology of *C. ulmi* in the feeding grooves.

The second is in diseased elm bark. In summer and autumn the vector beetles enter the bark of diseased elms to breed, bringing with them the spores of *C. ulmi* which colonises the bark around the breeding galleries. The fungus now enters an overwintering saprophytic phase which may last for anything up to 8–10 months before the next generation of beetles emerges in the following spring. This long saprophytic phase undoubtedly provides the greatest opportunity for *C. ulmi* mycelial interactions to occur, and is also the phase about which most is known of the structure of the *C. ulmi* mycelial population. Lea (1977) recently examined the bark population of *C. ulmi* in an epidemic area of Britain (resulting from the arrival of the NAN aggressive) by taking samples of bark tissue from points on a 9 × 9 cm grid overlying the bark surface. The samples were taken at three different levels: the inner bark/xylem interface; from the inner bark; and from the outer bark. In addition, he sampled synnemata developing on the inner bark/xylem interface. The isolates obtained from each point were then characterised into different phenotypic variants (within the NAN race) on the basis of morphological characteristics, growth-rate and mating-type.

Lea's results showed that the bark population contained a mosaic of different genotypes, some of which occupied many square centimetres (Fig. 2, levels 1 and 2). He also showed that the mosaic was three-dimensional, the space occupied by a single phenotype often extending across several layers of the same piece of bark (Fig. 2 levels 3, 4). Although this may be a somewhat static picture, it provides valuable insight into the type of mycelial population structure that is maintained and regulated by the following recognition systems.

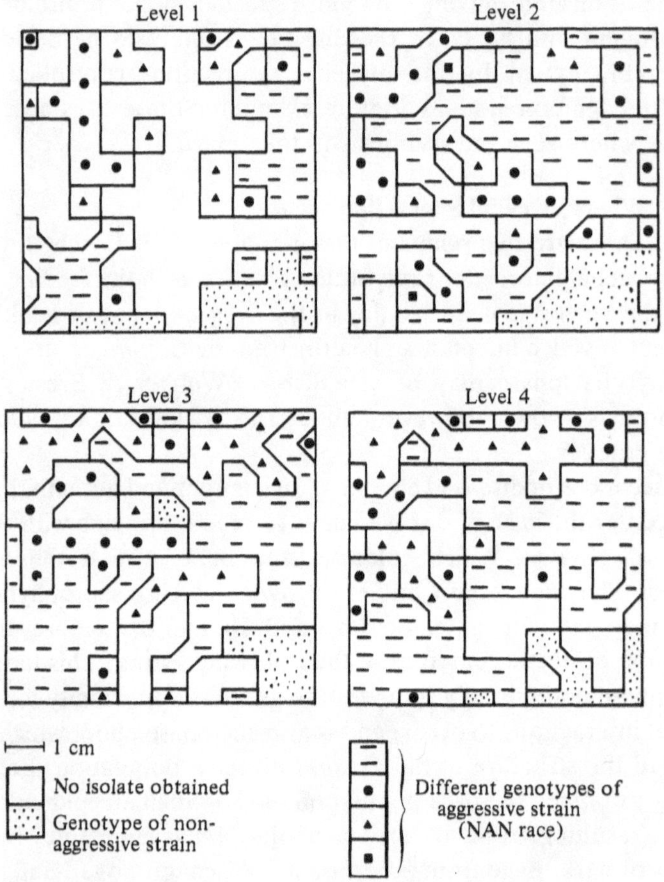

Fig. 2. The three-dimensional mosaic structure of the *C. ulmi* bark population. The *C. ulmi* genotypes obtained as a result of sampling a piece of elm bark *c.* 9 × 9 × 4 cm. Level 1, sample from synnemata on inner bark surface; Level 2, sample from inner bark surface; Level 3, sample from within inner bark; Level 4, sample from within outer bark. Note the approximately corresponding positions at each level of the same genotype e.g. the non-aggressive genotype, stippled. (After Lea, 1977.)

The recognition systems of *C. ulmi*

Three inter-mycelial recognition systems occur in *C. ulmi*. The mating type (homogenic) system was the first such system to be identified in *C. ulmi*. Two other systems have been identified only very recently: vegetative (= somatic) incompatibility, which occurs between individual mycelia and/or spores; and sub-group incompatibility, which operates between individuals of the different sub-groups (aggressive and non-aggressive, EAN and NAN) via the mating system. In a typical laboratory pairing of genetically different *C. ulmi* isolates in a Petri dish, the normal sequence of events following mycelial contact would be vegetative recognition (compatibility or incompatibility), followed by mating-type recognition and then sub-group recognition. The recognition or compatibility/incompatibility systems of *C. ulmi* will therefore be considered in this same ontogenetic sequence.

Vegetative incompatibility

Pairings of *C. ulmi* isolates in culture have revealed five different classes of vegetative reaction type. These are the wide or 'w' reaction; the narrow, or 'n' reaction; the line or 'l' reaction; the line-gap or 'lg' reaction; and the compatible or 'c' reaction. The general visual features of these reactions are shown in Fig. 3 and summarised in Table 1 in order of *decreasing* genetic complexity: the 'w' being the fully vegetatively incompatible reaction, the 'lg' the least incompatible reaction, and the 'c' the fully vegetatively compatible reaction. In practice the 'w' reaction is the common reaction obtained in pairings of wild isolates; 'n' reactions occurring only occasionally, and 'l' and 'lg' reactions being relatively scarce.

Number of vegetative-compatibility groups

The number of vegetative compatibility (v-c) groups in *C. ulmi* was examined by pairing samples of 11–13 wild isolates of a given sub-group in all possible combinations and scoring the resulting reaction types. Isolates of the same v-c group will be expected to give a 'c' reaction, and those of different v-c groups, w, n, 1, or 1g reactions. The results are summarised in Fig. 4.

In the non-aggressive strain, both a 'world-wide' sample of isolates (Fig. 4a) and a regional population sample from Sicily (not illustrated) gave a similar result, in that almost every isolate was found to be of a different v-c group, there being only one repeat in each case. In contrast, in a 'world-wide' sample of EAN aggressive isolates (Fig. 3b)

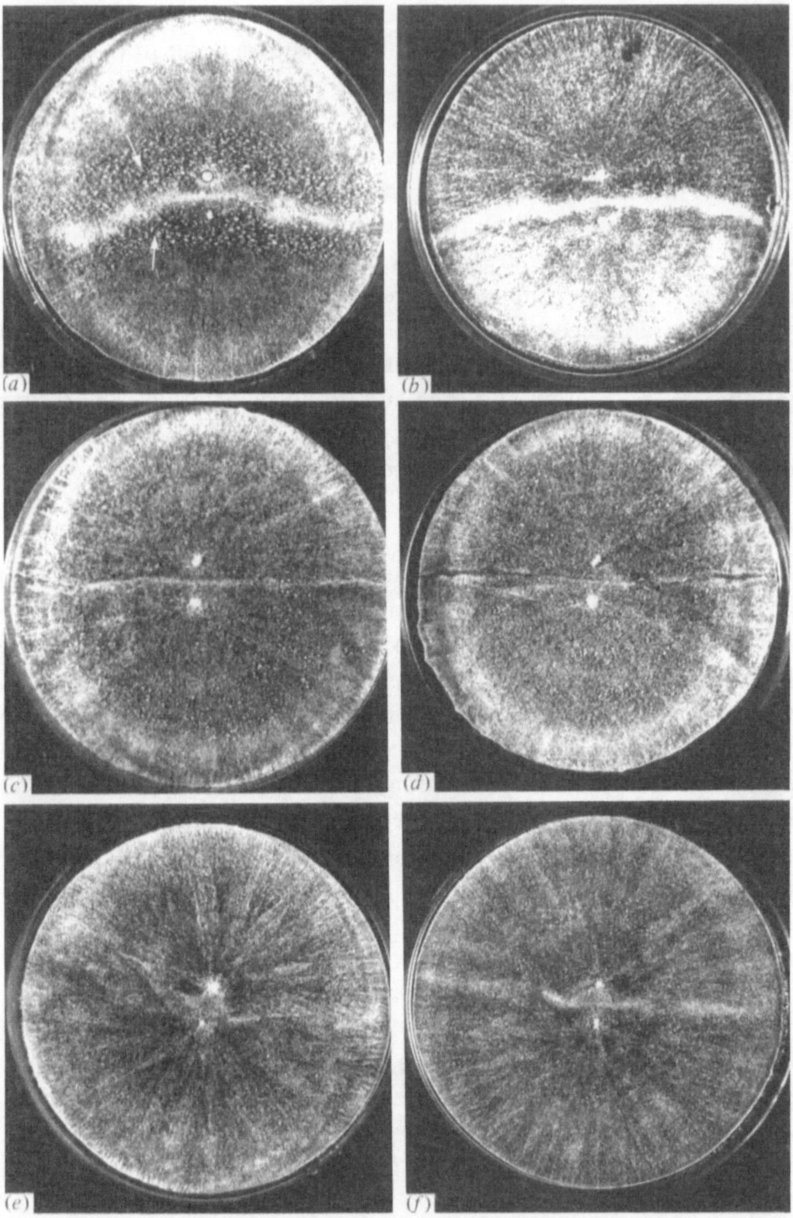

Fig. 3. Vegetative incompatibility reaction types in *C. ulmi.* (a) wide or w-reaction (see also Fig. 6a–d); note synnematal lines (arrowed). (b), narrow or n-reaction; (c), line or l-reaction; (d), line-gap or lg-reaction; (e, f), compatible or c-reaction. For descriptions see Table 1 and text. N.B. all the reactions are variable: the barrage zone of the n-reaction (b) is commonly narrower than shown. Cultures of aggressive strain NAN race (B-mating type) paired on elm sapwood agar.

Table 1. *Frequency and visual features of vegetative incompatibility reaction types in* C. ulmi

Reaction type	Relative frequency[a]	Visual features[b]
'w' or 'wide' reaction, Fig. 3a; Fig. 6a–d	common	Wide (*c.*1–1.5 cm) diffuse white mycelial barrage with synnematal lines of varying intensity parallel to barrage, often extending several cm in one or both directions
'n' or 'narrow' reaction, Fig. 3b	occasional	Narrow (*c.*0.4–0.7 cm) denser white mycelial barrage with some synnematal lines close to barrage zone
'l' or 'line' reaction, Fig. 3c	scarce	Thin white mycelial barrage line *c.* 1–3 mm wide; no synnemata associated with barrage
'lg' or 'line-gap' reaction, Fig. 3d	scarce	A gap or barrage of *c.* 1–3 mm between the two colonies; no synnemata
'c' or 'compatible' reaction, Fig. 3e, f	variable[c]	Either no reaction evident or slight diffuse mycelial thickening at junction line of mycelia

[a] Frequency in random pairings between wild isolates (excluding reactions involving the EAN and NAN 'super-groups').
[b] Pairings on elm sapwood agar (Brasier, 1981). Cultures grown initially for 7 days at 20 °C in darkness (27 °C for non-aggressive) followed by 30 or more days in diffuse daylight. Based mainly on observations of NAN and EAN aggressive isolates.
[c] Usually scarce unless v-c 'super-groups' are present.

one v-c group predominated, accounting for 60% of the isolates. Each of the remaining isolates, however, was of a different v-c group. Similarly, in a 'world-wide' sample of NAN aggressive isolates (Fig. 3c) one 'super v-c group' accounted for 40% of the sample, but among the remaining isolates there were no repeated v-c groups.

The number of v-c groups in a regional NAN population sample of 61 isolates from around Chichester, England (not illustrated) was examined in two stages. First, all the isolates were paired with a representative of the world-wide NAN 'super v-c group'. Ten of the 61 isolates, or 16%, fell into this group (i.e. giving a 'c' reaction; of the 51 remaining isolates 43 (i.e. 70%) gave a 'w'; 5 (8%) gave an 'n' and 3 (5%) gave an l/lg reaction with the 'super group' representative). Second, 12 of the 51 non 'super-group' isolates were selected at random and paired in all possible combinations. There was only one repeat v-c group among the 12 isolates (i.e. of the 78 possible reactions one was a 'c' reaction; 74 were 'w' and 3 were 'n' reactions). It is therefore likely that there were as many as 44 v-c groups among the 51 non-'super group' isolates.

Taking the results obtained with the non-aggressive and NAN aggressive samples as representative, and putting aside the matter of the 'super' v-c groups in the EAN and NAN, the frequency of repeat groups

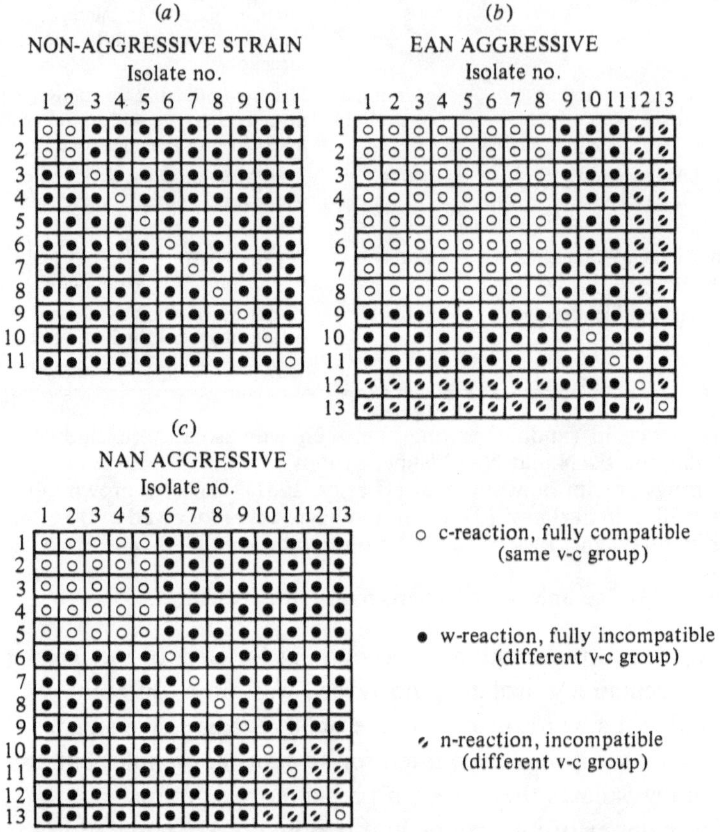

Fig. 4. Frequency of vegetative incompatibility groups in a 'world-wide' sample of each of the three *C. ulmi* sub-groups. (a), non-aggressive strain; (b), aggressive strain, EAN race; (c), aggressive strain, NAN race. Note the frequent v-c group repeats or 'super-groups' in the EAN and NAN races, and the patterns of n-reactions in the EAN and NAN. Non-aggressive isolates from 1, Italy; 2, Ireland; 3, 4, 5, Holland; 6, 7, Great Britain; 8, Poland; 9, Yugoslavia; 10, 11, Turkey. EAN isolates from 1, Ireland; 2, Turkey; 3, Yugoslavia; 4, 5, Denmark; 6, Italy; 7, West Germany; 8, 9, Iran; 10, Ireland; 11, 12, Bulgaria; 13, Soviet Union. NAN isolates from 1, 2, Great Britain; 3, 4, Ireland; 5, 6, France; 7, Denmark; 8, Belgium; 9, Spain; 10, Italy; 11, Yugoslavia; 12, 13, Holland.

at the regional level is very low (<1 in 10). This not only indicates that the total number of v-c groups in *C. ulmi* is likely to be extremely large, but gives an expectation of vegetative incompatibility between any given pair of isolates (in a sample of 10–12 isolates) of well over 95% (e.g. 99% in the case of the Chichester sample).

Regarding the 'super-groups' found in the EAN and NAN samples, pairings between representative isolates of the EAN 'super-group' with several NAN 'super-group' isolates all resulted in 'w' reactions (i.e. the two 'super-groups' are different v-c groups in each case). One possible explanation for the occurrence of the 'super-groups' is a founder effect due to the chance dispersal and spread of a particular group during the current epidemic. To test whether the NAN 'super-group' was as equally widely represented in North America as in Britain and Europe, a representative NAN 'super-group' isolate was paired with 19 NAN isolates from localities across the United States and their v-c reaction types scored. Only one of the 19 isolates belonged to the same v-c group as the 'super-group' showing that the latter is not as over-represented in the American NAN population. The significance of this will be discussed later.

Genetic control

Vegetative incompatibility in *C. ulmi* parallels that in other Ascomycotina in being heterogenic, that is incompatibility results from genetic differences (Esser & Blaich, 1973). The large number of v-c groups suggests multi-allelic and/or polygenic control. In the latter case, the more v-c genes there are in common, the greater will be the degree of vegetative compatibility and vice versa.

That the nature of genetic control of v-c in *C. ulmi* is indeed polygenic is illustrated by the results of a back-cross experiment, summarised in Fig. 5. Two NAN aggressive isolates (H321 and H326) which together give a fully incompatible 'w' reaction, were mated. Twenty of the resulting F_1 progeny were then paired with each parent (H321 and H326) and the resulting v-c reactions scored. Two F_1 isolates of appropriate mating type giving the *weakest* v-c reactions with one or other parent were then back-crossed to that parent, giving rise to two separate back-cross lines and an F_2 generation. Twenty of each F_2 line were then paired with H321 and H326 and the resulting v-c reactions scored. An F_2 isolate giving the *weakest* v-c reaction against its original parent was then selected as that parents' next back-cross partner, and this was repeated with the F_3.

As might be expected, progressively weaker sets of v-c reactions were obtained after each back-cross as more differences at v-c genes were eliminated (Fig. 5). Thus there were entirely w and n reactions in the F_1; mainly n-lg reactions in the F_2; and l-c reactions in the F_3. With progeny samples being restricted to 20 at each generation, it required four back-crosses to eliminate vegetative incompatibility completely in the two lines, that is to achieve all 'c' reactions. Assuming random assortment of v-c genes, even if only one v-c locus were eliminated at each generation, such a result would suggest that at least three loci are involved in vegetative incompatibility. If, as seems likely, the deliberate selection for back-crossing of an isolate giving the weakest v-c reaction at each generation would probably result in more than a single v-c gene being eliminated on each occasion, then the results suggest that the number of loci involved may be rather larger than three.

The results of this experiment also indicate that in *C. ulmi*, the v-c loci

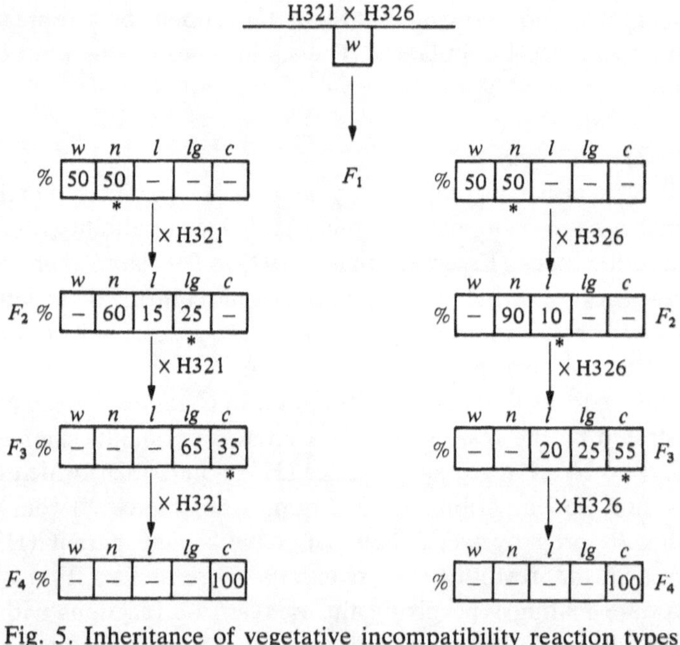

Fig. 5. Inheritance of vegetative incompatibility reaction types in *C. ulmi*: proportion of reaction types resulting when F_1–F_4 backcross lines derived from the cross H321 × H326 are paired with their respective parent NAN isolates H321 (left) and H326 (right). The parental combination H321 × H326 gives a w-reaction. *, reaction type from which progeny isolate was selected for the next backcross × the parent. Note gradual change from all w or n-reactions to all c-reactions over the four backcross generations. For details see text.

are functionally independent of the mating-type locus. Thus mating-type segregated regularly (1:1) in all four back-cross generations. In the F_4 generation, in which all the reactions were fully compatible 'c' reactions, a normal mating reaction (i.e. perithecial formation) occurred at the junction of parent and F_4 colonies where the two were of opposite mating type. In this respect *C. ulmi* is similar to *Endothia parasitica* (see Anagnostakis, Chapter 22) but contrasts with *Neurospora crassa* (Sansome, 1946; Garnjobst & Wilson, 1956) where the mating type locus is involved in the expression of vegetative incompatibility.

Another important point illustrated by this experiment is the genetic control of the 'w' versus 'n' reactions. The F_1 generation each give a 1:1 segregation ratio of 'n' to 'w' reactions against each parent (Fig. 5). In fact, an F_1 isolate giving an n-reaction against H321 gave a w-reaction against H326 and vice versa (not shown) indicating firstly that the 'n' reaction differed from the 'w' reaction by a single locus, which will be termed the '*w*' locus; and secondly that a w-reaction resulted when two isolates possessed different alleles (w^1 or w^2) at the putative *w* locus, and that an n-reaction resulted when they possessed the same *w* allele. This was confirmed by pairing the F_1 progeny against each other (the pattern obtained being $w^1 \times w^1$ or $w^2 \times w^2 \rightarrow$ n-reaction and $w^1 \times w^2 \rightarrow$ w-reaction). Similar 'n' versus 'w' reaction patterns have been obtained in pairings between F_1 progeny of other NAN × NAN crosses and their parents (*cf.* also reaction patterns between wild isolates e.g. in Fig. 4b where EAN isolates 12 and 13 evidently share the same *w* allele with the eight 'super-group' isolates and with each other; and Fig. 4c where NAN isolates 10–13 evidently share the same *w* allele). Taken together with evidence that 'n' reactions are generally scarce and 'w' reactions common in random pairings between wild isolates of different v-c groups (*cf.* the Chichester isolates above) the data also suggest that the 'w' locus is multiallelic (w^1, w^2, w^3, w^4 . . . etc). Hence the *w* locus may be a good example of multiallelic heterogenic incompatibility, with w-reactions being epistatic to n-reactions.

Evidence has also come, from pairing the F_3 progeny against each other, that the lg-reaction differs from a c-reaction by only a single v-c gene, the '*lg*' locus, such that $lg^1 \times lg^1$ and $lg^2 \times lg^2$ give a c-reaction and $\mathrm{lg}^1 \times \mathrm{lg}^2$ give an lg-reaction. The possibility of multi-allelism at the *lg* locus would be difficult to demonstrate.

The final expression of such a polygenic/multi-allelic vegetative incompatibility system is likely to be exceedingly complex given the

Table 2. *Apparent physiological and genetical features of vegetative incompatibility reaction types in* C. ulmi[a]

Reaction type	Physiological features	Genetical features[b]
w	Full vegetative incompatibility. Strongly restricted cytoplasmic transfer (of d-factors). Strong mycelial penetration effects.	All or most vegetative incompatibility genes different; different *w* allele
n	Strong vegetative incompatibility. Restricted cytoplasmic transfer. Some mycelial penetration effects.	Most vegetative incompatibility genes different; common *w* allele
l	Slightly restricted cytoplasmic transfer. No nuclear migration; little or no mycelial penetration.	A few (>1) vegetative incompatibility genes different
lg	Probably unrestricted cytoplasmic transfer. No nuclear migration or mycelial penetration.	Difference at only one vegetative incompatibility locus: different *lg* allele?
c	Full vegetative compatibility. Free cytoplasmic transfer (of d-factors). No nuclear migration or mycelial penetration	All vegetative incompatibility genes the same.

[a] Based on present observations, mainly with NAN aggressive isolates.
[b] Probable genetic similarities or differences between two paired isolates.

additional possibility of qualitatively different interactions between different v-c alleles at the same locus, interactions between different v-c loci, and the likely influence of modifiers and of background genotype. Thus, even in the simple case of an lg-reaction type in which, in terms of their vegetative incompatibility system, two isolates apparently differ by only a single v-c locus, the width of the 'line-gap' (Fig. 3d) varies considerably from one lg-reaction to another, suggesting either allelic interactions or the influence of secondary genetic factors. The morphology of l, n and w-reactions also varies greatly from one isolate combination to another.

Transfer of nuclei and cytoplasm

Only a limited amount of work has been done so far on the movement of cytoplasm and nuclei across the different types of vegetative incompatibility reactions, much of it concerned either with the transmission of the d-factor (see below) and most of it using isolates of

Table 3. *Transmission of d-factors across vegetative incompatibility reaction types in* C. ulmi

	A. Wild isolates[(a)]				B. Back-cross isolates[(b)]			
Reaction type with d-infected donor	w	n	l/lg	c	w	n	l/lg	c
No. of isolates	79	8	3	21	5	6	10	9
No. giving d-reaction as recipient	3	4	2	21	1	2	10	9
% d-factor transmission	<4	50	66	100	20	33	100	100

(a) 111 wild NAN recipient isolates paired against d-infected donor W2 to $111d^2$ (belonging to the NAN 'super-group').

(b) Selected F_1–F_4 back-cross progeny of varying relatedness to H321 (see Fig. 5) paired as recipients against d-infected $H321d^1$ as donor. Pairings carried out on elm sapwood agar (see Brasier, 1983*b*).

the NAN sub-group. The main physiological and genetical features of the different reaction types are summarised in Table 2.

The d-factor is a cytoplasmically transmissible disease of *C. ulmi* which spreads from an infected donor mycelium to a healthy recipient mycelium via hyphal anastomoses (see Brasier, 1983*b*). In doing so, it causes change or d-reaction in the newly infected mycelial segments of the recipient. Two experiments have been carried out examining the transmission of d-factors across vegetative incompatibility barriers in *C. ulmi.* In the first experiment (Table 3A), transmission of a d-factor to 111 wild NAN recipient isolates was investigated. The majority (79) of the isolates were of a different v-c group from the d-infected donor and hence gave a w-reaction with it. Since the donor isolate was a representative of the NAN 'super-group', a substantial number (21) of the remaining wild isolates gave a fully compatible or c-reaction with the donor. Based on the observation of a visible d-reaction between donor and recipient, only 4% of the w-reactions allowed the transmission of the d-factor, compared with 50% of the n-reactions and 100% of the c-reactions. In the second experiment (Table 3B), the donor d-infected isolate $H321d^1$ was paired with a sample of backcross isolates of varying genetic relatedness to it (see Fig. 7), i.e. giving either w, n, l, lg or c-reactions with $H321d^1$ respectively. The results again demonstrated restricted d-factor transfer in w and n-reactions, and showed unrestricted transfer in l, lg and c-reactions.

With isolates of the same v-c group, therefore, there appears to be ready transfer of cytoplasmic material from one mycelium to another, as demonstrated by the free migration of d-factors, indicating that there are free hyphal anastomoses between the mycelia (*cf.* Anagnostakis: Chapters 16, 22). In contrast, in fully incompatible w-reactions transfer of cytoplasm appears to be strongly restricted. The l and lg-reactions show similar cytoplasmic transfer characteristics to the c-reaction, and the n-reactions similar characteristics to the w-reactions, although there may be slightly more frequent cytoplasmic transfer in n than in w-reactions, and slightly less frequent cytoplasmic transfer in l and lg than in c-reactions.

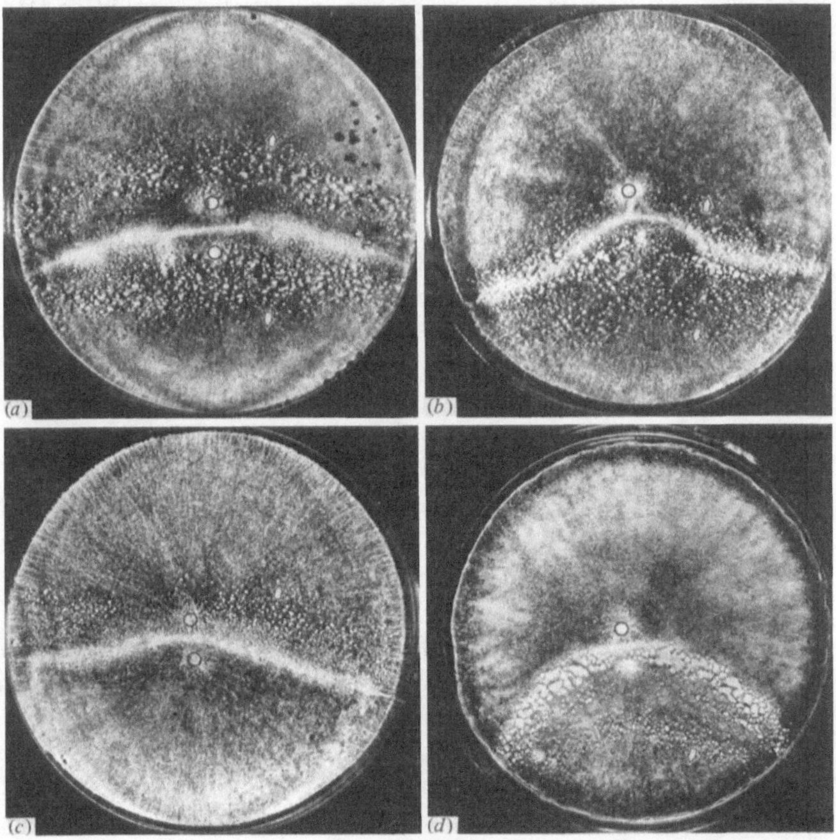

Fig. 6. The penetration effect in w-reactions. (a), equal bi-directional penetration; (b), unequal bi-directional penetration; (c), unidirectional penetration with penetrated isolate (above) faster growing; (d), unidirectional penetration with penetrated isolate (below) slower growing. Open arrows indicate extent of synnematal lines. Paired NAN B-mating type isolates on elm sapwood agar after 7 days darkness and 30 days diffuse daylight at 20 °C.

However, although cytoplasmic movement may be unrestricted there appears to be little or no migration of nuclei from one mycelium to another in l, lg and c-reactions as indicated by failure of movement of nuclei carrying markers such as MBC (fungicide) tolerance in pairings between MBC-tolerant and MBC-sensitive isolates, illustrated in Fig. 7B. This applies both to experiments involving pairings between different NAN isolates and between EAN and NAN isolates. No direct

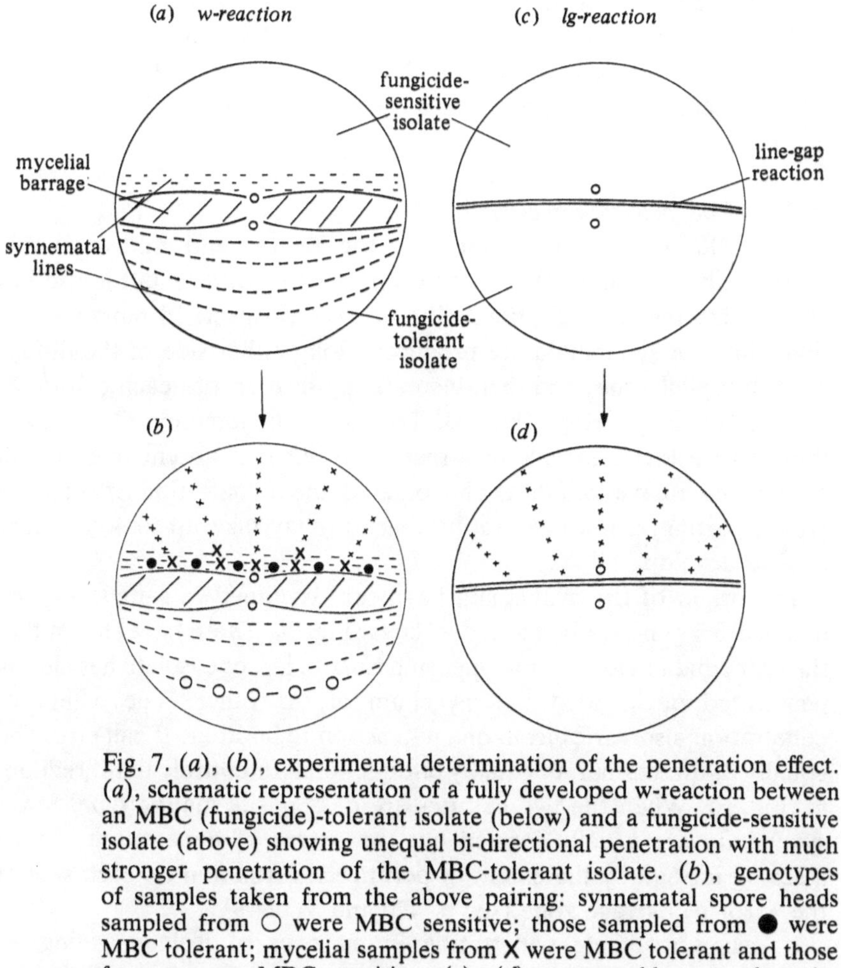

Fig. 7. (*a*), (*b*), experimental determination of the penetration effect. (*a*), schematic representation of a fully developed w-reaction between an MBC (fungicide)-tolerant isolate (below) and a fungicide-sensitive isolate (above) showing unequal bi-directional penetration with much stronger penetration of the MBC-tolerant isolate. (*b*), genotypes of samples taken from the above pairing: synnematal spore heads sampled from ○ were MBC sensitive; those sampled from ● were MBC tolerant; mycelial samples from X were MBC tolerant and those from x were MBC sensitive. (*c*), (*d*), comparable test of an lg-reaction. (*c*), schematic representation of an lg-reaction between an MBC-tolerant isolate (below) and an MBC-sensitive isolate (above). (*d*), samples of mycelium from x were all MBC sensitive, indicating that no significant mycelial penetration/nuclear migration had occurred. Similar results were obtained with l and c-reactions.

evidence is available for or against nuclear migration in w-reactions, but it is most unlikely that nuclear migration will occur when cytoplasmic movement is already so restricted.

Thus vegetative incompatibility in *C. ulmi* appears to restrict the formation of heteroplasmons, presumably as a result of some form of cell breakdown following hyphal fusion (*cf.* Anagnostakis: Chapter 22; and Croft and Dales: Chapter 20). Although the possibility of free nuclear exchange arises between isolates of the same v-c group but otherwise of different background genotype, it does not seem to occur. Apparently *C. ulmi* may not readily form unforced heterokaryons, a point of considerable importance for interpreting its behaviour in nature.

The 'penetration effect'

The w-reaction is not only the typical and full vegetative incompatibility reaction between two mycelia of *C. ulmi* but it also has another interesting and potentially important aspect. In most w-reactions, lines of synnemata are produced along either side of the diffuse white mycelial zone, and then increasingly further into each culture as the w-reaction develops (Fig. 6). The same phenomenon also occurs, though to a lesser extent, in n-reactions, but it is absent in l, lg and c-reactions. I have called the phenomenon the 'Penetration Effect'. In a typical pairing, a full penetration reaction may take up to 40 or more days to develop.

The origin of the synnemata has been examined in sample experiments using genetically marked isolates (Fig. 7). These have shown that the synnemata belong to the opposing isolate, i.e. one isolate has clearly penetrated or invaded the mycelium of the other. The extent of penetration also varies from one w-reaction to another. It can either be equally bi-directional, unequally bi-directional or entirely uni-directional (Fig. 6). When the two cultures are of opposite mating type, fewer synnemata are usually formed but the extent of penetration can be assessed instead by the extent of perithecial formation on each side of the w (or n) barrage zone (see p. 470 and Table 4).

If, when assessing penetration in matrices of isolates giving w-reactions, a relative numerical score is allotted to an isolate for its penetration reaction in a given pairing (on the basis: entirely penetrating, +2; predominantly penetrating, +1; equally penetrating, 0; predominantly invaded, −1; entirely invaded, −2) the isolates can be ranked in order of overall penetrating ability, high-scoring isolates being

Table 4. *Relative perithecial formation (fertility) in different vegetative incompatibility reaction types of* C. ulmi

	Experiment II[a]				Experiment 2					Experiment 3[b]				
Reaction type	w	n	l/lg	c	w	n	l	lg	c	w	n	l	lg	c
No. of pairings assessed	(4)	(4)	(4)	(4)	(26)	(14)	(1)	(–)	(3)	(–)	(39)	(2)	(4)	(–)
Mean distance (mm) of perithecial formation either side of junction line	18.1	10.8	2.0	2.1	20.4	12.1	6.0	–	1.3	–	11.0	6.5	4.0	–
Approximate area (cm^2) of perithecial formation[c]	32.6	19.4	3.6	3.8	36.8	21.8	10.8	–	2.4	–	19.8	11.8	7.2	
Relative fertility (by area) as fraction of 'w' reaction	(1.0)	0.6	0.1	0.1	(1.0)	0.6	0.3	–	0.07					
Mean distance (mm) of synnematal formation each side of junction line[d]	16.7	9.3	NA	NA	19.8	10.9	NA	NA	NA	(–)	13.4	NA	NA	NA

[a] Each experiment involved pairing F_1–F_4 progeny of an NAN × NAN cross or back-cross against parent isolates on elm sapwood agar and assessing perithecial formation in the different reaction types after 40 days.
[b] Experiment 3 involved two parents carrying the same *w* allele: hence there were no 'w' reaction types.
[c] Assuming Petri dish width of 9.0 cm.
[d] Synnematal penetration was assessed from a similar number of pairings but of the same rather than opposite mating type.
NA not applicable (synnemata not formed by these reaction types).

the strong penetrators and those with a low score being weak penetrators. The results of such a matrix calculation involving pairings of 14 different NAN aggressive isolates is shown in Fig. 8. A feature which emerges is that penetrating ability seems to be strongly hierarchical in that, for example, strong × strong or moderate × moderate isolates tend to be equally matched, that is equally bi-directional, whereas strong × moderate pairings result in penetration of the moderate isolate, and moderate × weak pairings result in penetration of the weak isolate, and so on.

Another important feature to emerge was that penetrating ability appeared to vary between members of a v-c group. This was tested by

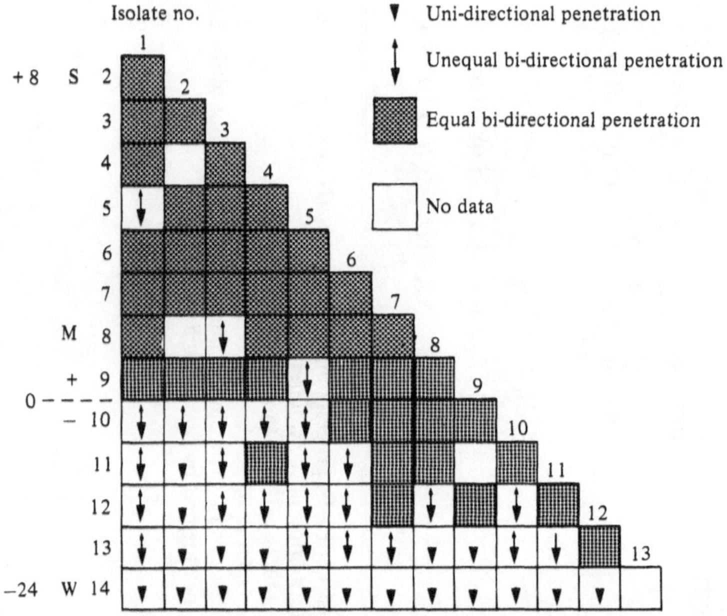

Fig. 8. Apparent hierarchical nature of penetrating ability. Ranking of 14 NAN aggressive isolates of different v-c groups according to the mean penetrating ability of each isolate (based on its 14 possible score combinations in a matrix, see text). S, strong penetrators; M, moderate penetrators; W, weak penetrators. e.g. isolate 1 penetrates isolate 14 unidirectionally; isolate 1 penetrates isolate 13 more strongly than 13 penetrates 1; isolate 1 and 9 penetrate each other equally. '+8' and '−24' indicate the highest and lowest overall penetration scores achieved by the isolates. The fourteen NAN B-type isolates (from Chichester, United Kingdom) were paired in all possible combinations on elm sapwood agar for 7 days in darkness and 30 days in diffused daylight before penetration effects were scored.

pairing nine isolates of the same v-c group with six other isolates, each of a different v-c group, scoring the resulting penetration effects and ranking the isolates on the above numerical basis. The former isolates were found to have different penetrating abilities (Fig. 9). Some were strong penetrators; others were weak penetrators and tended, instead, to be penetrated. Penetrating ability must therefore be influenced or controlled by some factor other than the v-c genes. It appears to be influenced to some extent by mycelial vigour, since slower-growing d-infected isolates tend to be more strongly penetrated than their healthy counterparts. However, some fast-growing isolates such as RDT-38 (Fig. 6c and Fig. 9. no 9) seem to be consistently penetrated even when paired against a slower-growing isolate. Whether penetrating ability is therefore controlled by a separate and specific polygenic genetic system, by the general effect of nuclear genes influencing growth vigour, by cytoplasmic influences on vigour or a combination of all three needs to be determined.

Also to be confirmed is whether penetration consists of true mycelial introgression or results from nuclear migration. Since penetration is an aspect of full vegetative incompatibility, mycelial introgression seems much the most likely explanation since nuclear migration would probably be largely prevented by the self-same incompatibility. Another question surrounds the stimulus underlying synnematal formation. The

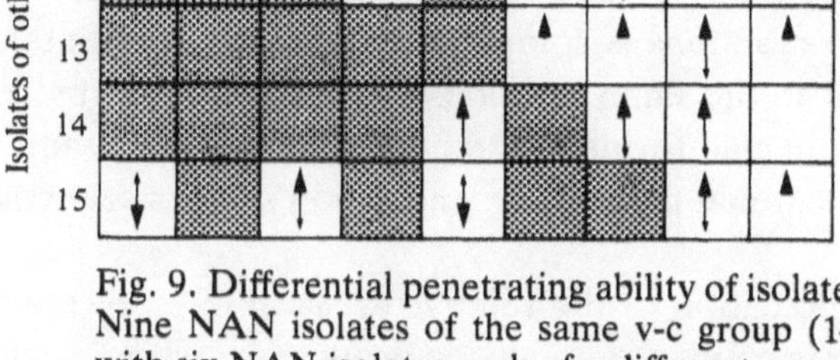

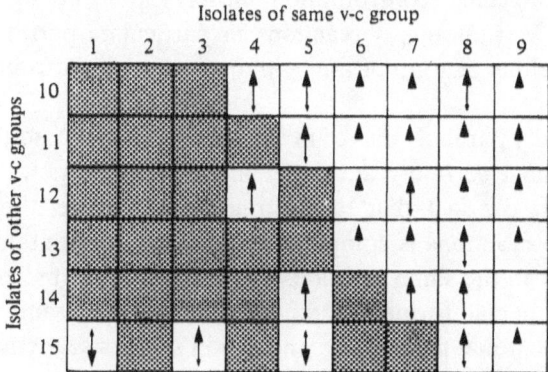

Fig. 9. Differential penetrating ability of isolates of the same v-c group. Nine NAN isolates of the same v-c group (1–9, across) were paired with six NAN isolates each of a different v-c group (10–15, down) and each isolate ranked according to its mean penetrating ability (based on its seven or ten possible score combinations in the matrix, see text). Note that the nine isolates of the same v-c group show different penetrating abilities. All isolates were NAN isolates of B-mating type paired on elm sapwood agar. Key as in Fig. 8.

latter is at its most striking in uni-directional penetration where synnemata are commonly produced in parallel arcs or lines, often accompanied by pigment production, suggesting successive waves of attack and defence. Are they perhaps formed in response to nutrients released by lysis of the penetrated isolates' mycelium; or alternatively, formed in response to an attack by the mycelium of the penetrated isolate on that of the partner? In view of the potential ecological significance of the penetration effect these problems are well worthy of further investigation.

Perithecial formation and vegetative incompatibility reaction type

Another interesting feature of vegetative incompatibility in *C. ulmi* is the different extent of perithecial formation in w, n, l, lg and c reactions (when the two isolates involved are of different mating type). The results of experiments assessing the average width of the zone of perithecial formation (in mm) on either side of the junction-line in the different reaction types are shown in Table 4.

In w-reactions, perithecia were formed deeply into each culture, that is well on either side of the junction-line. In the n-reaction their formation was a little and in l-reactions much more restricted. In the lg- and c-reactions, perithecial formation was more or less confined to the area of immediate mycelial confrontation closely adjacent to the junction-line. In the lg ('line-gap') reactions in particular, perithecia were confined along either margin of the 'gap' between the confronting mycelia.

This situation broadly parallels that of mycelial penetration described above. That it is almost certainly another feature of the penetration effect is illustrated by the fact that the average extent of perithecial formation in w and n reactions is comparable to the average extent of synnematal formation in the same reactions (Table 4). Furthermore, in a given pairing perithecial formation tends to follow the same bi-directional or uni-directional pattern of penetration as does synnematal formation.

The data give an indication of the relative area within each reaction type in which perithecia could potentially be produced. Assuming that a constant number of ascogonia cm^{-2} are available for fertilisation throughout a culture, then potential perithecial formation in w-reactions will be nearly twice that in n-reactions, between three and ten times that in l-reactions and around ten times that in c-reactions (Table 4). In terms of numbers of perithecia produced, therefore, w-reactions are likely to be the most fertile and l, lg and c-reactions the least fertile reaction types.

Mating type recognition

A second system of inter-mycelial recognition, which usually comes into operation after vegetative recognition, is that of mating-type recognition. This system either enables or prevents ascogonial initials differentiated in a vegetative mycelium to develop into fully functional perithecia, depending upon the genotypes of the two mycelia involved.

Buisman (1932) first showed that *C. ulmi* was heterothallic, with perithecial formation requiring the interaction of two mating types which she termed + and −. Schafer & Lyming (1950) substituted the terms A and B-type for Buisman's + and − types, and these have since become the accepted terminology. Work up to the early 1970s, in particular that of Holmes (1965, 1977), has confirmed that the mycelium of *C. ulmi* exists either as the A-type or B-type (apart from a few anomalies which will be considered later) and that only A-type × B-type pairings normally result in perithecial formation. As in most other Ascomycotina the two mating types of *C. ulmi* are bi-sexual, able to function either as a female recipient via formation of ascogonia, or as a male donor via spores acting as spermatia. Holmes (1977) has underlined the importance of distinguishing between compatibility type (A- or B-type) and sex (maleness or femaleness) by demonstrating occasional loss in laboratory cultures of ability to act as a recipient (female) irrespective of mating type.

The discovery of the different sub-groups of *C. ulmi* in the 1970s soon led to evidence of differences in mating behaviour between them, and stimulated further studies on a number of aspects of mating recognition.

Genetic control of mating type

Many A-type × B-type crosses have been carried out by the author in recent years during investigations of genetical differences between the sub-groups of *C. ulmi*, and a number of these have yielded information on the inheritance of mating type. These data are summarised in Table 5. Resulting F_1 progeny have always been of either A-type or B-type. Where an uncertain or a mixed result has been obtained, the isolate has usually proved to be of multiple ascospore origin. Furthermore, in most crosses, whether within a sub-group (e.g. non-aggressive × non-aggressive) or between sub-groups (e.g. NAN aggressive ×

Table 5. *Inheritance of mating type in* C. ulmi

Cross (A type × B type) Isolate no.	Source[(a)]	No. of progeny A type	No. of progeny B type	Ratio (A-type: B-type)
Non-aggressive × non-aggressive				
W6 × W7	(GB × GB)	45	46	
W6 × W9	(GB × GB)	42	54	
Gol A1[(b)]	(IR)	21	27	
Total		108	127	1:1.18
EAN × EAN				
V9 × V8	(SU × SU)	27	23	
AST-20 × AST-27	(IR × IR)	28	28	
Total		55	51	1:0.93
NAN × NAN				
H107 × H151	(F × USA)	20	30	
MM2/1 × H175	(GB × USA)	26	33	
Total		46	63	1:1.37
NAN × Non-aggressive				
W10 × G36	(GB × GB)	6	16	
EIS = BV2	(GB × GB)	10	7	
RR10 A4/43 × W7	(CA × GB)	12	9	
RR10 A4/43 × W9	(CA × GB)	8	12	
Total		36	44	1:1.22
NAN × EAN				
MM2/1 × AST-27	(GB × IR)	22	19	
H170 × AST-27	(USA × IR)	22	26	
AST-20 × H175	(IR × USA)	18	26	
H107 × V8	(USA × SU)	26	24	
V9 × H151	(SU × USA)	17	33	
I20 × I12	(I × I)	29	21	
Total		134	149	1:1.11
Total overall		379	434	1:1.15

[(a)] GB, Great Britain; IR, Iran; F, France; USA, United States of America; SU, Soviet Union; CA, Canada; I, Italy.
[(b)] 'Wild' perithecium from Golestan, Iran (Brasier & Afsharpour, 1979).

non-aggressive) a ratio of A : B-types reasonably close to 1 : 1 has usually resulted, the few exceptions probably being due to small sample sizes and to a slight but fairly consistent numerical bias towards B-types which is in turn thought to be due to a pleiotropic effect of the A mating type resulting in slower growth rate of A-type germlings (see below). The effects of the latter are also reflected in the overall ratio of

A:B-types for the crosses which deviates slightly from 1:1 in favour of B-types (Table 5).

These results are otherwise consistent with control of mating type by two alleles at a single locus, that is a homogenic system in which incompatibility occurs when the alleles are the same. In order to maintain the existing mating-type terminology in *C. ulmi* therefore, it is suggested that the two alleles should be termed the *A* and *B* alleles of the mating-type or *mt* locus.

Mating-type frequency

Samples of wild isolates of *C. ulmi* taken from different geographical locations during the current epidemics have revealed striking differences in the frequency of the two mating types in the different sub-groups of the fungus, as shown in Table 6. In the non-aggressive strain, the two mating types occur in a near 1:1 ratio. In both races of the aggressive strain, however, the B-type is predominant over the A-type, the latter accounting for only *c.* 15% of the population. The proportion of A-types present also varies considerably between geographical locations, and possibly with epidemic status, ranging from around 8% in old epidemic areas such as parts of Britain (NAN aggressive) and in Romania (EAN aggressive) to around 26% in current epidemic front areas such as Holland (mainly NAN aggressive) and Poland (EAN aggressive). Some implications of these differences will be considered in the discussion. A part of their cause may be found in the differential effects of the *A* and *B mt* alleles on mycelial development, which will now be described.

Influence of mating-type on mycelial development

The mating-type alleles not only function in recognition, but also influence mycelial development, especially in the EAN and NAN races of the aggressive strain. Thus, on malt extract agar the NAN aggressive A-type is much more fertile than the B-type, usually producing numerous protoperithecia throughout the culture, often with an associated production of dark pigment (Brasier & Gibbs, 1975*b*). In contrast with B-types only the much smaller ascogonia are normally formed, and these usually only in low frequency around the inoculum plug. In addition, NAN A-types generally tend to grow a little more slowly in culture than B-types and when inoculated into elm are generally a little less pathogenic (Brasier & Gibbs, 1975*b*; Brasier, Lea & Rawlings, 1981). Similar A-type/B-type polymorphism occurs in the

Table 6. *Mating type frequency in the* C. ulmi *sub-groups*

Sample source and date[a]		No. of isolates A type	B type	% A type
Non-aggressive strain				
Great Britain	(1972)	13	25	27.1
Iran	(1977)	53	46	53.5
Ireland	(1978)	17	16	51.5
Italy (Sicily)	(1979)	10	17	37.0
Holland	(1980)	3	5	
Turkey	(1980)	15	12	55.6
Total		111	121	47.8%
Aggressive strain EAN race				
Iran	(1977)	4	14	22.2
Soviet Union	(1979)	2	21	8.6
Italy	(1979)	4	68	5.5
Poland	(1980)	38	107	26.2
Yugoslavia	(1980)	20	134	13.0
Romania	(1980)	10	117	7.9
Turkey	(1980)	29	81	26.4
Total		107	542	16.5%
Aggressive strain NAN race				
Great Britain	(1972)	2	47	4.1
United States	(1977)	6	41	12.8
Italy	(1979)	2	43	4.4
Holland	(1980)	24	78	23.5
Total		34	209	14.0%

[a] Results of sample surveys of isolates from diseased elm twig material collected by the author.

EAN aggressive at least with regard to ascogonial production and growth rate but the overall level of fecundity of the EAN is less than that of the NAN. In the non-aggressive strain, on the other hand, ascogonia are not usually produced on malt extract agar by either mating type, and there are no detectable differences between the two mating types for any of the above characters.

Thus the *A* and *B* alleles exert different effects according to the genotype of the mycelium they occupy. This phenomenon reaches its most extreme expression in the NAN aggressive, where the *A* allele apparently exerts a pleiotropic effect on three major fitness characters. In particular, from the point of view of mating-type recognition between mycelia, the NAN A-type is a highly productive protoperithecial and hence perithecial producer and indeed probably functions as such during the bark phase of *C. ulmi* (Brasier & Gibbs, 1975*b*). The non-aggressive strain, on the otherhand, is generally much less fertile than the aggressive. Even on elm sapwood agar (Brasier, 1981) or on sterilised elm twigs where it does produce some ascogonia or protoperithecia, these are usually formed in very much smaller numbers than by, for example, an NAN A-type. The EAN aggressive appears to occupy a somewhat intermediate position in this respect between the non-aggressive strain and the NAN aggressive.

Mating-type mutation and pseudoselfing

During studies on the mating types of NAN aggressive isolates, a small proportion of B-type isolates of single spore origin, instead of being self-sterile, were found to produce scattered perithecia when grown by themselves on sterilised elm twigs. The elm twigs had been inoculated by dipping them in shake cultures of the respective isolates. Samples of ascospore progeny taken from these unexpectedly produced perithecia segregated 1:1 for the A and B mating-types i.e. A-type progeny had somehow arisen from a B-type mycelium.

The explanation given by Brasier & Gibbs (1975*b*) for this phenomenon is that mutation occurs from the *B* to the *A* allele at the *mt* locus, and that whereas in a mycelially growing culture nuclei carrying such a mutation would probably be suppressed, in liquid shake culture, in which *C. ulmi* grows as a yeast-like budding phase, cells carrying the *A* mutation can develop freely, resulting (after twig dipping) in fertilisation between independent A and B-type mycelia and a 1:1 segregation of the two types among progeny of the resulting perithecia. Since the fungus was in effect mating with itself, the only genetic difference between the partners probably being at the *mt* locus, Brasier & Gibbs (1975) termed the phenomenon 'pseudoselfing' to distinguish it from true selfing.

Further observations showed that the reverse pseudoselfing phenomenon involving an $A \rightarrow B$ mutation also occurred with A-type shake cultures. Moreover, since the NAN A-type is intrinsically much more fertile than the B-type (producing more ascogonia or protoperithecia available for fertilisation by B-type spores) pseudoselfing was generally much more frequently expressed in A-types than in B-types. Pseudoselfing has since been shown to occur in the EAN aggressive (again more readily in A-types) but has yet to be confirmed in the non-aggressive strain.

Perithecia resulting from pseudoselfing are often smaller than normally outcrossed perithecia, tend to be malformed and to produce fewer ascospores. In addition, they often appear only after prolonged incubation (3–6 weeks) and are frequently associated with the appearance of patches of felty white mycelium on the twig surface. The origin of the latter and its relationship to pseudoselfing is unknown. Indeed, the exact mechanism of pseudoselfing is also yet to be demonstrated, but mutation must remain the most likely explanation. A transposon or 'cassette' mating-type mutation mechanism comparable to that recently described in yeast (Hicks, Strathern & Klar, 1979) is another possibility (*cf.* also *mt* mutation in *Chromocrea spinulosa*, Mathieson, 1952; and *Sclerotinia trifoliorum*, Uhm & Fujii, 1983).

The ecological role of pseudoselfing will be considered later. If not recognised, its occurrence in mating-type tests with *C. ulmi* isolates may lead to difficulty in the interpretation of results. Indeed pseudoselfing may account for the 'Ab', 'AB', and 'aB' mating-types of *C. ulmi* reported by Schafer & Lyming (1950) and Ansina & Hubbes (1980).

The sub-group fertility barriers

A third form of recognition between mycelia of *C. ulmi* is that which occurs between the three sub-groups of the fungus. This system operates via the mating system when mycelia or spores of A and B types of different sub-groups are brought together, but is independent of mating type. The system first became apparent when difficulty was experienced in obtaining perithecia in crosses between the NAN aggressive and the non-aggressive strain. It was soon clear that mating between these two sub-groups was in some way restricted. Fertilisation experiments were therefore made by applying donor conidia (♂) of either NAN aggressive or non-aggressive isolates in patches on the surface of an NAN aggressive culture acting as the recipient (♀) and the resulting number of perithecia cm^{-2} was taken as an indication of the relative fertility of a given sub-group combination. The results confirmed the existence of the suspected reproductive barrier (Brasier, 1977). Thus, in one experiment involving fertilisation of three NAN A-type recipients (♀) by six different NAN and six different non-aggressive B-type donors (♂), the NAN ♀ × NAN ♂ patches produced a total of 11 968 perithecia; the equivalent NAN ♀ × non-aggressive ♂ patches produced only 650 perithecia, an 18-fold difference in relative fertility.

Subsequent tests involving many isolates of all three sub-groups of the fungus in all possible reciprocal combinations have revealed an intricate system of fertility barriers between the sub-groups, which is summarised in Fig. 10.

The non-aggressive strain when a recipient (♀) is a universal acceptor, accepting donor NAN, EAN and also non-aggressive mating partners equally. In contrast, both the NAN and EAN, as recipients, strongly reject the non-aggressive as a mating partner. In other words there is a uni-directional barrier to hybridisation between the aggressive and non-aggressive sub-groups. Another unidirectional barrier operates between the EAN and NAN sub-groups, in that the EAN, as recipient, partially rejects the NAN, whereas the NAN accepts the EAN equally with itself. EAN ♀ × NAN ♂ combinations are around 10–30 times less fertile than control EAN ♀ × EAN ♂ combinations (the magnitude of difference varying from one experiment to another).

Another feature of the non-aggressive → aggressive barrier is that, even where perithecia are produced, these tend to be poorly developed, often abnormal. Similar but less-marked perithecial abnormality occurs with the NAN → EAN barrier.

The barriers are independent of mating type. Thus the NAN → EAN barrier operates equally strongly in both EAN A-type ♀ × NAN B-type

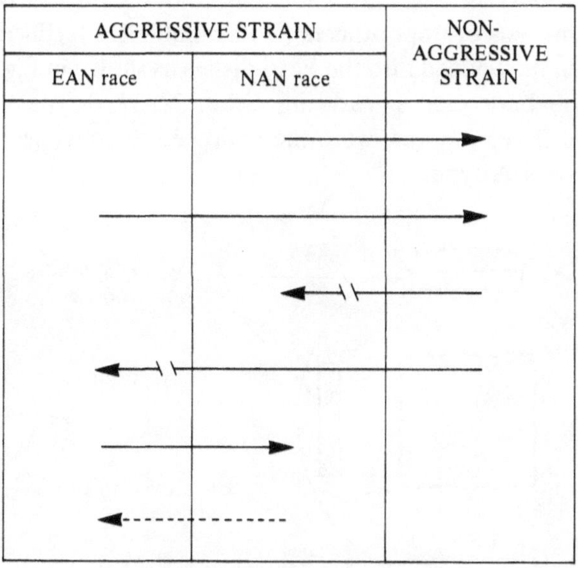

Fig. 10. The fertility barriers between the sub-groups of *C. ulmi*. ⟶, no barrier; ←⟍⟍— virtually complete barrier; ← – – –, strong partial barrier.

♂ and EAN B-type ♀ × NAN A-type ♂ combinations. The same applies to the non-aggressive→aggressive barrier.

Operation of the fertility barriers in nature

An important ecological question is whether the sub-group fertility barriers observed on agar also operate under natural conditions such as in the mycelial mosaic in elm bark illustrated in Fig. 2. It was postulated that in elm bark fertilisation of *C. ulmi* might be greatly enhanced by the action of the usually plentiful microfauna, especially mites, in brushing the sticky spores of *C. ulmi* onto ascogonia and protoperithecia formed in and around the beetle breeding galleries (Brasier, 1978). Laboratory experiments were therefore devised to test the effectiveness of the sub-group fertility barriers using mites as fertilising agents, as shown in Fig. 11.

To test the non-aggressive→aggressive barrier, pieces of sterilised elm bark previously dipped in spore suspensions of either NAN aggressive or non-aggressive B-type isolates, were allowed to develop a good mycelium and each piece was placed in the centre of a Petri dish. Mites were then introduced onto the bark pieces, on which they quickly multiplied. At the same time, sterile elm twigs were dipped into spores of an A-type NAN isolate and when the colonies on the twigs had developed numerous protoperithecia (♀) available for fertilisation, the twigs were then introduced into the Petri dishes as shown in Fig. 11. The mites from the bark pieces, carrying either NAN B-type spores or non-aggressive B-type spores, were then allowed to crawl across to the twigs of the NAN A-type.

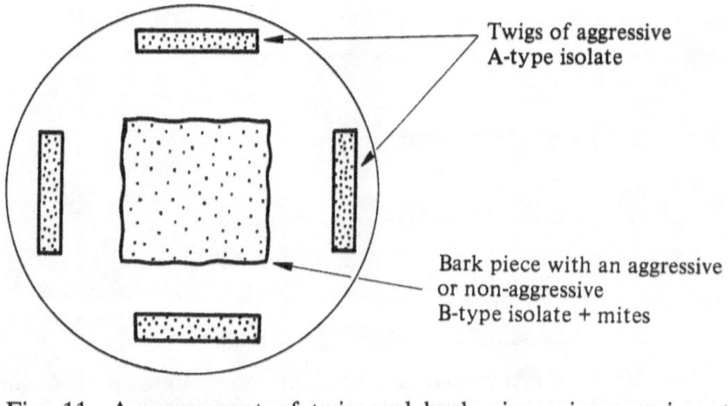

Fig. 11. Arrangement of twig and bark pieces in experiment using mites as fertilising agents.

After several weeks, the mites had eaten most of the mycelium from the twigs, and, in the NAN (twig) × NAN (bark) combinations perithecia were now plentiful on the twig surfaces (Fig. 12d). In contrast, perithecia were rare or absent on the twigs in the NAN (twig) × non-aggressive (bark) combinations, the protoperithecia remaining largely unfertilised in spite of considerable mite activity (Fig. 12c).

Assuming that this experiment is a reasonable model of natural conditions (see Brasier, 1978) then it is clear from the results that the non-aggressive → aggressive fertility barrier is indeed likely to act as a very strong inhibitor of hybridisation between the sub-groups in nature. Similar experiments to test the NAN → EAN barrier, on the other hand, have yielded comparable numbers of perithecia in both EAN (twig) × NAN (bark) and EAN (twig) × EAN (bark) combinations. The latter barrier therefore is unlikely to be so effective as the former in nature.

Genetic control of fertility barriers

The genetic control of the sub-group fertility barriers has so far been studied only in the NAN → EAN context. The aim has been both to examine the barrier mechanism and to produce information useful in the identification of any naturally occurring NAN × EAN hybrids in nature.

The behaviour of F_1 progeny of two crosses has been investigated: (i) a NAN A-type ♀ × EAN B-type ♂ cross; and (ii) an EAN A-type ♀ × NAN B-type ♂ cross. The progeny of each cross segregated 1 : 1 for mating type. The A and B-type progeny were each subjected to a separate type of fertility test as shown, together with the results, in Fig. 13.

The progeny showed a continuous distribution of behaviour types between the parental types rather than a discrete distribution, and each progeny sample showed a bias towards NAN. Thus, most of the A-type progeny, which were tested as recipients (♀), tended to accept NAN donors, some were intermediate, and only a few tended to reject NAN donors. Equally, many of the B-types which were tested as donors to an EAN recipient, tended to be rejected by the EAN as if they were NAN isolates, some were intermediate, and some were accepted by the EAN.

Since reciprocal crosses gave similar results, the NAN → EAN fertility barrier appears to be under nuclear gene control. It is apparently inherited on a polygenic basis rather than at a single locus, and the bias towards NAN phenotype in the progeny suggest that it is not a simple

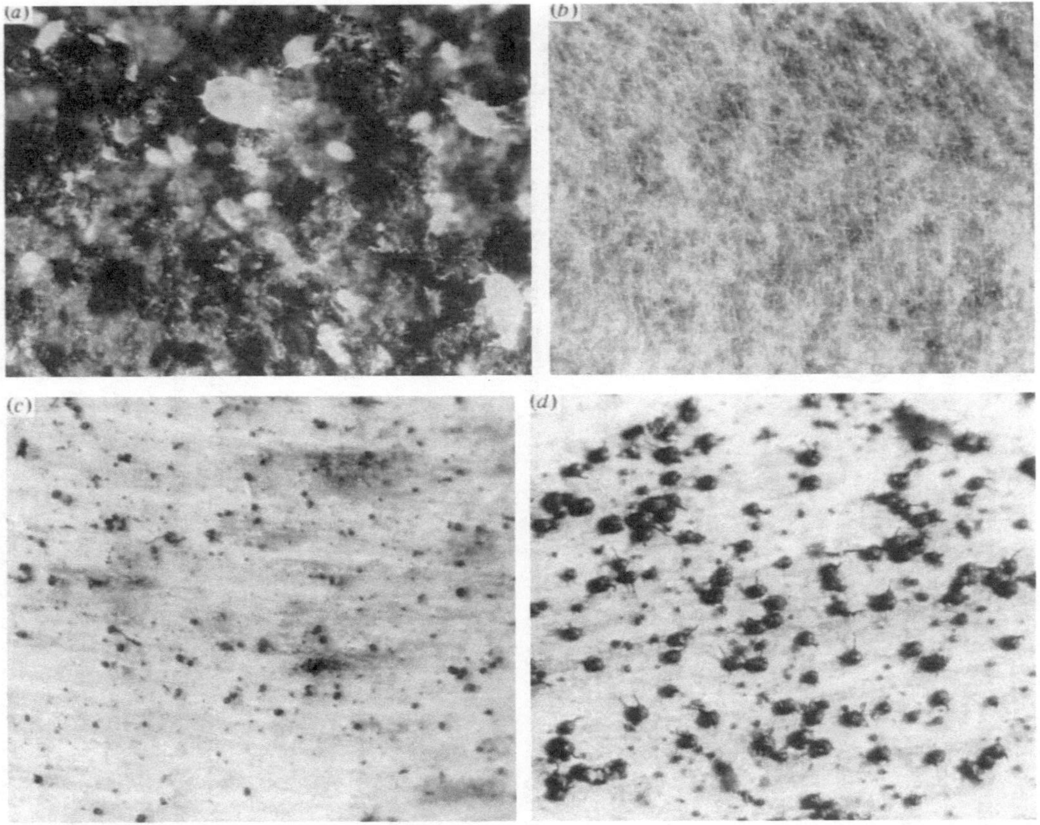

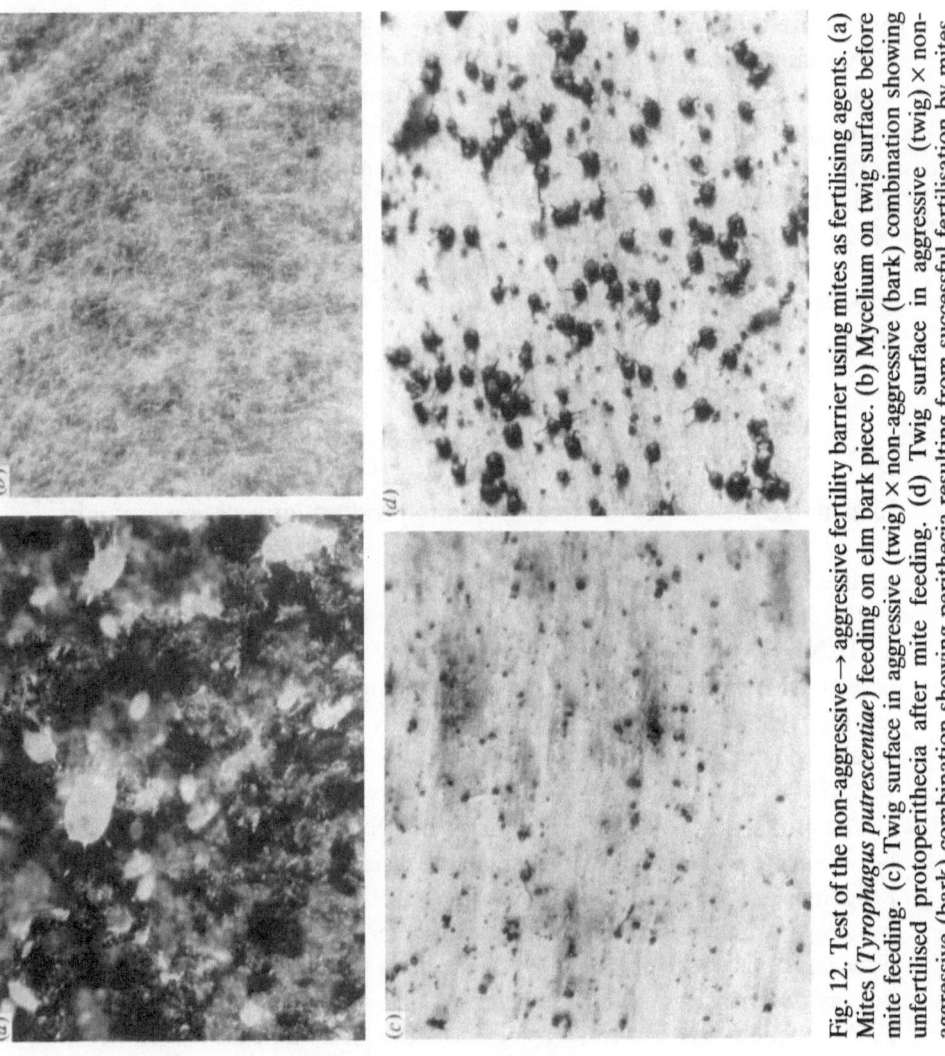

Fig. 12. Test of the non-aggressive → aggressive fertility barrier using mites as fertilising agents. (a) Mites (*Tyrophagus putrescentiae*) feeding on elm bark piece. (b) Mycelium on twig surface before mite feeding. (c) Twig surface in aggressive (twig) × non-aggressive (bark) combination showing unfertilised protoperithecia after mite feeding. (d) Twig surface in aggressive (twig) × non-aggressive (bark) combination showing perithecia resulting from successful fertilisation by mites. (From Brasier, 1978.)

additive system, but that some form of gene interaction may be involved. Any NAN × EAN hybrids formed in nature would apparently be more likely to behave, in fertility tests, like NAN isolates or intermediates than to behave like EAN isolates.

The physiological basis of the sub-group barriers

The NAN→EAN and non-aggressive→aggressive fertility barriers could conceivably result from recognition failure or incompatibility at any one of a number of steps in the mating and fertilisation processes. Unfortunately, since little is known of the mechanics or chemistry of fertilisation processes in *C. ulmi*, the point of incompatibility can only be a matter of speculation. Thus it is still not clear whether *C. ulmi* possesses a trichogyne, although the existing evidence points to the ascocarp being a simple coil of cells without such a specialised organ (Rosinski, 1961). In this case mating recognition may occur between the ascogonium and conidia acting as spermatia, the latter perhaps either carried to the ascocarp surface or growing chemotactically towards it via a germ tube, possibly guided by a hormonal system of the sort so elegantly demonstrated by Bistis & Georgopoulos (1979) for *Nectria*. In such a system, the sub-group fertility barriers could result from incompatibility at any of a number of points: slight differences in chemical structure of hormones could lead to failure of chemotropism; differences in cell-surface recognition molecules (e.g. differences in glycoprotein configuration) could result in failure of plasmogamy, and so on. Another possibility is that the barriers are due to comparable disruption of post-plasmogamy processes such as karyogamy and gametic fusion. The polygenic control of the fertility barriers (indicated above) would be consistent with any number of such mechanisms.

The function and ecological importance of recognition systems

Each of the three systems described above is an elegant physiological mechanism facilitating the recognition of one mycelial genotype by another. It is now appropriate to consider the role each system may play, jointly or in isolation, in regulating the population structure and ecological behaviour of *C. ulmi*.

Vegetative incompatibilility

Several authors in this volume emphasise the probability that vegetative incompatibility systems are widespread, possibly nearly universal in fungi, at least in the Ascomycotina and Basidiomycotina. It

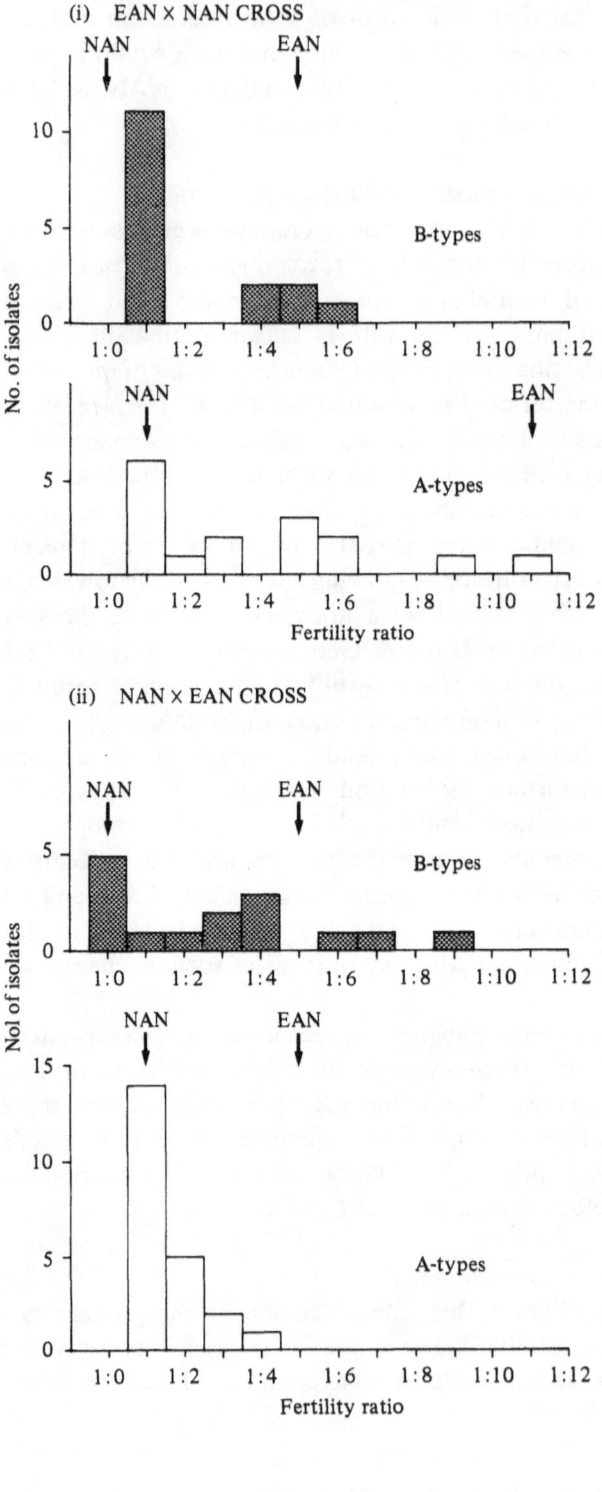
(i) EAN × NAN CROSS
NAN
EAN
B-types
NAN
EAN
A-types
No. of isolates
Fertility ratio
(ii) NAN × EAN CROSS
NAN
EAN
B-types
NAN
EAN
A-types
Nol of isolates
Fertility ratio
1:0
1:2
1:4
1:6
1:8
1:10
1:12
0
5
10
15

is evident therefore that such systems are likely to have a central function in mycelial biology. The possible consequences of this are considered in general terms, and for the Basidiomycotina in particular, by Rayner *et al.* (Chapter 23). Here, the consequences of such systems for *C. ulmi* and by analogy for many other outbreeding Ascomycotina, will be discussed.

Territorial defence and territorial invasion. Under the prevailing view of only a decade ago, it is likely that the genetic mosaic population structure of *C. ulmi* identified in elm bark by Lea (see above) would have been interpreted by many, including the author, in terms of unit mycelium concept, with each genotype a part of a physiological mosaic in which heterokaryosis and nuclear migration might play a significant part. Such an interpretation is no longer tenable in the light of present observations on vegetative incompatibility in the fungus which show that the majority of mycelia in such a mosaic would be of a different v-c group liable to produce fully vegetatively incompatible ('w') reactions

---

Fig. 13. Inheritance of the NAN → EAN fertility barrier. (i) Fertility behaviour of 32 F_1 progeny of EAN A-type ♀ × NAN B-type ♂ cross (AST-20 × W4) relative to EAN and NAN controls (arrowed). Above distribution of B-type progeny; below, distribution of A-type progeny. (ii) Fertility behaviour of 37 F_1 progeny of NAN A-type ♀ × EAN A-type ♂ cross (MM 2/1 × AST-27) relative to EAN and NAN controls (arrowed). Above, distribution of B-type progeny; below, distribution of A-type progeny.

Methods. (i) The relative fertility of B-type progeny was tested as follows: colonies of an EAN A-type were grown on elm sapwood agar; patches of each culture were fertilised (as recipient, ♀) with conidia of the different F_1 B-type donor isolates (♂) (three replicates). Conidia of standard NAN and EAN B-type donor isolates were also applied in patches to each EAN A-type recipient plate as controls. The mean number of perithecia cm^{-2} developing on the cultures of the EAN A-type was calculated for each F_1 B-type donor isolate and for the EAN and NAN control B-type isolates. The relative fertility of the F_1 isolates was calculated as the ratio of the EAN A-type × the NAN B-type control result to the individual EAN A-type × F_1 result.

(ii) The relative fertility of A-type progeny was tested as follows: each progeny isolate was grown on elm sapwood agar plates; patches of each culture were fertilised (as recipients ♀) with conidia of three EAN A-type and three NAN B-type donor isolates (♂) (three replicates) and control plates of NAN and EAN A-type isolates were fertilised in the same manner. The relative fertility of each F_1 isolate was calculated as the ratio of the mean no of perithecia cm^{-2} of the F_1 × NAN donors to the F_1 × EAN donors.

The 'patch' fertilisation method was carried out as described in Brasier (1977, 1979, 1981).

with each other. Indeed, a more logical deduction of the present observations (still to be demonstrated) is that the mosaic is itself a product of the vegetative incompatibility system, i.e. it is in some way initiated and maintained by it. Translated into behavioural terms, taking into account the apparently essentially antagonistic aspect of vegetative incompatibility interactions, the mosaic population structure could be better interpreted in terms of defence of an individual mycelium's territory. The latter concept is of course that which has been so convincingly presented by Rayner, Todd and colleagues (Todd & Rayner, 1980; Rayner *et al.*: Chapter 23) from their observations of Basidiomycotina. It would also appear to be in line with the 'selfish-gene' concept (see Dawkins, 1976).

The penetration effect observed in fully vegetatively incompatible reactions in *C. ulmi* (in which hyphae from one mycelium apparently penetrate the mycelium of another) extends the territorial concept of vegetative incompatibility beyond that of simple defence to that of territorial invasion, especially if, on further investigation, it is shown to result ultimately in replacement of one genotype by another. Thus it remains to be seen whether genetic mosaics of the sort illustrated in Fig. 2 result not only from territory obtained during primary bark colonisation but also reflect territory subsequently captured from one individual by another. Indeed, if penetration is also found to be consistently hierarchical, with some isolate combinations giving bi-directional penetration and others tending to be uni-directional penetrators, or to be penetrated, then some intra-specific fungal interactions may parallel the 'deadlock' and 'replacement' phenomena already known in inter-specific interactions (*cf.* Rayner & Webber: Chapter 18). Certainly the extent to which intra-specific mycelial penetration comparable to that in *C. ulmi* occurs in other fungi needs to be investigated, particularly in the many fungi which are likely to form genetic mosaics on solid substrata. The ecological implications of such phenomena would be far reaching, especially with plant pathogens, where competition between and replacement of one genotype by another would have enormous significance for our understanding of their population dynamics (see above) and might be exploited for biological control purposes.

At first sight, there appear to be strong parallels between the penetration effect in *C. ulmi* and, for example, the 'bow-tie' reaction in *Stereum hirsutum* and lytic crescent reaction in *Stereum gausapatum* (Coates, Rayner & Todd, 1981; Boddy & Rayner, 1982), or the 'killer' reaction in *Physarum polycephalum* (Lane & Carlile, 1979; Carlile,

1972). However, the latter are probably all associated with protoplast fusion and migration of nuclei or cytoplasm, and hence with a degree of vegetative compatibility (and therefore perhaps more analogous to the d-factor phenomenon in *C. ulmi*) rather than with full vegetative incompatibility as is thought to be the case with penetration. A further distinction is that with penetration, replacement will most often be of a genetically unrelated 'non-self' mycelium, whereas some of the above systems may tend to operate against a genotypically related mycelium (e.g. the same v-c group or homogenically incompatible sibs; see also inbreeding restriction, below).

Restriction of inbreeding. Another striking consequence of the mycelial penetration effect is that, where two isolates are of different mating type, penetration promotes the fertility of vegetatively incompatible as opposed to vegetatively compatible pairings by greatly extending the area in which sexual mating occurs. Thus, it was estimated that around ten times as many perithecia were produced in fully vegetatively incompatible w-reactions as in fully compatible c-reactions (Table 3). On the reasonable assumption that there is free recombination of v-c loci within a polygenic-heterogenic vegetative incompatibility system, then in a local population of *C. ulmi* fully vegetatively compatible isolates are likely to be the most closely related and fully incompatible isolates the least closely related as regards other loci in their genome. It follows therefore that the penetration effect will tend to restrict mating frequency between more similar genotypes and promote mating frequency between more diverse genotypes. As mycelial penetration is a particular feature of full vegetative incompatibility, it also follows that the vegetative incompatibility system in *C. ulmi* restricts inbreeding and promotes outbreeding in local populations. The significance of this interaction between vegetative incompatibility and mating-type recognition is considered in more general terms on page 491.

Since somatic or vegetative incompatibility systems have previously been reported to restrict inbreeding in a remarkably wide variety of animals and plants (see Jones & Partridge, 1983; Rayner *et al.* Chapter 23) it would perhaps be surprising if, amongst the Ascomycotina, this effect was confined to *C. ulmi* alone. Moreover, mechanisms promoting non-sib selection have recently been identified in Basidiomycotina, but here because of the lack of the specialised sexual structures found in Ascomycotina, a rather different type of somato-sexual interaction occurs. This involves the regulation of internal nuclear migration rather than an external mycelial penetration effect as in *C. ulmi* (see the 'access phenomenon,' Rayner *et al.*: Chapter 23). Thus the restriction of inbreeding may prove to be a major aspect of vegetative incompatibility in regularly outbreeding fungi, although the mechanisms by which this is achieved are likely to vary from one fungal group to another.

Prevention of cytoplasmic infection. Another attractive role for vegetative incompatibility put forward by Caten (1972) and Day (1970) is that it acts to prevent transfer of cytoplasmic infection between genetically different mycelia, i.e. to promote genetic isolation of the cytoplasm. This conclusion was derived in particular from observations on the restriction of cytoplasmic infection in *Aspergillus amstelodami* (Caten, 1972) and *Endothia parasitica* (Grente & Sauret, 1969; see also Anagnostakis: Chapters 16, 22; Croft & Dales: Chapter 20). It is now strongly supported by evidence from *C. ulmi* in which transfer of d-factors was found to be restricted to some 4% of fully vegetatively incompatible w-reactions compared with 100% transfer in fully compatible reactions and close to 100% transfer in the marginally incompatible 'l' and 'lg' reactions (see Chapter 16). In view of the extremely deleterious effects which d-factors may have on recipient mycelia (see Brasier, 1983*b*, and *loc. cit.*), the vegetative incompatibility system is likely to be a crucial factor in restricting d-factor transfer within the mycelial mosaics of *C. ulmi* found in elm bark.

It is difficult to imagine whether prevention of infection, territorial defence or the restriction of inbreeding could have been the more important to the fungus in evolutionary terms. This problem is further complicated by the potential interaction of these three functions. Thus, there is evidence that cytoplasmic infection may influence the extent of mycelial penetration, since d-infected isolates of *C. ulmi* appear to be more strongly penetrated than the equivalent 'healthy' isolates. Are they, and their 'territory', therefore, more likely to be eliminated? (See later.) Furthermore, the transmission of d-factors could be viewed not only as the spread of a disease, but also as a weapon of biological warfare between mycelia of the same or similar v-c groups. The latter argument would be strengthened if it could be shown that some isolates of *C. ulmi* were more resistant to d-factors than others, and not in themselves susceptible to the effects (of reduced fitness and survival) that they caused to their 'victims'.

Heterokaryosis and parasexuality: evidence for the case against. Another role frequently attributed to vegetative incompatibility in Ascomycotina, originating from comparisons with sexual mating systems, is that of regulating the occurrence of heterokaryosis between mycelia, and hence also of regulating the occurrence of any resulting parasexual recombination. The concept of heterokaryosis is particularly appealing because of the theoretical advantage it could provide a mycelium in terms of genotypic flexibility (e.g. Hansen, 1938; Beadle & Coonradt, 1944; Buxton, 1960).

However, Caten & Jinks (1966) have highlighted several reasons why the significance of heterokaryosis in fungi may have been greatly overestimated, including the probability that, owing to the nature of heterokaryon (vegetative) incompatibility systems, heterokaryon formation will usually be restricted to pairs of isolates with more or less identical genotypes. The present results with *C. ulmi* support this view. Where potential heterokaryotic partners are of widely differing genotype they would commonly be of different v-c groups. Hence the opportunity for heterokaryosis would be largely suppressed by vegetative incompatibility. Moreover, the present evidence also suggests that where two isolates are of the same v-c group, but otherwise of different background genotype, nuclear transfer does not occur although the cytoplasms of the two mycelia are apparently confluent. Thus, in *C. ulmi*, and perhaps in many other out-crossing Ascomycotina, the opportunity for formation of a balanced heterokaryon may not be taken up when it occurs.

Here, the observation of Croft & Dales (Chapter 20) that heterokaryon incompatibility in *Aspergillus* sometimes operates in the absence of a distinctive mycelial reaction, and hence may be genetically independent of vegetative incompatibility, could be due to the fact that the mycelial reactions are not always visually expressed on the media used. Thus in *C. ulmi*, while all v-c reactions are expressed on elm sapwood agar, many cannot be detected on other media such as malt extract, potato dextrose or carrot agars.

Caten (1971) has also pointed out that a role for vegetative incompatibility in regulating heterokaryosis is not consistent with the demonstrated occurrence of similar vegetative incompatibility systems in association with very diverse sexual breeding systems. Certainly it seems unlikely that an outbreeding fungus such as *C. ulmi* would gain any substantial benefit from heterokaryosis or parasexuality, since enough variation is likely to be available from sexual reproduction to meet the need to respond to natural selection. Clearly, specialised mating-type heterokaryons will be formed during karyogamy (see later) and transient 'intra-isolate' heterokaryons must result from mutation, especially from neutral mutations. However, even an advantageous mutation is perhaps more likely to survive if it occurs in a nucleus at the margin of a growing colony and can develop as a free-growing sector (or if it 'escapes' from the mycelium via a spore) than to persist as a component of a balanced heterokaryon.

Self-non-self recognition and the origin of the v-c 'super-groups'. The term 'self/non-self' recognition is sometimes applied to vegetative incompatibility systems. However it is important to realise that it is theoretically possible for two mycelia to have all v-c genes in common, and hence be compatible, yet be of quite different background genotypes. Thus vegetative compatibility can also mean failure of identity of 'non-self'. With a large number of v-c loci and a regularly out-crossing population lacking numerically dominant genotypes, the chance of two genetically different but vegetatively compatible mycelia coinciding is generally remote. However, EAN and NAN isolates of *C. ulmi* have been identified which are of the same v-c group, yet are otherwise, by definition, widely genetically separated entities. Similarly many isolates falling within the NAN and EAN v-c 'super groups' (see above) are probably of different background genotype.

This raises the question of why the EAN and NAN super-groups occur at all given the otherwise large number of potential groups and generally low frequency of v-c group repeats. A likely cause is a founder effect resulting from the recent rapid spread into Europe from outside of these two forms of the fungus (Brasier, 1983*a*). A particular v-c group in each case may have been, by chance, transported a long distance by man (from within the North American *C. ulmi* population in the case of the NAN race) forming the major part of a founder population which subsequently spread rapidly over a wide area aided by asexual dispersal. The number of additional v-c groups may have increased gradually as a result of mutation and sexual recombination and the 'super-group' itself may have become more genetically heterogeneous by the same means. The apparent absence of the NAN super-group in North America and its high frequency in Britain and Europe would support this theory. More information is required on the local frequency of the EAN and NAN super-groups at sites across Europe in order to confirm or reject this view. An alternative explanation is that the super-groups are favoured by natural selection as a result of the pleiotrophic effect of v-c genes on fitness. This seems less likely given the large number of potential v-c gene combinations on which selection would have to act, the expected high frequency of production of new v-c groups via recombination and the evidence that super-group isolates are otherwise probably genetically fairly heterogeneous (e.g. from evidence of their differential penetrating ability, see above).

Mating-type recognition and ascospore formation

Dispersal and survival. Mating type recognition serves to initiate perithecial formation and ascospore production in *C. ulmi* and the latter play a major part in the mycelial biology of the fungus in elm bark and on the vector beetles. In particular, an intense flush of perithecial formation has been recorded in the bark in Britain during November–February and the resulting ascospores are believed to contribute to an overwintering spore inoculum of *C. ulmi*, and to assist recolonisation of the bark in the spring (Lea & Brasier, 1983). Other perithecia formed in the pupal chambers of late-developing beetle larvae probably contribute ascospores to beetle inoculum (Webber & Brasier, 1983). An important function of mating type recognition, therefore, is to provide ascospore inoculum (possibly of a more durable nature than conidia) to facilitate survival and dispersal of the fungus at different stages of the life cycle. This is, of course, a common function of the sexual spores of most fungi.

Reduction of cytoplasmic infection levels: rejuvenation. There is evidence that where the two isolates involved in a mating are d-infected, perithecial formation is reduced, and that when perithecia are formed the d-factors may not be transmitted through to the ascospores. Another important role of ascospores, therefore, may be to maintain population fitness by ensuring that a certain proportion of the population is free from d-factor or similar infection, the proportion depending on the relative contribution of ascospores and conidia to the inoculum at different stages of the life cycle. This vital phenomenon of cytoplasmic 'clean-up' at meiosis is apparently common to many living organisms, and has been discussed in detail by Mather & Jinks (1958). In fungi, it has already been demonstrated for hypovirulence in *Endothia parasitica* (Chapter 16) and for dsRNA and DNA proviruses in *Geaumannomyces graminis* (McFadden, Buck & Rawlinson, 1983) and may be a common feature of Ascomycotina.

Outbreeding efficiency, mating competence and genetic recombination. The two allele single locus mating recognition system in *C. ulmi* is a classic homogenic Ascomycete incompatibility system (Esser & Blaich, 1973). Two important functional effects of this system are well known: that of preventing selfing (homothallism) and hence of enforcing obligatory out-crossing; and that of regulating outbreeding between individual mycelia to *c.* 50% when two alleles are present in a population in roughly equal frequency.

Probably too much emphasis is placed, however, in many discussions of this and other mating systems in fungi on the theoretical numerical inbreeding or outbreeding advantages which might arise from the different systems and not enough emphasis has been placed on the effect that other factors such as spore dispersal mechanisms, fitness effects of mating-type alleles, fertilisation mechanisms and selection intensities are likely to exert on mating competence and on the balance of sexual versus asexual reproduction. Thus in nature mating type frequencies in a two allele system may often diverge from 1:1 as a result of the selective advantage of one mating type over another. This is certainly the case in *C. ulmi* where mating competence appears to be regulated in different ways in the different sub-groups of the fungus. Although in the non-aggressive strain the two *mt* alleles do occur in near equal frequency, in the aggressive strain (both races) the B-type is predominant, the A-type only accounting on average for some 15% of the population (Table 6). In theory this reduces the chances of a compatible pairing between any two adjacent mycelia to around only 25%. However the low frequency of the A-type in the aggressive strain may be greatly offset by its intrinsically much higher level of fecundity (protoperithecial formation). The aggressive strain, therefore, may have balanced the slightly reduced pathogenic and growth fitness of the A-mating type against higher fecundity and so maintained a high level of mating competence.

The predominance of the B-type in the EAN and NAN aggressive populations may sometimes result in a local absence of the A-type (see Brasier & Gibbs, 1975). In these circumstances the mating type mutation and pseudoselfing mechanism of *C. ulmi* (see p. 475) could well supply the missing A-type, and indeed pseudoselfing if it occurs in nature might also contribute a small element of inbreeding. However, A-type protoperithecia formed after such a mating-type mutation would probably tend to be fertilised by B-types of different genetic background where the latter were present in the adjacent elm bark.

Indeed, in a mycelial mosaic mating competence will also be greatly increased by the fact that any one individual will often be surrounded by not one but several potential mating partners; by the fact that mites and other microfauna acting as fertilising agents may effectively intermingle many different genotypes; and also by the fact that a trichogyne may be able to grow selectively towards a spore of opposite mating type and vice versa. Thus mating competence is likely to be the product of a great many factors, few of which can be satisfactorily quantified. Judging by the large numbers of perithecia often formed in beetle breeding galleries by the NAN aggressive; (Lea & Brasier, 1983) a high level of mating competence is normally attained.

Ultimately, out-crossing systems produce, via genetic recombination, a sufficient reservoir of variation in the population to meet the need to respond to natural selection and yet maintain optimal population fitness. Polymorphism in vegetative incompatibility groups and differences in cultural characteristics and pathogenicity in samples of *C. ulmi* isolates all indicate extensive genetic variation within each of the three sub-groups of the fungus. Apart from the mating-type and vegetative incompatibility systems, however, the extent of allelic variation both at the local level and in world-wide populations has yet to be assessed.

The interplay of vegetative incompatibility and mating competence. At the cellular level there is clearly a theoretical functional conflict between the roles of vegetative incompatibility and mating type recognition, in that the expression of vegetative incompatibility between mycelia of both different v-c group and different mating type could potentially disrupt plasmogamy and karyogamy. That such disruption does not normally occur is presumably because the fungus can in some way switch off or bypass the vegetative incompatibility system when this situation arises. This is discussed in detail for Basidiomycetes by Rayner *et al.* (see Chapter 23). In Ascomycetes the expression of vegetative incompatibility may already be suppressed in cells differentiated as ascogonia, or *mt* allele complementation between an ascogonium and spermatium may switch off the vegetative incompatibility genes in each, for example by repression.

The present studies on *C. ulmi* also illustrate a variety of other ways in which the efficiency of the mating system may affect that of vegetative incompatibility and vice versa. For the vegetative incompatibility system to operate effectively, whether in terms of territorial behaviour or the prevention of cytoplasmic infection, there must be a suitably large number of different v-c groups within a given population size. This requirement is facilitated by the polygenic (and possibly multi-allelic) genetic system by which vegetative incompatibility is controlled (see page 459), since new v-c groups can arise with high frequency through genetic recombination. It is important to the system, therefore, that ascospores are a prime source of inoculum for host colonisation, since asexual dispersal would tend to favour clonal spread and smaller numbers of v-c groups (as may have occurred with the 'super-groups', see above). The winter flush of perithecia in beetle breeding galleries (Lea & Brasier, 1983) is therefore likely to be of great importance in generating and maintaining novel v-c groups in the *C. ulmi* population.

The efficiency of the mating system and the rate of sexual versus asexual reproduction is in turn likely to be influenced by vegetative incompatibility, since, where mating occurs between adjacent mycelia in a mosaic situation, more perithecia may be formed in mating-type compatible but vegetatively incompatible 'w' reactions in which mycelial penetration is involved than in mating-type compatible 'l', 'lg' or 'c' reactions. The net effect of this will be to promote fertilisation between mycelia of different v-c groups and to restrict it between mycelia of the same or a closely similar group, i.e. to promote outbreeding and to restrict inbreeding (see page 485).

The potential complexity of the interplay between vegetative incompatibility and mating can be illustrated further by considering the likely added influence of other mycelial fitness parameters such as cytoplasmic (d-factor) infection. Once an isolate is being effectively penetrated by the hyphae of another isolate of opposite mating type its main chance of survival (apart from conidial formation) may lie in passing on its genes via ascospores from any resulting perithecia. If it is d-infected it is apparently even more likely to be penetrated. However, the tendency for penetration to promote fertilisation and ascospore production may be offset by the tendency for d-infection to suppress perithecial formation! A further twist to this curious equation is that any resulting ascospores might be free from the d-factor and hence, in one sense, fitter than the d-infected parent.

The sub-group fertility barriers

Although mediated at the level of the individual mycelium via the mating system the sub-group fertility barriers effectively operate at the population level by restricting gene flow between and hence maintaining the genetic integrity of the sub-group mycelial populations.

Similar fertility barriers are well documented for other fungi, classical examples being the breeding groups of *Phellinus (Fomes) pinicola, Hirschioporus (Polyporus) abietinus* and *Nectria haematococca* (MaCrae 1967; Matuo & Snyder, 1973; for review see Burnett, 1983). Their prime role or effect is evolutionary: when the barrier to gene flow is total each sub-group is effectively an incipient species even though it may be largely morphologically indistinguishable from another group.

In *C. ulmi* the evolutionary status of the barriers should be seen in the context of other genetic differences between the sub-groups. The aggressive and non-aggressive strains have very different morphological and physiological characteristics (see Brasier, 1982*a,b*). When force-mated in the laboratory their progeny are extraordinarily variable morphologically and show negative interactions for continuous characters such as growth-rate and pathogenicity, indicating that the parental genomes are not only divergent but also somewhat incompatible. Since such hybrid types (which being morphologically unusual are not difficult to recognise) have not been found among thousands of wild isolates examined from nature, they either do not occur or do not survive (Brasier, 1983*a*). The genetical evidence, therefore, points to a considerable degree of evolutionary divergence between the aggressive and non-aggressive strains. This divergence is apparently effectively maintained by a combination of the fertility barrier and natural selection.

The extent to which the same may apply to the barrier between the NAN and EAN races is not yet clear. The two races are more closely related, and the barrier is only partial. Moreover any hybrids occurring in nature are likely to be difficult to identify without biochemical markers. Nevertheless, the extent to which even a small reduction in gene flow between sub-populations can promote genetic isolation is, I believe, frequently underestimated.

An outstanding question is whether the barriers are a consequence of evolutionary divergence of the sub-groups, or whether they were a contributing factor. The answer may depend in turn upon whether the evolution of the sub-groups has been sympatric, resulting from ecological specialisation in the same environment, or allopatric resulting from geographical isolation and genetic drift. In the former case the sub-group fertility barriers could have reinforced the divergence; in the latter they would simply be a consequence of it. Presently, I favour the former to explain the origin of the aggressive/non-aggressive barrier and the latter to explain the origin of the EAN/NAN barrier.

Recognition systems and the behaviour of Dutch elm disease

Changes in the pathogenic and saprophytic behaviour of plant pathogen populations will be understood better when studied in conjunction with pathogen recognition systems than when studied in isolation, since the former must in many ways be regulated by the latter. Nowhere perhaps could this be more vividly illustrated than by the present situation with Dutch elm disease.

Within the terms of this paper, the current epidemics of Dutch elm disease are a result of the 'territory' (i.e. elms) of the 'old' non-aggressive strain of *C. ulmi* being invaded, from east and west, by the EAN and NAN races of the aggressive strain. In many locations in Europe all three sub-groups of the fungus are, quite suddenly, cohabiting (see Brasier, 1983*a*). In North America and South West Asia the non-aggressive strain and either the NAN or the EAN aggressive respectively are now confronting each other. The initial confrontations will probably result in the destruction of the greater part of the mature field elm population across these regions. But what of the future?

The three sub-groups of *C. ulmi* have, in a sense, been thrown together into a melting pot which, as a result of the sudden and increasing destruction of so many of the larger elms, is getting rapidly smaller. The future of the elm depends upon what form of *C. ulmi* survives into the post-epidemic period and returns to attack the new generations of elm seedlings and suckers. The form of *C. ulmi* that will survive depends upon the product of firstly the interplay of pathogenic fitness and natural selection (aspects of which have been discussed by Brasier, 1983*a,b*), and secondly the outcome of mycelial interactions between the *C. ulmi* sub-groups.

The mycelial interactions between the sub-groups will, as already indicated, take place mainly in the bark phase of the fungus during which, in order to survive, a mycelium or genotype must pass through a long and complex series of ecological processes and ultimately succeed in contributing to the beetle inoculum in the pupal chambers (Webber & Brasier, 1983). During these processes vegetative incompatibility interactions are likely to be closely involved in determining territorial success. In this, the relative penetrating or replacement abilities of, for example, the aggressive and non-aggressive strains (as yet undetermined) could be crucial. The frequency of v-c groups within the sub-groups will influence the rate of spread of cytoplasmic infection both within and between them and here the existence of the 'super-groups' in the EAN and NAN might well enhance the spread of infection. The frequency of mating types in the sub-groups will influence the rate of ascospore formation and hence the level of genetic heterogeneity including the number of v-c groups, the proportion of inoculum free from cytoplasmic infection and the resistance of inoculum to the environment. Last, but by no means least, the sub-group fertility barriers will determine the extent of hybridisation between the sub-groups. Further mycelial interactions may occur in the beetle feeding grooves. Here vegetative incompatibility interactions could strongly influence, through inter-mycelial antagonism or through d-factor transfer, the success with which different mycelial genotypes achieve host infection (*cf.* Webber & Brasier, 1983).

Other genetical properties, and in particular relative pathogenic ability, will of course be closely involved in the final equation of this 'battle for territorial possession of the elm'. The first round of this battle may now have been fought in many areas, since there are strong indications of the probable demise of one of the three main contestants: the non-aggressive strain (Brasier, 1983*a*). The particular interaction phenomena which have most contributed to its downfall should be investigated.

I wish to thank Maxine Rawlings, Susan Kent and Susan Kirk, who over succeeding years have given superb technical support.

References

Anagnostakis, S. A. (1982). Biological control of chestnut blight. *Science*, **215**, 466–71.

Ansina, S. & Hubbes, M. (1980). Compatibility in *Ceratocystis ulmi. European Journal of Forest Pathology*, **10**, 201–8.

Beadle, G. W. & Coonradt, V. L. (1944). Heterocaryosis in *Neurospora crassa. Genetics*, **29**, 291–308.

Bistis, G. N. & Georgopoulos, S. G. (1979). Some aspects of sexual reproduction in *Nectria haematococca* var. *cucurbitae. Mycologia*, **70**, 127–43.

Boddy, L. & Rayner, A. D. M. (1982). Population structure, inter-mycelial interactions and infection biology of *Stereum gausapatum. Transactions of the British Mycological Society*, **78**, 337–51.

Brasier, C. M. (1977). Inheritance of pathogenicity and cultural characters in *C. ulmi*. Hybridisation of protoperithecial and non-aggressive strains. *Transactions of the British Mycological Society*, **68**, 45–52.

Brasier, C. M. (1978). Mites and reproduction in *Ceratocystis ulmi* and other fungi. *Transactions of the British Mycological Society*, **70**, 81–9.

Brasier, C. M. (1979). Dual origin of recent Dutch elm disease outbreaks in Europe. *Nature*, **281**, 78–9.

Brasier, C. M. (1981). Laboratory investigation of *Ceratocystis ulmi*. In *Compendium of Elm Diseases*, ed. R. J. Stipes & R. J. Campana, pp. 76–9. St Paul, Minnesota: American Phytopathological Society.

Brasier, C. M. (1982*a*). Genetics of pathogenicity in *Ceratocystis ulmi* and its significance for elm breeding. In *Resistance to Diseases and Pests in Forest Trees*, ed. H. M. Heybroek, B. R. Stephan & K. von Weissenberg, pp. 224–35. Wageningen, Netherlands: Pudoc.

Brasier, C. M. (1982*b*). Occurrence of three sub-groups within *Ceratocystis ulmi*. In *Proceedings of the Dutch elm Disease Symposium and Workshop*. Winnipeg, Manitoba, October 5–9, 1981, ed. E. S. Kondo, Y. Hiratsuka & W. B. G. Denyer, pp. 298–321. Manitoba, Canada: Manitoba Department of Natural Resources.

Brasier, C. M. (1983*a*). The future of Dutch elm disease in Europe. In *Research on Dutch Elm Disease in Europe*, ed. D. A. Burdekin, *Forestry Commission Bulletin*, **60,** 96–104.

Brasier, C. M. (1983*b*). A cytoplasmically transmitted disease of *Ceratocystis ulmi*. *Nature,* **305,** 220–3.

Brasier, C. M. & Afsharpour, F. (1979). The aggressive and non-aggressive strains of *Ceratocystis ulmi* in Iran. *European Journal of Forest Pathology*, **9,** 113–22.

Brasier, C. M. & Gibbs, J. N. (1975*a*). MBC tolerance in aggressive and non-aggressive isolates of *Ceratocystis ulmi. Annals of Applied Biology*, **80,** 231–5.

Brasier, C. M. & Gibbs, J. N. (1975*b*). Highly fertile form of the aggressive strain of *Ceratocystis ulmi. Nature*, **257,** 128–31.

Brasier, C. M. & Gibbs, J. N. (1976). Inheritance of pathogenicity and cultural characters in *C. ulmi*. I. Hybridisation of aggressive and non-aggressive strains. *Annals of Applied Biology*, **83,** 31–7.

Brasier, C. M., Lea, J. & Rawlings, M. K. (1981). The aggressive and non-aggressive strains of *Ceratocystis ulmi* have different temperature optima for growth. *Transactions of the British Mycological Society*, **76,** 213–18.

Buisman, C. J. (1932). Over het voorkomen van *Ceratostomella ulmi* (Schwarz) Buisman in de natur. *Tijdschrift over Plantenziekten*, **38,** 203–4.

Burnett, J. H. (1983). Speciation in fungi. *Transactions of the British Mycological Society*, **81,** 1–14.

Buxton, E. W. (1960). Heterokaryosis, saltation and adaptation. In *Plant Pathology, an Advanced Treatise,* vol. 2, ed. J. G. Horsfall & A. E. Dimond, pp. 359–405. London & New York: Academic Press.

Carlile, M. J. (1972). The lethal interaction following plasmodial fusion between two strains of the Myxomycete, *Physarum polycephalum. Journal of General Microbiology*, **71,** 581–90.

Caten, C. E. (1971). Heterocaryon incompatibility in imperfect species of *Aspergillus*. *Heredity*, **26,** 299–312.

Caten, C. E. (1972). Vegetative incompatibility and cytoplasmic infection in fungi. *Journal of General Microbiology*, **72,** 221–9.

Caten, C. E. & Jinks, J. L. (1966). Heterokaryosis: its significance in wild homothallic ascomycetes and fungi imperfecti. *Transactions of the British Mycological Society*, **49,** 81–93.

Coates, D., Rayner, A. D. M. & Todd, N. K. (1981). Mating behaviour, mycelial antagonism and establishment of individuals in *Stereum hirsutum. Transactions of the British Mycological Society*, **76,** 41–51.

Dawkins, R. (1976). *The Selfish Gene.* Oxford: Oxford University Press.

Day, P. R. (1970). The significance of genetic mechanisms in soil fungi. In *Root Diseases of Soil-borne Pathogens*, ed. T. A. Toussoun, R. V. Bega & P. E. Nelson, pp. 69–74. Berkeley: University of California Press.

Esser, K. & Blaich, R. (1973). Heterogenic incompatibility in plants and animals. *Advances in Genetics*, **17,** 107–52.

Garnjobst, L. & Wilson, J. F. (1956). Heterocaryosis and protoplasmic incompatibility in

Neurospora crassa. Proceedings of the National Academy of Sciences of the USA, **42,** 613–18.

Grente, J. & Sauret, S. (1969). 'L'hypovirulence exclusive' est-elle controlée par les determinants cytoplasmiques? *Compte rendu hebdomadaire des Séances de l'Académie des Sciences, Paris* (Série D) **268,** 3173–6.

Hansen, H. N. (1938). The dual phenomenon in imperfect fungi. *Mycologia*, **30,** 442–55.

Hicks, J., Strathern, J. N. & Klar, A. J. S. (1979). Transposable mating type genes in *Saccharomyces cerevisiae. Nature*, **282,** 478–83.

Holmes, F. W. (1965). Virulence in *Ceratocystis ulmi. Netherlands Journal of Plant Pathology*, **71,** 97–112.

Holmes, F. W. (1977). Distinction between sex and compatibility in *Ceratocystis ulmi* as shown by unisexual male cultures. *Mycologia*, **69,** 1149–61.

Jones, J. S. & Partridge, L. (1983). Tissue rejection: the price of sexual acceptance. *Nature*, **304,** 484–5.

Lane, E. B. & Carlisle, M. J. (1979). Post-fusion somatic incompatibility in *Physarum polycephalum. Journal of Cell Science*, **35,** 339–54.

Lea, J. (1977). A comparison of the saprophytic and parasitic stages of *Ceratocystis ulmi.* Ph.D. Thesis, University of London.

Lea, J. & Brasier, C. M. (1983). A fruiting succession in *Ceratocystis ulmi* and its significance for Dutch elm disease. *Transactions of the British Mycological Society*, **80,** 381–7.

MaCrae, R. (1967). Pairing incompatibility and other distinctions among *Hirschioporus (Polyporus) abietinus, H. fusco-violaceus* and *H. laricinus. Canadian Journal of Botany*, **45,** 1371–98.

Mather, K. & Jinks, J. L. (1958). Cytoplasm in sexual reproduction. *Nature*, **182,** 1188–90.

Mathieson, M. J. (1952). Ascospore dimorphism and mating type in *Chromocrea spinulosa. Annals of Botany, N.S.* **16,** 449–66.

Matuo, T. & Snyder, W. C. (1973). Use of morphology and mating populations in the identification of formae speciales in *Fusarium solani. Phytopathology*, **63,** 562–5.

McFadden, J. J. P., Buck, K. W. & Rawlinson, C. J. (1983). Infrequent transmission of double-stranded RNA virus particles, but absence of DNA proviruses in single ascospore cultures of *Geaumannomyces graminis. Journal of General Virology*, **64,** 927–37.

Rosinski, M. A. (1961). Development of the ascocarp in *Ceratocystis ulmi. American Journal of Botany*, **48,** 285–93.

Sansome, E. R. (1946). Heterokaryosis, mating-type factors and sexual reproduction in *Neurospora. Bulletin of the Torrey Botanical Club*, **73,** 397–409.

Schafer, T. & Lyming, O. N. (1950). *Ceratocystis ulmi* types in relation to development and identification of perithecia. *Phytopathology*, **40,** 1035–42.

Schwarz, M. B. (1922). *Des Zweigsterben der Ulmen, Traverweiden und Pfirschbaume, ein vergleichend-patologische Studie*, pp. 7–32. Utrecht: A. Osthoek.

Stebbins, G. L. (1950). *Variation and Evolution in Plants*. New York: Columbia University Press.

Swingle, R. V. (1936). A preliminary note on sexuality in *Ceratocystis ulmi. Phytopathology*, **26,** 925–7.

Todd, N. K. & Rayner, A. D. M. (1980). Fungal individualism. *Science Progress (Oxford)*, **66,** 331–54.

Uhm, J. Y. & Fujii, H. (1983). Heterothallism and mating type mutation in *Sclerotinia trifoliorum. Phytopathology*, **73,** 569–72.

Webber, J. F. & Brasier, C. M. (1984). The transmission of Dutch elm disease: a study of the processes involved. In *Invertebrate–Microbial Interactions*, ed. J. M. Anderson, A. D. M. Rayner & D. W. H. Walton. Cambridge, U.K.: Cambridge University Press.

22

The mycelial biology of *Endothia parasitica*. II. Vegetative incompatibility

SANDRA L. ANAGNOSTAKIS
Department of Plant Pathology and Botany, The Connecticut Agricultural Experiment Station, P. O. Box 1106, New Haven, CT 06504 USA

In *Endothia parasitica* two systems of incompatibility, sexual and vegetative (= somatic), provide us with natural genetic markers for basic research and with frustrating barriers to our attempts at biological control. I will discuss these two systems using terminology based on Ainsworth (1961), and the CMI 'Terms' Guide (1973). Thus, an *isolate* is the first culture of the fungus derived from spores or from host tissue, and this becomes a *strain* when it has been transferred and is maintained in a culture collection. My use of the word strain does not imply 'physiologic race'.

Sexual incompatibility

Sexual incompatibility in *Endothia parasitica* is homogenic. That is, strains are sexually compatible only if they have different alleles at the (single) mating type locus. There seem not to be any serious barriers to crossing, as occurs between the sub-groups of *Ceratocystis ulmi* described by Brasier (Chapter 21). Thirty-four strains from North America, France, Italy, Greece, and China were all compatible with one or the other of my American tester strains for *A* and *a* mating types (Anagnostakis, 1979). Fourteen were mating type *A* and twenty mating type *a*.

Each *E. parasitica* protoperithecium contains at least two female nuclei. Furthermore, more than one conidium can fertilise a single protoperithecium and gene reassortment between the male-parent nuclei is possible after plasmogamy and before karyogamy (Anagnostakis, 1982*b*). Further complicating progeny assessment, once a few perithecia resulting from cross-fertilisation have formed, production of selfed perithecia may be stimulated. An analogous situation occurs in 'heterothallic' *Phytophthora* species, where selfed oospores develop in crosses between mating types (Sansome, 1970). A perithecium can also be the result of one of the female nuclei in a protoperithecium fusing with an unlike partner and the other simply doubling or fusing with a like partner. Thus, a perithecium may be crossed, selfed, or half-crossed and half-selfed. I have been collecting data on this phenomenon for several years but no pattern is yet clear.

In host chestnut trees in the forest, the pathogen usually reproduces sexually every year. The resulting ascospores are shot out of the perithecia in wet weather and start new infections (Anderson, 1914; Rankin, 1914). Sexual reproduction in populations of pathogens is potentially important since it may lead to recombination of genes conferring virulence, and hence affect the balance of the host parasite interactions, as with *Venturia inaequalis* infecting apples (Day, 1974). However, there are few nuclear virulence genes in the natural population of *E. parasitica*, and, in any case, small differences in virulence are of little importance since host American chestnut trees (*Castanea dentata*) are very susceptible. This is often the case when hosts are exposed to an introduced pathogen. In the 80 years that *E. parasitica* has been present in the USA, no resistant *C. dentata* have been found (Jaynes & Elliston, 1982). However, sexual recombination in *E. parasitica* does have important implications in relation to reassortment of the genes for vegetative incompatibility.

Vegetative incompatibility

Vegetative incompatibility in *E. parasitica*, as in other fungi, (see Chapters 16, 20, 21, 23) is heterogenic and allelic. That is, viable anastomoses are formed only between strains which have identical alleles at all of the vegetative compatibility (v-c) loci. Andes (1961) reported that *E. parasitica* strains paired on potato dextrose agar either merged (Fig. 1a); formed a zone of inhibition (Fig. 1b); or formed a ridge of pycnidia where they met (Fig. 1c). Microscopic examination of such regions reveals, respectively, hyphae with many cross-bridges; hyphae with few cross-bridges and with some dead cells associated with

---

Fig. 1. Single ascospore colonies of *Endothia parasitica* on agar media (a) merge as their mycelia grow when they are all in the same vegetative-compatibility group. When some of the colonies are in different vegetative-compatibility groups they may form a weak barrage zone (b) or a strong barrage zone with a ridge of pycnidia (c).

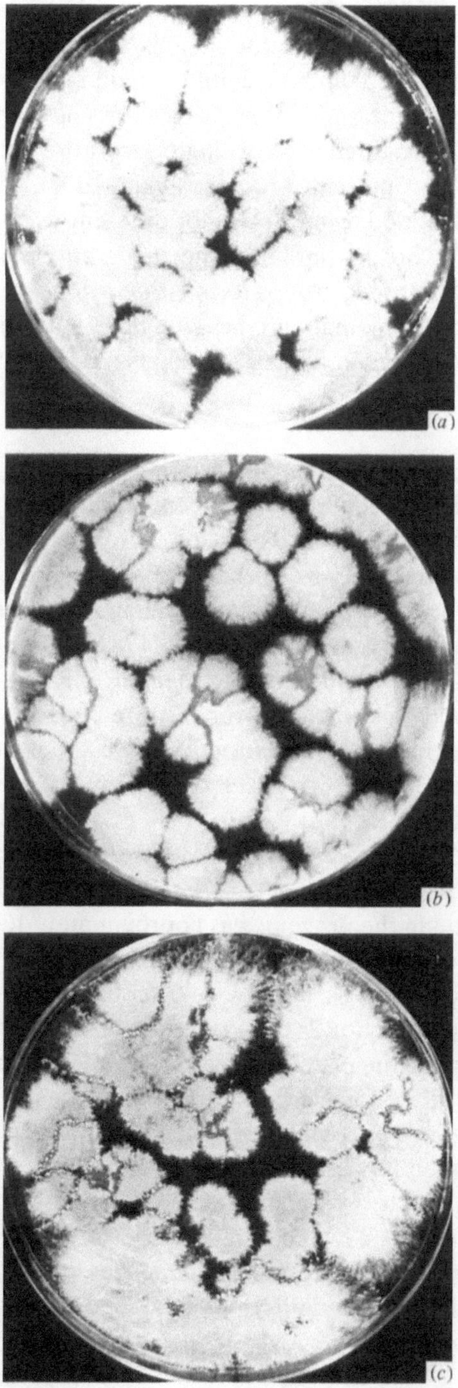
(a)
(b)
(c)

those few; or hyphae with few cross-bridges and dead cells associated with most of them. Fortunately these interactions are easily recognised macroscopically in our most common culture medium (Difco potato dextrose agar with methionine and biotin). This allows us to make large numbers of pairings of field isolates or of single ascospore progeny and assess them as 'same' or 'different' (Anagnostakis, 1977). Using such procedures, we have identified tester strains for different v-c types in our *E. parasitica* collection. Strains in the same v-c group merge when paired, and those in different v-c groups form either a zone of inhibition or two ridges of pycnidia. The mating type gene does not function as a v-c gene as it does in *Neurospora crassa* (Perkins & Barry, 1977).

Genetics of vegetative incompatibility

Some crosses have been made between *E. parasitica* strains in different v-c groups and progeny analysed for v-c type. These crosses allowed us to estimate the number of genes for v-c that were different between pairs of v-c types. We have found no linkage between these or any other tested genes (Anagnostakis, 1980, 1982*a*). For instance, crossing a v-c 5 strain with a v-c 39 strain yielded only v-c 5 and 39 progeny. Therefore, the types 5 and 39 probably are phenotypes of two alleles at the same gene locus. A cross of v-c 5 × v-c 8 yielded random ascospore progeny which were v-c 5, 8, 39, or 71. Thus, the v-c types 5 and 8 are different at two v-c genes.

Perhaps the most informative crosses were those made between two v-c 5 strains and two v-c 10 strains (Anagnostakis, 1982*a*). The total number of v-c types among the progeny was approximately 106. If all of the incompatibilities were caused by allelic differences (i.e. different alleles at the same locus) then v-c 5 and v-c 10 should be different at 7 v-c genes (which would allow for 2^7 total types, or 128). However, the male parental types among the progenies were 11/263 (1/24) and 37/973 (1/26). Since the number of female parental types is usually increased by perithecia which are half or completely selfed, it cannot be relied on. If the ratios of male parental v-c types to the total number of v-c types were really 1/32 (i.e. $1/2^5$), then v-c 5 and v-c 10 are different at 5 loci. In that case, the extra 70 or so v-c types found could mean that linkages exist, or might result from interactions between alleles at *different* genes (non-allelic heterogenic incompatibility) as described by Esser (1971) for *Podospora anserina.*

Vegetative incompatibility and biological control

Why do we care? It is more than intellectual curiosity. In Chapter 16, I have described curative strains of *E. parasitica* that have dsRNA molecules in their cytoplasm (see also Anagnostakis, 1982*c*). Not only can these abnormal strains cure existing blight when they are inoculated into the cankers, but they have spread all over Italy and Switzerland (Bonifacio & Turchetti, 1973; Bazzigher, Kanzler & Kübler, 1981) and seem to have effectively brought about a biological control of the disease. We have some information about the distribution of curative strains in Italy (Palenzona, 1978) but no thorough study has been done of their spread. Grente has used curative strains to control blight in French orchards, and says that he sees spread of biological control in those sites, as well (Grente, 1981).

Grente sent us curative strains in 1972. We then went through several stages of testing; in the laboratory, the greenhouse, and The Experiment Station Farm, to convince ourselves and the United States Plant Quarantine Office that the release of curative strains of *E. parasitica* would not cause any harm. We started our forest experiments in 1977 with high hopes which were based on the reports of biological control in Europe. We soon found, however, that only about 40% of Connecticut cankers treated with single curative strains were clearly cured when observed two years later (Jaynes & Elliston, 1980). Our failure to control all of the treated cankers with a single curative strain was reminiscent of the report by Grente & Sauret (1969), who found that only 6 of 50 pairs (12%) of virulent and curative strains from the same region formed cankers when inoculated together in *Castanea sativa.* By contrast, cankers were formed by 124 of 170 pairs (73%) originating from different regions.

These results prompted the laboratory experiments which produced all of our genetic information, discussed above, about vegetative incompatibility in *E. parasitica* (Anagnostakis, 1977, 1980, 1982*a*). We also started examining the interactions of curative strains with canker-causing strains. When curative hypovirulent (H) strains are paired with killing virulent (V) strains on agar media, the dsRNA from the curative strain can be transferred to the killing strain via anastomoses, converting virulent to hypovirulent; this is accompanied by morphological changes (see Fig. 5, Chapter 16). When V and H strains were in different v-c groups, the conversion did not always occur (Anagnostakis & Day, 1979) (see also Fig. 6, Chapter 16).

We have also found that vegetative incompatibility is an important factor in cure of cankers in the host (Anagnostakis & Waggoner, 1981). We started cankers with virulent strains in four v-c groups and then treated them two weeks later with curative strains in four v-c groups. The groups were chosen to make possible pairs different at 0, 1, 2 and 5 v-c genes. With increasing difference, there was decreasing cure success, until with strains different at five genes, the rate of cure was insignificant.

Laboratory tests

The strength of vegetative incompatibility reactions on agar medium in the laboratory also increases with the number of v-c gene differences, and strains different at only one gene sometimes produce a zone of inhibition that is hard to distinguish from merging. In the course of testing strains for v-c type over the years, we have kept records of which v-c pairings produced these weak reactions. Among the 4656 combinations of our tester strains for 97 different v-c types, 87 pairs (70 strains) reacted weakly. We tested 72 of these weakly-barraging pairs to see whether the cytoplasmic determinants for curative morphology could be easily transferred between them. We found that 62 such pairs allowed rapid (≤3 days) conversion in both directions (see Chapter 16). Furthermore, these strains form networks which provide the possibility of transferring the cytoplasmic genes for hypovirulence from one strain to another throughout a population (Anagnostakis, 1983). Although transfer was not possible between, for example strains A and D, cytoplasmic genes could be passed from strain A to B to C to D. Kuhlman (unpublished) has identified similar networks of *E. parasitica* strains in what he calls conversion clusters. The *E. parasitica* strains in his forest test areas in North Carolina and Virginia also fall into many v-c groups.

It is possible that this conversion across v-c types is not simply due to v-c relatedness. There is no evidence that the determinants for hypovirulence in the curative strains might influence these interactions, but tests with other cytoplasmic markers must be carried out before we can be sure that they do not. We are now trying to isolate *E. parasitica* strains with oligomycin-resistant mitochondria to make such tests.

Vegetative incompatibility and pathogen death in the host

Jaynes & Elliston (1980) also raised another problem. Why should cankers appear cured and then later start expanding again? One possibility is that inoculation with curative strains in very different v-c groups causes retardation of canker expansion via anastomoses leading to cell death, that is, vegetative incompatibility. To test this, V strains different at at least 5 v-c genes (v-c groups 5 and 10 discussed above) were inoculated into host chestnut trees alone, side by side, or as two different strains thoroughly mixed (Waggoner & Anagnostakis, unpublished). During the first 30 days, cankers usually develop slowly (primary rate) and then expand more rapidly (secondary rate) (Anagnostakis & Aylor, unpublished). There was very little difference in secondary expansion rate between strains in the same or different v-c group inoculated side by side. Mechanical blending of strains in the same v-c group reduced secondary rate slightly. However, mixing strains in v-c groups 5 and 10 significantly reduced canker-expansion rate. Although only strains in two unlike v-c groups have been thoroughly tested, we can probably generalise that killing due to anastomoses between unlike v-c groups can explain the temporary expansion halt noted by Jaynes & Elliston (1980) after treatment with curative strains. If there was no conversion, the canker-causing strain(s) would eventually grow away from or around the introduced strain, and the canker would start expanding again (cf. discussion of escape from somatic incompatibility, Rayner *et al.*, Chapter 23).

Population structure of *E. parasitica* in relation to vegetative incompatibility

Our 97 v-c types represent ascospore progeny and field isolates, but we need to know how many types are present in each area in nature.

Grente (1981) made tests with his European stock collection for vegetative compatibility and reported that most of 141 tested strains fit one of 22 v-c groups. I have tested his 22 European v-c types and found 9 to be the same as v-c types in our collection, and 13 not like any of our v-c types.

Kuhlman & Bhattacharyya (unpublished) examined the distribution of v-c types in natural *E. parasitica* cankers on American chestnut trees in Virginia. They found that although 12 out of 41 cankers yielded strains in only one v-c group, 29 cankers each contained 2 to 5 v-c types.

I have tested field isolates from all over North America (258 total samples), and found 73 v-c groups. In a survey of chestnut trees in Connecticut, we isolated 165 *E. parasitica* strains in 67 v-c groups (38 were each found only once). Thus, Connecticut has a very good sample of the diversity found in the whole country. On the other hand, 49 isolates from Italy fell into 9 v-c groups. Although our Italian sample is small, these data, together with those of Grente mentioned above, suggest that there may be more diversity of v-c in the *E. parasitica* population in North America than in the population of this fungus in Europe. There is no explanation for these differences at present, but they may underly the more-effective spread of hypovirulence in Europe.

Outlook

Does this mean that establishing a stable biological control, or balance tipped slightly in favour of the host, will be difficult or impossible in the US? We will have to have not only a high titre of curative strains in our forests, but also a genetically diverse population representing many points in the conversion network and capable of converting virulent strains in many v-c types.

We have seen some spread of cure in our test plots: here and there on the same tree, occasionally to a neighbouring tree. This inconsistent success is frustrating, but happens often enough to keep me optimistic. We continue to hope that if a vector is needed (insect, bird, or mammal) we have in the US the creature or creatures which promote spread in Europe. In addition, we keep trying new ways of putting curative strains of *E. parasitica* out into our chestnut trees to overcome the obstacle caused by vegetative incompatibility.

References

Ainsworth, G. C. (1961). *Ainsworth & Bisby's Dictionary of the Fungi*. 5th edn. Kew, Surrey, England: Commonwealth Mycological Institute.

Anagnostakis, S. L. (1977). Vegetative incompatibility in *Endothia parasitica*. *Experimental Mycology*, **1**, 306–16.

Anagnostakis, S. L. (1979). Sexual reproduction of *Endothia parasitica* in the laboratory. *Mycologia*, **71**, 213–15.

Anagnostakis, S. L. (1980). Notes on the genetics of *Endothia parasitica* in the laboratory. *Neurospora Newsletter*, **27**, 36.

Anagnostakis, S. L. (1982*a*). Genetic analyses of *Endothia parasitica*: linkage data for four single genes and three vegetative compatibility types. *Genetics*, **102**, 25–8.

Anagnostakis, S. L. (1982*b*). The origin of ascogenous nuclei in *Endothia parasitica*. *Genetics*, **100**, 25–8.

Anagnostakis, S. L. (1982*c*). Biological control of chestnut blight. *Science*, **215**, 466–71.

Anagnostakis, S. L. (1983). Conversion to curative morphology in *Endothia parasitica* and restriction by vegetative compatibility. *Mycologia*, **75**, 777–80.

Anagnostakis, S. L. & Day, P. R. (1979). Hypovirulence conversion in *Endothia parasitica*. *Phytopathology*, **69**, 1226–9.

Anagnostakis, S. L. & Waggoner, P. W. (1981). Hypovirulence, vegetative incompatibility, and the growth of cankers of chestnut blight. *Phytopathology*, **71**, 1198–1202.

Anderson, P. J. (1914). *The Morphology and Life History of the Chestnut Blight Fungus.* The Commission for the Investigation and Control of the Chestnut Tree Blight Disease in Pennsylvania Bulletin 7, State Printer, Harrisburg, Pennsylvania. 44 pp.

Andes, S. O. (1961). Cultural variation in *Endothia parasitica. Phytopathology*, **51,** 808.

Bazzigher, G., Kanzler, E. & Kübler, T. (1981). Irreversible Pathogenitätsverminderung bei *Endothia parasitica* durch übertragbare Hypovirulenz. *European Journal of Forest Pathology*, **11,** 358–69.

Bonifacio, A. & Turchetti, T. (1973). Differenze morphologiche e fisiologiche in isolati di *Endothia parasitica* (Murr.) And. *Annali Accademia Italiana di Scienze Forestali*, **2,** 111–31.

CMI and Fed. Br. Pl. Path. (1973). *Guide to the Use of Terms in Plant Pathology*. Phytopathological paper #17, C.A.B.

Day, P. R. (1974). *Genetics of Host–Parasite Interaction*. San Francisco: W. H. Freeman and Co.

Esser, K. (1971). Breeding systems in fungi and their significance for genetic recombination. *Molecular and General Genetics*, **110,** 86–100.

Grente, J. (1981). Les variants hypovirulents de l'*Endothia parasitica* et la lutte biologique contra le chancre du châtaignier. Ph.D. Thesis, Université de Bretagne Occidentale, Brest, France. 195 pp.

Grente, J. & Sauret, S. (1969). L'hypovirulence exclusive, phénomène original in pathologie végétale. *Compte Rendu Hebdomadaire des Seánces de l'Académie des Sciences, Paris, Série D,* **268,** 2347–50.

Jaynes, R. A. & Elliston, J. E. (1980). Pathogenicity and canker control by mixtures of hypovirulent strains of *Endothia parasitica* in American chestnut. *Phytopathology*, **70,** 453–6.

Jaynes, R. A. & Elliston, J. E. (1982). Hypovirulent isolates of *Endothia parasitica* associated with large American chestnut trees. *Plant Disease*, **66,** 769–72.

Palenzona, M. (1978). *Programma di studio ed interventi per la utilizzazione, la rigenerazione e la trasformazione delle foreste Castanili Piemontesi.* Technical report, Instituto Nazionale per Plante de Legno 'Giacomo Piccarolo', Corso Casale 476, Torino, Italy.

Perkins, D. D. & Barry, E. G. (1977). The cytogenetics of *Neurospora.* In *Advances in Genetics*, ed. E. W. Caspari, pp. 133–285. New York: Academic Press.

Rankin, W. H. (1914). Field studies on the *Endothia* canker of chestnut in New York state. *Phytopathology*, **4,** 233–61.

Sansome, E. (1970). Selfing as a possible cause of disturbed ratios in *Phytophthora* crosses. *Transactions of the British Mycological Society*, **54,** 101–7.

23

The biological consequences of the individualistic mycelium

A. D. M. RAYNER*, D. COATES*, A. M. AINSWORTH*, T. J. H. ADAMS†, E. N. D. WILLIAMS† and N. K. TODD†

** School of Biological Sciences, University of Bath, Claverton Down, Bath BA2 7AY, UK and †Department of Biological Sciences, University of Exeter, Washington Singer Laboratories, Perry Road, Exeter EX4 4PS, UK*

Introduction – somatic incompatibility and fungal individualism

As will be apparent from other contributions to this Symposium volume, sexual and vegetative development within species of higher fungi (Ascomycotina and Basidiomycotina) are affected by two distinct types of incompatibility which serve contrasting, but complementary, roles. Homogenic incompatibility (which more logically ought to be called *heterogenic compatibility*) promotes zygote formation between individuals with unlike, but not like, mating-type alleles, and therefore results in outcrossing between genetically different lines. Heterogenic incompatibility (*sensu* Esser & Blaich, 1973) may be either sexual in consequence, preventing zygote formation, or somatic (= vegetative), preventing somatic integration, between genetically non-alike individuals. However, as will be explained later, somatic reactions attributed to sexual heterogenic incompatibility may at least sometimes best be regarded as the consequence rather than cause of failure of sexual conjugation.

The widespread occurrence of somatic incompatibility mechanisms amongst higher fungi has been a consistent theme in this volume (see Chapters 1, 5, 9, 11, 17, 20, 21, 22) as well as the subject of several recent reviews (Rayner & Todd, 1979, 1982; Todd & Rayner, 1980; Lane, 1981), and it must now be recognised that collectivism and heterokaryosis between genetically different mycelia may have a much more restricted significance in natural populations than has previously been supposed (see also Gregory: Chapter 1). The value of somatic incompatibility studies as a practical tool in elucidating population structure, dynamics, origins and vulnerability of specific, economically-important fungi has already been emphasised by Anderson (Chapter

17), Brasier (Chapter 21) and Anagnostakis (Chapter 22). The aim of this chapter is to explore, in more general terms, the fundamental biological implications of intraspecific somatic incompatibility in relation to the behaviour of mycelial systems. First, however, it is necessary to obtain an overview of what is known of the occurrence, timing and mechanisms of the phenomenon (for further information see the reviews listed above).

Occurrence and timing

Manifestation in culture. As has been seen in other chapters, the most direct manifestation of somatic incompatibility in culture is the development of a reaction zone between paired mycelia of different genetic origin, which maintains them clearly demarcated from one another (see Chapter 5: Fig. 4; Chapter 17: Figs. 1, 2; Chapter 21: Fig. 3; Chapter 22: Fig. 1). Typically this reaction zone contains relatively sparse mycelium, with evidence of hyphal disruption, often associated with production of pigment within the hyphae and/or medium. However, certain forms of somatic incompatibility (e.g. heterokaryon incompatibility in *Aspergillus nidulans* and *Cochliobolus heterostrophus*, see Croft & Dales: Chapter 20; Leach & Yoder, 1983) may sometimes occur in the absence of a visible mycelial interaction, and vice versa. Also, the reaction is very variable, both within and between species, notably in relation to width of the interaction zone, intensity and hue of pigment production, and development of aerial mycelium on either side of, overarching, or even within the interaction zone. In some cases such aerial mycelium may be heterokaryotic, but unless mechanisms exist (see later) to maintain its stability and allow 'escape' from the interaction zone, this may not be capable of an independent existence. In pairings between different sibs (ascospore-derived progeny from the same perithecial stroma) of *Hypoxylon serpens*, a ridge of dense white mycelium characteristically developed in the interaction zone. This ridge widened very gradually, and as it did so, mycelium of the original isolates was apparently destroyed – even where pigmented crust-like patches had been formed. Outgrowth following subculture of strips cut from across the interaction zone was very slow and dense from the region of white mycelium, but relatively rapid from non-replaced regions (Fig. 1). The dense mycelium was unstable on subculture, reverting to one or other of the original homokaryotic types. Similar, less dramatic developments occur in sib-pairings of other xylariaceous fungi (Dowson, 1982), and the dense mycelium may, or may not, be analogous to the heterokaryotic tufts which have been described for *Rhizoctonia solani* (Anderson: Chapter 17) and the ascomycete *Pyricularia oryzae* (Fatemi & Nelson, 1977). Presently little information is available to account for these variations in expression of somatic incompatibility except in so far as they may result from the interplay of different genetic loci (see Brasier: Chapter 21).

Whatever the pattern of expression of somatic incompatibility, a common feature, in so far as the relevant evidence is available, is that it appears to restrict exchange of materials and organelles between the participants, consistent with maintenance of their functional and genetic integrity as discrete individuals. Since the most direct assault on the integrity of a mycelial individual is via its capacity to anastomose with

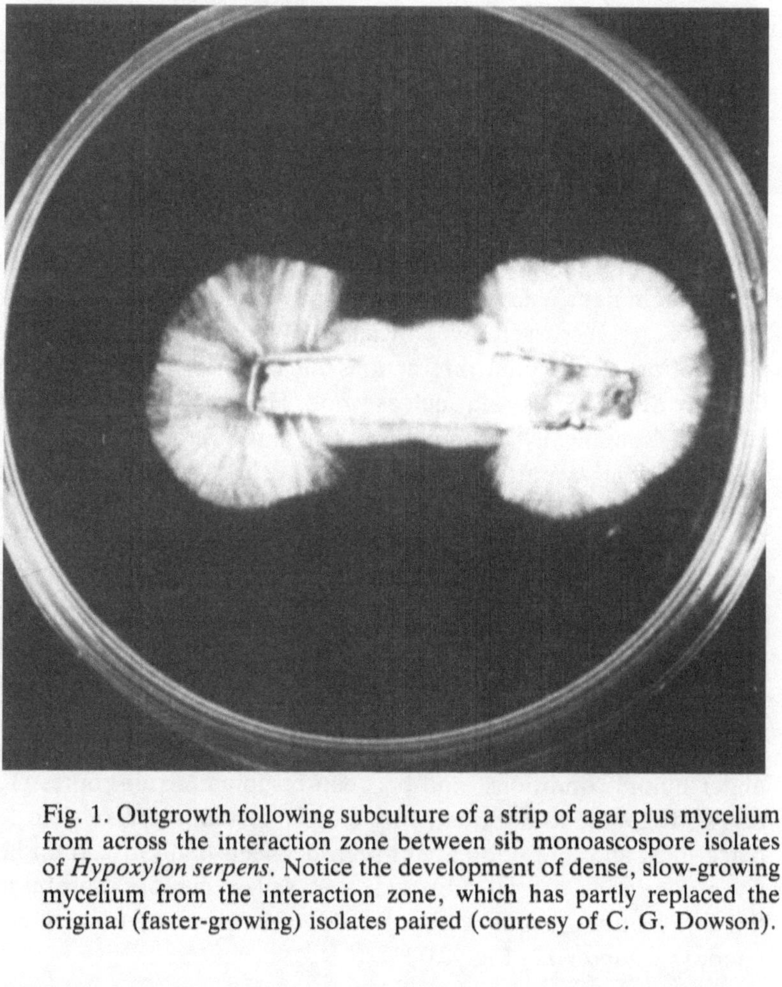

Fig. 1. Outgrowth following subculture of a strip of agar plus mycelium from across the interaction zone between sib monoascospore isolates of *Hypoxylon serpens*. Notice the development of dense, slow-growing mycelium from the interaction zone, which has partly replaced the original (faster-growing) isolates paired (courtesy of C. G. Dowson).

others, it is not surprising that somatic incompatibility reactions have been reported only for higher fungi, which have this capacity for anastomosis. Furthermore, in those cases which have been adequately examined, the events are directly associated with hyphal fusion. Usually, it seems, a post-fusion event is involved. This may directly follow fusion, resulting in rapid death of the compartments in continuity, as with *Rhizoctonia solani* (Anderson: Chapter 17), *Sclerotium rolfsii* (Punja & Grogan, 1983*a,b*) and *Phanerochaete velutina* (Phipps, 1980). Alternatively, it may follow, or be associated with, abnormal development of fusion segments, such as the spindle cells in *Coriolus versicolor* (Aylmore & Todd: Chapter 4) and moniliform development in *Piptoporus betulinus* (Adams, Todd & Rayner, 1981). However, caution is necessary here, since if the nucleus replacement reaction described by Aylmore & Todd (Chapter 4) predominates in fusions between genotypes, expression of somatic incompatibility may either be delayed, or restricted to relatively few fusion segments; this could account for the surprisingly normal appearance of many of the hyphae which is often a feature of interaction zones.

Manifestation in nature. Despite fairly frequent observations of somatic incompatibility phenomena in culture, demonstrations of their actual operation in nature appear to have been virtually absent until relatively recently. As discussed by Rayner & Webber (Chapter 18), direct observation of mycelial interactions in nature is facilitated in bulky, spatially defined substrata such as wood. Here, the boundaries between adjacent, somatically incompatible individuals of decay fungi are often readily visible as narrow, relatively undecayed regions of a different colour (usually darker) from the adjacent decay, and appearing as 'lines' ('interaction zone lines') in cross-section. By examining the distribution of interaction zone lines, it is possible to map, in three dimensions, the spatial distribution of individuals; and by using this to direct sampling points for isolation into culture, it becomes possible to investigate the spatial and genetic structure of populations simultaneously (Rayner & Todd, 1979, 1982). Somatic rejection can also often be observed between mycelial mats which develop over the surfaces of substrata under humid conditions, and between resupinate fruit-bodies (Rayner & Todd, 1982), as well as being almost certainly responsible for the black lines and inhibition zones seen between thalli of certain lichens. Even in relatively small substrata, such as leaf laminae, careful inspection will often reveal the presence of apparent interaction zones (Boddy: Chapter 12, Fig. 1d).

Timing in the life cycle. To understand its significance, it is crucial to know when somatic incompatibility occurs in relation to life cycle events. In Ascomycotina and ascomycetous Deuteromycotina somatic incompatibility is expressed directly between homokaryons, that is, mycelia derived from single ascospores or homokaryotic conidia, regardless of whether or not the species concerned is sexually self-fertile (homomictic/primary homothallic) or outcrossing (heteromictic/heterothallic) (see Croft & Dales: Chapter 20; Brasier: Chapter 21; Anagnostakis: Chapter 22). These fungi do not produce an independent secondary mycelial phase as an integral part of their sexual cycle, and also reproduce sexually predominantly via sex organs or cells (ascogonia, trichogynes, antheridia, spermatia, conidia), rather than unspecialised somatic hyphae (although some Ascomycotina have been reported to be somatogamous, evidence, or otherwise, of somatic incompatibility patterns in them is urgently required).

By contrast, in sexually outcrossing populations of Basidiomycotina, production, via somatic fusion between homokaryons, of a secondary mycelial phase (typically a dikaryon) *is* an essential feature of the life cycle (e.g. Nguyen & Niederpruem: Chapter 4). Here, somatic incompatibility is typically readily expressed between genetically different secondary mycelia, but *not* or *only transiently*, between homokaryons with complementary mating type alleles. On occasions, however, somatic incompatibility reactions may be expressed between homokaryons with common mating-type alleles (i.e. homogenically incompatible combinations) – providing they are different at the requisite genetic loci, and even between homogenically compatible homokaryons. Furthermore, as will be emphasised later, in non-outcrossing species or populations, rejection phenomena are expressed directly between different, homokaryotic, clonal lines.

Mechanisms

Genetic mechanisms. Probable genetic mechanisms underlying somatic incompatibility in Ascomycotina have been referred to elsewhere in this volume (see Croft & Dales: Chapter 20; Brasier: Chapter 21; Anagnostakis: Chapter 22). In summary, from the few examples available, they appear predominantly to be polygenic, and although they may sometimes be non-allelic, there is evidence for at least bi-allelic and probably multi-allelic incompatibility loci.

In Basidiomycotina, the virtually invariable occurrence of incompatibility between different secondary mycelia from outcrossing populations of common species, such as *Coriolus versicolor* and *Stereum hirsutum*, suggests that control must be either polygenic, multi-allelic or both. However, in our attempts to resolve the genetic loci responsible, several features have complicated analysis, notably the tendency for expression of somatic incompatibility primarily between secondary mycelia as opposed to homokaryons, and variability of the response. Thus, pairings between sib-composed, sib-related dikaryons (i.e. dikaryons synthesised from sexually compatible basidiospore progeny from the same basidiocarp) of tetrapolar species, such as *Coriolus versicolor* and *Flammulina velutipes*, showed no obvious pattern in the occurrence of somatic incompatibility pointing to simple genetic control. They showed instead a decrease in the intensity of the response compared with that seen between unrelated dikaryons, indicative of polygenic control (cf. Brasier: Chapter 21). Likewise, observations of somatic incompatibility between homogenically incompatible homokaryons, have as yet mostly not yielded any simple patterns (cf. Brasier: Chapter 21). However, in *Piptoporus betulinus*, which is bipolar, we have obtained evidence for the involvement of a single, major, multi-allelic locus ('h'-factor) in the production of a ridge of mycelium developing from highly-branched aerial knots associated with the interaction zone. This ridge develops both between dikaryons, and between homogenically incompatible (but *not* sexually compatible) monokaryons. However, detection of this apparently single locus depended on pairing experiments using inbred lines (Adams, 1982), suggesting that its expression may be conditioned by other loci, and vice versa (cf. Brasier: Chapter 21).

Physiological mechanisms. Knowledge of the physiological and biochemical processes associated with somatic incompatibility is, again, limited at present. That the interaction zone has a distinctive biochemistry is intuitively obvious from its appearance, such as production of pigment, and this is reinforced by observations such as the intriguing selective attraction of fungus gnats to interaction zones on culture plates, in a variety of fungi (Fig. 2; Boddy, Coates & Rayner, 1983). The most detailed biochemical studies have been in *Podospora anserina* where hyphal fusions between strains carrying the non-allelic incompatibility genes C/D, C/E or R/V result in disintegration of the fusion compartments and production of an unpigmented interaction zone (Labarère, Bégueret & Bernet, 1974). Studies using the temperature-dependent R/V system have revealed that the disintegration is due to specific proteolytic activity resulting from *de novo* synthesis on stable messenger RNAs. Concomitant with this is the reduction of both normal polypeptide synthesis (Boucherie, Dupont & Bernet, 1981) and RNA synthesis, together with the production of a laccase exoenzyme, as a result of de-repression of a dominant repressor gene. It is postulated that a close relationship exists between certain phenoloxidase enzymes and the specific protease responsible for lysis (Boucherie & Bernet, 1977). Under normal circumstances both species of enzyme are repressed, at least until ageing allows the onset of natural cell death. It is therefore of interest that there is induced phenoloxidase activity and pigmentation following activation of somatic incompatibility in *Phellinus tremulae* (Hiroth, 1965) and *Phellinus weirii* (Li, 1981). In *Podospora*, as might be expected, addition of a protease inhibitor (β-phenyl pyruvic acid) to the medium appears to prevent lysis, and similar observations have been made in *Coriolus versicolor* (R. Aylmore, personal communication).

Biological consequences

The inference from the above discussions, is that somatic incompatibility serves to delimit somatic individuals from one another in natural populations of higher fungi, but does not, in itself, *necessarily*

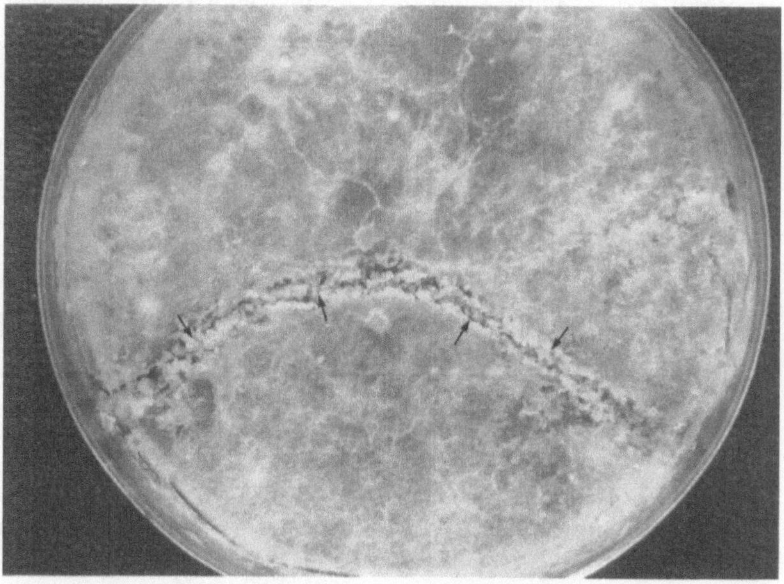

Fig. 2. Selective grazing by fungus gnat larvae of a somatic-incompatibility zone (arrowed) between two isolates of *Phlebia rufa*. (From Boddy, Coates & Rayner, 1983.)

preclude sexual conjugation between them. In attempting to assign biological consequences to this, our approach will be to follow an interconnected sequence of questions. We will start by asking what is the outcome in immediate functional terms, then we will consider what are the socio-evolutionary consequences of this outcome, from which a variety of benefits and advantages of individualistic behaviour will become apparent. In turn this will raise the issue of mechanisms which regulate the occurrence or non-occurrence of individualism, from which the functional outcome and socio-evolutionary consequences of operation of such mechanisms can in turn be considered.

Territoriality and the spatial dynamics of populations

The immediate functional outcome of somatic incompatibility is that the space-invading higher mycelial fungi emerge as highly territorial organisms, in which access to domain captured by one individual in a population is denied to another. This has fundamental implications for the way in which populations of these organisms become established and function in resource utilisation.

Individualism and population establishment

Modes of arrival and invasion. Associated with their invasiveness of solid or semi-solid substrata, a crucial, yet curiously neglected phase in the development of mycelial fungi involves establishment of individuals at or from a resource surface. In very real terms, the surface of a suitable substratum is the site of a barrier, characterised by prevailingly hostile competitive and abiotic conditions, which must be breached if access to the relative security of the interior is to be gained. This was recognised by Garrett (1960, 1970) in his concept of *inoculum potential* as the energy of growth of a fungus available for invasion at the surface to be invaded (see also Jennings, 1982: Chapter 7; Frankland: Chapter 11).

Important within this context is the extent to which synergism is possible between separate colonising units. Here it is necessary to distinguish three basic modes of arrival at a surface: genetically identical propagules; genetically different propagules (e.g. ascospores or basidiospores arising from heterozygotic ascus or basidium initials); vegetative mycelium. Whilst synergism may be possible between genetically identical colonisation units, antagonism may be expected either directly (as between homokaryons of Ascomycotina) or indirectly (as between secondary mycelia arising from sexually compatible homokaryons of

Basidiomycotina) between genetically different ones. Furthermore, not only is synergism possible between hyphae in a mycelial inoculum, but as has been explained by other authors (e.g. Jennings: Chapter 7; Watkinson: Chapter 8; Thompson: Chapter 9; Read: Chapter 10) the mycelium itself may be capable of bringing in via translocation vital resources such as nutrients and water, obviating the requirement to absorb these from the immediate environment during establishment. The likely outcome of these features is exemplified by differences between patterns resulting from air- and soil-borne colonisation of cut beech logs placed upright, and partly buried in the ground (Coates, 1984; see Rayner & Webber: Chapter 18, Figs. 10, 11). Colonisation of the base by mycelial-cord-forming decay fungi, such as *Tricholomopsis platyphylla* resulted in occupation of a considerable volume by a single individual. By contrast, airborne colonisation by decay fungi such as *Coriolus versicolor, Stereum hirsutum* and *Xylaria hypoxylon* typically resulted in establishment of numerous individuals. A further interesting feature of the latter fungi was the tendency for certain of their decay columns to expand in width with distance from the surface, as though a select few individuals with a positional advantage had 'escaped' the intense competition operating near the surface, attaining dominance lower down. This relates to the suggestion that temporary retardation of canker development due to *Endothia parasitica* following inoculation of curative strains may be explained by initially increased incidence of somatic incompatibility followed by renewed outgrowth from peripheral regions (Anagnostakis: Chapter 22).

Another important consideration is the degree to which selective stresses may govern entry into a substratum. A generalisation which is emerging concerning wood decay fungi is that those colonising the standing tree often show marked preferences for particular tree species, coincident with which the fungi form relatively few (often only one) very extensive individuals in any one trunk or branch, for example, *Phaeolus schweinitzii* in sitka spruce (Barrett & Uscuplic, 1971), *Daldinia concentrica* in ash (Dowson, 1982) and *Piptoporus betulinus* in birch (Adams, 1982). This contrasts with the behaviour of ubiquitous airborne fungi, such as *S. hirsutum* and *C. versicolor*, which are common on broken or cut timber, and often form numerous individuals within a single resource unit.

As has been mentioned by Boddy (Chapter 12), one feature of the large individuals in standing trees is that they often appear to form more rapidly than can be explained simply in terms of mycelial invasion from a colonisation court. This gives rise to the concept of latent invasion mechanisms (see Cooke & Rayner, 1984), whereby a fungus adopts a strategy of producing colonisation units or 'modules' (buds, conidia, mycelial fragments etc.), under conditions militating against mycelial development, reverting to mycelial development from such units when the stress conditions are alleviated. Synergism may then be possible between genetically identical separate modules, which combine to produce an extensive mycelial individual and hence highly effective primary resource capture. Fungi causing systemic infections of plants and animals probably frequently behave in this way (Cooke & Rayner, 1984), and the significance of the mycelial-yeast dimorphism which they often exhibit can readily be appreciated in this context.

Homokaryon–dikaryon transitions. The 'module-mycelium' transition is only one of several mycelial transitions, or, in the terminology of Gregory (Chapter 1), changes of mode, which occur during colonisation processes. Another well-known, but ecologically neglected, transition occurs between primary and secondary mycelia in Basidiomycotina. This is important, because, depending on its duration, the homokaryon may be principally responsible for primary resource capture. The final distribution of mutually antagonistic secondary mycelia may therefore not be so dependent on their own intrinsic properties as on the growth rates and patterns of nucleus exchange of homokaryons. One way of modelling this experimentally is to inoculate Petri plates with arrays of sexually compatible homokaryons and to examine the spatial development of the resulting secondary mycelia. Sample results for *Coriolus versicolor* are illustrated in Fig. 3. These were obtained using non-sib-related monokaryons, and differences between the areas occupied by the resultant dikaryons were relatively slight – taking into account the timing of the inoculations. However, studies using *Stereum hirsutum*, in which arrays of sib- and non-sib-related homokaryons were inoculated, revealed that the domains subsequently occupied by non-sib-composed

---

Fig. 3. Distribution of dikaryons of *Coriolus versicolor* formed following inoculation of five non-sib-related sexually compatible monokaryons onto a 14-cm diameter 3% malt agar plate. (From Williams, Todd & Rayner, 1981.) (a) Following simultaneous inoculation. (b) Following sequential, clockwise inoculation at 2-day intervals, starting with 06. The closed circles represent the monokaryon inocula, the dotted lines interaction zones formed between monokaryons, and the shaded areas the dikaryons formed.

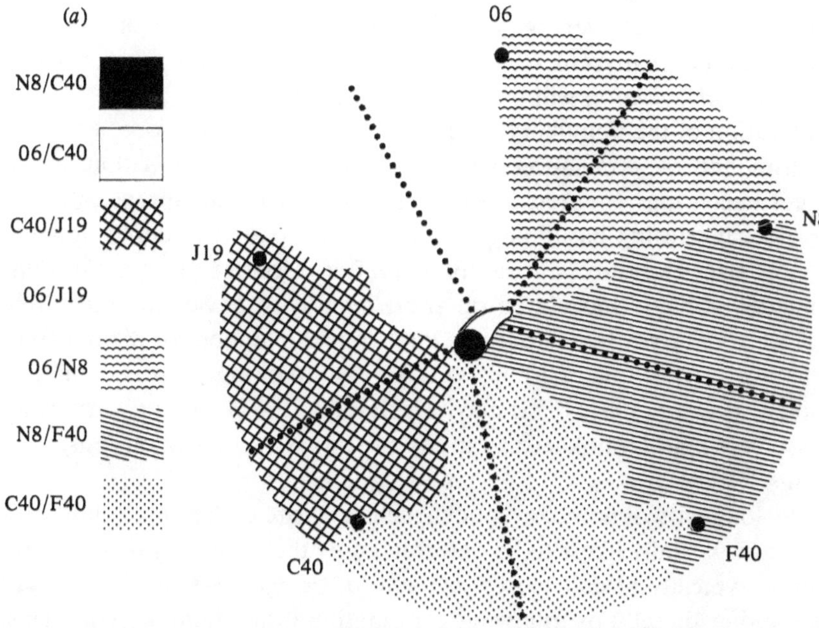
(a)
N8/C40
06/C40
C40/J19
06/J19
06/N8
N8/F40
C40/F40
06
N8
J19
C40
F40

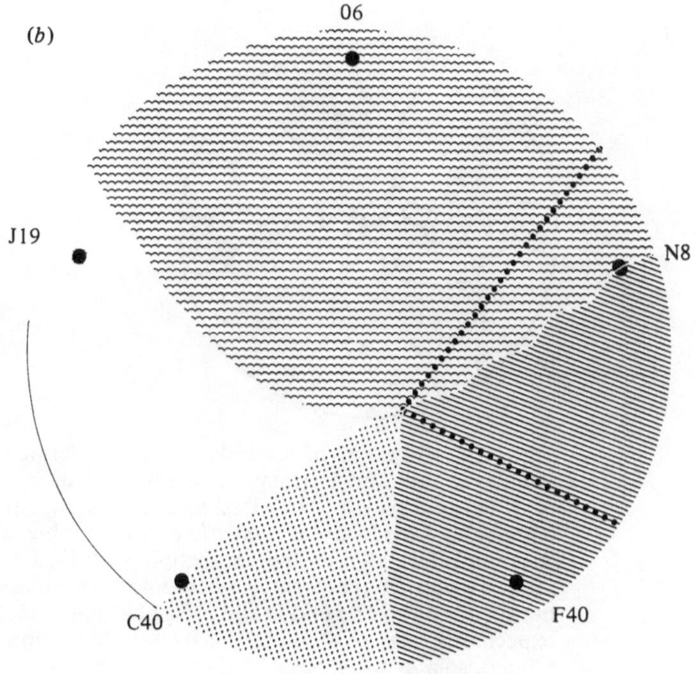
(b)
06
J19
N8
C40
F40

secondary mycelia were substantially greater than those of sib-composed secondary mycelia (Fig. 4). The possible basis and significance of such *non-sib-selection* will be returned to later.

Duration of the homokaryotic phase is obviously, therefore, a crucial determinant of population structure. Observations on colonisation of cut beech logs (Coates, 1984) indicated that homokaryons of common decay-causing basidiomycetes could be isolated at depth up to 6 months after exposure, but rarely thereafter. Monokaryons of *Coriolus versicolor* directly inoculated on short lengths of dowelling into birch logs were dikaryotised 12 months later, but the number of dikaryotisation events was low, perhaps one or two per inoculum (Williams, Todd & Rayner, 1981). This may be related to the fact that dikaryotisation of a single homokaryon rapidly precludes further dikaryotisation events.

The longevity of homokaryotic mycelia at a site under field conditions is at least partially likely to be determined by the frequency of arrival of spores, typically basidiospores, capable of bringing about dikaryotisation, either directly or following germination into a homokaryon. This idea prompted us to devise a species-specific technique using a homokaryotic mycelium as a selective surface, to investigate the dissemina-

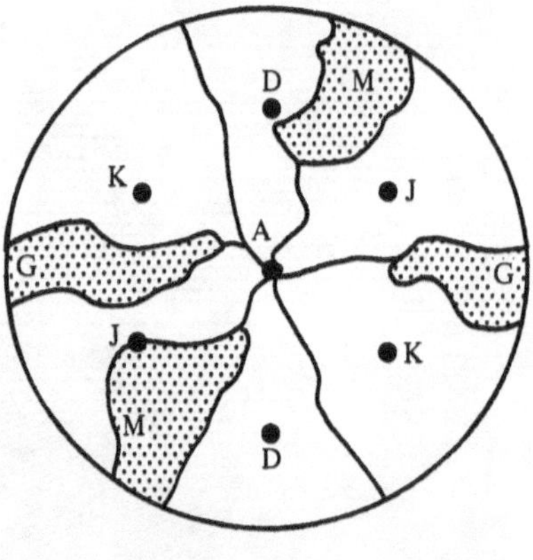

Fig. 4. Distribution of secondary mycelia formed following simultaneous equidistant inoculation of four homokaryons onto 2% malt agar. (From Coates, 1984.) D, J and K are sib-related, with J sexually compatible with D and K, which are incompatible. The central isolate A is non-sib-related to and compatible with D, J and K. The shaded areas G and M represent sib-composed secondary mycelia formed between K and J or J and D and are significantly smaller than would be expected if rates of secondary mycelium formation between non-sib-related isolates were equal to that between sibs.

tion and deposition of viable spores of heteromictic basidiomycetes (Adams, Williams, Todd & Rayner, 1984). The technique depends on the assumption that actively growing homokaryotic mycelia will limit mycelial spread from spores of another species, whereas nuclei from conspecific spores, providing they are of appropriate mating-type, are able to enter, divide and migrate through the mycelium to establish a secondary mycelium whose presence can be verified by simple tests (such as, in appropriate cases, presence of clamp-connections). The particular procedure we followed (refinements are possible) involved the use of square Petri-dishes, 10 cm × 10 cm, subdivided into 25 chambers, each 2 cm × 2 cm. Each chamber was partly filled with 1 ml of sterile 2.5% malt agar and inoculated with a small piece of a homokaryotic culture. After incubation until mycelium just covered the medium in each chamber, the dishes were taken to the field site and exposed to the atmosphere. After exposure they were incubated for a few days before examination of the mycelium in each compartment to detect whether or not it had been dikaryotised. In one instance, working with *Coriolus versicolor*, virtually all compartments were dikaryotised following a 6-h exposure during which a shower of rain occurred (Williams, Todd & Rayner, 1984).

Effects on resource utilisation

The development of numerous, mutually antagonistic individuals within a resource might well be expected to inhibit the overall performance of the population in utilisation of the resource. That this is indeed probable, has been indicated by studies of cut beech logs (Coates, 1984; see above), whose aerially exposed cut surfaces were inoculated with a large number (*ca* 10^3) of basidiospores of *Coriolus versicolor, Bjerkandera adusta, Stereum hirsutum* or *Hypholoma fasciculare.* Colonisation of these logs was compared with that occurring on ones which were allowed to become colonised naturally, and others for which the upper surface had been re-cut several times allowing build up of an artificially high surface inoculum load. The typical appearance of each of these three types of log two years after cutting is illustrated in Fig. 5. In all cases both decay and production of basidiocarps of decay fungi were markedly inhibited in inoculated logs, as opposed to ones colonised naturally, with the re-cut ones intermediate between the two. Isolations onto selective media from stained wood near the surface of inoculated logs often yielded numerous antagonistic individuals from very small volumes of wood. In a laboratory experiment (Coates, 1984) cut sections of beech inoculated with arrays of mycelium-plus-agar discs containing secondary mycelia of *Stereum hirsutum* of different genotype, decayed significantly less than ones inoculated with similar arrays of the same genotype.

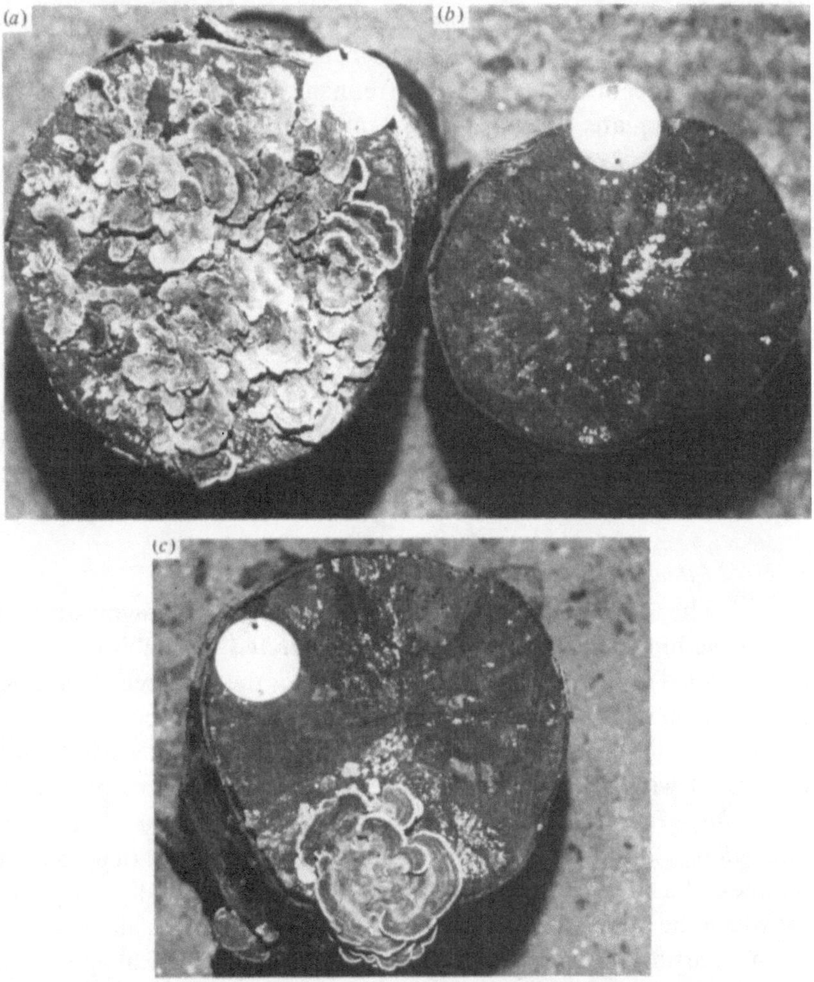

Fig. 5. Typical appearance, two years after cutting, of the upper surface of beech logs placed upright with their bases buried in the ground at a woodland site. (After Coates, 1983.) (a) Allowed to become colonised naturally. (b) Inoculated with 10^3 basidiospores of a common decay species. (c) The surface was recut three times during the first 12 weeks of exposure, presumably allowing an artificially high inoculum load to develop.

Relationships between resource size and distribution, territoriality, and reproductive behaviour

The previous section draws attention to the important, but complex interrelationship which exists between mycelial domain, resource pool and reproductive commitment. Put at its simplest, the greater the resource pool on which a mycelium can draw, the greater, in energy terms, can be its commitment to reproduction. In turn, the size of the resource pool will be affected, at least partially, by the size of the domain occupied by the mycelium, which in turn will be affected by the frequency and outcome of interactions with its neighbours. It follows that an approximate idea of mycelial distribution and resource relationships may be gained simply by observation of the characteristics of reproductive stages. Thus one can argue that it is no coincidence that amongst wood-decay fungi, those causing heart-rots, such as various *Ganoderma, Fomes* and *Phellinus* species, should form some of the most substantial basidiocarps, for here a single individual, may, as already mentioned, occupy substantial volumes of the whole or part of the tree. By contrast, fungi such as *Coriolus versicolor* and *Stereum hirsutum* typically colonise felled, broken or fallen timber under less-selective conditions than in the standing tree, but in the presence of larger numbers of competitive individuals of the same and different species, and form much smaller basidiocarps. Amongst leaf-litter decomposers, non-component-restricted species of basidiomycetes, that is ones forming large diffuse-spreading mycelia, form either substantial numbers of basidiocarps, or as in large *Clitocybe* species such as *C. nebularis* and *C. geotropa*, ones of considerable size. At the other end of the spectrum are some of the minute agarics of component-restricted species of *Mycena* and *Marasmius* such as those mentioned by Boddy (Chapter 12) respectively inhabiting the laminae and petioles of beech leaves. In this connection it is worth noting the general principle, mentioned by Rayner & Webber (Chapter 18) that because of the proportionate increase in surface to volume ratio with decrease in substratum size, similar numbers of mycelial individuals may often be present in substrata of markedly different sizes.

In order to establish a general point, the above discussion has been deliberately simplistic, and it must be recognised that the issues are complicated by a variety of factors. Thus the relation between mycelial domain and resource pool is not necessarily absolute or direct; the resource pool available to a mycelium is also dependent on ability to assimilate available substrates – including those which may be of a refractory nature – ability to mobilise resources and the nature of the resource supply. With respect to the latter, if nutrient resources are continually brought to a mycelium, rather than having to be sought out – as might apply in the case of certain mycorrhizal fungi for example – then considerable commitment to reproduction may be possible without development of extensive vegetative mycelium (cf. Jennings: Chapter 7; Read: Chapter 10). Here too the reciprocal relationship between vegetative and reproductive development must be remembered, in that one activity often occurs at the expense of the other (cf. Raudaskoski: Chapter 13; Lysek: Chapter 14). Thus transition to the reproductive mode often occurs after completion of phases of primary resource capture (see Rayner & Webber: Chapter 18) either because of rhythmic growth, attainment of the boundary of a substratum (Lysek: Chapter 14) or contact with an antagonistic neighbour. In culture, stimulation of reproductive differentiation is often associated with expression of somatic incompatibility, as in the production of pycnidia in interaction zones by *Endothia parasitica* (Anagnostakis: Chapter 22, Fig. 1c) and of synnemata and perithecia in *Ceratocystis ulmi*, where the increase in fertility with increased intensity of the interactions is of particular interest (Brasier: Chapter 21). Underpinning all these considerations is the concept of *r*- and *K*-strategies for mycelial fungi (see Rayner & Webber: Chapter 18).

Socio-evolutionary consequences of individualism and collectivism

Individualism versus collectivism

Many mycologists, brought up to the tradition of widespread and unrestricted heterokaryosis and mosaic mycelium formation (see Gregory: Chapter 1), and the supposed advantages they confer, are perplexed by somatic incompatibility in fungi, seeing in it only the negation of potentially harmonious relationships. By the same token, general ecologists and evolutionists, having ideas of selfish genes, kin-selection and the operation of selection pressures on individuals rather than groups, would be amazed by the widespread collectivism which has been proposed for fungi.

Here it is worth reflecting that collectivism (concerted action between individuals) and individualism (independent action of individuals) represent not only the extremes of the human socio-political dilemma, but also consistent themes operating at every level of biological organisation. An effective statement of this principle is found in the Open

Hierarchical Systems theory of Koestler (1976). Amongst other things, this holds that

> 'the organism is to be regarded as a multi-levelled hierarchy of semi-autonomous sub-wholes, branching into sub-wholes of a lower order, and so on. Sub-wholes on any level of the hierarchy are referred to as *holons* . . . Every holon has the dual tendency (*Janus Effect*) to preserve and assert its individuality as a quasi-autonomous whole, and to function as an integrated part of an existing or evolving larger whole. This polarity between the Self-Assertive and Integrative tendencies is inherent in the concept of hierarchic order; and a universal characteristic of life'.

An *indication* of some of the biologically important consequences of individualism and collectivism *within a species* (mutualistic and unilateral symbioses between different species are outside the scope of this article) is provided in Table 1. No doubt the list is incomplete, but it emphasises the conflicting roles of the two strategies. A consistent outcome of this conflict in eukaryotic populations is the maintenance of somatic individualism between different genotypes via rejection mechanisms which can operate at *cellular, tissue*, or *whole organism* (*behavioural*) levels. Somatic collectivism is, by contrast, operated typically between genetically identical or related individuals (kin-selection) – herein the significance of Gregory's comment (Chapter 1) that 'membership of the co-op is restricted to members of the clone'. The major occasions on which collectivism between *genetically* non-alike individuals comes into play in eukaryote populations undoubtedly occur in connection with sexual reproduction.

The override imperative

Sexual reproduction, then, represents the major occasion on which the individualistic response to reject is brought into direct challenge with the collectivistic response to accept. In many instances, the conflict may be minimised via production of specialised sexual nuclei, cells or organs which mediate the collectivistic act. However, in other cases, interaction between somatic structures is partially or wholly involved during sexual reproductive processes, and for the sake of argument we shall describe these as 'somato-sexual' interactions. A feature of somato-sexual interactions is that the degree to which acceptance is achieved is determined by the extent to which rejection is

Table 1. *Socio-evolutionary consequences of individualism and collectivism in relation to some important biological issues*

Issue	Collectivism	Individualism
Variation and natural selection	Between different genotypes allows complementation and flexibility. Loss of resolution of selective pressures. Masking of recessive by dominant traits. Reproduction of some components at expense of others. Destabilisation of favourable gene combinations	Selection acts directly on individuals. Limitation of complementation and flexibility. Stabilisation and maintenance of favourable gene combinations
Resource capture and utilisation	Synergism and sharing. Loss of collective if resources are inadequate	Independent capture and utilisation. Direct feedback on population size
Susceptibility to adverse influences	'Safety in numbers' – e.g. against predators; but collectives make larger targets. Control of external environment and development of an internal one – homoeostasis. Susceptibility to transmissible 'internal' pathogens – e.g. viruses and toxins	Increased exposure and susceptibility to biotic and abiotic external forces. Lessened susceptibility to internal pathogens
Differentiation and division of labour	Possible, leading to greater 'efficiency', but with loss of functional independence of the parts	Impossible – self-sufficiency imperative

expressed, and vice versa. In other words, acceptance involves *override* of rejection, or vice versa.

Principles and consequences. There are numerous examples of somatosexual interactions amongst eukaryotes (Table 2) and examination of what is known about these indicates a number of apparent generalisations. The latter include the fact that the recognition systems which mediate rejection or acceptance of non-self appear often to be based on multi-allelic loci; that where a 'choice' is available, non-sib partners are

Table 2. *Examples of somato-sexual interactions amongst eukaryotes, and some recent references*

Fertilisation in colonial tunicates (Scofield, Schlumpberger, West & Weissman, 1982)
Myxomycete plasmogamy (Collins, 1979)
Conjugation in algae (Wiese & Wiese, 1977)
Conjugation in ciliates (Miyake, 1974)
Pollen-stigma-style interactions (Pandey, 1981)
Mate selection in birds, mammals and insects (Getz, 1981)
Courtship behaviour (Bateson, 1983)
Materno-foetal interactions in mammals (Faulk *et al.*, 1978)
Secondary mycelium formation in Basidiomycotina

often favoured over related ones (non-sib-selection); that there is an increased tendency to reject between 'aged' or 'differentiated' systems and a corresponding tendency for acceptance between 'undifferentiated' or 'juvenile' systems.

A very important feature is that both acceptance and rejection may effectively be operated by the same trigger, genetic difference between self and non-self, with the outcome that they are in a delicate balance with one another. Relatively small genetic or physiological differences or changes may then be sufficient to swing the balance away from one mechanism towards the other. It is thus possible to envisage, if the tendencies to reject and accept are *not in phase*, that with increasing genetic difference between participants, the balance between acceptance and rejection may *alternate* through one or more cycles, as shown in Fig. 6. The particular model illustrated is based on the possibility that at approaching zero genetic difference there is full acceptance, even though the rate of recognition is slow, since there is insufficient difference to trigger the rejection response. With increased genetic difference, rejection becomes possible and if the rate of operation of the override mechanism allowing acceptance is slower than that of rejection, rejection will be the net result. With further genetic difference, rate of recognition and operation of acceptance become rapid, until ultimately the difference is so overwhelming as to result in complete rejection. This type of model accounts for rejection of sib and non-sib preference, and explains several otherwise puzzling observations, a notable example being the problem of repeated abortions in couples found to have *common* HLA antigens (Komlos, Zamir, Joshua & Halbrecht, 1977). Such abortions are believed to be due possibly to the mother's immune system rejecting foreign trophoblast antigens on the blastocyst (Faulk *et al.*, 1978). Such rejection is believed to be prevented (overridden) during normal pregnancies by a class of antibodies (blocking factors) produced in response to recognition of trophoblasts invading the maternal circulation, and it has been suggested that failure of such non-self recognition leads to non-production of blocking factors and consequent immune rejection of the foetus (Faulk *et al.*, 1978). Accordingly, a novel treatment prescribed involved transfusions of incompatible white cells during pregnancy. In one case this resulted in the birth of a healthy daughter to a woman who had previously suffered ten consecutive miscarriages (Taylor & Faulk, 1981).

Override of somatic incompatibility in Basidiomycotina

Somatogamy and multi-allelic mating systems

Summarising, heteromictic Basidiomycotina (hemibasidiomycetes excepted) are unique amongst Eumycota in possessing a mating (= homogenic incompatibility) system based on multi-allelic rather than bi-allelic genetic loci (diaphoromixis as opposed to dimixis; see Burnett, 1975, 1976). They are further unique in that outcrossing is dependent on *somatic fusion* and establishment of a stable secondary mycelium between genetically different homokaryons; this is a somato-sexual interaction. Somatic incompatibility is typically evident between different secondary mycelia – even where these contain a common nucleus type – and between certain homogenically incompatible homokaryons.

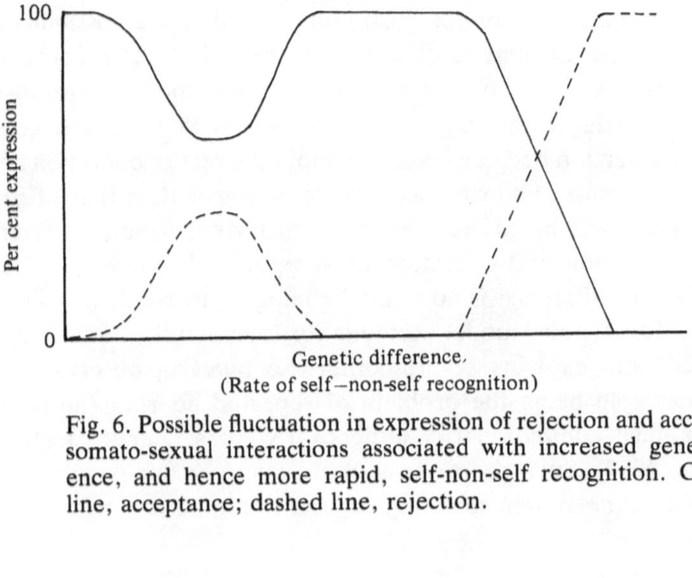

Fig. 6. Possible fluctuation in expression of rejection and acceptance in somato-sexual interactions associated with increased genetic difference, and hence more rapid, self-non-self recognition. Continuous line, acceptance; dashed line, rejection.

However, somatic incompatibility is not *usually* expressed between mating-type-compatible homokaryons (see below). This suggests that an *override* system, based on non-self recognition and mediated by multi-allelic genetic loci, allows acceptance between sexually compatible homokaryons and obviates the latent rejection response between them. To us, this hypothesis also provides a basis for understanding a wide range of other facts and observations concerning Basidiomycotina. These include the diverse patterns of interaction and nucleus exchange between homokaryons and between homokaryons and secondary mycelia, evolutionary links with the Ascomycotina, the relative infrequency of homomixis and the origins and consequences of reproductive isolation.

Interactions between homogenic and somatic incompatibility

At first sight, the variations, both within and between species, in expression and timing of homogenic and somatic incompatibility are perplexing. However, following studies of a range of species with various types of nucleus behaviour (Boidin, 1971; see Butler: Chapter 3), we have found that each in its idiosyncratic way conforms to a pattern of behaviour which can be predicted from the override hypothesis just mentioned.

First, let us consider the situation with *Stereum hirsutum* which seems to conform to the pattern predicted in Fig. 6. When first-generation sibs from basidiocarps collected from the field are paired in all combinations, it is common to find that a large number of pairings result in the mutually antagonistic reaction which is characteristic of somatic incompatibility between different secondary mycelia. Also pairings between mating-type complementary homokaryons sometimes fail to show a sexually compatible reaction, instead mutual antagonism or bow-tie formation (see below) occurs. By contrast, second-generation sibs (i.e. produced from a basidiocarp produced by a secondary mycelium obtained by pairing sexually compatible first-generation sibs) show less antagonism, generally interact much more slowly and less overtly, and full homogenic incompatibility is normally expressed. By the same token, non-sibs from the field are generally fully interfertile (Coates, 1984). This is the situation with the British population. However, in pairings between British and Finnish homokaryons, we have found mutual antagonism (or even replacement of the British by the Finnish) in all cases, with no interfertility; this may, however, be partly related to the fact that the Finnish population appears to be non-outcrossing (A.

M. Ainsworth, unpublished; see below). In contrast to the behaviour of *S. hirsutum*, but still readily interpretable in terms of override, is that of *Rhizoctonia solani* (Anderson: Chapter 17) where expression of heterogenic incompatibility at the expense of homogenic incompatibility appears to increase directly with increased genetic difference. At a further extreme, *Athelia (Sclerotium) rolfsii* showed expression of antagonism among a sample of 50 homokaryons from five parent-field heterokaryons of 80–92% between sibs and 86% in non-sib pairings. Associated with this was a very low incidence of secondary mycelium formation in both cases (Punja & Grogan, 1983*b*). In both *R. solani* and *A. rolfsii*, somatic incompatibility is expressed by direct lysis of fusion segments, and this rapid expression may underlie the above patterns. However, in *Phanerochaete velutina* where the same direct mechanism operates, sometimes very strikingly, between secondary mycelia, somatic incompatibility seems not to be expressed readily between homokaryons; associated with this, full interfertility between mating-type complementary strains occurs (A. M. Ainsworth, unpublished).

It is unfortunate that, with respect to the actual mechanisms whereby homogenic incompatibility functions, emphasis has been placed on *Schizophyllum commune* and *Coprinus cinereus*, both tetrapolar and with clamp-connections only on the dikaryon, as vehicles for detailed study. First, the tendency to produce inbred laboratory lines, which although valuable for genetic and biochemical analysis, can eliminate heterogeneity, and hence produce interaction patterns unlike those which might occur in strains isolated directly from nature. Secondly, and perhaps more importantly there is a very wide range of mating and nuclear behaviour amongst Basidiomycotina. Thus there may be homomixis, or homodiaphoromixis or unifactorial, bifactorial and possibly trifactorial control of homogenic incompatibility; presence or absence of clamp-connections on secondary mycelia, primary mycelia or both; and normal, subnormal, heterocytic, astatocoenocytic and holocoenocytic nuclear behaviour (see Burnett, 1976; Boidin, 1971; Kühner, 1977; Butler: Chapter 3). In order to establish whether there are any common principles underlying homogenic incompatibility mechanisms and hence possible interrelations between them, it is important to investigate, in reasonable detail, species representing a range of these different types of behaviour. Unfortunately, outside *S. commune* and *C. cinereus*, studies of mating interactions, although very extensive (Boidin, 1971; Kühner, 1977), have, with some notable exceptions, often gone no further than scoring the presence or absence of clamp-connections and detection of bi- and tetrapolarity and multiple alleles. Since the work of

Vandendries & Brodie (1933), there has, in particular, been little published material readily available concerning actual patterns of mycelial interaction, nucleus exchange and establishment of the secondary mycelium.

Kemp (1975; personal communication) has made a particular study of *Coprinus* in which there is considerable variation between species with respect to presence or absence of clamps, bipolarity, tetrapolarity, homomixis or homodiaphoromixis. He suggests that the two central functions of homogenic incompatibility relate to *nuclear migration* and *fertility*. In *C. cinereus* heterozygosity for both factors is necessary for 'full fertility', whilst nuclear migration can occur when just the B factor is heterozygous. Bipolar species such as *C. gonophyllus* and *C. congregatus* have migration and fertility controlled by a single factor. In addition there is evidence that in some cases two factors are present, one controlling migration and the other fertility; here junction-line dikaryons develop between homokaryons homozygous for the migration factor, but heterozygous for the fertility factor. In other cases (in the *C. patouillardii* group) migration is controlled by two factors, one of which also controls fertility.

We have examined the interactions of both tetrapolar species, notably *Coriolus versicolor, Flammulina velutipes* and *Pleurotus cornucopiae*, and bipolar ones, notably *Piptoporus betulinus, Phlebia radiata* and *P. rufa*. We have taken a particular interest in the genera *Stereum* (*sensu stricto*), *Phanerochaete* and *Coniophora* which exhibit holocoenocytic nuclear behaviour (Boidin, 1971). These holocoenocytic fungi were originally thought all to be 'homothallic' due to the presence of whorled clamp-connections on the larger hyphae of both monospore and polyspore cultures (Boidin, 1971). However, evidence for a unifactorial homogenic incompatibility system governing outcrossing and secondary mycelium formation has been found in *Stereum gausapatum* (Boddy & Rayner, 1982), *S. hirsutum* (Coates, Rayner & Todd, 1981) *S. rugosum* (Rayner & Turton, 1982), *Phanerochaete velutina* and *Coniophora puteana* (A. M. Ainsworth unpublished).

An interesting feature of several of the unifactorial species is that, in addition to the presence of a single multi-allelic compatibility locus controlling secondary mycelium formation, there is evidence for another multi-allelic locus, which, *if expressed*, gives rise to appressed mycelium containing hyphae with abnormal branching and/or swollen cells. In *Piptoporus betulinus* heterozygosity at this locus between paired mating-type-incompatible monokaryons leads to moniliform morphology. This is restricted to the interaction zone in the presence of heterozygous 'h'-factors (see above), but in the presence of homozygous 'h'-factors may spread unilaterally or bilaterally through the interacting colonies, associated with marked inhibition of radial extension. This seems analogous to the 'flat' reaction in common-A-factor heterokaryons of *Schizophyllum commune* (Papazian, 1950) and some other tetrapolar species, indicating a possible link between bi- and tetrapolarity.

In *Stereum hirsutum*, a remarkable reaction between certain sexually incompatible homokaryons, has been termed the 'bow-tie' and is shown in Fig. 7. It consists of a band of appressed, sparse, mycelium, widest at the ends, bounded by zones of watery droplets, and characterised by unusual hyphal branching patterns (Coates, Rayner & Todd, 1981). Crossing experiments have confirmed that this reaction is due to heterozygosity at a multi-allelic locus, which may be subdivided, and is unlinked to another locus conferring full interfertility. The shape of the bow-tie on either side of the interaction interface is often asymmetric, and sibs exhibiting the reaction can be ranked in a hierarchic series from 'net donators' to 'net acceptors' based on the degree to which they are penetrated by the bow-tie front (Coates, 1984). The reaction appears to involve migration of nuclei, in the course of which acceptor nuclei are often replaced, followed by regeneration of the donor mycelium; in this way partial or complete replacement of one genotype may occur. Similar reactions, but segregating non-independently of the mating-type locus, and resulting in differently shaped zones of appressed mycelium have been seen in *Stereum gausapatum* (Boddy & Rayner, 1982). There appear to be some strong parallels with the d-factor and 'penetration' reactions in *Ceratocystis ulmi* (Brasier: Chapter 21); whether this is due to more than coincidence remains to be seen. In *S. hirsutum*, whilst a form of nuclear migration appears to be involved in the bow-tie reaction, this occurs at fundamentally different, and more variable, rates than that following implantation of a compatible donor into an acceptor homokaryon (Coates, 1984). Observations of this type lead us to suggest that three fundamental functions are associated with secondary mycelium formation in Basidiomycotina:

> *Access migration*: Ingress of donor nuclei into an acceptor, at a rate dependent on the donor-acceptor capacities of the participants. When 'unstabilised' produces 'flat' morphologies and mutual or unilateral inhibition reactions. Probably involves R-glucanase activity and dissolution of dolipore septal swellings.

Acceptor migration: Occurs after access has been gained. Probably effected via a contractile mechanism operated entirely by the acceptor, so that different donor nuclei migrate at the same rate in a common acceptor homokaryon.

Stabilisation: Production of a stable secondary mycelium following access and/or acceptor migration, in which there is maintenance of a 1:1 nuclear ratio in dikaryotic or multikaryotic hyphal compartments.

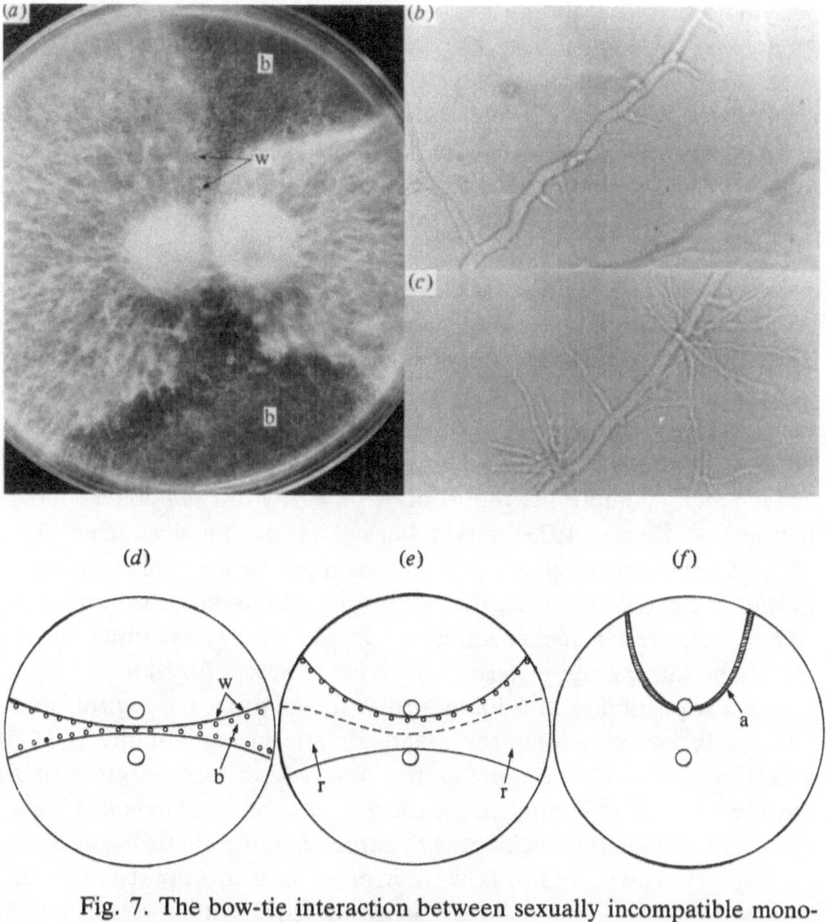

Fig. 7. The bow-tie interaction between sexually incompatible monobasidiospore isolates of *Stereum hirsutum*. (From Coates, Rayner & Todd, 1981.) (a) General appearance during early development. (b, c) Abnormal branching of hyphae in the bow-tie region. (d, e, f) Stages in the progression of a typical bow-tie interaction. a, zone of mutual antagonism (= somatic incompatibility); b, bow-tie region; r, renewed aerial growth; w, watery exudation.

The precise way in which these three functions are co-ordinated and controlled genetically will vary between species. In some tetrapolar species, such as *S. commune*, it may be that separate loci for access migration (B-factor), and acceptor migration plus stabilisation (A-factor) are required for full fertility. In certain bipolar species, such as *Stereum hirsutum* and *Piptoporus betulinus*, it may be that the absolute requirement for a separate access locus for secondary mycelium formation is obviated, whilst the locus itself may still be retained. In other bipolar species, it may be that there is no acceptor migration, so that secondary mycelium formation is dependent entirely on effective access migration plus stabilisation. This might explain the behaviour of *Mycena galopus*, where the dikaryon develops in bow-tie shaped regions (Frankland: Chapter 11) and *Rhizoctonia solani* (Anderson: Chapter 17). In the latter, heterokaryotic tufts arise from zones in which abundant mycelial knots are formed, a feature characteristic of access reactions. Often, in what seem to be fully compatible pairings, there are two rows of mycelial knots; in others there is only a single row, possibly related to unilateral exchange of nuclei or cytoplasm (N. A. Anderson; personal communication). This relates to our view that unilateral or asymmetric nuclear migration is determined by the extent of access function operated by an acceptor. Since we have identified three particular functions in secondary mycelium formation, it might be possible, theoretically, for all three to be controlled by independent genetic loci, and here the report of octopolarity in *Psathyrella coprobia* (Jurand & Kemp, 1973) might be significant. Finally, since access appears to be associated with development of unusual hyphal morphology and branching patterns, the occurrence of moniliform morphology, spindle cells and mycelial knots within somatic interaction zones can readily be interpreted in terms of restricted access function.

The last possibility also follows from the fact that the logical site for interaction between homogenic and somatic incompatibility is at the access step. Of interest here is the observation that progress of the bow-tie interaction – putatively an access reaction – in *Stereum hirsutum* is commonly halted coincident with expression of somatic incompatibility (Fig. 7). Furthermore, bow-tie expression is a property of freshly-obtained isolates, old cultures being directly mutually antagonistic. Bowtie expression is normally retained if fresh cultures are stored under oil (Coates, 1984).

In *S. hirsutum*, somatic incompatibility between homokaryons thus excludes the possibility of access between them. However, it is possible for nuclei to transgress an existing somatic incompatibility barrier, via first becoming established into a secondary mycelium, that is via a 'di-mon' interaction. We have tried several designs based on this principle of which two are illustrated in Fig. 8. Salient features are the

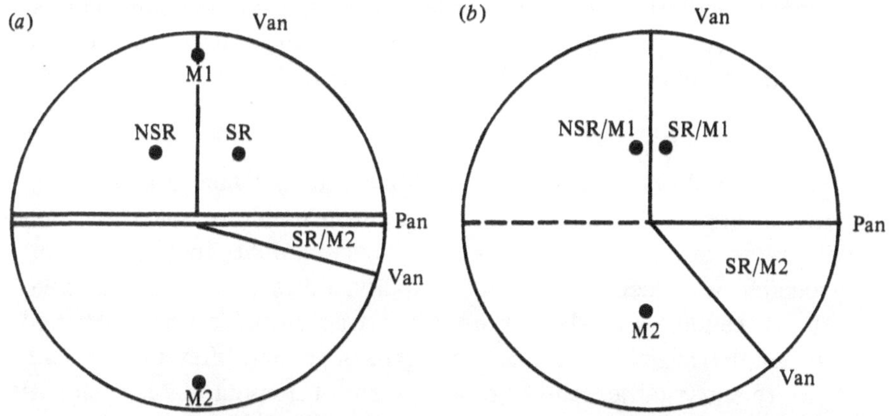

Fig. 8. Typical results of experiments with *Stereum hirsutum* designed to follow transgression of a somatic-incompatibility barrier by compatible donor nuclei into an acceptor homokaryon.(After Coates, 1984) (a) Transgression of a pre-established barrier. Two somatically and sexually incompatible homokaryon acceptors, M1 and M2 were paired opposite one another, and a pigmented zone of mutual antagonism (Pan) allowed to develop between them. Two weeks after inoculation of M1 and M2, inocula of two homokaryotic donors, respectively non-sib-related (NSR) and sib-related (SR) to and compatible with both acceptors, were inoculated into one of them. Rates of nucleus migration and patterns of secondary mycelium formation were followed on both sides of the barrier. Whilst the areas occupied by NSR and SR delimited by a zone of mutual antagonism (Van) on the inoculated side of the barrier were similar, the area occupied by SR on the uninoculated side was smaller than that containing NSR in all 39 reciprocal combinations tested. Since rates of nucleus migration within the acceptors were found to be equal, this corresponded to a relative delay in access of SR across the barrier (80 h in one example) compared with NSR at 25 °C. (b) Transgression of a developing barrier. Here secondary mycelia NSR/M1 and SR/M1 were synthesised separately and then paired directly against M2, so that the somatic incompatibility barrier putatively due to the reaction between M1 and M2 develops during the course of secondary mycelium formation. NSR again occupied a greater area beyond Pan, and associated with its access into M2, Pan formation was, in the combination studied, apparently *suppressed* relative to that associated with access of the SR. The latter occupied a greater proportion of M2 than during access across a pre-established barrier, corresponding with a lower calculated relative delay of 38 h at 25 °C.

stronger non-sib selection across a pre-established rather than developing somatic-incompatibility barrier and the apparent relative *suppression* of somatic incompatibility by migrating non-sib as opposed to sib-nuclei.

Thus the characteristic features of somato-sexual interactions (see previous section), that is multi-allelic loci, non-sib-selection and increased rejection with ageing or differentiation all seem to be present, to a greater or lesser extent, in Basidiomycotina.

Breakdown of override

It follows from the above discussion that failure of override between genetically different homokaryons capable of somatic fusion will allow development of somatic incompatibility between them, associated with lack of outcrossing ability, that is, reproductive isolation. Depending on whether or not a true sexual cycle is retained between the original lines, and the degree of genetic difference between them, this may either lead to development of non-outcrossing (clonal) subpopulations within a species, or to speciation itself.

Clonal subpopulations within a species. If outcrossing is essential for the generation of variation which fuels evolutionary progress, then the relative rarity of homomixis in the Basidiomycotina (Burnett, 1975), where outcrossing is dependent on somatogamy – and hence overridden – is readily understood (Rayner & Turton, 1982). This raises the question of the origin and maintenance of non-outcrossing within Basidiomycotina populations.

In *Stereum sanguinolentum*, sibs from the same basidiocarp typically appear more or less identical morphologically, and intermingle in culture, suggesting genetic identicality and origin from a homokaryotic parent. Non-sibs either appear similar and intermingle, or are directly mutually antagonistic, the isolates being assignable to interaction groups, such that members of the same group intermingle, those of different groups are antagonistic (Rayner & Turton, 1982). This population structure is remarkably similar to that of *Aspergillus nidulans* (Croft & Jinks, 1977; Croft: Chapter 20) in which each interaction group (in their terminology, h-c group) is seen as a clonally related group of strains – in the case of *S. sanguinolentum* fully reproductively isolated, in which evolution occurs independently. In *S. sanguinolentum*, this apparently complete absence of outcrossing between genetically significantly different lines, is associated with probable lack of meiosis in the basidium (Robak, 1942), so that it might be best to regard the fungus as amictic (Burnett, 1975) or apomictic, rather than homomictic or homothallic (Rayner & Turton, 1982). Either way, the consequence of non-outcrossing in development of clonal subpopulations is the same. Twenty-three different interaction groups have been discovered so far in *S. sanguinolentum*, of which seven have been found in more than one location (A. M. Ainsworth, unpublished). *Stereum rameale* appears to behave similarly (Rayner & Turton, 1982; A. M. Ainsworth unpublished).

Stereum hirsutum is particularly interesting here, since as already indicated the British population appears to be outcrossing, with a functional homogenic incompatibility system, whilst the Finnish population behaves like *S. sanguinolentum*, being divisible into apparently clonal interaction groups. Such occurrence of non-outcrossing and outcrossing subpopulations within what in morphological terms appears to be the same species, associated with full expression of somatic incompatibility between different homokaryotic lines, is, we think, strong evidence for the override hypothesis. The occurrence of reproductively isolated homomictic, unifactorial and bifactorial diamoromictic groups has been reported in *Sistotrema brinkmanii* (for summary, see Burnett, 1975) and 'homothallic races' have been found in *Hyphoderma tenue*, *H. setigerum* and certain species of *Phlebia*, *Mycoacia* and *Coriolellus*, of which the other representatives are bipolar (Boidin, 1971). Boidin suggests from this that 'bipolarity seems to be . . . the access to homothallism'. Our paraphrasing of this would be 'loss of access leads to loss of homogenic incompatibility'.

Speciation. The whole question of speciation in Basidiomycotina and its origins from reproductive isolation has been considered in depth by Burnett (1983). To us, a significant feature is that homokaryons of closely related species-pairs often show what we would regard as a strong somatic-incompatibility reaction in culture. In some cases, as between *Phlebia radiata* and *P. rufa* (Boddy & Rayner, 1983) this reaction may follow hyphal anastomosis. Whilst the occurrence of this interaction has sometimes been interpreted as the *cause* it may actually be the *consequence* of intersterility, due to breakdown of override.

Concluding comment

We hope to have demonstrated that in terms of the genetic structure of populations and interactions between individuals, higher fungi show much in common with higher organisms. However, they are distinguished from the latter by their mycelia, which, whilst lacking powers of locomotion, are indeterminate in extent. This must be accounted for in all discussion of fungal individualism, its consequences and comparisons with other organisms. By the same token the relative simplicity and culturability of the mycelium facilitates experimentation. This offers the mycologist an opportunity to contribute significantly to the development of general biological principles.

We thank the Science and Engineering Research Council, and the Natural Environment Research Council, for provision of financial support.

References

Adams, T. J. H. (1982). *Piptoporus betulinus*, some aspects of population biology. PhD thesis, University of Exeter.

Adams, T. J. H., Todd, N. K. & Rayner, A. D. M. (1981). Antagonism between dikaryons of *Piptoporus betulinus*. *Transactions of the British Mycological Society*, **76,** 510–13.

Adams, T. J. H., Williams, E. N. D., Todd, N. K. & Rayner, A. D. M. (1984). A species-specific method of analysing populations of basidiospores. *Transactions of the British Mycological Society* (in press).

Barrett, D. K. & Uscuplic, M. (1971). The field distribution of interacting strains of *Polyporus schweinitzii* and their origin. *New Phytologist*, **70,** 581–98.

Bateson, P., ed. (1983). *Mate Choice*. Cambridge: Cambridge University Press.

Boddy, L., Coates, D. & Rayner, A. D. M. (1983). Attraction of fungus gnats to zones of intraspecific antagonism on agar plates. *Transactions of the British Mycological Society*, **81,** 149–51.

Boddy, L. & Rayner, A. D. M. (1982). Population structure, inter-mycelial interactions and infection biology of *Stereum gausapatum*. *Transactions of the British Mycological Society*, **78,** 337–51.

Boddy, L. & Rayner, A. D. M. (1983). Mycelial interactions, morphogenesis and ecology of *Phlebia radiata* and *Phlebia rufa* from oak. *Transactions of the British Mycological Society*, **80,** 437–48.

Boidin, J. (1971). Nuclear behaviour in the mycelium and the evolution of the Basidiomycetes. In *Evolution in the Higher Basidiomycetes*, ed. R. H. Petersen, pp. 129–48. Knoxville: University of Tennessee Press.

Boucherie, H. & Bernet, J. (1977). Intracellular and extracellular phenoloxidases in the fungus *Podospora anserina*: effect of a constitutive mutation in a gene involved in their posttranscriptional control. *Molecular and General Genetics*, **157,** 53–9.

Boucherie, H., Dupont, C. H. & Bernet, J. (1981). Polypeptide synthesis during protoplasmic incompatibility in the fungus *Podospora anserina*. *Biochimica et Biophysica Acta*, **653,** 18–26.

Burnett, J. H. (1975). *Mycogenetics*. London, New York, Sydney, Toronto: John Wiley & Sons.

Burnett, J. H. (1976). *Fundamentals of Mycology*, 2nd edn. London: Edward Arnold.

Burnett, J. H. (1983). Speciation in fungi. *Transactions of the British Mycological Society*, **81,** 1–14.
Coates, D. (1984). Biology of intraspecific antagonism in wood decay fungi. PhD thesis, University of Bath.
Coates, D., Rayner, A. D. M. & Todd, N. K. (1981). Mating behaviour, mycelial antagonism and the establishment of individuals in *Stereum hirsutum*. *Transactions of the British Mycological Society*, **76,** 41–51.
Collins, O. R. (1979). Myxomycete biosystematics: some recent developments and future research opportunities. *Botanical Reviews*, **45,** 145–201.
Cooke, R. C. & Rayner, A. D. M. (1984). *Ecology of Saprotrophic Fungi.* London & New York: Longman.
Croft, J. H. & Jinks, J. L. (1977). Aspects of the population genetics of *Aspergillus nidulans*. In *Genetics and Physiology of Aspergillus*, ed. J. E. Smith & J. A. Pateman, pp. 339–60. London, New York, San Francisco: Academic Press.
Dowson, C. G. (1982). Mycelial ecology of the Xylariaceae. Project report, University of Bath.
Esser, K. & Blaich, R. (1973). Heterogenic incompatibility in plants and animals. *Advances in Genetics*, **17,** 107–52.
Fatemi, J. & Nelson, R. R. (1977). Intra-isolate heterokaryosis in *Pyricularia oryzae*. *Phytopathology*, **67,** 1523–5.
Garrett, S. D. (1960). Inoculum potential. In *Plant Pathology*, vol. 3, ed. J. G. Horsfall & A. E. Dimond, pp. 23–56. New York: Academic Press.
Garrett, S. D. (1970). *Pathogenic Root-Infecting Fungi.* Cambridge: Cambridge University Press.
Getz, W. M. (1981). Genetically based kin recognition systems. *Journal of Theoretical Biology*, **92,** 209–26.
Faulk, W. P., Temple, A., Lovins, R. & Smith, N. C. (1978). Antigens of human trophoblast: a working hypothesis for their role in normal and abnormal pregnancies. *Proceedings of the National Academy of Sciences of the USA*, **75,** 1947–51.
Hiroth, J. (1965). The phenoloxidase and peroxidase activities of two culture types of *Phellinus tremulae* (Bond.) Bond. et Borriss. *Norwegian Forest Research Institute*, **20,** 255–72.
Jennings, D. H. (1982). The movement of *Serpula lacrimans* from substrate to substrate over nutritionally inert surfaces. In *Decomposer Basidiomycetes, Their Biology and Ecology*, ed. J. C. Frankland, J. N. Hedger & M. J. Swift, pp. 91–108. Cambridge: Cambridge University Press.
Jurand, M. K. & Kemp, R. F. O. (1973). An incompatibility system determined by three factors in a species of *Psathyrella* (Basidiomycetes). *Genetical Research, Cambridge*, **22,** 125–34.
Kemp, R. F. O. (1975). Breeding biology of *Coprinus* species in the section *Lanatuli*. *Transactions of the British Mycological Society*, **65,** 375–88.
Koestler, A. (1976). *The Ghost in the Machine.* Danube edition. London: Hutchinson.
Komlos, L., Zamir, R., Joshua, H. & Halbrecht, I. (1977). Common HLA antigens in couples with repeated abortions. *Clinical Immunology and Immunopathology*, **7,** 330–5.
Kühner, R. (1977). Variation of nuclear behaviour in the Homobasidiomycetes. *Transactions of the British Mycological Society*, **68,** 1–16.
Labarère, J., Bégueret, J. & Bernet, J. (1974). Incompatibility in *Podospora anserina*: comparative properties of the antagonistic cytoplasmic factors of a nonallelic system. *Journal of Bacteriology*, **120,** 854–60.
Lane, E. B. (1981). Somatic incompatibility in fungi and myxomycetes. In *The Fungal*

Nucleus, ed. K. Gull & S. G. Oliver, pp. 239–58. Cambridge: Cambridge University Press.

Leach, J. & Yoder, O. C. (1983). Heterokaryon incompatibility in the plant pathogenic fungus, *Cochliobolus heterostrophus. Journal of Heredity*, **74,** 149–52.

Li, C. Y. (1981). Phenoloxidase and peroxidase activities in zone lines of *Phellinus weirii. Mycologia*, **73,** 811–21.

Miyake, A. (1974). Cell interaction in conjugation of ciliates. *Current Topics in Microbiology and Immunology*, **64,** 49–77.

Pandey, K. K. (1981). Evolution of unilateral incompatibility in flowering plants: further evidence in favour of twin specificities controlling intra- and interspecific incompatibility. *New Phytologist*, **89,** 705–28.

Papazian, H. (1950). Physiology of the incompatibility factors in *Schizophyllum commune. Botanical Gazette*, **112,** 143–63.

Phipps, B. C. (1980). A study of vegetative hyphal fusion among fungi. Project report, University of Bath.

Punja, Z. K. & Grogan, R. G. (1983*a*). Basidiocarp induction, nuclear condition, variability, and heterokaryon incompatibility in *Athelia (Sclerotium) rolfsii. Phytopathology*, **73,** 1273–8.

Punja, Z. K. & Grogan, R. G. (1983*b*). Hyphal interactions and antagonism among field isolates and single-basidiospore strains of *Athelia (Sclerotium) rolfsii. Phytopathology*, **73,** 1279–83.

Rayner, A. D. M. & Todd, N. K. (1979). Population and community structure and dynamics of fungi in decaying wood. *Advances in Botanical Research*, **7,** 333–420.

Rayner, A. D. M. & Todd, N. K. (1982). Population structure in wood-decomposing basidiomycetes. In *Decomposer Basidiomycetes: Their Biology and Ecology*, ed. J. C. Frankland, J. N. Hedger & M. J. Swift, pp. 109–28. Cambridge: Cambridge University Press.

Rayner, A. D. M. & Turton, M. N. (1982). Mycelial interactions and population structure in the genus *Stereum: S. rugosum, S. sanguinolentum* and *S. rameale. Transactions of the British Mycological Society*, **78,** 483–93.

Robak, H. (1942). Cultural studies in some Norwegian wood-destroying fungi. *Meddelelse fra Vestlands Forstlige Forsøksstation*, **7,** 1–248.

Scofield, V. L., Schlumpberger, J. M., West, L. A. & Weissman, I. L. (1982). Protochordate allorecognition is controlled by a MHC-like gene system. *Nature*, **295,** 499–502.

Taylor, C. & Faulk, W. P. (1981). Prevention of recurrent abortion with leucocyte transfusions. *The Lancet*, July 11, 68–9.

Todd, N. K. & Rayner, A. D. M. (1980). Fungal individualism. *Science Progress (Oxford)*, **66,** 331–54.

Vandendries, R. & Brodie, H. J. (1933). Nouvelles investigations dans la domaine de la sexualité des Basidiomycètes et l'étude expérimentale des barrages sexuels. *La Cellule*, **42,** 165–209.

Wiese, L. & Wiese, W. (1977). On speciation by evolution of gametic incompatibility: a model case in *Chlamydomonas. American Naturalist*, **111,** 733–42.

Williams, E. N. D., Todd, N. K. & Rayner, A. D. M. (1981). Spatial development of populations of *Coriolus versicolor. New Phytologist*, **89,** 307–19.

Williams, E. N. D., Todd, N. K. & Rayner, A. D. M. (1984). Characterisation of the spore rain of *Coriolus versicolor* and its ecological significance. *Transactions of the British Mycological Society*, **82,** 323–6.

Index of generic and specific names

For organisms with a well-known common name the reader is also advised to consult the subject index.
Numbers in **bold type** refer to pages on which relevant figures or tables appear.

Subject index

Numbers in **bold type** refer to pages on which relevant figures or tables appear. *passim* denotes scattered references.

For EU product safety concerns, contact us at Calle de José Abascal, 56–1°,
28003 Madrid, Spain or eugpsr@cambridge.org.

www.ingramcontent.com/pod-product-compliance
Lightning Source LLC
LaVergne TN
LVHW091526060526
838200LV00036B/503
9780521106269